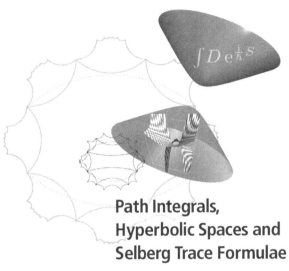

Path Integrals,
Hyperbolic Spaces and
Selberg Trace Formulae

2nd Edition

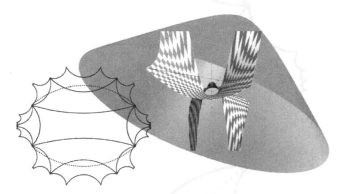

Path Integrals, Hyperbolic Spaces and Selberg Trace Formulae

2nd Edition

Christian Grosche

Universität Hamburg & Stadtteilschule Walddörfer
Germany

World Scientific

NEW JERSEY · LONDON · SINGAPORE · BEIJING · SHANGHAI · HONG KONG · TAIPEI · CHENNAI

Published by

World Scientific Publishing Co. Pte. Ltd.

5 Toh Tuck Link, Singapore 596224

USA office: 27 Warren Street, Suite 401-402, Hackensack, NJ 07601

UK office: 57 Shelton Street, Covent Garden, London WC2H 9HE

British Library Cataloguing-in-Publication Data
A catalogue record for this book is available from the British Library.

ISBN   978-981-4460-07-1

Printed in Singapore by B & Jo Enterprise Pte Ltd

Contents

List of Tables

List of Figures

Preface

Motivation.

In this monograph I want to give an overview and summary of two lines of research I have carried out: These are the theory of path integrals on the one hand, and Selberg trace formulæ on the other. The first topic, the study of path integrals I started with my Diploma Thesis which was entitled "Das Coulombpotential im Pfadintegral" [195]. I calculated the radial path integral for the Coulomb potential, however, in a somewhat complicated way: I used a two-dimensional analogue of the Kustaanheimo-Stiefel transformation. My Diploma Thesis was the starting point for an intensive investigation of path integral formulations on curved manifolds.

Later on, I could generalize these results to the path integral formulation for some *specific* coordinate systems in spaces of constant curvature on the sphere and the pseudosphere, for general hyperbolic spaces of rank one, for hermitian spaces (later also for hermitian hyperbolic spaces), and for single-sheeted hyperboloids. I started a systematic investigation of the path integral formulation (and evaluation if possible) in spaces of constant curvature, where all coordinate systems which separate the Schrödinger equation or the path integral, respectively, were taken into account.

Motivated by string theory, in particular by the Polyakov approach to string perturbation theory which is a path integral formulation, and quantum mechanics in spaces of constant negative curvature, I started an investigation in the theory of the Selberg trace formula, i.e., quantum field theory on Riemann surfaces. The first principal achievement I presented in my Dissertation [203]. It included a thorough discussion of the Selberg super-trace formula on super-Riemann surfaces, and I could derive the trace formula for super-automorphic forms of integer weight. Analytic properties of the Selberg super-zeta-functions could be discussed by a proper choice of testfunctions in the trace formula, and super-determinants of Laplacians on super-Riemann surfaces could be expressed in terms of the Selberg super-zeta-functions, thus giving well-defined expressions for all genera in the integrand of the Polyakov partition function. It is interesting to note that the Selberg trace formula can be derived by a path integration. This is true for the usual as well as the super-hyperbolic plane.

I developed the theory further, first by including elliptic and parabolic conjugacy classes in the Selberg super-trace formula, and secondly by the incorporation of bordered super-Riemann surfaces. In the latter case I could generalize results from a joint paper with Jens Bolte, where we merged results from Venkov, Hejhal and Bolte and Steiner to derive a Selberg trace formula on bordered Riemann surfaces for automorphic forms of integer weight.

Also, the asymptotic distribution for the lengths of geodesics and the eigenvalues of Laplacians, i.e. analogies for Huber's and Weyl's law, respectively, on super Riemann surfaces could be stated.

In the first edition of this monograph I successfully solved the path integral representations for many coordinate systems on two- and three-dimensional pseudo-Euclidean space, Euclidean space, the sphere and the pseudosphere. These coordinate systems separate the Schrödinger equation, respectively the path integral. However, several path integral representations could not be evaluated, which is mainly due to that they are parametric systems. Surprisingly several path integral representations for elliptic and spheroidal coordinate systems could be derived which is due to expansion-theorems for elliptic and spheroidal coordinates in three-dimensional Euclidean space.

In the second edition of this monograph I was able to extend this study to more spaces: The two- and three-dimensional complex spheres, which are, however, somewhat only a formal level, because the complex sphere is a complex space. However, parametric coordinate systems cannot be treated generally.

Furthermore I can include path integral representations and solution on the hermitian hyperbolic space HH(2), and for two- and three-dimensional Darboux spaces. Darboux spaces have the interesting property that they are spaces of *non-constant* curvature, which contain as limiting case spaces of constant curvature, though, i.e. spheres and hyperboloids. In two dimensions four cases of Darboux spaces are studied, and in three dimensions two cases of Darboux spaces. Another interesting property of Darboux spaces is that in some of them there is a *zero-point* energy for the free motion in the continuous spectrum $(D_{II}, D_{IV}, D_{3d-II})$. Also, for the free motion, a discrete spectrum can arise (D_{III}). Another feature in three-dimensional Darboux spaces is the fact that a proper quantum theory must be set up. The incorporation of the third dimension changes the corresponding momentum operators (and the Hamiltonian) in such a way that without a proper additional quantum potential no calculations at all would be possible.

A further extension to the so-called Koenig-spaces I have omitted here. Including them would require more space without giving more substantial information and insight.

Also omitted are the discussions of super-integrable potentials, also called Smorodinsky-Winternitz potentials. Including them would extend this work far too much. In the future it is planned to publish another monograph about the topic of super-integrable potentials in spaces of constant and non-constant curvature. Also complex spaces will be included, as well as the investigations on the O(2, 2) and SU(2, 2) hyperboloids.

In the path integral representation on the single-sheeted hyperboloid no new results have been added. This is due to the fact that an extensive study of the free motion of the $O(2,2)$ hyperboloid is still missing. In Chapter 11 I only provide some more information about the coordinate systems on the two-dimensional single-sheeted hyperboloid.

In the investigation of billiard systems and periodic orbit theory I have added several more integrable systems in two and three dimensions, this includes billiard systems in flat space as well in hyperbolic space (e.g. in the Poincaré disc), for instance triangles, circles, parallelepipeds, and spheres. The numerical investigations nicely confirm the semiclassical theory.

In the theory of the Selberg trace formula I have added a study on asymptotic distributions on super-Riemann surfaces, i.e. Huber's and Weyl's law. Here, not surprisingly, the feature is recovered that in super-symmetric quantum mechanics there are as many bosonic as fermionic states, except for a (zero-energy) ground state. In some sense, Weyl's law for compact super-Riemann surfaces can be quite boring.

Acknowledgements for the Second Edition.

I would like to thank the publisher, World Scientific Publishing, for giving me the possibility to publish a second edition of this monograph. I could include a lot of new material.

I would like thank my friend and colleague George Pogosyan. We wrote several papers an integrable coordinate systems and superintegrability in spaces of constant and non-constant curvature. I enjoyed very much several visits in Dubna (Russia) and Yerewan (Armenia) with him and his family. It was always a great pleasure for me. Im memoriam I also thank Prof. Alexei Sissakian who passed away too early in 2010 for the collaboration in our joint work and given me the possibilty for visiting Russia.

I also thank the "Stadtteilschule Walddörfer, Hamburg" for giving me the possibility to compile and finish this second edition of this monograph.

Furthermore I would like to thank in addition to the first edition Yota Ardamerina, Jochen Bartels, Cornelia Brandl-Hoff Sensei and Feliks Hoff Sensei, Wolf Beiglböck and his wife Urda Beiglböck, Christian Caron, Inken Eckelmann, Rafael Gil-Brand, Stefan Gößling and Verena Reisemann, Dieter Greschok, Dieter Haidt, Burkhard Hofmann, Gertrude Huber, Ernie Kalnins, Hagen Kleinert, Angela Lahee, Ole Nydal, Elisabeth Thölke, Pavel Winternitz, for support, fruitful discussions, useful hints, friendship, and many more.

Last but not least I would like to thank my son Urs, who gives me a lot of joy.

Acknowledgements for the First Edition.

First of all I would like to thank my beloved friend Claudia Renner for her endurance and patience.

I am grateful to my parents, Emmy and Dr. Heinz Grosche, for their support, and also my brother Karl Rudolf Grosche and his wife Gloria for reading a big manuscript on short notice.

I would like to thank my teacher Frank Steiner for his constant encouragement and support for completing this Habilitationsschrift. I particularly appreciated his attitude to be open for discussions all the time, his fruitful hints and suggestions, and the friendly atmosphere during all the years of our collaboration, and I thank him for a critical reading of the manuscript.

I am very grateful to George Pogosyan who drew my attention to the subject of interbasis expansions and coordinate systems in spaces of constant curvature.

I would like to thank Akira Inomata and Georg Junker for fruitful discussions concerning path integration on group manifolds, in particular during my visit to the State University of New York at Albany, N.Y., which was made possible by the invitation of A. Inomata and financial support of the Deutsche Forschungsgemeinschaft. I also thank Pavel Winternitz for his kind invitation to the University of Montreal during my stay in Albany, N.Y., and I thank Dennis Hejhal for the encouragement for studying Selberg trace formulæ.

Furthermore I would like to thank my friends and colleagues at the II. Institut für Theoretische Physik, Hamburg, for their friendship and help in discussing issues arising in the work; among them in particular Ralf Aurich, Arnd Bäcker, Guido Bennecke, Jens Bolte, Thomas Hesse, Claudia Matthies, Jens Marklof, Holger Ninnemann, Christer Oldhoff, and Oliver Rudolph.

I am also grateful to several people in the computing center at DESY for software support, among them Erwin Deffur, Jan Hendrik Peters, and Katherine Wipf.

This work was supported by Deutsche Forschungsgemeinschaft under contract number GR 1031/2–1.

Chapter 1

Introduction

Path Integrals.

Contrary to common believe, the proper quantum potential in the path integral in quantum mechanics is in general not just a constant proportional to the curvature. There may be a formulation, where this is the case, but not necessarily. In particular, if the path integral is formulated in terms of the classical Lagrangian, thus giving rise to an effective Lagrangian, then the quantum potential is explicitly coordinate-dependent.

Our first paper [248] was followed by other instructive examples of path integrals which could be treated by this theory in a consistent way. Among them were the path integral on the Poincaré upper half-plane [247], and its related conformally equivalent formulations, the Poincaré disc and the hyperbolic strip [202], and the pseudosphere [249]. Some potential problems [198, 204] and the incorporation of magnetic fields [197, 200] could also be discussed in this context, among them the Kepler problem on the pseudosphere [201]. Here a useful lattice formulation of the path integral was extensively used, which I have called "product form" [196]. In comparison to the often used (arithmetic) mid-point formulation, this lattice prescription is basically a geometric mid-point formulation. Also the already in [248] improved space-time transformation (also Duru-Kleinert transformation) technique could be further developed in [218] by the incorporation of explicitly time-dependent transformations.

The part of this volume concerned with path integrals is designed as follows: In Chapter 2 I review the definition of path integrals on curved manifolds. This includes the explicit construction of the path integral in its lattice definition. The two most important lattice prescriptions, mid-point and product-form are presented with the emphasize on the latter. Other lattice representations are not discussed, and neither the Vielbein approach of Kleinert [352]. Furthermore, transformation techniques are outlined. This includes point canonical transformations, space-time transformations, pure time transformations, and separation of variables. Some of the path integral investigations were done in joint work with Frank Steiner [248]. It must be noted that in recent years several review articles and textbooks following the classical books of Feynman and Hibbs [164] and Schulman [460] on exactly solvable path integrals with many examples have been published, e.g., Albeverio et al. [5, 3], Dittrich and Reuter

[131], Glimm and Jaffe [182], Inomata et al. [289], Khandekar and Lawande [337], Kleinert [352], Roepstorff [454]. Simon [470], and Wiegel [519].

In Chapter 3 I give a summary of the classification of coordinate systems in spaces of constant curvature. This includes some remarks about the physical significance concerning separation of variables and breaking of symmetry, a general classification scheme, and an overview of the coordinate systems in Euclidean and Minkowski spaces and on spheres and hyperboloids.

In the next nine Chapters the path integral representations in several classes of (homogeneous) spaces are discussed. It includes the two- and three-dimensional Minkowski or pseudo-Euclidean spaces (Chapter 4), the two- and three-dimensional Euclidean spaces (Chapter 5), the two- and three-dimensional spheres (Chapter 6) and hyperboloids (Chapter 7), the two- and three-dimensional complex sphere (Chapter 8), hermitian hyperbolic space (Chapter 9), and Darboux spaces (Chapter 10). Additional results for the two-dimensional single-sheeted hyperboloid are presented in Chapter 11 and for more general homogeneous and hyperbolic spaces in Chapter 12. This includes the single-sheeted two-dimensional hyperboloid, the hyperbolic space corresponding to $SO(p,q)$ and $SU(p,q)$, and the case of hyperbolic spaces of rank one.

In comparison to the first edition of this monograph, the Chapters 8, 9 and 10 are entirely new. They contain results, which have been achieved in the course of studying more general cases as just real spaces, respectively real spaces of constant curvature, respectively spaces whose curvature is not constant, i.e. the Darboux spaces. However, I will not discuss the so-called Koenig-spaces which emerge from, say, usual flat space by multiplying the metric with a super-integrable potential in this flat space. This can be done in two- and three-dimensional Euclidean space, respectively. In two dimensions these potentials are the isotropic harmonic oscillator, the Holt-potential, and the Coulomb potential, e.g. [241]. The quantum motion then can be analyzed in the usual way by path integration [235, 237]. However, the quantization conditions for the energy-levels turn out to be rather complicated. They require the solution of an equation of eighth order in the energy E. Surprisingly, special cases of the Koenig-spaces turn to be Darboux spaces, spaces of constant (negative) curvature, and, of course, Euclidean space.

Koenig spaces which emerge form an analogous way from, say, Minkowski space, or spaces with constant (positive and negative) curvature with their corresponding super-integrable potentials have not been discussed yet. Their construction should be straightforward, though, including a path integral evaluation.

The cases of two- and three-dimensional Darboux spaces have been partly done in collaboration with George Pogosyan [244], in particular in the context of super-integrable potentials in these spaces. This was an extension of earlier work of super-integrable potentials on spaces of constant curvature [241]–[243].

Generally, I denote by "u" coordinates with indefinite metric, and by "**q**" coordinates with a positive definite metric. I start with the case of the pseudo-Euclidean space, because the proposed path integral solutions are entirely new. Some of the path integral solutions in the remaining three other spaces have been already reviewed in [223], and I do not discuss all the solutions in detail once more. Only the new solutions are treated more explicitly. In particular, I concentrate on the path integral solutions which can be obtained by means of interbasis expansions. This includes the case of elliptic coordinates in two-dimensional Euclidean space, on the sphere and on the pseudosphere, the case of spheroidal coordinates in three-dimensional Euclidean space, and some cases of ellipsoidal coordinates in spaces of constant curvature in three dimensions. As we will see, all the developed path integral techniques will come into play. I have cross-checked the solutions with the ones available in the literature achieved by other means. My hope is that my presentation will serve as a table for path integral representations in homogeneous spaces. I did not intend a discussion according to Camporesi [94] with a generalization to higher dimensions, applications of my results in cosmology, zeta-function regularization, etc. In addition, some results of path integration on generalized hyperbolic spaces are given.

Periodic Orbit Theory and Selberg Trace Formulæ.

For about twenty years trace formulæ have played a major role in mathematical physics and string theory. Generally, trace formulæ relate the classical and the quantum properties of a given system to each other. This can be very easily visualized with a simple example, the drum: The periodic orbits are the classical trajectories on the drum, and the energy eigenvalues are related to its modes (hence the question "Can one hear the shape of a drum?" [296]). This is true for every system one wants to study, however, trace formulæ become particularly important and useful when the system under consideration is classically chaotic. A necessary condition that a classical system can be solved exactly is that the phase space separates into invariant tori. If this is not the case, the usual tools of a perturbative approach break down and because of the exponentially diverging distance of initially nearby trajectories described by the Lyapunov exponent, no statement about the long-time behaviour of the system can be made. In the mathematical literature explicit statements of this feature were first made by Hadamard [261] and Poincaré [445]. They considered classical motion on spaces of constant negative curvature.

Surprisingly enough it was Einstein [147] who pointed out that any attempt to quantize a generic classical system runs into trouble if there are not enough constants of motion in this system. The existence of constants of motion in conservative classical systems - the energy E being just one constant of motion among others - cause that the phase space corresponding to this system separates into invariant tori. Einstein made this observation in connection to the "old" quantum theory, and he considered the problem under which conditions the quantization rule $\oint \mathbf{p} \cdot d\mathbf{q} = n\hbar$ makes sense. It makes sense if one can find in $\mathbb{R}^D$, say, a coordinate system such that for any generalized coordinate q_a one can find a generalized conjugate momentum p^a with p^a a conserved quantity. In other words, we must find in a D-dimensional space $(D > 1)$ at least one coordinate system which separates the classical equations of motions or the Laplacian, respectively. Finding such a coordinate system is equivalent in finding a set of observables. If this is not the case a quantization procedure *cannot* be found in the usual way by introducing position and momentum operators and impose commutation relations among them. A minimum of two dimensions is required in order that this feature can occur. Therefore the only systems which can be quantized semiclassically are those whose classical phase space consists of D-fold separating invariant tori. Among them are many well-known standard systems as the harmonic oscillator, the hydrogen atom, anharmonic oscillators like the Morse- or the Pöschl–Teller oscillators, and all one-dimensional systems. Excluded are the motion on spaces of constant negative curvature, billiard systems with boundaries which have defocusing properties, and many others like a hydrogen atom in a uniform magnetic field or the anisotropic Kepler problem. All these systems are classically chaotic.

In the 1960's Gutzwiller [258] was the first who developed by means of path integrals a semiclassical theory for systems, which are classically chaotic and cannot be quantized semiclassically, because no set of invariant tori exists. What do exist, however, are *periodic orbits* and *energy levels*, the very things physicists are interested in. Gutzwiller discovered his *periodic orbit formula* in the study of the problem of the semiclassical quantization of separable and non-separable (chaotic) systems, where the most important separable system under consideration was the hydrogen atom [258]. The semiclassical trace formula of Gutzwiller has been applied successfully to many physical systems, like the anisotropic Kepler problem [260], the Helium-atom [522], the hydrogen-atom in crossed electric and magnetic fields [493], and billiard systems, e.g., [23, 467, 469]. Later on, in the study of more general systems, in particular for the classical and quantum motion on a Riemann surface, he realized that he had rediscovered the Selberg trace formula [259]. A sound mathematical footing for a large class of systems is also due to Albeverio et al. [2].

The second part of this volume is therefore devoted to periodic orbit theory, the theory of the Selberg (super-) trace formula, and some of its applications in mathematical physics. It is important to keep in mind that contrary to one's first impression, these topics do have a relation to path integral techniques indeed. First of all, for the *derivation* of the periodic orbit formula the *path integral is essential*. Only the path integral gives in its semiclassical (stationary phase etc.) approximation all the necessary information for a proper set up for a correct and comprehensive periodic orbit formula. This is due to the property of the path integral that it represents in quantum mechanics not just only a summation over paths, but a summation *over all paths*. This huge amount of information goes into the periodic orbit formula if one studies more and more refinements and improvements of it, i.e., information about the Maslov-indices, caustics, discontinuities in the van Vleck-Pauli-Morette determinant, and many more. Another line of reasoning is valid in the case of the Selberg trace formula. Considering a path integral formulation on a Riemann surface represents just but a special case of the Selberg trace formula, i.e., the Selberg trace formula for the heat kernel: Usually the propagator or the Green's function, respectively, in a hyperbolic space can be evaluated in closed form. Applying the composition law for a path integral on a coset space yields an expansion over group elements for the propagator. In the case of Riemann surfaces the summation is over the elements of a Fuchsian group. Taking the trace gives the Selberg trace formula.

In Chapter 13 I start with an elementary introduction into the periodic orbit theory and the periodic orbit trace formula of Gutzwiller [258]–[260] and Sieber and

Steiner [467, 469]. Aspects of mathematical rigour are set aside, e.g., Albeverio et al. [2]. I give a simple derivation of the periodic orbit formula, the regularized periodic orbit formula is stated without proof, and I give some arguments that the periodic orbit theory is the proper semiclassical quantization procedure for classically chaotic systems. These remarks are supposed to be on an informal level, and in Chapter 14 I give some more details of the theoretical background. This emphasizes the importance of trace formulæ, and in particular the importance of Selberg trace formulæ. The remainder of Chapter 13 is then devoted several integrable billiard system. One of them I have called a billiard system in a "rectangle in the hyperbolic plane". A numerical analysis has been presented in [219], and some of the most important features are reported here. Furthermore, I give some additional analysis concerning the fluctuations of the number of energy levels about the mean number of energy levels. The other two- and three-dimensional integrable billiard systems have been investigated in [238], which included the numerical evaluation of the corresponding energy-levels and the statistics.

These studies of Selberg trace formulæ in the connection with periodic orbit theory and quantum chaos led to the investigation of a separable billiard system in the hyperbolic plane. This system was analyzed in [211]. The most important results are reported in Chapter 13, including some new studies concerning the statistic of the fluctuations of the number of energy levels about the mean number of levels.

A conjecture of Steiner et al. [18, 480] predicts for classically chaotic systems a Gaussian distribution of the fluctuations of the spectral staircase about the mean number of energy levels (Weyl's law), whereas the fluctuations for classically integrable systems should be *all* non-Gaussian. Within the error margins, the conjecture is supported by the investigated systems. Whereas the original attempt of periodic orbit theory was to determine from the knowledge of the classical periodic orbits the quantal energy levels, it can also be used to do the analysis the other way round: Take the energy levels and determine the lengths of the periodic orbits. This analysis is done for the rectangular billiard system in the hyperbolic plane, and it is found that the two shortest periodic orbits (and their multiplies) can be indeed determined. However, a systematic investigation of the periodic orbits of this system has not been done yet; this and a desirable extended and refined determination of the energy levels have not been the topic of this volume.

In the next two Chapters I deal with the Selberg trace formula on Riemann surfaces and with the Selberg super-trace formula on super-Riemann surfaces, respectively. Selberg trace formulæ in symmetric space forms of rank higher than one as, e.g., considered by Efrat [145] and Wallace [516], are not taken into account. In Chapter 14 I give a summary of the theory of the Selberg trace formula. I start with an overview of several applications in Mathematical Physics, i.e., cosmology, string theory, relation with the Riemann zeta functions, and higher dimensional generalizations, i.e., hyperbolic space forms of rank one. I am concerned mainly with the statement of the trace formulæ, including the incorporation of elliptic and parabolic conjugacy classes. After using the trace formula for the determination of the analytic properties of the Selberg zeta-function, I show how determinants of Laplacians on

Riemann surfaces can be evaluated by means of the Selberg trace formula and can be expressed by means of the Selberg zeta-functions. The discussion is then repeated for bordered Riemann surfaces, and I rely on results as derived in [70]. The presentation is in the form of theorems, they are given without proofs with only the necessary explanations. The boundary conditions for the bordered Riemann surfaces are either Dirichlet or Neumann boundary conditions.

In Chapter 15 I present the theory of the Selberg super-trace formula for super-Riemann surfaces. I consider only the case of $N = 1$ super-symmetry. In my Dissertation [203] I have started this investigation by extending earlier results of Baranov et al. [41]–[43] and have continued the development of this theory and its application in several published papers [213, 215, 221] which I rely on. In order to make the presentation self-contained I review some elementary results from the quantum theory on super-Riemann manifolds, the super-uniformization theorem for super-Riemann surfaces, and super-automorphic forms. In the sequel the results concerning the statement of the Selberg super-trace formula for super-Riemann surfaces and the analytic properties of the Selberg super-zeta-functions are presented in the form of theorems given without proof. Here partly some results from [199, 203] are repeated, and partly the results from [215] concerning the incorporation of elliptic and parabolic conjugacy classes are reported. Similarly as in the usual Selberg theory, I also show how super-determinants of Laplacians on super-Riemann surfaces are evaluated by means of the Selberg super-trace formula and expressed by means of the Selberg super-zeta-functions. In the second part of Chapter 15 the discussion is repeated for the case of bordered super-Riemann surfaces and the results of [221] are reported.

The last Chapter is devoted to a summary, critical discussion and an outlook. Concerning path integration it includes the enumeration of the most important path integral results and identities as achieved in this work. Concerning the theory of the Selberg (super) trace formula I summarize my results, and discuss open problems and questions. Furthermore, I can give a general formula for determinants of Laplacians on hyperbolic space forms of rank one. Mainly due to lack of time, it was not possible for me to undergo a systematic study of Selberg super-trace formula for extended super-symmetry, for instance to generalize my theory of the Selberg super-trace formula for super-Riemann surfaces for $N = 1$ super-symmetry to $N = 2$ super-symmetry [400], or a Selberg super-trace formula on analogues of higher dimensional hyperboloids. Such generalizations will be studied in the future.

Chapter 2

Path Integrals in Quantum Mechanics

2.1   The Feynman Path Integral

The invention of the path integral by Feynman [160] is one of the major achievements in theoretical physics. In its 70 years of history it has become an indispensable tool in field theory, cosmology, molecular physics, condensed matter physics and string theory as well.

Originally developed as a "space-time approach to non-relativistic quantum mechanics" [159, 160] with the famous solution of the harmonic oscillator, it became soon of paramount importance in quantum electrodynamics (QED), especially in the development of the nowadays so-called "Feynman rules". It did not take long and Feynman succeeded in discussing problems not only in QED but also in the theory of super-fluidity [162], and solid state physics [163]. However, it took a considerable time before it was generally accepted by most physicists as a powerful tool to analyze any physical system, giving non-perturbative global information. The advantage in comparison to the operator approach lies in the fact that the path integral is more comprehensive. The path integral gives a global point of view of the problem in question, in comparison to the operator approach which gives only a local one. Not only the propagator or the Green's function can be often evaluated explicitly, but also the energy spectrum and the correctly normalized wavefunctions are automatically given once the path integral has been calculated. Perturbation theory can be incorporated in a straightforward way.

Eventually, a satisfying theory should be based on a field theory formulation, let it be the second quantization of the Schrödinger or the Dirac equation, or let it be a field theory path integral. However, the path integral is quite a formidable and difficult functional-analytic object. The very early field theory formulations by a path integral, first by Feynman and in the following by, e.g., Matthews and Salam [175] remained only on a formal level, however with well-described rules to extract the relevant information, say, for Feynman diagrams. Indeed, in field theory the path integral is cursed by several pathologies which cause some people now and then to state that "the path integral does not exist".

What does exist, however, is the original Feynman path integral, i.e., Feynman's "space-time approach to non-relativistic quantum mechanics". Thanks to the work of many mathematicians and physicists as well, the theory of the Feynman path integral can be considered as quite comprehensively developed. Actually, the theory of the "Wiener-integral" [175, 520] existed some twenty years right before Feynman published his ideas, and was developed in the theory of diffusion processes. The Wiener integral itself represents some sort of an "imaginary time" version of the "real time" Feynman path integral. This particular feature of the Wiener integral makes it a not-too-complicated and convenient tool in functional analysis, mostly because convergence properties are easily shown. These convergence properties are absent in the Feynman path integral and the emerging challenge attracted many mathematicians and mathematical physicists, c.f. the references given in [175]. Let us in addition mention Nelson [420] concerning the Feynman path integral in cartesian coordinates, DeWitt concerning curvilinear coordinates [123], Morette-DeWitt et al. [414], and Albeverio et al. [3, 5] who developed a theory of "pseudomeasures" appropriate to the interference of probability amplitudes in the path integral. This interference of probabilities, in particular in the lattice formulation of the path integral leads to the very interpretation of the Feynman path integral. One encounters for finite lattice spacing, i.e., finite N, a complex number $\Phi(\mathbf{q}_1, \ldots, \mathbf{q}_{N-1})$ which is a function of the variables $\mathbf{q}_j$ defining a path $\mathbf{q}(t)$, and the path integral can be interpreted as a "sum over all paths" or a "sum over all histories"

$$K(\mathbf{q}'', \mathbf{q}'; T) = \sum_{\substack{\text{over all paths} \\ \text{from } \mathbf{q}' \text{ to } \mathbf{q}''}} \Phi[\mathbf{q}(t)] = \sum_{\substack{\text{over all paths} \\ \text{from } \mathbf{q}' \text{ to } \mathbf{q}''}} e^{iS[\mathbf{q}(t)]/\hbar} \ . \tag{2.1}$$

The path integral then gives a prescription how to compute the important quantity Φ for each path: "The paths contribute equally in magnitude, but the phase of their contribution is the classical action (in units of $\hbar$). ... That is to say, the contribution $\Phi[\mathbf{q}(t)]$ from a given path $\mathbf{q}(t)$ is proportional to $\exp(\frac{i}{\hbar} S[\mathbf{q}(t)])$, where the *action* is the time integral of the classical Lagrangian taking along the path in question" [160]. All possible paths enter and interfere which each other in the convolution of the probability amplitudes. In fact, the nowhere differentiable paths span the continuum in the set of all paths, the differentiable ones are being a set of measure zero (the quantity $\Delta\mathbf{q}_j/\Delta t$ does not exist, whereas $(\Delta\mathbf{q}_j)^2/\Delta t$ does).

A tabulation which represents the state of the art of path integral representations in spaces of constant curvature will be presented in Chapters four to nine, i.e., path integral representations on two- and three-dimensional pseudo-Euclidean (Minkowski) space, Euclidean space, sphere and the pseudosphere (or hyperboloid), the complex sphere. and hermitian hyperbolic spaces. However, comprehensive path integral evaluations for single-sheeted hyperboloids and the $O(2,2)$, are not possible yet. Some results will be presented in Chapters 11 and 12, respectively. Path integral evaluations in spaces of non-constant curvature (Darboux spaces in two and three dimensions) are given in Chapter 10. Koenig spaces will be not discussed, see e.g. [235, 237] for some details.

In addition to the tabulation some comments will be given on how to calculate the path integrals. The results concerning the pseudo-Euclidean plane and the pseudo-Euclidean space are entirely new. The path integral solutions in the other three spaces are partly new and partly have been calculated in previous publications. The already known results will be cited with only little comment, whereas the new ones will be commented. I restrict myself to the discussion of the two- and three-dimensional cases, because the most relevant features appear in these low-dimensional cases; in the higher dimensional ones either these features are repeated, or matters become too complicated anyway. In Chapter 12 some miscellaneous results are presented. Due to the very complicated structure of the path integrals in these spaces, only a very limited number of path integral evaluations seem possible.

I am able to present all path integral solutions corresponding to all coordinate systems which separate the Hamiltonian in spaces of constant curvature in two dimensions. The explicit construction of these coordinate systems is omitted, and I refer to the literature (sphere and pseudosphere c.f. [425], Euclidean space c.f. [416], and pseudo-Euclidean space [300, 405]). My results are roughly summarized in table 2.1. As can be seen the case of the two-dimensional homogeneous spaces is (more or less) settled. In the case of the two-dimensional pseudosphere it remains only to present a comprehensive discussion and set-up of the relevant interbasis expansions for the elliptic, hyperbolic and semi-hyperbolic coordinate systems. Methods and formal treatments are already well-defined and will be taken for granted in this volume. The case of the single-sheeted pseudospheres is far more involved and complicated and will not be thoroughly discussed here (see below and Chapter 11). Only some selected results will be given.

The results in three dimensions are less satisfying. I have nevertheless included the case of $\mathbb{R}^3$ and $S^{(3)}$ as complete. The ellipsoidal and paraboloidal coordinate systems in $\mathbb{R}^3$ and the ellipsoidal in $S^{(3)}$ are two-parametric and therefore the most complicated ones. However, the wavefunctions can in principle be constructed, see e.g., in $\mathbb{R}^3$ Arscott [9, 11] and Miller [405] for the wavefunctions in ellipsoidal coordinates, and Arscott and Urwin [10, 500, 501] for the wavefunctions in paraboloidal coordinates; and see, e.g., Karayan et al. [240], Harnad and Winternitz [262, 263], and Kuznetsov et al. [306, 356, 365, 366] for the ellipsoidal wavefunctions on $S^{(3)}$. The corresponding interbasis expansions are defined and could be worked out in principle in a tedious way. This is postponed [240], and the *fact that this is possible in principle* will be taken for granted. Matters are far more complicated in the cases of the parametric coordinate systems on $\Lambda^{(3)}$, S_{3C}, and the pseudo-Euclidean space. Here no explicit solutions seem to be known, let alone interbasis expansions relating the various systems which each other, and the corresponding path integral solutions, respectively.

A discussion of these systems is completely omitted. It is obvious that things are becoming even worse in higher dimensions. Of course, it is possible for many systems to consider the various subgroup coordinate systems, and the corresponding path integral solutions can then be easily constructed from the two- and three-dimensional cases. However, this is not very instructive. Some remarks concerning this and path

Table 2.1: Path Integration on Homogeneous Spaces

Homogeneous Space	Number of Coordinate Systems	Number of Systems in which Path Integral Representation can be given
Two-Dimensional Pseudo-Euclidean Space	10	10 (Chapter 4)
Three-Dimensional Pseudo-Euclidean Space	54	32 (Chapter 4)
Two-Dimensional Euclidean Space	4	4 (Chapter 5)
Three-Dimensional Euclidean Space	11	11 (Chapter 5)
Two-Dimensional Sphere	2	2 (Chapter 6)
Three-Dimensional Sphere	6	6 (Chapter 6)
Two-Dimensional Pseudosphere	9	9 (Chapter 7)
Three-Dimensional Pseudosphere	34	24 (Chapter 7)
D-Dimensional Pseudosphere	(not given)	3 (Chapter 12)
Two-Dimensional Complex Sphere	5	5 (Chapter 8)
Three-Dimensional Complex Sphere	21	12 (Chapter 8)
Three-Dimensional O(2,2) Hyperboloid	74	3 (Chapter 2)
Two-Dimensional One-Sheeted Pseudosphere	9	2 (Chapter 11)
Hermitian Hyperbolic Space	12	6 (Chapter 9)
Hyperbolic Rank-One Spaces	(not given)	4 (Chapter 12)

integral representations in $\mathbb{R}^4$ and $\mathbb{R}^{(1,3)}$ can be found in the summary, though. The only exception will the hermitian hyperbolic space HH(2), which is four-dimensional. In fact, a path integral evaluation is possible in 6 out of 12 coordinate systems.

A special role play the Darboux spaces in two and three dimensions. Whereas in two dimensions almost all path integral representations can be derived, matters are not so satisfying in three dimensions, which is due to the parametric, parabolic, prolate and ellipsoidal coordinate systems.

In my consideration I do not treat the spaces of the two- and three-dimensional single-sheeted hyperboloid (with nine and 34 coordinates systems separating the free Schrödinger equation), and the $O(2,2) = SO(2,1) \times SO(2,1) \simeq SU(1,1) \times SU(1,1)$ hyperboloid with 74 orthogonal and three non-orthogonal coordinates systems [317] (note $SO_0(2,1) \simeq SU(1,1)/Z_2$, $SO_0(2,2) \simeq SO_0(2,1) \times SU_0(2,1)$). Actually the free quantum motion on one sheet of the $O(2,2)$ hyperboloid is isomorphic to the free quantum motion on the group manifold corresponding to $SU(1,1)$, a property which will be used in deriving a new path integral representation. I do not present a table of all the separating coordinate systems on the $O(2,2)$ hyperboloid.

Table 2.2: Path Integration on Darboux Spaces

Darboux Space	Number of Coordinate Systems	Number of Systems in which Path Integral is possible
Two-Dimensional Darboux Space D_{I}	3	2 (Chapter 10)
Two-Dimensional Darboux Space D_{II}	4	4 (Chapter 10)
Two-Dimensional Darboux Space D_{III}	5	4 (Chapter 10)
Two-Dimensional Darboux Space D_{IV}	4	4 (Chapter 10)
Three-Dimensional Darboux Space $D_{3d-\mathrm{I}}$	7	5 (Chapter 10)
Three-Dimensional Darboux Space $D_{3d-\mathrm{II}}$	12	8 (Chapter 10)

I do not treat non-orthogonal coordinate systems [300, 315, 324], and integrable Hamiltonian systems with velocity-dependent potentials, i.e., with magnetic fields [133].

I do not consider so-called R-separable coordinate systems [82, 311]–[315, 319, 320, 326]. Quantum mechanically they correspond to the very restricted case $E = 0$. There are some treatments of this kind of problem, in particular for power potentials [478] and the Holt-potential [241]. In free space things are either trivial, as in two dimensions where one has an infinite number of possible systems whose $E = 0$ solutions go over into the already well-known systems [412], or they are not solvable at all.

I do not go into the details of path integrals for potential problems. Some systematic investigation have been reviewed in [209] for the Coulomb problem (with the additional papers [212, 214] for even more complicated Coulomb-like problems which are only separable in parabolic coordinates), and in [241]–[243] for Smorodinsky-Winternitz potentials, i.e., super-integrable potentials, in Euclidean space, on the sphere, and on the hyperboloid, respectively. Two other applications of the path integral technique have been the Dirac monopole [206], and the Kaluza-Klein monopole [207].

I also do not go into details of point perturbations and boundary conditions in the path integral. Boundary conditions are of particular importance in the path integral. Some "usual" boundary conditions are automatically contained in any path integral formulation: These are boundary conditions at infinity, i.e., the vanishing of the wavefunctions at infinity for the flat space path integrals; boundary conditions connected with singular point potentials are taken into account by using a functional weight formulation. The single valuedness conditions for the quantum motion on spheres is taken into account by periodic boundary conditions; and finally Dirichlet and Neumann boundary conditions can be incorporated into the path integral by considering the infinite strength limit of point interactions. As discussed in [205, 216, 217] point interactions in turn can be incorporated in the path integral by, e.g., a simple δ-function perturbation. The path integral with this perturbation can be evaluated by the summation of a perturbation expansion, giving in the general case the energy-dependent Green's function only instead of the propagator. The latter can be obtained only in specific cases, e.g., for the free motion subject to a point interaction. The whole procedure can be repeated to incorporate arbitrarily many point interactions in order to model band structures [4]. The limit of infinitely repulsive δ-function perturbations gives Dirichlet boundary conditions at the location of the point interaction. Repeating the procedure, one can state the Green's function for a particle in a box, where an otherwise arbitrary well-behaving potential may be included!

It is also possible to discuss δ'-interactions and two- and three-dimensional point perturbations in the path integral. In these cases, however, the problem is more complicated and an ultraviolet regularization has to be done. For two- and three-dimensional point interactions this regularization prescription corresponds to a simultaneous smoothing [53] of the "δ-function" and the coupling: the coupling has to be zero in a "suitable way" [4]. In the case of the δ' point interaction the ultraviolet regularization can be performed by considering a point interaction for the one-dimensional Dirac particle, and then taking the non-relativistic limit [225, 226]. Consequently, Neumann boundary conditions are then obtained by making the δ' interaction infinitely repulsive. These examples show in a nice way the importance of regularization procedures for singular interactions in the path integral. They do work also for step-potentials [216].

In [251] we presented an up-to-date "Handbook of Feynman Path Integrals". This Table will contains in its first part an introduction into the theory of Feynman path integrals in quantum mechanics, the presentation of the Basic Path Integral Solutions, a summary about some perturbative and approximation methods, i.e., effective potentials or the semiclassical expansion, an outline of the periodic orbit theory, and coherent state path integrals. In the second part we will list over 300 exactly solvable path integrals which will thus present a reference guide for path integrals in quantum mechanics. Included are many "master formulæ", like the formulæ for the general quadratic Lagrangian, how to implement boundary conditions, point interactions, separate variables, etc. In the bibliography we cite almost 1000 titles relevant path integral references. This allows to treat almost all known problems in quantum mechanics by path integration.

2.2 Defining the Path Integral

In order to set up our notation for path integrals on curved manifolds we proceed in the canonical way (see, e.g., DeWitt [123], D'Olivio and Torres [132], Feynman [160], Gervais and Jevicki [179], [196, 248], McLaughlin and Schulman [384], Mayes and Dowker [398], Mizrahi [408], and Omote [426]). To avoid unnecessary overlap with our Table of Path Integrals [251] I give in the following only the essential information required for the path integral representation on homogeneous spaces. For more details concerning ordering prescriptions, transformation techniques, perturbation expansions, point interactions, and boundary conditions I refer to [251], where also listings of the application of Basic Path Integrals will be presented. In the following $\mathbf{x}$ denotes D-dimensional cartesian coordinates, $\mathbf{q}$ some D-dimensional coordinates, and x, y, z etc. one-dimensional coordinates. We start by considering the classical Lagrangian corresponding to the line element $ds^2 = g_{ab}dq^a dq^b$ of the classical motion in some D-dimensional Riemannian space

$$\mathcal{L}_{Cl}(\mathbf{q}, \dot{\mathbf{q}}) = \frac{m}{2}\left(\frac{ds}{dt}\right)^2 - V(\mathbf{q}) = \frac{m}{2}\sum_{ab} g_{ab}(\mathbf{q})\dot{q}^a \dot{q}^b - V(\mathbf{q}) \ . \tag{2.2}$$

The quantum Hamiltonian is *constructed* by means of the Laplace-Beltrami operator

$$H = -\frac{\hbar^2}{2m}\Delta_{LB} + V(\mathbf{q}) = -\frac{\hbar^2}{2m}\sum_{ab}\frac{1}{\sqrt{g}}\frac{\partial}{\partial q^a}g^{ab}\sqrt{g}\frac{\partial}{\partial q^b} + V(\mathbf{q}) \tag{2.3}$$

as a *definition* of the quantum theory on a curved space [442]. Here are $g = \det(g_{ab})$ and $(g^{ab}) = (g_{ab})^{-1}$. The scalar product for wavefunctions on the manifold reads $(f, g) = \int d\mathbf{q}\sqrt{g}f^*(\mathbf{q})g(\mathbf{q})$, and the momentum operators which are hermitian with respect to this scalar product are given by

$$p_a = \frac{\hbar}{i}\left(\frac{\partial}{\partial q^a} + \frac{\Gamma_a}{2}\right) \ , \qquad \Gamma_a = \frac{\partial \ln\sqrt{g}}{\partial q^a} \ . \tag{2.4}$$

In terms of the momentum operators (2.4) we can rewrite $\underline{H}$ by using a product according to $g_{ab} = h_{ac}h_{cb}$ [196]. Then we obtain for the Hamiltonian (2.3) (PF - Product-Form)

$$\underline{H} = -\frac{\hbar^2}{2m}\Delta_{LB} + V(\mathbf{q}) = \frac{\hbar^2}{2m}\sum_{abc} h^{ac}p_a p_b h^{bc} + \Delta V_{PF}(\mathbf{q}) + V(\mathbf{q}) \ , \tag{2.5}$$

and for the path integral

$$K(\mathbf{q}'', \mathbf{q}'; T)$$

$$= \int_{\mathbf{q}(t')=\mathbf{q}'}^{\mathbf{q}(t'')=\mathbf{q}''} \mathcal{D}_{PF}\mathbf{q}(t)\sqrt{g(\mathbf{q})} \exp\left\{\frac{i}{\hbar}\int_{t'}^{t''}\left[\frac{m}{2}\sum_{abc}h_{ac}(\mathbf{q})h_{bc}(\mathbf{q})\dot{q}^a\dot{q}^b - V(\mathbf{q}) - \Delta V_{PF}(\mathbf{q})\right]dt\right\}$$

$$= \lim_{N\to\infty}\left(\frac{m}{2\pi i\epsilon\hbar}\right)^{ND/2}\prod_{k=1}^{N-1}\int d\mathbf{q}_k\sqrt{g(\mathbf{q}_k)}$$

$$\times \exp\left\{\frac{i}{\hbar}\sum_{j=1}^{N}\left[\frac{m}{2\epsilon}\sum_{abc}h_{ac}(\mathbf{q}_j)h_{bc}(\mathbf{q}_{j-1})\Delta q_j^a\Delta q_j^b - \epsilon V(\mathbf{q}_j) - \epsilon\Delta V_{PF}(\mathbf{q}_j)\right]\right\}. \quad (2.6)$$

ΔV_{PF} denotes the well-defined quantum potential

$$\Delta V_{PF}(\mathbf{q}) = \frac{\hbar^2}{8m}\sum_{abc}\left[4h^{ac}h^{bc}{}_{,ab} + 2h^{ac}h^{bc}\frac{h_{,ab}}{h}\right.$$

$$\left. + 2h^{ac}\left(h^{bc}{}_{,b}\frac{h_{,a}}{h} + h^{bc}{}_{,a}\frac{h_{,b}}{h}\right) - h^{ac}h^{bc}\frac{h_{,a}h_{,b}}{h^2}\right] \quad (2.7)$$

arising from the specific lattice formulation (2.6) of the path integral or the ordering prescription for position and momentum operators in the quantum Hamiltonian, respectively. Here we have used the abbreviations $\epsilon = (t'' - t')/N \equiv T/N$, $\Delta\mathbf{q}_j = \mathbf{q}_j - \mathbf{q}_{j-1}$, $\mathbf{q}_j = \mathbf{q}(t' + j\epsilon)$ $(t_j = t' + \epsilon j, j = 0, \ldots, N)$ and we interpret the limit $N \to \infty$ as equivalent to $\epsilon \to 0$, T fixed. The lattice representation can be obtained by exploiting the composition law of the time-evolution operator $U = \exp(-iHT/\hbar)$, respectively its semi-group property. The classical Lagrangian is modified into an effective Lagrangian via $\mathcal{L}_{eff} = \mathcal{L}_{Cl} - \Delta V$. In cartesian coordinates ordering problems do not appear in the Hamiltonian, and the path integral takes on the simple form (with obvious lattice discretization)

$$K(\mathbf{x}'', \mathbf{x}'; T) = \int_{\mathbf{x}(t')=\mathbf{x}'}^{\mathbf{x}(t'')=\mathbf{x}''} \mathcal{D}\mathbf{x}(t)\exp\left\{\frac{i}{\hbar}\int_{t'}^{t''}\left[\frac{m}{2}\dot{\mathbf{x}}^2 - V(\mathbf{x})\right]dt\right\}. \quad (2.8)$$

Actually, only to the product "$(m/2\pi i\epsilon\hbar)^{N/2} \cdot \prod_j dx_j \cdot \exp\left[(im/2\epsilon\hbar)\sum_j(\Delta x_j)^2\right]$" the meaning of a measure which is well-defined in the limit $N \to \infty$ can be given, e.g. [454].

The necessity of the quantum potential in order to obtain the correct Schrödinger equation from the short-time kernel of the path integral by means of the time evolution equation was observed quite early by several authors; among them were DeWitt [123], Gutzwiller [258], and Arthurs [12]. However, a systematic derivation of the proper quantum potential corresponding to a specific lattice prescription (as, e.g., the midpoint prescription) in the path integral was done only later on. In the Weyl ordering prescription we have for instance

$$\underline{H}(\mathbf{p}, \mathbf{q}) = \frac{1}{8m}\sum_{ab}(g^{ab}p_ap_b + 2p_ag^{ab}p_b + p_ap_bg^{ab}) + V(\mathbf{q}) + \Delta V_W(\mathbf{q}), \quad (2.9)$$

with the quantum potential (W = Weyl)

$$\Delta V_W = \frac{\hbar^2}{8m}\left(\sum_{abcd} g^{ab}\Gamma^d_{ac}\Gamma^c_{bd} - R\right) = \frac{1}{8m}\sum_{ab}\left[g^{ab}\Gamma_a\Gamma_b + 2(g^{ab}\Gamma_a)_{,b} + g^{ab}{}_{,ab}\right] . \quad (2.10)$$

The Weyl-ordering prescription leads to a path integral formulation in the mid-point formulation. However, I do not go into the details of the corresponding path integral formulation any further, and refer to the literature instead, e.g., [248, 251] and references therein.

I only use the lattice formulation (2.6) in this volume unless otherwise and explicitly stated. If the metric tensor is diagonal, i.e., $g_{ab} = f_a^2\delta_{ab}$, the quantum potential simplifies into

$$\Delta V_{PF}(\mathbf{q}) = \frac{\hbar^2}{8m}\sum_{ab}\frac{1}{f_a^2}\left[\left(\frac{f_{b,a}}{f_b}\right)^2 - 4\frac{f_{a,aa}}{f_a}\frac{\delta_{bb}}{D} + 4\frac{f_{a,a}}{f_a}\left(2\frac{f_{a,a}}{f_a}\frac{\delta_{bb}}{D} - \frac{f_{b,a}}{f_b}\right) + 2\left(\frac{f_{b,a}}{f_b}\right)_{,a}\right] . \quad (2.11)$$

Let us assume that g_{ab} is proportional to the unit tensor, i.e., $g_{ab} = f^2\delta_{ab}$. Then ΔV_{PF} simplifies further into

$$\Delta V_{PF} = \hbar^2\frac{D-2}{8m}\sum_a\frac{(D-4)f_{,a}^2 + 2f\cdot f_{,aa}}{f^4} . \quad (2.12)$$

This implies, that if the dimension of the space is $D = 2$, the quantum correction ΔV_{PF} vanishes.

Let us consider the special case that the metric is of the form ($\mathbf{q} = (a, b, z)$ are some three-dimensional coordinates)

$$ds^2(\mathbf{q}) = h^{-2}(da^2 + db^2) + u^2dz^2 , \quad (2.13)$$

with $h = h(a, b), u = u(a, b)$. Then the quantum potential is of the form

$$\Delta V_{PF} = \frac{\hbar^2}{8m}\frac{h^2}{u^2}\sum_{ab}\left[2u(u_{,aa} + u_{,bb}) - (u_{,a}^2 + u_{,b}^2)\right] . \quad (2.14)$$

In the special case that the metric is of the form ($\mathbf{q} = (a, b, z, w)$ are some four-dimensional coordinates)

$$ds^2(\mathbf{q}) = h^{-2}(da^2 + db^2) + u^2(dz^2 + dw^2) , \quad (2.15)$$

with $h = h(a, b), u = u(a, b)$, the quantum potential reads

$$\Delta V_{PF} = \frac{\hbar^2}{2m}h^2\sum_{ab}\frac{u_{,aa} + u_{,bb}}{u} . \quad (2.16)$$

These specific examples of ΔV_{PF} will be useful in the sequel.

2.3   Transformation Techniques

2.3.1   Point Canonical Transformations

Indispensable tools in path integral techniques are transformation rules. In order to avoid cumbersome notation, we restrict ourselves to the one-dimensional case. For the general case we refer to DeWitt [123], Duru and Kleinert [141, 142], Fischer, Leschke and Müller [167, 168], Gervais and Jevicki [179], Refs. [209, 218, 248, 250, 251], Ho and Inomata [278], Inomata [283], Junker [295], Kleinert [350, 351, 352], Pak and Sökmen [429], Pelster and Wunderlin [440], Steiner [478] and Storchak [484], and references therein. Implementing a transformation $x = F(q)$, one has to keep all terms of $O(\epsilon)$ in (2.8). Expanding about midpoints, the result is

$$K(F(q''), F(q'); T) = \left[F'(q'') F'(q') \right]^{-1/2} \lim_{N \to \infty} \left(\frac{m}{2\pi i \epsilon \hbar} \right)^{1/2} \prod_{k=1}^{N-1} \int dq_k \cdot \prod_{l=1}^{N} F'(\bar{q}_l)$$

$$\times \exp \left\{ \frac{i}{\hbar} \sum_{j=1}^{N} \left[\frac{m}{2\epsilon} F'^2(\bar{q}_j)(\Delta q_j)^2 - \epsilon V(F(\bar{q}_j)) - \frac{\epsilon \hbar^2}{8m} \frac{F''^2(\bar{q}_j)}{F'^4(\bar{q}_j)} \right] \right\} \tag{2.17}$$

$$\equiv \left[F'(q'') F'(q') \right]^{-\frac{1}{2}} \int_{q(t')=q'}^{q(t'')=q''} \mathcal{D}q(t) F' \exp \left\{ \frac{i}{\hbar} \int_{t'}^{t''} \left[\frac{m}{2} F'^2 \dot{q}^2 - V(F(q)) - \frac{\hbar^2}{2m} \frac{F''^2}{F'^4} \right] dt \right\} .$$

$$\tag{2.18}$$

2.3.2   Space-Time Transformations

It is obvious that the path integral representation (2.18) is not completely satisfactory. Whereas the transformed potential $V(F(q))$ may have a convenient form when expressed in the new coordinate q, the kinetic term $\frac{m}{2} F'^2 \dot{q}^2$ is in general nasty. Here the so-called "time transformation" comes into play which leads in combination with the "space transformation" already carried out to general "space-time transformations" in path integrals. The time transformation is implemented [141, 142, 248, 278, 283, 352, 440, 478, 484] by introducing a new "pseudo-time" s''. In order to do this, one first makes use of the operator identity

$$\frac{1}{H - E} = f_r(x, t) \frac{1}{f_l(x, t)(H - E) f_r(x, t)} f_l(x, t) , \tag{2.19}$$

where H is the Hamiltonian corresponding to the path integral $K(t'', t')$, and $f_{l,r}(x, t)$ are functions in q and t, multiplying from the left or from the right, respectively, onto the operator $(H - E)^{-1}$. Secondly, one introduces a new pseudo-time s'' and assumes that the constraint

$$\int_0^{s''} ds f_l(F(q(s)), s)) f_r(F(q(s)), s)) = T = t'' - t' \tag{2.20}$$

has for all admissible paths a unique solution $s'' > 0$ given by

$$s'' = \int_{t'}^{t''} \frac{dt}{f_l(x,t)f_r(x,t)} = \int_{t'}^{t''} \frac{ds}{F'^2(q(s),s)} \ . \tag{2.21}$$

Here one has made the choice $f_l(F(q(s),s)) = f_r(F(q(s),s)) = F'(q(s),s))$ in order that in the final result the metric coefficient in the kinetic energy term is equal to one. A convenient way to derive the corresponding transformation formulæ uses the energy dependent Green's function $G(E)$ of the kernel $K(T)$ defined by

$$G(q'',q';E) = \left\langle q'' \left| \frac{1}{H - E - \mathrm{i}\epsilon} \right| q' \right\rangle = \frac{\mathrm{i}}{\hbar} \int_0^\infty dT \mathrm{e}^{\mathrm{i}(E+\mathrm{i}\epsilon)T/\hbar} K(q'',q';T) \ . \tag{2.22}$$

For the path integral (2.18) one obtains the following transformation formula

$$K(x'',x';T) = \int_{\mathbb{R}} \frac{dE}{2\pi\mathrm{i}} \mathrm{e}^{-\mathrm{i}ET/\hbar} G(q'',q';E) \ , \tag{2.23}$$

$$G(q'',q';E) = \frac{\mathrm{i}}{\hbar} \left[F'(q'')F'(q') \right]^{1/2} \int_0^\infty ds'' \hat{K}(q'',q';s'') \ , \tag{2.24}$$

with the transformed path integral $\hat{K}$ having the form

$$\hat{K}(q'',q';s'') = \lim_{N\to\infty} \left(\frac{m}{2\pi\mathrm{i}\epsilon\hbar} \right)^{1/2} \prod_{k=1}^{N-1} \int dq_k$$

$$\times \exp\left\{ \frac{\mathrm{i}}{\hbar} \sum_{j=1}^N \left[\frac{m}{2\epsilon} (\Delta q_j)^2 - \epsilon F'^2(\bar{q}_j) \big(V(F(\bar{q})) - E \big) - \epsilon \Delta V(\bar{q}_j) \right] \right\} \tag{2.25}$$

$$\equiv \int_{q(0)=q'}^{q(s'')=q''} \mathcal{D}q(s) \exp\left\{ \frac{\mathrm{i}}{\hbar} \int_0^{s''} \left[\frac{m}{2}\dot{q}^2 - F'^2(q)\big(V(F(q)) - E\big) - \Delta V(q) \right] ds \right\} \ , \tag{2.26}$$

and with the quantum potential ΔV given by

$$\Delta V(q) = \frac{\hbar^2}{8m} \left(3 \frac{F''^2}{F'^2} - 2 \frac{F'''}{F'} \right) \ . \tag{2.27}$$

Note that ΔV has the form of a Schwarz derivative of F. A rigorous lattice derivation is far from being trivial and has been discussed by some authors. Recent attempts to put it on a sound footing can be found in Castrigiano and Stärk [101], Fischer et al. [167, 168], and Young and DeWitt-Morette [530].

Let us consider a pure time transformation in a path integral. Let

$$G(\mathbf{q}'',\mathbf{q}';E) = \sqrt{f(\mathbf{q}')f(\mathbf{q}'')} \frac{\mathrm{i}}{\hbar} \int_0^\infty ds'' \big\langle \mathbf{q}'' \big| \exp\big(-\mathrm{i}s''\sqrt{f}\,(H - E)\sqrt{f}/\hbar \big) \big| \mathbf{q}' \big\rangle \ , \tag{2.28}$$

which corresponds to the introduction of the "pseudo-time" $s'' = \int_{t'}^{t''} ds/f(\mathbf{q}(s))$ and we assume that the Hamiltonian H is product ordered. Then

$$G(\mathbf{q}'',\mathbf{q}';E) = \frac{\mathrm{i}}{\hbar} (f'f'')^{\frac{1}{2}(1-D/2)} \int_0^\infty \tilde{K}(\mathbf{q}'',\mathbf{q}';s'') ds'' \tag{2.29}$$

with the path integral

$$
\tilde{K}(\mathbf{q}'', \mathbf{q}'; s'') = \int\limits_{\mathbf{q}(0)=\mathbf{q}'}^{\mathbf{q}(s'')=\mathbf{q}''} \mathcal{D}\mathbf{q}(s) \sqrt{\tilde{g}}
$$

$$
\times \exp\left\{ \frac{i}{\hbar} \int_0^{s''} \left[\frac{m}{2} \sum_{abc} \tilde{h}_{ac}\tilde{h}_{cb}\dot{q}^a\dot{q}^b - f\Big(V(\mathbf{q}) + \Delta V_{PF}(\mathbf{q}) - E \Big) \right] ds \right\} \ . \quad (2.30)
$$

Here $\tilde{h}_{ac} = h_{ac}/\sqrt{f}$, $\sqrt{\tilde{g}} = \det(\tilde{h}_{ac})$ and (2.30) is of the canonical product form.

2.3.3   Separation of Variables

Separation Formula for the Path Integral.

By the same technique also the separation of variables in path integrals can be stated, c.f. [204]. Let us consider a $D = d + d'$ dimensional system, where $\mathbf{x}$ represents the d-dimensional coordinate and $\mathbf{z}$ the d'-dimensional coordinate. For simplicity we consider the special case where the metric tensor for the $\mathbf{x}$ coordinates is equal to $f^2(\mathbf{z})\mathbb{1}$, and the metric tensor for the $\mathbf{z}$ coordinates is diagonal and denoted by $\mathbf{g} = \mathbf{g}(\mathbf{z})$ with elements $g_i = g_{ii}(\mathbf{z})$, $i = 1, \dots, d'$. Furthermore, we incorporate a potential of the special form $\widehat{W}(\mathbf{x}, \mathbf{z}) = W(\mathbf{z}) + V(\mathbf{x})/f^2(\mathbf{z})$ which also includes all quantum potentials arising from metric terms. Then $(g = \prod g_i^2)$

$$
\int\limits_{\mathbf{z}(t')=\mathbf{z}'}^{\mathbf{z}(t'')=\mathbf{z}''} \mathcal{D}\mathbf{z}(t) f^d(\mathbf{z})\sqrt{g} \int\limits_{\mathbf{x}(t')=\mathbf{x}'}^{\mathbf{x}(t'')=\mathbf{x}''} \mathcal{D}\mathbf{x}(t)
$$

$$
\times \exp\left\{ \frac{i}{\hbar} \int_{t'}^{t''} \left[\frac{m}{2}\Big((\mathbf{g}\cdot\dot{\mathbf{z}}^2) + f^2(\mathbf{z})\dot{\mathbf{x}}^2 \Big) - \left(\frac{V(\mathbf{x})}{f^2(\mathbf{z})} + W(\mathbf{z}) \right) \right] dt \right\}
$$

$$
= [f(\mathbf{z}')f(\mathbf{z}'')]^{-d/2} \int dE_\lambda \Psi_\lambda^*(\mathbf{x}')\Psi_\lambda(\mathbf{x}'') \int\limits_{\mathbf{z}(t')=\mathbf{z}'}^{\mathbf{z}(t'')=\mathbf{z}''} \mathcal{D}\mathbf{z}(t)\sqrt{g}
$$

$$
\times \exp\left\{ \frac{i}{\hbar} \int_{t'}^{t''} \left[\frac{m}{2}(\mathbf{g}\cdot\dot{\mathbf{z}}^2) - W(\mathbf{z}) - \frac{E_\lambda}{f^2(\mathbf{z})} \right] dt \right\} \ . \quad (2.31)
$$

Here we assume that the d-dimensional $\mathbf{x}$-path integration has the special representation

$$
\int\limits_{\mathbf{x}(t')=\mathbf{x}'}^{\mathbf{x}(t'')=\mathbf{x}''} \mathcal{D}\mathbf{x}(t) \exp\left\{ \frac{i}{\hbar} \int_{t'}^{t''} \left[\frac{m}{2}\dot{\mathbf{x}}^2 - V(\mathbf{x}) \right] dt \right\} = \int dE_\lambda \Psi_\lambda^*(\mathbf{x}')\Psi_\lambda(\mathbf{x}'') e^{-iE_\lambda T/\hbar} \ . \quad (2.32)
$$

Transformation Formula for Separable Coordinate Systems.

We want to look for coordinate systems which separate the relevant partial differential equations, i.e., the Hamiltonian, and more important from our point of view, the path integral. In order to develop a separation formula we consider according to [416] the Lagrangian $\mathcal{L} = \frac{m}{2} \sum_{i=1}^{D} h_i^2(\boldsymbol{\xi}) \dot{\xi}_i^2$ and the Laplacian Δ_{LB} acting on wavefunctions Ψ, respectively, in the following way and only orthogonal coordinate systems are taken into account

$$\Delta_{LB} = \sum_{i=1}^{D} \frac{1}{\prod_{j=1}^{D} h_j(\boldsymbol{\xi})} \frac{\partial}{\partial \xi_i} \left(\frac{\prod_{k=1}^{D} h_k(\boldsymbol{\xi})}{h_i^2(\boldsymbol{\xi})} \frac{\partial}{\partial \xi_i} \right) \tag{2.33}$$

$\boldsymbol{\xi}$ denotes the set of variables $(\xi_1, \ldots, \xi_D)$. As was shown by Moon and Spencer [411] the necessary and sufficient condition for simple separability in a D-dimensional Riemannian space with an orthogonal coordinate system $\boldsymbol{\xi}$, is the factorization of the h_i according to

$$\frac{\prod_{j=1}^{D} h_j(\boldsymbol{\varrho})}{h_i^2(\boldsymbol{\varrho})} = M_{i1} \prod_{j=1}^{D} f_j(\varrho_j) \tag{2.34}$$

such that

$$M_{i1}(\varrho_1, \ldots, \varrho_{i-1}, \varrho_{i+1}, \ldots, \varrho_D) = \frac{\partial S}{\partial \Phi_{i1}} = \frac{S(\boldsymbol{\varrho})}{h_i^2(\boldsymbol{\varrho})} \ , \qquad \frac{h^{1/2}}{S(\boldsymbol{\varrho})} = \prod_{i=1}^{D} f_i(\varrho_i) \ , \tag{2.35}$$

$$S(\boldsymbol{\xi}) = \begin{vmatrix} \Phi_{11}(\xi_1) & \Phi_{12}(\xi_1) & \cdots & \Phi_{1D}(\xi_1) \\ \Phi_{21}(\xi_2) & \Phi_{22}(\xi_2) & \cdots & \Phi_{2D}(\xi_2) \\ \vdots & \vdots & \ddots & \vdots \\ \Phi_{D1}(\xi_D) & \Phi_{D2}(\xi_D) & \cdots & \Phi_{DD}(\xi_D) \end{vmatrix} = \det(\Phi_{ij}(\xi_i)) = \prod_{i=1}^{D} \frac{h_i(\boldsymbol{\xi})}{f_i(\xi_i)} \ , \tag{2.36}$$

where $h = \prod_{i=1}^{D} h_i(\boldsymbol{\varrho})$, M_{i1} is called the cofactor of Φ_{i1}, and S is the Stäckel determinant [301, 412, 416]. Note the property of the Stäckel-matrix [416]

$$\sum_{i=1}^{D} \frac{\Phi_{ij}(\xi_i)}{h_i^2(\boldsymbol{\xi})} = \frac{1}{S} \sum_{i=1}^{D} \Phi_{ij}(\xi_i) M_{1i} = \delta_{1j} \ . \tag{2.37}$$

A tabulation of the Stäckel-matrix for the coordinate systems in $\mathbb{R}^3$ can be found in [416]. We abbreviate $\Gamma_i = f_i'/f_i$, introduce the (new) momentum operators $P_{\xi_i} = \frac{\hbar}{i}(\partial_{\xi_i} + \frac{1}{2} f_i'/f_i)$, and write the Legendre transformed Hamiltonian [247] as follows

$$\begin{aligned} H - E &= -\frac{\hbar^2}{2m} \Delta_{LB} - E \\ &= -\frac{\hbar^2}{2m} \sum_{i=1}^{D} \frac{1}{\prod_{j=1}^{D} h_j} \frac{\partial}{\partial \xi_i} \left(\frac{\prod_{k=1}^{D} h_k}{h_i^2} \frac{\partial}{\partial \xi_i} \right) - E \\ &= -\frac{\hbar^2}{2m} \frac{1}{S} \sum_{i=1}^{D} \left[\frac{1}{f_i} \frac{\partial}{\partial \xi_i} \left(f_i \frac{\partial}{\partial \xi_i} \right) \right] - E = -\frac{\hbar^2}{2m} \frac{1}{S} \sum_{i=1}^{D} M_{1i} \left(\frac{\partial^2}{\partial \xi_i^2} + \Gamma_i \frac{\partial}{\partial \xi_i} \right) - E \\ &= \frac{1}{S} \sum_{i=1}^{D} M_{1i} \left[\frac{1}{2m} P_i^2 - E h_i^2 + \frac{\hbar^2}{8m} \left(\Gamma_i^2 + 2\Gamma_i' \right) \right] \\ &= \frac{1}{S} \sum_{i=1}^{D} M_{1i} \left[\frac{1}{2m} P_i^2 - \frac{\hbar^2}{2m} \sum_{j=1}^{D} k_j^2 \Phi_{ij}(\xi_i) + \frac{\hbar^2}{8m} \left(\Gamma_i^2 + 2\Gamma_i' \right) \right] \ . \end{aligned} \tag{2.38}$$

The k_i^2 are the separation constants [473] with $E = \frac{\hbar^2}{2m} k_1^2$ the energy, c.f. (2.37). We obtain according to the general theory by means of a space-time transformation the following identity in the path integral [223] ($g = \prod h_i^2$)

$$
\int\limits_{\boldsymbol{\xi}(t')=\boldsymbol{\xi}'}^{\boldsymbol{\xi}(t'')=\boldsymbol{\xi}''} \mathcal{D}\boldsymbol{\xi}(t) \sqrt{g} \exp\left\{ \frac{i}{\hbar} \int_{t'}^{t''} \left[\frac{m}{2}(\mathbf{h} \cdot \dot{\boldsymbol{\xi}})^2 - \Delta V_{PF}(\boldsymbol{\xi}) \right] dt \right\}
$$

$$
= \int\limits_{\boldsymbol{\xi}(t')=\boldsymbol{\xi}'}^{\boldsymbol{\xi}(t'')=\boldsymbol{\xi}''} \mathcal{D}\boldsymbol{\xi}(t) \prod_{i=1}^{D} \sqrt{\frac{S}{M_{1i}}} \exp\left\{ \frac{i}{\hbar} \int_{t'}^{t''} \left[\frac{m}{2} S \frac{\dot{\xi}_i^2}{M_{1i}} - \Delta V_{i,PF}(\boldsymbol{\xi}) \right] dt \right\}
$$

$$
= (S'S'')^{\frac{1}{2}(1-D/2)} \int_{\mathbb{R}} \frac{dE}{2\pi\hbar} e^{-iET/\hbar} \int_0^\infty ds'' \prod_{i=1}^{D} (M_{i1}' M_{i1}'')^{1/4}
$$

$$
\times \int\limits_{\xi_i(0)=\xi_i'}^{\xi_i(s'')=\xi_i''} \mathcal{D}\xi_i(s) \exp\left\{ \frac{i}{\hbar} \int_0^{s''} \left[\frac{m}{2} \dot{\xi}_i^2 + \frac{\hbar^2}{2m} \sum_{j=1}^{D} k_j^2 \Phi_{ij}(\xi_i) - \frac{\hbar^2}{8m} \left(\Gamma_i^2 + 2\Gamma_i' \right) \right] ds \right\} .
$$

$$(2.39)$$

Therefore we have obtained a complete separation of variables in the $\boldsymbol{\xi}$-path integral.

2.4 Group Path Integration

One may ask, if it is possible to analyze the path integral of a potential problem in terms of its dynamical symmetry group [289]. In order to look at such a path integral formulation we consider in a not-necessarily positive definite space with signature

$$
(g_{ab}) = \text{diag}(\underbrace{+1, \ldots, +1}_{p \text{ times}}, \underbrace{-1, \ldots, -1}_{q \text{ times}}) \tag{2.40}
$$

the generic Lagrangian $\mathcal{L}(\mathbf{x}, \dot{\mathbf{x}}) = \frac{m}{2} g_{ab} \dot{x}^a \dot{x}^b - V(\mathbf{x})$ ($\mathbf{x} \in \mathbb{R}^{p+q}$) and its corresponding short-time kernel $K(\mathbf{x}_j, \mathbf{x}_{j-1}; \epsilon)$. The short-time kernel is evaluated by harmonic analysis with respect to the symmetry group of the Lagrangian. This is usually a Lie group. In order to do this one seeks for an expansion of $e^{z\mathbf{x}_{j-1} \cdot \mathbf{x}_j}$ in terms of representations of the group. This may be done in generalized polar coordinates involving generalized spherical harmonics. We assume that we can introduce a generalized polar variable τ and a set of generalized angular variables $\boldsymbol{\vartheta}$ such that $x_\nu = \tau \hat{e}_\nu(\vartheta_1, \ldots, \vartheta_{p+q-1})$ ($\nu = 1, \ldots, p+q$), where the $\hat{e}$'s are unit vectors in some suitably chosen (timelike, spacelike or lightlike) set with $V(\mathbf{x}) = V(\tau)$ [66]. To perform the integration over the spherical harmonics the scalar product $\mathbf{x}_{j-1} \cdot \mathbf{x}_j$ must be rewritten in terms of a group element, say a function $f(g_{j-1}^{-1} g_j)$, such that $e^{z\mathbf{x}_{j-1} \cdot \mathbf{x}_j} = e^{zf(g_{j-1}^{-1} g_j)}$. Since $g_{j-1}^{-1} g_j$ is a group element we set $F(g) = e^{zf(g)}$. The expansion then yields

$$
F(g) = \int dE_\lambda d_\lambda \sum_m \hat{F}_m^\lambda(\tau) D_m^\lambda(g) , \qquad \hat{F}_m^\lambda = \int_G F(g) D_m^{\lambda*}(g^{-1}) dg , \tag{2.41}
$$

where dg is the invariant group (Haar) measure . $\int dE_\lambda$ stands for a Lebesgue-Stieltjes integral to include discrete ($\int dE_\lambda \to \sum_\lambda$) as well as continuous representations. The summation index m may be a multiindex. d_λ denotes (in the compact case) the dimension of the representation; otherwise we take

$$d_\lambda \int_G D_m^\lambda(g) D_{m'}^{\lambda'}{}^*(g) dg = \delta(\lambda, \lambda') \delta_{m,m'} \tag{2.42}$$

as a definition for d_λ. $\delta(\lambda, \lambda')$ can denote a Kronecker delta or a δ-function, depending on whether the variable λ is a discrete or continuous parameter. The path integration over the group elements can be performed due to their orthonormality. Choosing a particular basis in the group fixes the matrix elements of the representation and makes it possible to expand the $D_m^\lambda(g'^{-1}g'')$ in terms of the wavefunctions $\Psi_{mn}^\lambda(\{\vartheta\})$, corresponding to the eigenfunctions of the Casimir operator of the group. This can be, e.g., done by means of the group (composition) law

$$D_{mn}^\lambda(g_a^{-1} g_b) = \sum_k D_{kn}^{\lambda}{}^*(g_a) D_{km}^\lambda(g_b) \ . \tag{2.43}$$

In [66] the authors concentrated on the cases where the harmonic analysis can be performed either with the $D_m^\lambda(g)$ as the characters of the group or the zonal spherical functions. In the case of SO(D) the Casimir operator is the Legendre operator and the wavefunctions are the hyperspherical harmonics $S_l(\mathbf{\Omega})$ which are products of Gegenbauer polynomials. However, the method is of course more general, c.f. Böhm and Junker [67], Dowker [134, 135], Marinov and Terentyev [393], and Picken [441]. These spherical functions are eigen-functions of the corresponding Laplace-Beltrami operator on a homogeneous space, and the entire Hilbert space is spanned by a complete set of associated spherical functions D_{0m}^λ (Gelfand et al. [176, 177] and Vilenkin [510]).

Dowker [135] and Marinov and Terentyev [393] have considered group path integration on unitary groups. In particular, Dowker [134] has observed that the exact propagator equals its corresponding semiclassical approximation. A general consideration of the propagator on a group manifold is due to Picken [441].

However, whereas a path integral representation in polar coordinates may be in most cases sufficient and convenient, it is not the only possibility. In particular in the case of a homogeneous space, the corresponding path integral allows as many spectral expansions in coordinate space representations as there are coordinate systems in this space which separate the Laplace-Beltrami operator. To our knowledge, the path integral representations in polar coordinates have allowed to derive several of the Basic Path Integrals, among them the path integral of the radial harmonic oscillator, the Pöschl–Teller and the modified Pöschl–Teller potential (see below in the discussion of the path integration on the SU(2) and SU(1, 1) group manifolds). In the next section we discuss interbasis expansions which allow to switch from one coordinate space representation into an other in the path integral. These new coordinate space representations of a path integral in a homogeneous space then can give rise to new and more complicated path integral identities.

In the following we concentrate on two principal possibilities of harmonic analysis of a path integral on a group space. They are

1. *H_α is isomorphic to the group manifold G: $H_\alpha \simeq G$.*
   $H_\alpha \simeq G$ is a quite strong constraint, so it is not surprising that there are only four groups which satisfy it. They are $SO(2), SO(1,1), SU(2)$, and $SU(1,1)$. The harmonic analysis in these cases is performed by the characters of the group $\chi^l(g)$. For the one-parameter groups $SO(2)$ and $SO(1,1)$ the irreducible representations are one-dimensional and in fact the general Fourier transformation (2.41) is reduced to the usual Fourier and Laplace transformation, respectively. Therefore we are left as the only nontrivial examples with $SU(2)$ and $SU(1,1)$.

2. *H_α is given by a group quotient: $H_\alpha = G/H$.*
   This case describes motion on quotient space group manifolds, i.e., on a homogeneous space [284, 287]. Examples are the motion in (pseudo-) Euclidean spaces, and on spheres and pseudospheres. The harmonic analysis in this case is usually performed by means of the zonal spherical harmonics $D^l_{00}(g)$.

Another aspect of group path integration is the so-called interbasis expansion for problems which are separable in more than one coordinate system. In the case of potential problems these potentials are called super-integrable. This aspect is very closely connected to the fact that such problems have more integrals of motions as necessary for complete separation of variables, and that the underlying dynamical symmetry group allows the representation of the problem in various coordinate space representations. Super-integrable systems can be found in Euclidean space, as well as in spaces of constant curvature. The basic formula is quite simple being

$$|\mathbf{k}> = \int dE_{\mathbf{l}} C_{\mathbf{l},\mathbf{k}} |\mathbf{l}> \, , \tag{2.44}$$

where $|\mathbf{k}>$ stands for a basis of eigenfunctions of the Hamiltonian in the coordinate space representation $\mathbf{k}$, and $\int dE_{\mathbf{l}}$ is the spectral-expansion with respect to the coordinate space representation $\mathbf{l}$ with coefficients $C_{\mathbf{l},\mathbf{k}}$ which can be discrete, continuous or both. The main difficulty is, in case one has two coordinate space representations in the quantum numbers $\mathbf{k}$ and $\mathbf{l}$, respectively, to find the expansion coefficients $C_{\mathbf{l},\mathbf{k}}$. Well-known are the expansions which involve cartesian coordinates and polar coordinates. In the simple case of free quantum motion in Euclidean space, this means that exponentials representing plane waves are expanded in terms of Bessel functions and spherical waves (a discrete interbasis expansion).

This general method of changing a coordinate basis in quantum mechanics can now be used in the path integral. We assume that we can expand the short-time kernel, respectively the exponential $e^{z\mathbf{x}_j - 1 \cdot \mathbf{x}_j}$ in terms of matrix elements of a group according to (2.41). Here a specific coordinate basis has been chosen. We then can change the coordinate basis by means of (2.44). Due to the unitarity of the expansion coefficients $C_{\mathbf{l},\mathbf{k}}$ the short-time kernel is expanded in the new coordinate basis, and the orthonormality of the basis allows to perform explicitly the path integral, exactly in the same way as in the original coordinate basis.

From the two (or more) different equivalent coordinate space representations, formulæ and path integral identities can be derived. These identities actually correspond to integral and summation identities, respectively, between special functions. The case of the expansion from cartesian coordinates to polar coordinates has been studied by Peak and Inomata [439] and they obtained the solution of the radial harmonic oscillator as well. The path integral solution of the radial harmonic oscillator in turn enables one to calculate numerous path integral problems related to the radial harmonic oscillator, actually problems which are of the so-called Besselian type, including the Coulomb problem.

2.5 Klein-Gordon Particle

The path integral formulation of a Klein-Gordon particle was already presented by Feynman [161] in one of his classical papers. It goes as follows: One considers the Green's function corresponding to the Klein-Gordon equation

$$(\Box + M^2)G(x'', x') = \delta(x'' - x') \ , \tag{2.45}$$

where $\Box = g^{\mu\nu}\nabla_\mu\nabla_\nu = \partial_t^2 - \Delta$ is the Klein-Gordon operator, and $\delta(x)$ the four-dimensional δ-function. M is the mass of the particle. According to [161, 456] we can now write $G(x'', x')$ as

$$G(x'', x') = \frac{\mathrm{i}}{2\hbar} \int_0^\infty d\tau \, \mathrm{e}^{-\mathrm{i}M^2\tau/2\hbar} K(x'', x'; \tau) \ , \tag{2.46}$$

where $s \in [0, \tau]$ is a new time-like variable. The new propagator $K(x'', x'; \tau)$ describes time evolution in τ from x' to x'' and can in turn be written as

$$K(x'', x'; \tau) = \int_{x(0)=x'}^{x(\tau)=x''} \mathcal{D}x(s) \exp\left(\frac{\mathrm{i}}{2\hbar}\int_0^\tau g_{\mu\nu}\dot{x}^\mu\dot{x}^\nu ds\right) \ . \tag{2.47}$$

This path integral is usually a path integral with an indefinite (Minkowski-like) metric, and satisfies a Schrödinger-like equation

$$\mathrm{i}\frac{\partial K(x'', x'; \tau)}{\partial \tau} = \Box_{x''} K(x'', x'; \tau) \ , \tag{2.48}$$

where τ serves as the time parameter, together with the initial condition $\lim_{\tau\to 0} K(x'', x'; \tau) = \delta(x'' - x')$. Therefore, the propagator can be seen as a usual quantum mechanical path integral, defined on a manifold with metric $g_{\mu\nu}$. Potentials and magnetic fields can be incorporated in an obvious way. The path integral representation of the Klein-Gordon particle can be kept in mind if we deal with path integral representations on pseudo-Euclidean spaces or Minkowski-spaces. Equation (2.46) always allows to obtain the Green's function from a propagator in such a geometry.

2.6 Basic Path Integrals

In this Section we present the path integrals which we consider as the Basic Path Integrals, and which will be useful in the sequel.

2.6.1 The Quadratic Lagrangian

The first elementary example is the path integral for the harmonic oscillator. It has been first evaluated by Feynman [159, 160]. We have the identity ($x \in \mathbb{R}$)

Basic Path Integral: Harmonic Oscillator:

$$
\int_{x(t')=x'}^{x(t'')=x''} \mathcal{D}x(t) \exp\left[\frac{im}{2\hbar} \int_{t'}^{t''} \left(\dot{x}^2 - \omega^2 x^2\right) dt\right]
$$
$$
= \sqrt{\frac{m\omega}{2\pi i\hbar \sin\omega T}} \exp\left\{\frac{im\omega}{2\hbar}\left[(x'^2 + x''^2)\cot\omega T - \frac{2x'x''}{\sin\omega T}\right]\right\} . \qquad (2.49)
$$

The path integral for general quadratic Lagrangians, i.e., Gaussian Path Integrals, can also be stated exactly ($\mathbf{x} \in \mathbb{R}^D$)

Basic Path Integral: Quadratic Lagrangian:

$$
\int_{\mathbf{x}(t')=\mathbf{x}'}^{\mathbf{x}(t'')=\mathbf{x}''} \mathcal{D}\mathbf{x}(t) \exp\left(\frac{i}{\hbar}\int_{t'}^{t''} \mathcal{L}(\mathbf{x}, \dot{\mathbf{x}})dt\right) = \left(\frac{1}{2\pi i\hbar}\right)^{D/2} \sqrt{\det\left(-\frac{\partial^2 S[\mathbf{x}_{Cl}]}{\partial x_a'' \partial x_b'}\right)} \exp\left(\frac{i}{\hbar}S[\mathbf{x}_{Cl}]\right).
$$
$$
\tag{2.50}
$$

Here $\mathcal{L}(\mathbf{x}, \dot{\mathbf{x}})$ denotes any classical Lagrangian at most quadratic in $\mathbf{x}$ and $\dot{\mathbf{x}}$, and $S[\mathbf{x}_{Cl}] = \int_{t'}^{t''} \mathcal{L}(\mathbf{x}_{Cl}, \dot{\mathbf{x}}_{Cl})dt$ the corresponding classical action evaluated along the classical solution $\mathbf{x}_{Cl}$ satisfying the boundary conditions $\mathbf{x}_{Cl}(t') = \mathbf{x}'$, $\mathbf{x}_{Cl}(t'') = \mathbf{x}''$ (we assume that the classical dynamics allows only a single classical path). The determinant appearing in (2.50) is known as the van Vleck-Pauli-Morette determinant (see e.g. DeWitt [123], Morette [414], van Vleck [504] and references therein). The explicit evaluation of $S[\mathbf{x}_{Cl}]$ may have any degree of complexity due to complicated classical solutions of the Euler-Lagrange equations as the classical equations of motion.

Furthermore, the formula of the general quadratic Lagrangian (2.50) serves as a starting point for the semi-classical expansion, and the general moments formula in the path integral (DeWitt-Morette [124], Mizrahi [409], and Roepstorff [454]).

Based on the solution of the harmonic oscillator and the quadratic Lagrangian, it is possible to derive expressions for the generating functional [369] in a perturbative approach which is also applicable in quantum field theory (Feynman graphs!). They are based on the moments formula for arbitrary functionals F of positions and momenta (the analogue of Wick's theorem in quantum mechanics) [409]. Some important moments formulæ can be found in [454].

Moreover, very satisfying expressions exist for the trace of the Euclidean time-evolution kernel, i.e., the partition function in terms of an effective potential (Feynman and Hibbs [164], Feynman and Kleinert [165], Giachetti et al. [180], and Kleinert [352]). For details we refer to the literature [164, 165, 352, 369], and to our Table of Path Integrals [251].

2.6.2 The Radial Harmonic Oscillator

In order to evaluate the path integral for the radial harmonic oscillator, one has to perform a separation of the angular variables, see [183, 439]. In the following I have also displayed the case where for the modified Bessel functions in the functional weight the asymptotic expansion for small ϵ according to

$$I_\nu(z) \simeq (2\pi z)^{-\frac{1}{2}} e^{z-(\nu^2-\frac{1}{4})/2z} \qquad (|z| \gg 1, \quad \Re(z) > 0) \ . \tag{2.51}$$

($z = mr_j r_{j-1}/i\epsilon\hbar$) has been used to give a radial potential $V_\lambda = \hbar^2(\lambda^2 - \frac{1}{2})/2mr^2$. But this expansion is valid only for $\Re(z) > 0$ and the emerging path integral violates the boundary conditions at the origin [476]. However, particularly in space-time transformations such an expansion is indispensable in numerous calculations, and it turns out that it can be rigorously justified for this kind of transformations [167]. But one has to keep in mind that the Besselian functional weight in the lattice approach [167, 183, 248, 439, 478] is necessary for the explicit evaluation of the radial harmonic oscillator path integral. One has ($r > 0$, $I_\lambda(z)$ is a modified Bessel function)

Basic Path Integral: Radial Harmonic Oscillator:

$$\int_{r(t')=r'}^{r(t'')=r''} \mathcal{D}r(t) \exp\left\{\frac{i}{\hbar}\int_{t'}^{t''}\left[\frac{m}{2}\left(\dot{r}^2 - \omega^2 r^2\right) - \hbar^2\frac{\lambda^2-\frac{1}{4}}{2mr^2}\right]dt\right\}$$

$$\equiv \int_{r(t')=r'}^{r(t'')=r''} \mathcal{D}r(t)\mu_\lambda[r^2]\exp\left[\frac{im}{2\hbar}\int_{t'}^{t''}\left(\dot{r}^2 - \omega^2 r^2\right)dt\right]$$

$$= \lim_{N\to\infty}\left(\frac{m}{2\pi i\epsilon\hbar}\right)^{N/2}\prod_{k=1}^{N-1}\int_0^\infty dr_k \cdot \prod_{l=1}^N \mu_\lambda[r_l r_{l-1}] \cdot \exp\left\{\frac{i}{\hbar}\sum_{j=1}^N\left[\frac{m}{2\epsilon}(\Delta r_j)^2 - \epsilon V(r_j)\right]\right\}$$

$$= \sqrt{r'r''}\frac{m\omega}{i\hbar\sin\omega T}\exp\left[-\frac{m\omega}{2i\hbar}(r'^2 + r''^2)\cot\omega T\right]I_\lambda\left(\frac{m\omega r'r''}{i\hbar\sin\omega T}\right) \ , \tag{2.52}$$

and the functional weight μ_λ is given by $\mu_\lambda[r_j r_{j-1}] = \sqrt{2\pi z_j}\, e^{-z_j} I_\lambda(z_j)$ with $z_j = mr_j r_{j-1}/i\epsilon\hbar$.

2.6.3 The Pöschl–Teller Potential

There are two Basic Path Integral Solutions based on the SU(2) (Böhm and Junker [66], Duru [139], Fischer et al. [168], and Inomata and Wilson [290]) and SU(1,1) [66] group path integration, respectively. The first yields the following path integral identity for the Pöschl–Teller potential ($0 < x < \pi/2$)

Basic Path Integral: Pöschl–Teller Potential:

$$
\int_{x(t')=x'}^{x(t'')=x''} \mathcal{D}x(t) \exp\left\{ \frac{i}{\hbar} \int_{t'}^{t''} \left[\frac{m}{2}\dot{x}^2 - \frac{\hbar^2}{2m}\left(\frac{\alpha^2 - \frac{1}{4}}{\sin^2 x} + \frac{\beta^2 - \frac{1}{4}}{\cos^2 x} \right) \right] dt \right\}
$$

$$
\equiv \int_{x(t')=x'}^{x(t'')=x''} \mathcal{D}x(t)\mu_{\alpha,\beta}[\sin x, \cos x] \exp\left(\frac{im}{2\hbar} \int_{t'}^{t''} \dot{x}^2 dt \right) \tag{2.53}
$$

$$
= \int_{\mathbb{R}} \frac{dE}{2\pi i} e^{-iET/\hbar} \frac{m}{2\hbar^2} \sqrt{\sin 2x' \sin 2x''} \frac{\Gamma(m_1 - L_E)\Gamma(L_E + m_1 + 1)}{\Gamma(m_1 + m_2 + 1)\Gamma(m_1 - m_2 + 1)}
$$

$$
\times \left(\frac{1 - \cos 2x'}{2} \cdot \frac{1 - \cos 2x''}{2} \right)^{(m_1 - m_2)/2} \left(\frac{1 + \cos 2x'}{2} \cdot \frac{1 + \cos 2x''}{2} \right)^{(m_1 + m_2)/2}
$$

$$
\times {}_2F_1\left(-L_E + m_1, L_E + m_1 + 1; m_1 - m_2 + 1; \frac{1 - \cos 2x_<}{2} \right)
$$

$$
\times {}_2F_1\left(-L_E + m_1, L_E + m_1 + 1; m_1 + m_2 + 1; \frac{1 + \cos 2x_>}{2} \right) \tag{2.54}
$$

$$
= \sum_{n \in \mathbb{N}_0} e^{-iE_N T/\hbar} \Phi_n^{(\alpha,\beta)}(x')\Phi_n^{(\alpha,\beta)}(x'') , \tag{2.55}
$$

$$
\Phi_n^{(\alpha,\beta)}(x) = \left[2(\alpha + \beta + 2n + 1) \frac{n!\Gamma(\alpha + \beta + n + 1)}{\Gamma(\alpha + n + 1)\Gamma(\beta + n + 1)} \right]^{1/2}
$$

$$
\times (\sin x)^{\alpha + 1/2}(\cos x)^{\beta + 1/2} P_n^{(\alpha,\beta)}(\cos 2x) , \tag{2.56}
$$

$$
E_n = \frac{\hbar^2}{2m}(\alpha + \beta + 2n + 1)^2 , \tag{2.57}
$$

with $m_{1/2} = \frac{1}{2}(\lambda \pm \kappa)$, $L_E = -\frac{1}{2} + \frac{1}{2}\sqrt{2mE}/\hbar$. $x_{<,>}$ the larger, smaller of x', x'', respectively. The $P_n^{(\alpha,\beta)}(z)$ are Jacobi polynomials, and $_2F_1(a, b; c; z)$ is the hypergeometric function. Note that I have displayed also a functional weight formulation similarly as in the radial path integral with the functional weight $\mu_{\alpha,\beta}$ defined by [198]

$$
\mu_{\alpha,\beta}[\sin x, \cos x] := \lim_{N \to \infty} \prod_{j=1}^{N} \mu_{\alpha,\beta}[\sin x_j, \cos x_j]
$$

$$
= \lim_{N \to \infty} \prod_{j=1}^{N} \frac{2\pi m}{i\epsilon\hbar} \widehat{\sin x_j}\widehat{\cos x_j} \exp\left[\frac{m}{i\epsilon\hbar}(\widehat{\sin^2} x_j + \widehat{\cos^2} x_j) \right]
$$

$$
\times I_\alpha\left(\frac{m}{i\epsilon\hbar}\widehat{\sin^2} x_j \right) I_\beta\left(\frac{m}{i\epsilon\hbar}\widehat{\cos^2} x_j \right) . \tag{2.58}
$$

This functional weight formulation is necessary to guarantee a consistent lattice definition of the path integral for the Pöschl-Teller potential, i.e., to guarantee the correct boundary conditions.

2.6.4 The Modified Pöschl–Teller Potential

Similarly one can derive a path integral identity for the modified Pöschl–Teller potential. One gets [219, 353] ($\eta, \nu > 0$, $r > 0$)

Basic Path Integral: Modified Pöschl–Teller Potential:

$$
\int\limits_{r(t')=r'}^{r(t'')=r''} \mathcal{D}r(t) \exp\left\{\frac{i}{\hbar} \int_{t'}^{t''} \left[\frac{m}{2}\dot{r}^2 - \frac{\hbar^2}{2m}\left(\frac{\eta^2 - \frac{1}{4}}{\sinh^2 r} - \frac{\nu^2 - \frac{1}{4}}{\cosh^2 r}\right)\right] dt\right\} \tag{2.59}
$$

$$
= \int_{\mathbb{R}} \frac{dE}{2\pi i} e^{-iET/\hbar} \frac{m}{\hbar^2} \frac{\Gamma(m_1 - L_\nu)\Gamma(L_\nu + m_1 + 1)}{\Gamma(m_1 + m_2 + 1)\Gamma(m_1 - m_2 + 1)}
$$
$$
\times (\cosh r' \cosh r'')^{-(m_1 - m_2)} (\tanh r' \tanh r'')^{m_1 + m_2 + 1/2}
$$
$$
\times {}_2F_1\left(-L_\nu + m_1, L_\nu + m_1 + 1; m_1 - m_2 + 1; \frac{1}{\cosh^2 r_<}\right)
$$
$$
\times {}_2F_1\left(-L_\nu + m_1, L_\nu + m_1 + 1; m_1 + m_2 + 1; \tanh^2 r_>\right) \tag{2.60}
$$

$$
= \sum_{n=0}^{N_M} e^{-iE_n T/\hbar} \Psi_n^{(\eta,\nu)*}(r')\Psi_n^{(\eta,\nu)}(r'') + \int_0^\infty dp\, e^{-iE_p T/\hbar} \Psi_p^{(\eta,\nu)*}(r')\Psi_p^{(\eta,\nu)}(r'') \tag{2.61}
$$

$[m_{1,2} = \frac{1}{2}(\eta \pm \sqrt{-2mE}/\hbar), L_\nu = \frac{1}{2}(\nu - 1)]$. The bound states are explicitly given by

$$
\Psi_n^{(\eta,\nu)}(r) = N_n^{(\eta,\nu)}(\sinh r)^{\eta+1/2}(\cosh r)^{n-\nu+1/2}{}_2F_1(-n, \nu - n; 1 + \eta; \tanh^2 r) , \tag{2.62}
$$
$$
N_n^{(\eta,\nu)} = \frac{1}{\Gamma(1+\eta)}\left[\frac{2(\nu - \eta - 2n - 1)\Gamma(n + 1 + \eta)\Gamma(\nu - n)}{\Gamma(\nu - \eta - n)n!}\right]^{1/2} ,
$$
$$
E_n = -\frac{\hbar^2}{2m}(2n + \eta - \nu - 1)^2 . \tag{2.63}
$$

Here denote $n = 0, 1, \ldots, N_M < \frac{1}{2}(\nu - \eta - 1)$. The continuous states are with $E_p = \hbar^2 p^2/2m$

$$
\Psi_p^{(\eta,\nu)}(r) = N_p^{(\eta,\nu)}(\cosh r)^{ip}(\tanh r)^{\eta+1/2}
$$
$$
\times {}_2F_1\left(\frac{\nu + \eta + 1 - ip}{2}, \frac{\eta - \nu + 1 - ip}{2}; 1 + \eta; \tanh^2 r\right) , \tag{2.64}
$$
$$
N_p^{(\eta,\nu)} = \frac{1}{\Gamma(1+\eta)}\sqrt{\frac{p \sinh \pi p}{2\pi^2}}\Gamma\left(\frac{\nu + \eta + 1 - ip}{2}\right)\Gamma\left(\frac{\eta - \nu + 1 - ip}{2}\right) . \tag{2.65}
$$

In the path integral formulation of the modified Pöschl–Teller potential again a functional weight $\mu_{\eta,\nu}$ can be used, c.f. [198], which is, however, omitted here.

2.6.5 Parametric Path-Integrals

In the course of evaluating the path integral representations on the two- and three-dimensional Euclidean space we encounter also three further Basic Path Integrals which can be derived from elliptic coordinates in $\mathbb{R}^2$ and spheroidal coordinates in $\mathbb{R}^3$ [223], respectively (see Chapter 5 for details):

Basic Path Integral: Elliptic Coordinate Path Integral:

$$
\int\limits_{\mu(t')=\mu'}^{\mu(t'')=\mu''} \mathcal{D}\mu(t) \int\limits_{\nu(t')=\nu'}^{\nu(t'')=\nu''} \mathcal{D}\nu(t)d^2(\sinh^2\mu + \sin^2\nu)
$$

$$
\times \exp\left[\frac{im}{2\hbar}d^2\int_{t'}^{t''}(\sinh^2\mu + \sin^2\nu)(\dot\mu^2 + \dot\nu^2)dt\right]
$$

$$
= \frac{1}{2\pi}\sum_{n\in\mathbb{Z}}\int_0^\infty pdp\, \mathrm{me}_n^*(\nu';\tfrac{d^2p^2}{4})\mathrm{me}_n(\nu'';\tfrac{d^2p^2}{4})M_n^{(1)*}(\mu';\tfrac{dp}{2})M_n^{(1)}(\mu'';\tfrac{dp}{2})\,\mathrm{e}^{-i\hbar p^2 T/2m}. \quad (2.66)
$$

Basic Path Integral: Prolate-Spheroidal Coordinate Path Integral:

$$
\int\limits_{\mu(t')=\mu'}^{\mu(t'')=\mu''} \mathcal{D}\mu(t) \int\limits_{\nu(t')=\nu'}^{\nu(t'')=\nu''} \mathcal{D}\nu(t)d^2(\sinh^2\mu + \sin^2\nu)
$$

$$
\times \exp\left\{\frac{i}{\hbar}\int_{t'}^{t''}\left[\frac{m}{2}d^2(\sinh^2\mu + \sin^2\nu)(\dot\mu^2 + \dot\nu^2) - \frac{\hbar^2}{2md^2}\frac{\lambda^2 - 1/4}{\sinh^2\mu\sin^2\nu}\right]dt\right\}
$$

$$
= d\sqrt{\sin\nu'\sin\nu''\sinh\mu'\sinh\mu''}\sum_{l=0}^\infty \frac{2l+1}{\pi}\frac{\Gamma(l-\lambda+1)}{\Gamma(l+\lambda+1)}\int_0^\infty p^2 dp\,\mathrm{e}^{-i\hbar p^2 T/2m}
$$

$$
\times \mathrm{ps}_l^{\lambda*}(\cos\nu';p^2d^2)\mathrm{ps}_l^\lambda(\cos\nu'';p^2d^2)S_l^{\lambda\,(1)*}(\cosh\mu';pd)S_l^{\lambda\,(1)}(\cosh\mu'';pd)\ . \quad (2.67)
$$

Basic Path Integral: Oblate-Spheroidal Coordinate Path Integral:

$$
\int\limits_{\xi(t')=\xi'}^{\xi(t'')=\xi''} \mathcal{D}\xi(t) \int\limits_{\nu(t')=\nu'}^{\nu(t'')=\nu''} \mathcal{D}\nu(t)d^2(\cosh^2\xi - \sin^2\nu)
$$

$$
\times \exp\left\{\frac{i}{\hbar}\int_{t'}^{t''}\left[\frac{m}{2}d^2(\cosh^2\xi - \sin^2\nu)(\dot\mu^2 + \dot\nu^2) - \frac{\hbar^2}{2md^2}\frac{\lambda^2 - 1/4}{\cosh^2\mu\sin^2\nu}\right]dt\right\}
$$

$$
= d\sqrt{\sin\nu'\sin\nu''\cosh\xi'\cosh\xi''}\sum_{l=0}^\infty \frac{2l+1}{\pi}\frac{\Gamma(l-\lambda+1)}{\Gamma(l+\lambda+1)}\int_0^\infty p^2 dp\,\mathrm{e}^{-i\hbar p^2 T/2m}
$$

$$
\times \mathrm{psi}_l^{\lambda*}(\cos\nu';p^2d^2)\mathrm{psi}_l^\lambda(\cos\nu'';p^2d^2)\mathrm{Si}_l^{\lambda\,(1)*}(\cosh\xi';pd)\mathrm{Si}_l^{\lambda\,(1)}(\cosh\xi'';pd)\ . \quad (2.68)
$$

2.6.6 The $O(2,2)$-Hyperboloid

The $O(2,2)$ hyperboloid is the simplest model for an Anti-de-Sitter gravity theory with the metric given by (e.g. [119, 354])

$$
ds^2 = dz_0^2 - \sum_{i=1}^{D-1} dz_i^2 + dz_D^2\ , \qquad z_0^2 - \sum_{i=1}^{D-1} z_i^2 + z_D^2 = 1\ , \qquad (2.69)
$$

where $D \geq 2$, with the two-dimensional single-sheeted hyperboloid as a special case. The O(2,2) hyperboloid has the metric $ds^2 = dz_0^2 + dz_1^2 - dz_2^2 - dz_3^2$. We want to study path integrals in this geometry, and in particular we are interested in three out of the 74 possible coordinate systems on the O(2,2) hyperboloid [317]. These are the spherical, the horicyclic and the equidistant coordinate systems. In the spherical system the coordinates have been chosen according to

$$\left.\begin{array}{ll} z_0 = \cosh \tau \cos \varphi_1 \ , & z_2 = \sinh \tau \cos \varphi_2 \ , \\ z_1 = \cosh \tau \sin \varphi_1 \ , & z_3 = \sinh \tau \sin \varphi_2 \end{array}\right\} \tag{2.70}$$

with the coordinate domains as indicated below. This is exactly the parameterization as used by Böhm and Junker [66] to calculate the SU(1,1) path integral and is not surprising since one sheet of the O(2,2) hyperboloid is isomorphic to the group manifold corresponding to SU(1,1).

Let us turn to the horicyclic system. It parameterizes the coordinates as follows ($v \in \mathbb{R}^{(1,1)}$, $a \cdot b = a_0 b_0 - a_1 b_1$, $u \in \mathbb{R}$):

$$\left.\begin{array}{ll} z_0 = \frac{1}{2}[e^{-u} + e^u(1 - v^2)] \ , & z_2 = v_1 e^u \ , \\ z_1 = v_0 e^u \ , & z_3 = \frac{1}{2}[-e^{-u} + e^u(1 + v^2)] \ . \end{array}\right\} \tag{2.71}$$

As known from the theory of harmonic analysis on this manifold one has a discrete and a continuous series in the variable u with wavefunctions corresponding to ($k^2 = k_0^2 - k_1^2 > 0$, the subspectrum is taken in the physical domain) [308, 367, 405, 496]

Discrete series ($n \in \mathbb{N}$, $0 < \alpha \leq 2$):

$$\Psi_{k_0,k_1,n}(v_0, v_1, u) = \frac{e^{ik_0 v_0 - ik_1 v_1}}{2\pi} \sqrt{2(2n + \alpha)} \, J_{2n+\alpha}(k \, e^{-u}) \ . \tag{2.72}$$

Continuous series ($p > 0$):

$$\Psi_{k_0,k_1,p}(v_0, v_1, u) = \frac{e^{ik_0 v_0 - ik_1 v_1}}{2\pi} \sqrt{\frac{p}{2 \sinh \pi p}} \left[J_{ip}(k \, e^{-u}) + J_{-ip}(k \, e^{-u}) \right] \ . \tag{2.73}$$

α is the parameter of the self-adjoint extension [405, p.54]. These wavefunctions are the matrix element expansions of the Titchmarch transformation ([385, p.150], [405, p.54], [496, p.93–95]). According to the general theory of the path integration on group manifolds we have to calculate the quantity

$$\hat{F}_m^\lambda = \int_G F(g) D_m^{\lambda *}(g^{-1}) dg \ . \tag{2.74}$$

Since this expression is actually independent of the representation one chooses we can take the result of Böhm and Junker [66] and we have in the limit $\epsilon \to 0$ the result

Discrete series ($n \in \mathbb{N}$): $\quad K_n(T) = e^{-iE_n T/\hbar} = \exp\left[\frac{i\hbar T}{2m}(4n^2 - 1)\right] \ , \tag{2.75}$

Continuous series ($p > 0$): $K_p(T) = e^{-iE_p T/\hbar} = \exp\left[-\frac{i\hbar T}{2m}(p^2 + 1)\right] \ . \tag{2.76}$

Note that the relevant spectrum we need emerging from the spectrum of the group manifold $SU(1,1)$ is of the form [318]

$$E_{\sigma,j_0} = -\frac{\hbar^2}{2m}[j_0^2 + \sigma(\sigma+2)] \ , \qquad \begin{array}{ll} \text{continuous spectrum:} & j_0 = 0, \sigma = -1 + ip \ , \\ \text{discrete spectrum:} & j_0 = 2n \ (n \in \mathbb{N}), \sigma = -1 \ . \end{array} \tag{2.77}$$

In the notation of [66] we have $E_n \to E_l = -\frac{\hbar^2}{2m} 2l(2l+2)$ with $l = -\frac{1}{2}, 0, \frac{1}{2} \dots$ which is equivalent to the discrete series (2.75). The wavefunctions of the subgroup systems corresponding to the coordinates v have been denoted by $\Psi_{\lambda,k}^{\mathbb{R}^{(1,1)}}$ with the continuous quantum number (λ, k), c.f. Chapter 4. They may be any of the wavefunctions of the free motion on $\mathbb{R}^{(1,1)}$. In particular, in (2.72, 2.73) we have chosen plane waves with quantum numbers $(k_0, k_1) \in \mathbb{R}^{(1,1)}$.

The equidistant coordinate system finally is given by ($\tau \in \mathbb{R}, u \in \Lambda^{(2)}$)

$$z_0 = \cosh\tau \ , \qquad \mathbf{z} = \mathbf{u} \cdot \sinh\tau \ . \tag{2.78}$$

We have therefore the following path integral representations which follow from the corresponding group spectral expansions

Spherical, $\tau > 0, \varphi_{1,2} \in [0, 2\pi)$:

$$\int_{\tau(t')=\tau'}^{\tau(t'')=\tau''} \mathcal{D}\tau(t)\sinh\tau\cosh\tau \int_{\varphi_1(t')=\varphi_1'}^{\varphi_1(t'')=\varphi_1''} \mathcal{D}\varphi_1(t) \int_{\varphi_2(t')=\varphi_2'}^{\varphi_2(t'')=\varphi_2''} \mathcal{D}\varphi_2(t)$$

$$\times \exp\left\{\frac{i}{\hbar}\int_{t'}^{t''}\left[\frac{m}{2}(\dot\tau^2 + \sinh^2\tau\dot\varphi_1^2 - \cosh^2\tau\dot\varphi_2^2) - \frac{\hbar^2}{8m}\left(4 - \frac{1}{\sinh^2\tau\cosh^2\tau}\right)\right]dt\right\}$$

$$= (\sinh\tau'\sin\tau''\cosh\tau'\cosh\tau'')^{-1/2} \sum_{\nu_{1,2}\in\mathbb{Z}} \frac{e^{i\nu_1(\varphi_1''-\varphi_1')+i\nu_2(\varphi_2''-\varphi_2')}}{(2\pi)^2}$$

$$\times\left\{\sum_{n\in\mathbb{N}} \Psi_n^{(\nu_1,\nu_2)}(\tau')\Psi_n^{(\nu_1,\nu_2)\,*}(\tau')\exp\left[\frac{i\hbar T}{2m}(2n+|\nu_2|-|\nu_1|)(2n+|\nu_2|-|\nu_1|+2)\right]\right.$$

$$\left. + \int_0^\infty dp\, e^{-i\hbar(p^2+1)T/2m}dt\Psi_p^{(\nu_1,\nu_2)}(\tau')\Psi_p^{(\nu_1,\nu_2)\,*}(\tau')\right\} \ , \tag{2.79}$$

$$\Psi_n^{(\nu_1,\nu_2)}(r) = N_n^{(\nu_1,\nu_2)}(\sinh r)^{|\nu_1|+1/2}(\cosh r)^{n-|\nu_2|+1/2}$$
$$\times {}_2F_1(-n, |\nu_2|-n; 1+|\nu_1|; \tanh^2 r) \ , \tag{2.80}$$

$$N_n^{(\nu_1,\nu_2)} = \frac{1}{\Gamma(1+|\nu_1|)}\left[\frac{2(|\nu_2|-|\nu_1|-2n-1)\Gamma(n+1+|\nu_1|)\Gamma(|\nu_2|-n)}{\Gamma(|\nu_2|-|\nu_1|-n)n!}\right]^{1/2} \ , \tag{2.81}$$

$$\Psi_p^{(\nu_1,\nu_2)}(r) = N_p^{(\nu_1,\nu_2)}$$
$$\times {}_2F_1\left(\frac{|\nu_2|+|\nu_1|+1-ip}{2}, \frac{|\nu_1|-|\nu_2|+1-ip}{2}; 1+|\nu_1|; \tanh^2 r\right) \ ,$$

$$N_p^{(\nu_1,\nu_2)} = \sqrt{\frac{p\sinh\pi p}{2\pi^2}} \frac{\Gamma[\frac{1}{2}(|\nu_1|+|\nu_2|+1-ip)]\Gamma[(|\nu_1|-|\nu_2|+1-ip)]}{\Gamma(1+|\nu_1|)} \ . \tag{2.82}$$

Horicyclic, $u \in \mathbb{R}, \mathrm{v} \in \mathbb{R}^{(1,1)}$:

$$\int\limits_{u(t')=u'}^{u(t'')=u''} \mathcal{D}u(t)\mathrm{e}^{2u} \int\limits_{\mathrm{v}(t')=\mathrm{v}'}^{\mathrm{v}(t'')=\mathrm{v}''} \mathcal{D}\mathrm{v}(t) \exp\left\{\frac{\mathrm{i}}{\hbar}\int_{t'}^{t''}\left[\frac{m}{2}(\dot{u}^2 - \mathrm{e}^{2u}\dot{\mathrm{v}}^2) - \frac{\hbar^2}{2m}\right]dt\right\}$$

$$= \mathrm{e}^{u'+u''}\int d\lambda \int_0^\infty dk\, \Psi_{\lambda,k}^{\mathbb{R}^{(1,1)}}(\mathrm{v}'')\Psi_{\lambda,k}^{\mathbb{R}^{(1,1)}\,*}(\mathrm{v}')$$

$$\times\left\{\sum_{n\in\mathbb{N}} 2(2n+\alpha)\,\mathrm{e}^{\mathrm{i}\hbar(4n^2-1)T/2m}J_{2n+\alpha}(k\,\mathrm{e}^{-u'})J_{2n+\alpha}(k\,\mathrm{e}^{-u''})\right.$$

$$\left.+\int_0^\infty \frac{p\,dp\,\mathrm{e}^{-\mathrm{i}\hbar(p^2+1)T/2m}}{2\sinh\pi p}\left[J_{\mathrm{i}p}(k\,\mathrm{e}^{-u''})+J_{-\mathrm{i}p}(k\,\mathrm{e}^{-u''})\right]\left[J_{\mathrm{i}p}(k\,\mathrm{e}^{-u'})+J_{-\mathrm{i}p}(k\,\mathrm{e}^{-u'})\right]\right\}. \quad (2.83)$$

Equidistant $\Lambda^{(2)}$, $\tau \in \mathbb{R}, \mathrm{u} = (u_0, \mathbf{u}) \in \Lambda^{(2)}$:

$$\int\limits_{\tau(t')=\tau'}^{\tau(t'')=\tau''} \mathcal{D}\tau(t)\cosh^2\tau \int\limits_{\mathbf{u}(t')=\mathbf{u}'}^{\mathbf{u}(t'')=\mathbf{u}''} \frac{\mathcal{D}\mathbf{u}(t)}{u_0}$$

$$\times \exp\left\{\frac{\mathrm{i}}{\hbar}\int_{t'}^{t''}\left[\frac{m}{2}(\dot{\tau}^2 + \cosh^2\tau\dot{\mathbf{u}}^2) - \frac{\hbar^2}{2m} - \frac{\Delta V(\mathrm{u})}{\cosh^2\tau}\right]dt\right\}$$

$$= (\cosh\tau'\cosh\tau'')^{-1}\int d\lambda \int_0^\infty dk \Psi_{\lambda,k}^{\Lambda^{(2)}}(\mathrm{u}'')\Psi_{\lambda,k}^{\Lambda^{(2)}\,*}(\mathrm{u}')$$

$$\times\left(\sum_{n\in\mathbb{N}}(n-k-\tfrac{1}{2})\frac{\Gamma(2k-n)}{n!}\exp\left\{\frac{\mathrm{i}\hbar T}{2m}[(n-k-\tfrac{1}{2})^2 - 1]\right\}\right.$$

$$\times P_{k-1/2}^{n-k+1/2}(\tanh\tau')P_{k-1/2}^{n-k+1/2}(\tanh\tau'')$$

$$\left.+\int_{\mathbb{R}} \frac{p\,dp\sinh\pi p}{\cos^2\pi k + \sinh^2\pi p}P_{k-1/2}^{\mathrm{i}p}(\tanh\tau'')P_{k-1/2}^{-\mathrm{i}p}(\tanh\tau')\mathrm{e}^{-\mathrm{i}\hbar T(p^2+1)/2m}\right). \quad (2.84)$$

In the horicyclic system $\{k, \lambda\}$ are the corresponding continuous quantum numbers of the path integral representations of the ten two-dimensional pseudo-Euclidean subsystems, i.e., in (2.83) ten path integral solutions are contained.

In the equidistant-$\Lambda^{(2)}$ system $\{k, \lambda\}$ are the corresponding quantum numbers of the path integral representations on $\Lambda^{(2)}$, i.e., in (2.84) nine path integral solutions are contained, c.f., Chapter 7 for more details. However, no new path integral identities can be derived from this system.

Our result enables us to derive two path integral identities. The first is the path integral representation of the modified Pöschl–Teller potential as discussed by Böhm and Junker [65, 66] and already stated in the last subsection, c.f. (2.61). The second is a path integral identity of the inverted Liouville problem. Separating off in (2.83) the variables $\mathrm{v} = (v_0, v_1, v_2)$ and performing $u = \mathrm{e}^y$ yields ($\kappa > 0$, $\alpha > 0$ is the parameter of the self-adjoint extension)

Basic Path Integral: Inverted Liouville Potential:

$$\int\limits_{y(t')=y'}^{y(t'')=y''} \mathcal{D}y(t) \exp\left[\frac{i}{\hbar}\int_{t'}^{t''}\left(\frac{m}{2}\dot{y}^2 + \frac{\hbar^2\kappa^2}{2m}e^{2y}\right)dt\right]$$

$$= \sum_{n\in\mathbb{N}} 2(2n+\alpha)\,e^{i\hbar T(2n+\alpha)^2/2m}\,J_{2n+\alpha}(\kappa\,e^{y'})J_{2n+\alpha}(\kappa\,e^{y''})$$

$$+ \int_0^\infty dp\frac{p e^{-i\hbar p^2 T/2m}}{2\sinh\pi p}\left[J_{ip}(\kappa\,e^{y''}) + J_{-ip}(\kappa\,e^{y''})\right]\left[J_{ip}(\kappa\,e^{y'}) + J_{-ip}(\kappa\,e^{y'})\right]\ . \quad (2.85)$$

2.6.7 δ-Functions and Boundary Problems

Inevitable in many path integral evaluations are the considerations of boundary problems in path integrals, i.e. motion in half spaces, respectively in boxes or rings with infinitely high walls [217, 222, 225].

The first problem consists of a δ-function perturbation at $x = a$ yielding
Basic Path Integral: δ-Function Perturbation:

$$\frac{i}{\hbar}\int_0^\infty dT\,e^{iET/\hbar}\int\limits_{x(t')=x'}^{x(t'')=x''}\mathcal{D}x(t)\exp\left\{\frac{i}{\hbar}\int_{t'}^{t''}\left[\frac{m}{2}\dot{x}^2 - V(x) + \gamma\delta(x-a)\right]dt\right\}$$

$$= G^{(V)}(x'',x';E) - \frac{G^{(V)}(x'',a;E)G^{(V)}(a,x';E)}{G^{(V)}(a,a;E) - 1/\gamma}\ . \quad (2.86)$$

The second problem consists of a δ'-function perturbation at $x = a$ yielding [225]
Basic Path Integral: δ'-Function Perturbation:

$$\frac{i}{\hbar}\int_0^\infty dT\,e^{iET/\hbar}\int\limits_{x(t')=x'}^{x(t'')=x''}\mathcal{D}x(t)\exp\left\{\frac{i}{\hbar}\int_{t'}^{t''}\left[\frac{m}{2}\dot{x}^2 - V(x) + \beta\delta'(x-a)\right]dt\right\}$$

$$= G^{(V)}(x'',x';E) - \frac{G^{(V)}_{,x'}(x'',a;E)G^{(V)}_{,x''}(a,x';E)}{\widehat{G}^{(V)}_{,x'x''}(a,a;E) + 1/\beta}\ , \quad (2.87)$$

$$\widehat{G}^{(V)}_{,xy}(a,a;E) = \left(\frac{\partial^2}{\partial x\partial y}G^{(V)}(x,y;E) - \frac{2m}{\hbar^2}\delta(x-y)\right)\bigg|_{x=y=a}\ . \quad (2.88)$$

The path integral solutions for the δ- and δ'-function perturbation enables one to perform the limit $\gamma\to\infty$ and $\beta\to\infty$, respectivley, yielding path integral repesentation for Dirichlet and Neumann-boundary conditions respectievly [225]:

Basic Path Integral: Dirichlet Boundary Condition at $x = a$:

$$\frac{i}{\hbar}\int_0^\infty dT\,e^{iET/\hbar}\int\limits_{x(t')=x'}^{x(t'')=x''}\mathcal{D}^{(D)}_{(x=a)}x(t)\exp\left\{\frac{i}{\hbar}\int_{t'}^{t''}\left[\frac{m}{2}\dot{x}^2 - V(x)\right]dt\right\}$$

$$= G^{(V)}(x'', x'; E) - \frac{G^{(V)}(x'', a; E)G^{(V)}(a, x'; E)}{G^{(V)}(a, a; E)} . \tag{2.89}$$

Basic Path Integral: Neumann Boundary Condition at $x = a$:

$$\frac{\mathrm{i}}{\hbar} \int_0^\infty dT \, \mathrm{e}^{\mathrm{i}ET/\hbar} \int\limits_{x(t')=x'}^{x(t'')=x''} \mathcal{D}_{(x=a)}^{(N)} x(t) \exp\left\{ \frac{\mathrm{i}}{\hbar} \int_{t'}^{t''} \left[\frac{m}{2}\dot{x}^2 - V(x) \right] dt \right\}$$

$$= G^{(V)}(x'', x'; E) - \frac{G_{,x'}^{(V)}(x'', a; E)G_{,x''}^{(V)}(a, x'; E)}{\widehat{G}_{,x'x''}^{(V)}(a, a; E)} , \tag{2.90}$$

$$\widehat{G}_{,xy}^{(V)}(a, a; E) = \left(\frac{\partial^2}{\partial x \partial y} G^{(V)}(x, y; E) - \frac{2m}{\hbar^2}\delta(x - y) \right)\Big|_{x=y=a} . \tag{2.91}$$

The procedure can be repeated to obtain the results for particle in boxes with boundaries at $x = a$ and $x = b$ [224, 225]:

Basic Path Integral:
Dirichlet-Dirichlet Boundary Conditions at $x = a$ and $x = b$:

$$\frac{\mathrm{i}}{\hbar} \int_0^\infty dT \, \mathrm{e}^{\mathrm{i}ET/\hbar} \int\limits_{x(t')=x'}^{x(t'')=x''} \mathcal{D}_{(a \leq x \leq b)}^{(DD)} x(t) \exp\left\{ \frac{\mathrm{i}}{\hbar} \int_{t'}^{t''} \left[\frac{m}{2}\dot{x}^2 - V(x) \right] dt \right\}$$

$$= \frac{\begin{vmatrix} G^{(V)}(x'', x'; E) & G^{(V)}(x'', b; E) & G^{(V)}(x'', a; E) \\ G^{(V)}(b, x'; E) & G^{(V)}(b, b; E) & G^{(V)}(b, a; E) \\ G^{(V)}(a, x'; E) & G^{(V)}(a, b; E) & G^{(V)}(a, a; E) \end{vmatrix}}{\begin{vmatrix} G^{(V)}(b, b; E) & G^{(V)}(b, a; E) \\ G^{(V)}(a, b; E) & G^{(V)}(a, a; E) \end{vmatrix}} . \tag{2.92}$$

Basic Path Integral:
Neumann-Neumann Boundary Conditions at $x = a$ and $x = b$:

$$\frac{\mathrm{i}}{\hbar} \int_0^\infty dT \, \mathrm{e}^{\mathrm{i}ET/\hbar} \int\limits_{x(t')=x'}^{x(t'')=x''} \mathcal{D}_{(a \leq x \leq b)}^{(NN)} x(t) \exp\left\{ \frac{\mathrm{i}}{\hbar} \int_{t'}^{t''} \left[\frac{m}{2}\dot{x}^2 - V(x) \right] dt \right\}$$

$$= \frac{\begin{vmatrix} G^{(V)}(x'', x'; E) & G_{,x'}^{(V)}(x'', b; E) & G_{,x'}^{(V)}(x'', a; E) \\ G_{,x''}^{(V)}(b, x'; E) & \widehat{G}_{,x'x''}^{(V)}(b, b; E) & G_{,x'x''}^{(V)}(b, a; E) \\ G_{,x''}^{(V)}(a, x'; E) & G_{,x'x''}^{(V)}(a, b; E) & \widehat{G}_{,x'x''}^{(V)}(a, a; E) \end{vmatrix}}{\begin{vmatrix} \widehat{G}_{,x'x''}^{(V)}(b, b; E) & G_{,x'x''}^{(V)}(b, a; E) \\ G_{,x'x''}^{(V)}(a, b; E) & \widehat{G}_{,x'x''}^{(V)}(a, a; E) \end{vmatrix}} . \tag{2.93}$$

Basic Path Integral: Dirichlet Boundary Conditions at $x = a$ and
Neumann Boundary Conditions $x = b$:

$$\frac{\mathrm{i}}{\hbar} \int_0^\infty dT \, \mathrm{e}^{\mathrm{i}ET/\hbar} \int\limits_{x(t')=x'}^{x(t'')=x''} \mathcal{D}_{(a \leq x \leq b)}^{(DN)} x(t) \exp\left\{ \frac{\mathrm{i}}{\hbar} \int_{t'}^{t''} \left[\frac{m}{2}\dot{x}^2 - V(x) \right] dt \right\}$$

$$
= \frac{\begin{vmatrix} G^{(V)}(x'', x'; E) & G^{(V)}_{,x'}(x'', b; E) & G^{(V)}(x'', a; E) \\ G^{(V)}_{,x''}(b, x'; E) & \widehat{G}^{(V)}_{,x'x''}(b, b; E) & G^{(V)}_{,x''}(b, a; E) \\ G^{(V)}(a, x'; E) & G^{(V)}_{,x'}(a, b; E) & G^{(V)}(a, a; E) \end{vmatrix}}{\begin{vmatrix} \widehat{G}^{(V)}_{,x'x''}(b, b; E) & G^{(V)}_{,x''}(b, a; E) \\ G^{(V)}_{,x'}(a, b; E) & G^{(V)}(a, a; E) \end{vmatrix}} . \tag{2.94}
$$

This concludes the discussion.

2.6.8   Miscellaneous Results

For the classification of solvable path integrals, one also requires a few additional formulæ which generalize the usual problems in quantum mechanics in a specific way. Here one has, e.g.,

1. Perturbation expansions (Feynman and Hibbs [164], Devreese et al. [184]–[186]), and point interactions (Bauch [49], Goovaerts et al. [184, 185] and Refs. [205, 216, 222, 217, 225]). Also $1/r$- [186, 370] and $1/r^2$-potentials [370] can be taken into account by such a formalism.

2. Partition functions and effective potentials for partition functions have been discussed, e.g., by Feynman and Hibbs [164], Giachetti et al. [180], and Feynman and Kleinert [165].

3. Explicitly time-dependent problems according to, e.g., $V(\mathbf{x}) \mapsto V(\mathbf{x}/\zeta(t))/\zeta^2(t)$ $(\mathbf{x} \in \mathbb{R}^D)$ [140, 218].

4. Point interactions in two and three dimensions [4, 89, 217].

5. Path integral representation of the one-dimensional Dirac particle according to Feynman and Hibbs [164], Ichinose and Tamura [282], Jacobson [293], and Jacobson and Schulman [294]. Simple applications are the free particle [164, 294], and point interactions [224, 225].

6. Fermionic path integrals based on coherent states according to, e.g., Faddeev and Slavnov [157], Klauder et al. [347, 348], Singh and Steiner [472], and references therein.

Of course, in the case of general quantum mechanical problems, more than just one of the Basic Path Integral Solutions is required. However, such problems can be conveniently put into a hierarchy according to which of the Basic Path Integrals is the most important one for its solution, see, e.g., [250, 251] for a more detailed enumeration. In the monography [252] a comprehensive list known path integrals is given.

Chapter 3

Separable Coordinate Systems on Spaces of Constant Curvature

3.1 Separation of Variables and Breaking of Symmetry

Attempts to classify separable potentials go back to Eisenhart [148], Aly and Spector [7], Bhattacharjie and Sudarshan [76], and Luming and Predazzi [378]. A systematic study was eventually undertaken by Smorodinsky, Winternitz and co-workers [171, 386, 527]. They looked systematically for potentials which are separable in more than one coordinate system, i.e., which have additional integrals of motions. In two dimensions [171, 527] it turns out that there are four potentials of the sought type which all have three constants of motion (including energy), i.e., there are two more operators commuting with the Hamiltonian. Smorodinsky, Winternitz et al. extended their investigations in a classical paper [386] to three dimensions by listing all potentials which separate in more than one coordinate system.

Let us shortly discuss the physical significance of the consideration of separation of variables in coordinate systems. The free motion in some space is, of course, the most symmetric one, and the search for the number of coordinate systems which allow the separation of the Hamiltonian is equivalent to the investigation how many inequivalent sets of variables can be found. The incorporation of potentials usually removes at least some of the symmetry properties of the space. Well-known examples are spherical systems, and they are most conveniently studied in spherical coordinates. For instance, the isotropic harmonic oscillator in three dimensions is separable in eight coordinate systems, namely in cartesian, spherical, circular polar, circular elliptic, conical, oblate spheroidal, prolate spheroidal, and ellipsoidal coordinates. The Coulomb potential is separable in four coordinate systems, namely in conical, spherical parabolic, and prolate spheroidal II coordinates (for a comprehensive review with the focus on path integration, e.g., [241]).

The separation of a particular quantum mechanical potential problem in more than one coordinate systems has the consequence that there are additional integrals of motion and that the spectrum is degenerate. The Noether theorem [424] connects the particular symmetries of a Lagrangian, i.e., the invariances with respect to the dynamical symmetries, with conservation laws in classical mechanics and with observables in quantum mechanics, respectively. In the case of the isotropic harmonic oscillator one has in addition to the conservation of energy and the conservation of the angular momentum, the conservation of the quadrupole moment; in the case of the Coulomb problem one has in addition to the conservation of energy and the angular momentum, the conservation of the Pauli-Runge-Lenz vector. In total these additional conserved quantities add up to five functionally independent integrals of motion in classical mechanics, respectively observables in quantum mechanics. It is even possible to introduce extra terms in the pure oscillator and Coulomb potential in such a way that one still has all these integrals of motion, however somewhat modified [155, 156].

The harmonic oscillator in various coordinate systems has been studied by Boyer et al. [79], Evans [155, 156], Kallies et al. [297], and Pogosyan et al. [444]; in spaces of constant curvature by Bonatsos et al. [75], and [242, 243]. The Coulomb-Kepler problem in various coordinate systems has been studied by many authors including, e.g., Carpio-Bernido et al. [96]–[99], Chetouani at al. [105, 106], Cisneros and McIntosh [108], Coulson et al. [110, 111], Davtyan et al. [116, 118, 390], Fock [169], Gerry [178], Granovsky et al. [188], [198, 209, 212, 214, 219, 241], Hodge [279], Guha and Mukherjee [255], Hartmann [264], Kibler et al. [339]-[343], [344], Lutsenko et al. [380], Vaidya and Boschi-Filho [503], Pauli [437], Sökmen [474], Teller [494], Zaslow and Zandler [531], and Zhedanov [533], and in spaces of constant curvature by Barut et al. [45, 46], Granovsky et al. [189, 190], [201, 242, 243], Kibler et al. [341, 345], Mardoyan et al. [389, 391], Izmest'ev et al. [292], Katayama [336], Otchik and Red'kov [428], and Vinitsky et al. [513].

As it turns out, the so-constructed modified harmonic oscillator and Coulomb problems, belong to a larger class of potentials which are called super-integrable: the maximally super-integrable potentials in three dimensions have five functionally independent integrals of motion, and the minimally super-integrable potentials have four functionally independent integrals of motion [155, 156]. All such systems have the special property that the energy-levels of the system are organized in representations of the non-invariance group which contain representations of the dynamical subgroup realized in terms of the wave-functions of these energy-levels [171]. In the case of the hydrogen it enabled Pauli [437], Fock [169] and Bargmann [44] to solve the quantum mechanical Kepler problem without explicitly solving the Schrödinger equation. Actually, the algebra of the dynamical symmetry of the hydrogen atom turns out to be a centerless twisted Kac-Moody algebra [114]. The additional integrals of motion also have the consequence that in the case of the super-integrable systems in two dimensions and maximally super-integrable systems in three dimensions all finite trajectories are found to be periodic; in the case of minimally super-integrable systems in three dimensions all finite trajectories are found to be quasi-periodic [339]. (The notion quasi-periodic means that they are periodic in each coordinate, but not necessarily periodic in a global way. They are periodic globally if the respective periods are commensurable, i.e., their quotients are rational numbers.) Of course, in the case of the pure Kepler or the isotropic harmonic oscillator all finite trajectories are periodic.

It is interesting to remark that the notion of "super-integrability" can also be introduced in spaces of constant curvature [242]. Whereas the general form of potentials which are "super-integrable" in some kind is not clear until now, one knows that the corresponding Higgs-oscillator (c.f. Higgs [277], Bonatsos et al. [75], Granovsky et al. [188], Katayama [336]), Leemon [371], Pogosyan et al. [444], and Nishino [423]) and Kepler problems (c.f. Granovsky et al. [190], Kibler et al. [341], Kurochkin and Otchik [363], Nishino [423], Otchik and Red'kov [428], Pervushin et al. [514], Vinitsky et al. [513], Zhedanov [532]) in spaces of constant curvature do have additional constants of motion: the analogues of the flat space. For the Higgs-oscillator this is the Demkov-tensor [121, 188, 423] and in the Kepler problem an analogue of the Pauli-Runge-Lenz vector on spaces of constant curvature can be defined, c.f. [190, 363, 423].

Disturbing the spherical symmetry usually spoils it. The first step consists of deforming the ring-shaped feature of the (maximally super-integrable) modified oscillator and Coulomb potential. One gets in the former a ring-shaped oscillator and in the latter the Hartmann potential, two-minimally super-integrable systems. The number of coordinate systems which allow a separation of variables drop from eight to four (namely spherical, circular polar, oblate spheroidal and prolate spheroidal coordinates), and from four to three, namely spherical, parabolic, prolate spheroidal II coordinates. The ring-shaped oscillator has been discussed by, e.g., Carpio-Bernido et al. [98, 99], Kibler et al. [339, 340, 342, 344], Lutsenko et al. [379], and Quesne [448]. The Hartmann system has been discussed by, e.g., Carpio-Bernido et al. [96]–[99], Chetouani at al. [105], Gal'bert et al. [172], Gerry [178], Granovsky et al. [188], [209], Guha and Mukherjee [255], Hartmann [264], Kibler et al. [339]-[342], [344], Lutsenko et al. [380], Vaidya and Boschi-Filho [503], and Zhedanov [533]; compare also the connection with a Coulomb plus Aharonov-Bohm solenoid, e.g., Chetouani et al. [106], Kibler and Negadi [343], and Sökmen [474].

Disturbing the system further, and one is left with, say, only one coordinate systems which still allows separation of variables. A constant electric field (Stark effect) allows only the separation in parabolic coordinates [209, 523]. Here it is interesting to remark that in the momentum representation of the hydrogen atom the bound state spectrum is described by the free motion on the sphere $S^{(3)}$. To be more precise, the dynamical group O(4) describes the discrete spectrum, and the Lorentz group O(3, 1) the continuous spectrum [39]. Now, there are six coordinate systems on $S^{(3)}$ which separate the corresponding Laplacian. The solution in spherical and cylindrical coordinates corresponds to the spherical and parabolic solution in the coordinate space representation. The elliptic cylindrical system is of special interest because it enables one to set up a complete classification for the energy-levels of the quadratic Zeeman effect (c.f. Herrick [276], Lakshmann and Hasegawa [368], Brown and Solov'ev [86]).

The separation in parabolic coordinates is also possible in the case of a perturbation of the pure Coulomb field with a potential force $\propto z/r$ which still allows an exact solution [212, 214]. The two-center Coulomb problem turns out to be separable only in spheroidal coordinates (Coulson and Josephson [110], Coulson et al. [111], [415]) as has been studied first in the connection with the hydrogen-molecule ion by Teller [494]. Let us also note that the Kaluza-Klein monopole separates in polar (Bernido [55] and Inomata and Junker [286]) and parabolic coordinates [207], in comparison to the Dirac monopole [206], respectively the Dyonium system (Chetouani et al. [107], Dürr and Inomata [138], Inomata et al. [288], and Kleinert [349]) which is separable only in spherical coordinates.

Another possibility to disturb the spherical symmetry is to remove the invariance to rotations with respect to some axis, e.g., about a uniform magnetic field. Usually this invariance is used to illustrate the azimuthal quantum number m of the L_z operator. The physical meaning of this quantum number then is that there exists a preferred axis in space. This symmetry can be broken if one considers a Hamiltonian of a nucleus with an electric quadrupole moment Q and spin J in a spatially varying electric field [374, 475]. Here sphero-conical coordinates are most convenient, and the projection of the terminus of the angular momentum vector traces out a cone of elliptic cross Section about the z-axis [475]. Also the problem of the asymmetric top (Kramers and Ittmann [360], Lukač [373], Smorodinsky et al. [376, 473, 525]), the symmetric oblate top [373], or the case of tensor-like potentials (Lukač and Smorodinsky [377]) can be treated best in sphero-conical coordinates. Therefore sphero-conical coordinates are most suitable for problems which have spherical symmetry but not a sphero-axial symmetry. Another system where elliptical coordinates are suitable are electromagnetic traps with electric quadrupole fields (c.f. Paul [436]), here one has to deal with Matthieu equations. The harmonic oscillator in elliptic coordinates has been discussed by Kalnins and Miller [312].

In order that a potential problem can be separated in ellipsoidal coordinates is that the shape of the potential resembles the shape of a ellipsoid. Of course, the anisotropic harmonic oscillator belongs to this class. Introducing quartic and sextic [502] interaction terms then eventually allows only a separation of variables in ellipsoidal coordinates. Another example is the Neumann model [257, 421] , which describes a particle moving on a sphere subject to anisotropic harmonic forces (Babelon and Talon [33] and MacFarlane [382]).

In the cases of coordinate systems on single- or two-sheeted hyperboloids and in pseudo-Euclidean spaces applications occur in the theory of gravity. Examples are gravity waves [83], Robertson-Walker-type space-times [327], and Kerr background spaces [307, 335]; the $O(2,2)$ hyperboloid is the simplest model for an Anti-de-Sitter gravity theory, e.g., De Bievre and Renaud [119] and Kleppe [354]. By means of an expansion to a larger space, $SU(2,2)$, one can construct super-integrable potentials on the $O(2,2)$ hyperboloid [120].

Historically very important were the investigations in particle scattering. The usual partial-wave expansions of a spinless scattering amplitude for the reaction

$1 + 2 \to 3 + 4$ can be written as

$$f(E, \vartheta) = \sum_{l=0}^{\infty} (2l + 1) a_l(E) P_l(\cos \vartheta) \ , \tag{3.1}$$

where E and ϑ are the center-of-mass-system energy and scattering angle. The variables E and ϑ are treated asymmetrically, and the energy is contained in the unknown partial-wave amplitude $a_l(E)$, whereas the dependence on ϑ is displayed explicitly. A more symmetric treatment, as well as a greater separation of kinematics and dynamics, is provided by the Lorentz group expansion using the chain $O(3, 1) \supset O(3) \supset O(2)$ and supplementing the above expansion according to [91]

$$f(E, \vartheta) = \sum_{l=0}^{\infty} \int_{\delta - i\infty}^{\delta + i\infty} d\sigma (\sigma + 1)^2 \frac{\Gamma(\sigma + 1)}{\Gamma(\sigma - l + 1)} \frac{A_l(\sigma)}{\sqrt{\sinh a}} P_{\sigma + 1/2}^{-l - 1/2}(\cosh a) P_l(\cos \vartheta) \ , \tag{3.2}$$

where

$$\cosh a = \frac{s + m_1^2 - m_2^2}{2 m_1 \sqrt{s}} \ ,$$

$$\sinh a = \frac{\sqrt{s - (m_1 + m_2)^2} \sqrt{s - (m_1 - m_2)^2}}{2 m_1 \sqrt{s}} \ , \tag{3.3}$$

$$s = (p_1 + p_2)^2 = E^2 \ .$$

Here $m_{1,2}$ are the masses of the particles 1 and 2, and $A_l(\sigma)$ are the Lorentz amplitudes (a systematic study of various kinds of subgroup chains can be, e.g., found in [50, 298, 406, 417, 430]–[435, 523]). Vilenkin and Smorodinsky [511] started a study of such scattering amplitudes in various coordinate space representations. It was further developed by Winternitz and Smorodinsky and coworkers, e.g., [381, 524]–[527, 534]. Here the two-sheeted hyperboloid is taken for the physical domain of the invariant mass, and the single-sheeted hyperboloid is taken in the unphysical domain. The incorporation of potentials can be found in [334].

Lobachevskian space has also attracted some attention in field theory as a model of a non-trivial field theory in a space of constant (non-zero) curvature by Boyer and Fleming [77] and Kerimov [338].

It is also observed that several potential problems can be put into connection with free motion on a space of constant curvature by symmetry arguments (Boyer and Kalnins [78] and Kalnins and Miller [310, 316]), and c.f. the Basic Path Integrals.

3.2 Classification of Coordinate Systems

In order to set up a systematic method of classifying coordinate systems which separate the Hamilton-Jacobi equations in classical mechanics or the Hamiltonian in quantum mechanics, one starts to consider symmetry operators. Let us consider the time-independent Schrödinger equation for the free motion

$$\underline{H}\Psi = -\frac{\hbar^2}{2m} \Delta_{LB} \Psi = E\Psi \tag{3.4}$$

in some homogeneous space and Δ_{LB} is the Laplace-Beltrami operator. The simplest case is, of course, the Euclidean space. Let us consider $\mathbb{R}^3$. A symmetry operator is a linear differential operator

$$L = \sum_{j=1}^{3} a_j(\mathbf{x})\partial_j + b(\mathbf{x}) \ , \tag{3.5}$$

where a_j and b are analytic functions of $\mathbf{x} \in D \subset \mathbb{R}^3$ such that $L\Psi$ is a solution of the time-independent Schrödinger equation in D for any Ψ which is a solution of $(H - E)\Psi = 0$. The set of all such symmetry operators is the Lie algebra $\mathfrak{L}$ under the operations of scalar multiplication and commutator brackets $[L_i, L_j] = L_iL_j - L_jL_i$. (This is equivalent to the assertion that $\mathbf{a}$ is a Killing field.) It consists of the momentum P_i and angular momentum operators L_i, $i = 1, 2, 3$.

Table 3.1: Operators and Coordinate Systems in $\mathbb{R}^3$

Commuting Operators S_1, S_2	Coordinate System
P_2^2, P_3^2	Cartesian
L_3^2, P_3^2	Circular Polar
$L_3^2 + d^2 P_1^2, P_3^2$	Circular Elliptic
$\{L_3, P_2\}, P_3^2$	Circular Parabolic
$\mathbf{L}^2, L_1^2 + k'^2 L_2^2$	Sphero-Conical
$\mathbf{L}^2, L_3^2$	Spherical
$\mathbf{L}^2 - d^2(P_1^2 + P_2^2), L_3^2$	Prolate Spheroidal
$\mathbf{L}^2 + d^2(P_1^2 + P_2^2), L_3^2$	Oblate Spheroidal
$\{L_1, P_2\} - \{L_2, P_1\}, L_3^2$	Parabolic
$\mathbf{L}^2 + (a_2 + a_3)P_3^2 + (a_1 + a_3)P_2^2 + (a_1 + a_2)P_1^2,$	Ellipsoidal
$\quad a_1^2 L_1^2 + a_2 L_2^2 + a_3 L_3^2 + a_2 a_3 P_3^2 + a_1 a_3 P_2^2 + a_1 a_2 P_1^2$	
$L_3^2 - d^2 P_3^2 + d(\{L_2, P_1\} + \{L_1, P_2\}),$	Paraboloidal
$\quad d(P_2^2 - P_1^2) + \{L_2, P_1\} - \{L_1, P_2\}$	

However, for the classification of coordinate systems, and hence for sets of inequivalent observables we need second-order differential operators S_i ($i \in I$, I an index set) which are at most quadratic in the derivatives. In order that they can characterize a coordinate system which separates the Hamiltonian we must require that they commute with the Hamiltonian and with each other, i.e., $[H, S_i] = [S_i, S_j] = 0$. This property characterizes them as observables (in classical mechanics as constants of motion). Because we consider $\mathbb{R}^3$ there are two of these operators S_1, S_2 which correspond to the two separation constants which appear for each coordinate system in consideration. Finding all inequivalent sets of $\{S_1, S_2\}$ is equivalent in finding all inequivalent sets of observables for the Hamiltonian of the free motion in $\mathbb{R}^3$. Because the operators $S_{1,2}$ commute with the Hamiltonian and with each other one can find simultaneously eigenfunctions of H, S_1, S_2.

The small table 3.1 illustrates this for the eleven coordinate systems in $\mathrm{I\!R}^3$ ($a, d >$ $0, 0 < r \leq 1$ parameters, $\{\cdot, \cdot\}$ is the anticommutator). Here $P_{1,2,3}$ and $L_{1,2,3}$ are the usual momentum and angular momentum operators taken in cartesian coordinates, the parameters are as defined in table 5.2, and for the ellipsoidal system according to [242, 297].

The method has been illustrated by Kalnins and Miller in [80] for the three-dimensional Euclidean space, in the book of Miller [405] for the two- and three-dimensional Euclidean and pseudo-Euclidean space (c.f. also [308]), for the time-dependent one- and two-dimensional Schrödinger (heat-) equation (c.f. also [310, 311]), and for the time-dependent radial Schrödinger (heat-) equation (c.f. also [310, 316]), by Kalnins [300] for two- and three-dimensional pseudo-Euclidean space, and by Kalnins and Miller for four-dimensional Euclidean and pseudo-Euclidean space [322].

Of course, the same method applies for spaces with non-zero constant curvature, as for the two- and three-dimensional sphere [373, 374, 432, 425], for the two- and three-dimensional pseudosphere [318, 425, 473, 512, 524, 525, 526], or for the O(2, 2) hyperboloid [317]. Also, the same method applies for potential problems [171, 386, 527].

3.3   Coordinate Systems in Spaces of Constant Curvature

In this Section we cite the general construction of coordinate systems in the most important spaces of constant curvature. These are the sphere, Euclidean space, the pseudosphere, and Minkowski space. For the sphere, Euclidean space and pseudosphere we follow Kalnins [301]. For the pseudo-Euclidean space we follow Kalnins [300], Kalnins and Miller [322], and Miller et al. [406]. In order to set up a convenient classification scheme for coordinate systems on spaces of constant curvature, we cite the classification scheme of [406]. Here one makes use of the notions of "ignorable coordinates" and "generic coordinates", respectively. If $\{x_1, \ldots, x_D\}$ are separable coordinates for the Hamiltonian in a D-dimensional space, the coordinate x_1 is said to be *ignorable* if for the operator L in the corresponding Lie algebra we have $L = \partial_1 \in \mathfrak{L}$ provided the metric tensor in these coordinates is independent of x_1. *Generic* coordinates on the other hand have at most one ignorable coordinate, and *parametric* coordinates at least one *parameter*. Examples for the latter are elliptic or spheroidal coordinates.

The method can, of course, be extended to the four-dimensional case, where in total 261 coordinate systems can be found for the pseudo-Euclidean space, and 42 coordinate systems for the Euclidean space [322]. Because we do not discuss path integrals in four-dimensional spaces of constant curvature in detail this is omitted. In Patera et al. [435] a discussion can be found for the four-dimensional pseudosphere, however only the subgroup coordinate systems have been discussed, there is no complete treatment of all coordinate systems; a similar analysis for the Euclidean space can be found in [482]. An incorporation of these spaces would not give any substantially new path integral solutions, on the contrary, we can only solve some of the corresponding three-dimensional subspace-systems, and no path integral for a genuine generic four-dimensional coordinate system can be found.

We cite the general construction of coordinate systems on spheres, pseudospheres and Euclidean spaces nevertheless for the sake of completeness. It is obvious from the presented methods to obtain in a systematic way the various degenerations, and therefore the explicit possible coordinate systems in each space for any dimension one is interested in. However, there seems to be no closed recurrence relation to determine the number of inequivalent coordinate systems for a given dimension. For each of the discussed spaces we will state at the end of each section the proper definition of the various coordinate systems.

3.3.1   Classification of Coordinate Systems

In the classification of three-dimensional coordinate systems in spaces of constant curvature which separate the Hamiltonian we follow [323, 406]. In each of the subsequent distinct types we will state the line element and the set $\{S_1, S_2\}$ of observables.

Three Ignorable Coordinates: Type I.

$$ds^2 = dx_1^2 + dx_2^2 + \epsilon dx_3^2 \ ,$$
$$S_1 = \partial_1^2, S_2 = \partial_2^2 \ . \tag{3.6}$$

$\epsilon = \pm 1$ must be taken into account due to non-definite metrics. Usually this kind of coordinates are associated with cartesian coordinates. $\mathfrak{L}$ contains a three-dimensional Abelian subalgebra generated by $P_{1,2,3}$, and the manifold is flat. The operator $S_3 = \partial_3^2$ is automatically diagonalized.

Two Ignorable Coordinates: Type II.

$$ds^2 = g_{ij}(x_3)dx^i dx^j \ ,$$
$$S_1 = \partial_1^2, S_2 = \partial_2^2 \ . \tag{3.7}$$

$\mathfrak{L}$ contains a two-dimensional Abelian subalgebra $\mathfrak{A}$ generated by $P_{1,2}$. The coordinates may be non-orthogonal. The subalgebra must be maximal Abelian since otherwise the system would be of type I. In three-dimensional Euclidean space this provides cylindrical coordinates, where e.g. z and φ are the ignorable coordinates.

One Ignorable Coordinate: Type III.

This case splits into four subtypes, for each of which $\mathfrak{L}$ contains P_1 and $S_1 = \partial_1^2$.

Centralizer Coordinates (orthogonal): Type III$_1$.

$$ds^2 = dx_1^2 + (\sigma_2 + \sigma_3)(dx_2^2 + \epsilon dx_3^2) \ ,$$
$$S_2 = \frac{1}{\sigma_2 + \sigma_3}(\sigma_3 \partial_2^2 - \epsilon \sigma_2 \partial_3^2) \ . \tag{3.8}$$

Here and in the following we denote $\sigma_i = \sigma_i(x_i)$, which is some function of the coordinates x_i. Under this type of coordinates we find elliptic and parabolic cylindrical coordinates in three-dimensional Euclidean space.

Centralizer Coordinates (non-orthogonal): Type III$_2$.

$$ds^2 = \sigma_2(\sigma_3 dx_2^2 + 2dx_1 dx_2 + dx_3^2) \ ,$$
$$S_2 = \partial_3^2 - \sigma \partial_1^2 \ . \tag{3.9}$$

In three-dimensional Euclidean space there are no coordinate systems of this type.

Subgroup Coordinates: Type III$_3$.

$$ds^2 = \sigma_2 \sigma_3(dx_1^2 + dx_2^2) + \sigma_3 dx_3^2 \ ,$$
$$S_2 = \frac{1}{\sigma_2}(\partial_1^2 + \epsilon \partial_2^2) \ . \tag{3.10}$$

Subgroup coordinates are for instance polar coordinates. The metric in coordinate systems of type III$_1$ and III$_3$ is of the form $ds^2 = dx_1^2 + d\omega^2(x_2, x_3)$ or $ds^2 = \sigma_3(dx_3^2 + d\omega^2(x_1, x_2))$, where $d\omega^2$ is the metric for a two-dimensional subspace of (positive or negative) constant curvature.

Generic Coordinates: Type III$_4$.

$$ds^2 = \epsilon_1 \sigma_2 \sigma_3 dx_1^2 + (\sigma_2 + \sigma_3)(dx_2^2 + \epsilon_2 dx_3^2) \ , \qquad \epsilon_{1,2} = \pm 1 \ ,$$
$$S_2 = \epsilon_1 \left(\frac{1}{\sigma_3} - \frac{1}{\sigma_2} \right) \partial_1^2 + \frac{1}{\sigma_2 + \sigma_3}(\sigma_3 \partial_2^2 - \epsilon_2 \sigma_2 \partial_3^2) + \frac{1/2}{\sigma_2 + \sigma_3}\left(\frac{\sigma_3 \sigma_2'}{\sigma_2} \partial_2 - \frac{\epsilon_2 \sigma_2 \sigma_3'}{\sigma_3} \partial_3 \right) \ . \tag{3.11}$$

Generic coordinates of type III$_4$ are, e.g., spheroidal and parabolic coordinates.

Semi-Subgroup Coordinates: Type IV$_1$.

$$ds^2 = \sigma_1 dx_1^2 + \sigma_1(\sigma_2 + \sigma_3)(dx_2^2 + \epsilon dx_3^2) \ ,$$
$$S_1 = \frac{1}{\sigma_2 + \sigma_3}(\partial_2^2 + \epsilon \partial_3^2) \ . \tag{3.12}$$

Semi-subgroup coordinates of type IV$_1$ coordinates are, e.g., sphero-conical coordinates.

Generic Coordinates: Type IV$_2$.

$$\left.\begin{aligned}
ds^2 &= (\sigma_1 - \sigma_2)(\sigma_1 - \sigma_3)dx_1^2 + (\sigma_2 - \sigma_1)(\sigma_2 - \sigma_3)dx_2^2 + (\sigma_3 - \sigma_1)(\sigma_3 - \sigma_2)dx_3^2 , \\
S_1 &= \frac{\sigma_2 + \sigma_3}{(\sigma_1 - \sigma_2)(\sigma_3 - \sigma_2)}\partial_1^2 + \frac{\sigma_3 + \sigma_1}{(\sigma_2 - \sigma_1)(\sigma_2 - \sigma_3)}\partial_2^2 + \epsilon\frac{\sigma_1 + \sigma_2}{(\sigma_3 - \sigma_1)(\sigma_3 - \sigma_2)}\partial_3^2 , \\
S_2 &= \frac{\sigma_2\sigma_3}{(\sigma_1 - \sigma_2)(\sigma_3 - \sigma_2)}\partial_1^2 + \frac{\sigma_3\sigma_1}{(\sigma_2 - \sigma_1)(\sigma_2 - \sigma_3)}\partial_2^2 + \epsilon\frac{\sigma_1\sigma_2}{(\sigma_3 - \sigma_1)(\sigma_3 - \sigma_2)}\partial_3^2 .
\end{aligned}\right\}$$
$$(3.13)$$

Generic coordinates of type IV$_2$ are all kinds of ellipsoidal, hyperboloidal and parabo-loidal coordinates. Coordinates of types III$_4$ and IV$_2$ are called generic because all others can be obtained by degenerations from the generic ones.

Parametric coordinate systems can be found in types III$_4$ and IV$_{1,2}$. It is only for the three-dimensional pseudo-Euclidean space that all kinds of coordinate types occur. Types I, III$_{1,2}$ do not appear in spaces of non-zero constant curvature, i.e., on spheres and pseudospheres. In our discussion we will not return to the non-orthogonal coordinate systems.

3.3.2   The Sphere

We denote the coordinates on the sphere $S^{(D-1)}$ by the vector $\mathbf{s} = (s_0, \ldots, s_{D-1})$. The basic building blocks of separable coordinates systems on $S^{(D-1)}$ are the $(D-1)$-sphere *elliptic* coordinates

$$s_j^2 = \frac{\prod_{i=1}^{D-1}(\varrho_i - e_j)}{\prod_{j\neq i}(e_i - e_j)} , \quad (j = 0, \ldots, D-1) , \quad \sum_{j=0}^{D-1} s_j^2 = \frac{1}{k} = R^2 , \quad (3.14)$$

corresponding to a metric

$$ds^2 = -\frac{1}{4k}\sum_{i=1}^{D-1}\frac{1}{P_D(\varrho_i)}\left[\prod_{j\neq i}(\varrho_i - \varrho_j)\right](d\varrho_i)^2 , \quad P_D(\varrho) = \prod_{i=0}^{D-1}(\varrho - e_i) \quad (3.15)$$

($k > 0$ curvature). In order to find the possible explicit coordinate systems one must pay attention to the requirements that, (i) the metric must be positive definite, (ii) the variables $\{\varrho_i\}_{i=1}^{D-1}$ should vary in such a way that they correspond to a coordinates patch which is compact. There is a unique solution to these requirements given by

$$e_0 < \varrho_1 < e_1 < \ldots < e_{D-1} < \varrho_D < e_D . \quad (3.16)$$

3.3.3   Euclidean Space

In D-dimensional Euclidean space we have first the coordinate system corresponding to the D-sphere elliptic (3.14)

$$x_j^2 = \frac{\prod_{i=1}^{D}(\varrho_i - e_j)}{\prod_{j\neq i}(e_i - e_j)} , \quad (j = 1, \ldots, D) \quad (3.17)$$

$(e_1 < \varrho_1 < \ldots < e_d < \varrho_D)$. In addition there is a second class of coordinate systems, namely the paraboloidal coordinates (c^2 constant)

$$
\begin{aligned}
x_1^2 &= \tfrac{c}{2}(\varrho_1 + \ldots + \varrho_D + e_1 + \ldots + e_{D-1}) , \\
x_j^2 &= -c^2 \frac{\prod_{i=1}^{D}(\varrho_i - e_j)}{\prod_{j \neq i}(e_i - e_j)} , \quad (j = 2, \ldots, D) .
\end{aligned}
\tag{3.18}
$$

3.3.4 The Pseudosphere

On the pseudosphere $\Lambda^{(D-1)}$ the complexity increases considerably. One starts by considering the line element

$$
ds^2 = -\frac{1}{4k} \sum_{i=1}^{D-1} \frac{1}{P_D(\varrho_i)} \left[\prod_{j \neq i}(\varrho_i - \varrho_j) \right] (d\varrho_i)^2 , \qquad u_0^2 - \mathbf{u}^2 = \frac{1}{k} = R^2
\tag{3.19}
$$

($k < 0$ curvature), and one must require that $ds^2 > 0$. It turns out that there are four classes of solutions determined by the character of the solutions of the characteristic equation $P_D(\varrho) = 0$.

The first class is characterized by $e_i \neq e_j$ for $i, j = 0, \ldots, D - 1$. If $D - 1 = n = 2p + 1$ is odd then

$$
\ldots < \varrho_{i-2} < e_{i-2} < \varrho_{i-1} < e_{i-1} < e_i < e_{i+1} < \varrho_i < e_{i+2} < \ldots < e_{2p+2} < e_{2p+1} ,
\tag{3.20}
$$

($i = 0, \ldots, p$) with the convention that $e_j, \varrho_j = 0$ for j a non-positive integer which give $p+1$ distinct possibilities. Using $E_i^{(j)} = e_{i+j+1}$ ($i = 1, \ldots, 2p+2, j = 1, \ldots, p+1$) and $e_r = e_s$ for $r = s \bmod(n+1)$, the coordinates on $\Lambda^{(D-1)}$ are written in the following way

$$
u_0^2 = \frac{\prod_{i=1}^{n}(\varrho_i - E_1^{(j)})}{\prod_{k \neq 1}(E_k^{(j)} - E_1^{(j)})} , \qquad u_l^2 = \frac{\prod_{i=1}^{n}(\varrho_i - E_{l+1}^{(j)})}{\prod_{k \neq l+1}(E_k^{(j)} - E_{l+1}^{(j)})} .
\tag{3.21}
$$

Similarly if $D - 1 = n = 2p + 2$ is even ($i = 0, \ldots, p$)

$$
\ldots < \varrho_{i-2} < e_{i-2} < \varrho_{i-1} < e_{i-1} < e_i < e_{i+1} < \varrho_i < \ldots < e_{2p+1} < e_{2p} .
\tag{3.22}
$$

The second class is characterized by the fact that there can be two complex conjugate zeros of $P_D(\varrho) = 0$ denoted by $e_1 = \alpha + i\beta$, $e_2 = \alpha - i\beta$ ($\alpha, \beta \in \mathbb{R}$), respectively. Together with the convention $e_{i+1} \equiv f_{i-1}$ for all other e_j there is the one possibility

$$
\varrho_1 < f_1 < \varrho_2 < f_2 < \ldots < \varrho_{n-1} < f_{n-1} < \varrho_n .
\tag{3.23}
$$

A suitable choice of coordinates is ($j = 2, \ldots, n$)

$$
(u_0 + iu_1)^2 = \frac{i}{\beta} \frac{\prod_{i=1}^{n}(\varrho_i - \alpha - i\beta)}{\prod_{i=1}^{n-1}(f_i - \alpha - i\beta)} , \qquad u_j^2 = \frac{-\prod_{i=1}^{n}(\varrho_i - f_{j-1})}{[(\alpha - f_{j-1})^2 + \beta^2] \prod_{i \neq j-1}(f_i - f_{j-1})} .
\tag{3.24}
$$

In the third class we have a two-fold root $e_1 = e_2 = a$. Let us denote $G_j^{(i)} = g_{j+1}$ $(j = 1, \ldots, n-1, i = 0, \ldots, p)$, where $e_j = g_{j-2}$ $(j = 3, \ldots, n+1)$ with $g_k \neq g_l$ if $k \neq l$ and $g_k \neq a$ for any k, $g_r = g_s$ for $r = s \bmod(n+1)$, $n = 2p+1$ for n odd, and $n = 2p$ for n even. This case divides into two families with coordinates varying in the ranges $(i = 0, \ldots, p)$

$$\ldots < \varrho_{i-1} < g_{i-1} < \varrho_i < g_i < a < \varrho_{i+1} < g_{i+1} < \ldots < g_{n-1} < \varrho_n \ , \quad (3.25)$$
$$\ldots < \varrho_{i-1} < g_{i-1} < \varrho_i < g_i < \varrho_{i+1} < a < g_{i+1} < \ldots < g_{n-1} < \varrho_n \ , \quad (3.26)$$

and in either case of n there are $p+1$ distinguishable cases to consider. A suitable choice of coordinates is

$$
\left.
\begin{aligned}
(u_0 - u_1)^2 &= \epsilon \frac{\prod_{i=1}^{n}(\varrho_i - a)}{\prod_{k=1}^{n-1}(G_j^{(i)} - a)} \ , \\[2mm]
(u_0^2 - u_1^2) &= \frac{\partial}{\partial a} \frac{\prod_{i=1}^{n}(\varrho_i - a)}{\prod_{k=1}^{n-1}(G_j^{(i)} - a)} \ , \\[2mm]
u_j^2 &= -\frac{\prod_{i=1}^{n}(\varrho_i - G_{j-1}^{(i)})}{(a - G_{j-1}^{(i)})^2 \prod_{l \neq j-1}(G_l^{(i)} - G_{j-1}^{(i)})} \ ,
\end{aligned}
\right\} \quad (3.27)
$$

$(j = 2, \ldots, n)$ and $\epsilon = +1$ in (3.25), $\epsilon = -1$ in (3.26).

The fourth case is characterized by $e_1 = e_2 = e_3 = b$. We set $e_j = h_{j-3}$ $(j = 4, \ldots, n+1)$ with $h_k \neq h_l$ for $k \neq l$ and $h_k \neq b$ for any k. Then

$$\ldots < \varrho_{i-1} < h_{i-1} < \varrho_i < \varrho_{i+1} < b < \varrho_{i+2} < h_{i+1} < \ldots < \varrho_{n-1} < h_{n-2} < \varrho_n \ , \quad (3.28)$$

$(i = 0, \ldots, p)$ and there are $p+1$ distinct cases. A suitable choice of coordinates is

$$
\left.
\begin{aligned}
(u_0 - u_1)^2 &= -\frac{\prod_{i=1}^{n}(\varrho_i - b)}{\prod_{k=1}^{n-2}(H_j^{(i)} - b)} \ , \\[2mm]
2u_2(u_0 - u_1) &= -\frac{\partial}{\partial b} \frac{\prod_{i=1}^{n}(\varrho_i - b)}{\prod_{k=1}^{n-2}(H_j^{(i)} - b)} \ , \\[2mm]
(u_0^2 - u_1^2 - u_2^2) &= -\frac{1}{2} \frac{\partial^2}{\partial b^2} \frac{\prod_{i=1}^{n}(\varrho_i - b)}{\prod_{k=1}^{n-2}(H_j^{(i)} - b)} \ , \\[2mm]
u_j^2 &= -\frac{\prod_{i=1}^{n}(\varrho_i - H_{j-2}^{(i)})}{\prod_{l \neq j-2}(H_l^{(i)} - H_{j-2}^{(i)})(b - H_{j-2}^{(i)})^3} \ ,
\end{aligned}
\right\} \quad (3.29)
$$

$(j = 3, \ldots, n)$ and $H_j^{(i)} = h_{j+1}$ $(j = 1, \ldots, n-2, i = 0, \ldots, p)$, and $h_r = h_s \bmod(n-2)$. Note that for all coordinate systems on the hyperboloid we have the constraint $u_0^2 - \mathbf{u}^2 = 1$, respectively $u_0^2 - \mathbf{u}^2 = R^2$ if $R \neq 1$.

3.3.5   Pseudo-Euclidean Space

The case of coordinate systems in pseudo-Euclidean space has been addressed by Kalnins [300], Kalnins and Miller [322], and Turakulov [498]. There seems to be no similar study as in the three cases before in order to construct coordinate systems in D-dimensional pseudo-Euclidean space. In order to illustrate a systematic construction of coordinate systems in pseudo-Euclidean spaces nevertheless we follow [300] for the three-dimensional pseudo-Euclidean space.

Three Ignorable Coordinates: Type I.

There is only one coordinate system in $\mathbb{R}^{(2,1)}$ with three ignorable coordinates. This is the (pseudo-) cartesian one. The prefix "pseudo" will be omitted in the following. Cartesian coordinates are of type I. In tables 4.1 and 4.2 this is coordinate system I.

$E(1,1)$-Cylindrical Coordinates: Types II and III$_1$.

The next set of coordinate systems consists of the $E(1,1)$ group reduction. There are nine of such systems, where we do not count the cartesian one. By $E(1,1)$ we mean the group of isomorphic operations in the pseudo-Euclidean plane. Similarly, the group $E(2)$ consists of the isomorphic operations in the usual Euclidean space, where the algebra is six-dimensional consisting of the momentum and angular momentum operators. We thus have the group chain $E(2,1) \supset E(1,1)$. The type II coordinate system is the cylindrical polar system (one system), all other eight are of type III$_1$. In table 4.2 the cylindrical $E(1,1)$ coordinates are the systems II.-X.

$E(2)$-Cylindrical Coordinates: Types II and III$_1$.

The next set of three coordinate systems consists of the $E(2)$-cylindrical type systems. Here one has the subgroup chain $E(2,1) \supset E(2)$. The type II system is the cylindrical polar system (one system), the other two are of class III$_1$. In table 4.2 the cylindrical $E(2)$ coordinates are the systems XI.-XIII.

(Semi-) Subgroup Coordinates: Types III$_3$ and IV$_1$.

The next set of nine coordinate systems is due to the subgroup chain $E(2,1) \supset SO(2,1)$, i.e., we have a polar variable and a set of variables of the nine different coordinate systems from the two-dimensional pseudosphere. Among them three are of class III$_3$, and six of type IV$_1$. In table 4.2 the subgroup coordinates (spherical $\Lambda^{(2)}$) are the systems XIV.–XXII.

Generic Coordinates: Type III$_4$.

In this class we find eight different coordinate systems, i.e., two parabolic, six spheroidal and two hyperbolic systems. In table 4.2 they are XXIII., XXIV., XXV.–XXX., and XXXI., XXXII., respectively.

Generic Coordinates: Type IV$_2$.

The generic class here consists of 22 types of coordinate systems, among them ellipsoidal, paraboloidal, hyperboloidal and others where the respective planes of constant ϱ_i do not have commonly accepted names. We list them in the second part of table 4.2. The various systems can be found in [322], and by related systems we mean that the coordinates have to be permutated in the indicated order, including possible factors of i. Very little seems to be known about these systems, let alone solutions of the Laplacian in these coordinate systems. Because we do not find any solution at all, and therefore also no interbasis expansions relating the wavefunctions on other coordinate system representations, no path integral representations can be found. The only path integral representations available consist of the solutions in terms of the coordinate systems I.–XXXII.

It must be noted that there is one non-orthogonal coordinate system in the pseudo-Euclidean plane [300], and there are two non-orthogonal coordinate systems in pseudo-Euclidean space [406]. These coordinate systems are of the type II and are called semi-hyperbolic in [300].

We see that pseudo-Euclidean space has a rich structure concerning the number of different coordinate systems which separate the Hamiltonian. We can either have the subgroup chains, e.g., for the four-dimensional pseudo-Euclidean space $E(3,1)$: $E(3,1) \supset E(2,1)$, $E(3,1) \supset E(3)$ or $E(3,1) \supset SO(3,1) \supset SO(2)$. Thus all three previous coordinate systems of the sphere, the pseudosphere and Euclidean space come into play, and in addition, there are many coordinate systems in four-dimensional pseudo-Euclidean space which are not of the subgroup type. All together add up to 261 different coordinate systems [322].

3.3.6   A Hilbert Space Model

Let us shortly mention a technique how to construct solutions of the Hamiltonian on spaces of constant curvature. This method is based on the Gel'fand-Graev [177] transform which is constructive harmonic analysis on the space. In D-dimensional Euclidean space one has for instance ($\mathbf{x} \in \mathbb{R}^D$)

$$\Psi(\mathbf{x}) = \text{l.i.m.} \int_{S^{(D-1)}} d\Omega(\mathbf{k})\, e^{\mathrm{i}p\mathbf{x}\cdot\mathbf{k}} h(\mathbf{k}) \ . \tag{3.30}$$

Here $\mathbf{k}$ is the (momentum) unit vector on the sphere $S^{(D-1)}$ with $d\Omega$ its integration measure. The notion l.i.m. says that the integral exists in the L^2-sense ($\equiv$ limes in medio). h is a complex valued function on the sphere, and $h \in L^2(S^{(D-1)})$. In order to apply this formula in the construction of separable eigenfunctions of the Hamiltonian, one must find a (normalized) basis function of the characterizing differential operators which corresponds to a particular coordinate system. These basis functions are eigenfunctions of observables which are peculiar for the coordinate system. Evaluating the integral yields the eigenfunctions of the Hamiltonian. In two-dimensional Euclidean space one has [405] $((x, y) \in \mathbb{R}^2)$

$$\Psi(x, y) = \text{l.i.m.} \int_{-\pi}^{\pi} d\varphi \, e^{ip(x \cos \varphi + y \sin \varphi)} h(\varphi) \ , \tag{3.31}$$

in three-dimensional Euclidean space [80] $((x, y, z) \in \mathbb{R}^3)$

$$\Psi(x, y, z) = \text{l.i.m.} \int_{-\pi}^{\pi} d\varphi \int_{0}^{\pi} d\vartheta \, e^{ip(x \sin \vartheta \cos \varphi + y \sin \vartheta \sin \varphi + z \cos \vartheta)} h(\vartheta, \varphi) \ , \tag{3.32}$$

in two-dimensional pseudo-Euclidean space [405] $((v_0, v_1) \in \mathbb{R}^{(1,1)})$

$$\Psi(v_0, v_1) = \text{l.i.m.} \int_{\mathbb{R}} dr \, e^{ip(v_0 \cosh r + v_1 \sinh r)} h(r) \ , \tag{3.33}$$

and analogously in three-dimensional pseudo-Euclidean space. In [80, 405] this method is used to construct explicitly the separating solutions of the Hamiltonian in these spaces.

In order to apply this method of harmonic analysis in spaces of (non-zero) constant curvature, the formula (3.30) must be generalized in such a way that the complex valued function h is still a basis function, i.e., an eigenfunction of the observables characteristic for the coordinate system in question. The harmonic analysis must then be performed with "plane waves" in the corresponding geometry, and the integral must be taken along a contour Γ on the cone $y^2 = 0$ in this geometry. This gives for pseudospheres $(u, y \in \Lambda^{(D-1)})$

$$\Psi(u) = \text{l.i.m.} \int_{\Gamma} dw(y)(u \cdot y)^{ip-(D-2)/4} h(y) \ . \tag{3.34}$$

dw is the invariant measure on the pseudosphere, and the value of the integral is independent of the contour. This method has been used in [309] for the two-dimensional two-sheeted pseudosphere, in [299] for the horicyclic system on the two-dimensional single-sheeted pseudosphere, and in [318] for various coordinate systems on the two- and single-sheeted three-dimensional pseudosphere. However, I do not use this method in the path integral evaluations.

Chapter 4

Path Integrals in Pseudo-Euclidean Geometry

4.1   The Pseudo-Euclidean Plane

We start with the two-dimensional pseudo-Euclidean, respectively two-dimensional Minkowski space. We first state the relevant path integral representations, and then discuss the methods of evaluating them. We have the following path integral representations in $\mathbb{R}^{(1,1)}$ ($a \cdot b = a_0 b_0 - a_1 b_1$, $p^2 > 0$ is taken in the physical domain):

I. Cartesian and General Form of the Propagator , $(v_0, v_1) = v \in \mathbb{R}^{(1,1)}$:

$$\int_{v(t')=v'}^{v(t'')=v''} \mathcal{D}v(t) \exp\left(\frac{im}{2\hbar}\int_{t'}^{t''}(\dot{v}_0^2 - \dot{v}_1^2)dt\right) = \frac{m}{2\pi\hbar T}\exp\left(\frac{im}{2\hbar T}|v'' - v'|^2\right) \tag{4.1}$$

$$= \int_{\mathbb{R}^{(1,1)}} \frac{dp}{4\pi^2}\exp\left[-\frac{i\hbar T}{2m}p^2 + ip\cdot(v'' - v')\right] . \tag{4.2}$$

II. Polar , $\varrho > 0, \tau \in \mathbb{R}$:

$$\int_{\varrho(t')=\varrho'}^{\varrho(t'')=\varrho''} \mathcal{D}\varrho(t)\varrho \int_{\tau(t')=\tau'}^{\tau(t'')=\tau''} \mathcal{D}\tau(t) \exp\left\{\frac{i}{\hbar}\int_{t'}^{t''}\left[\frac{m}{2}(\dot{\varrho}^2 - \varrho^2\dot{\tau}^2) + \frac{\hbar^2}{8m\varrho^2}\right]dt\right\}$$

$$= \int_{\mathbb{R}} \frac{dk}{2\pi}e^{ik(\tau''-\tau')}\int_0^\infty \frac{pdp}{\pi^2}K_{ik}(-ip\varrho'')K_{ik}(ip\varrho')\,e^{-i\hbar p^2 T/2m} . \tag{4.3}$$

III. Parabolic 1 , $\xi \in \mathbb{R}, \eta > 0$:

$$\int_{\xi(t')=\xi'}^{\xi(t'')=\xi''} \mathcal{D}\xi(t) \int_{\eta(t')=\eta'}^{\eta(t'')=\eta''} \mathcal{D}\eta(t)(\xi^2 - \eta^2) \exp\left[\frac{im}{2\hbar}\int_{t'}^{t''}(\xi^2 - \eta^2)(\dot{\xi}^2 - \dot{\eta}^2)dt\right]$$

$$= \int_{\mathbb{R}} d\zeta \int_{\mathbb{R}} \frac{dp}{32\pi^4} e^{-i\hbar p^2 T/2m}$$

$$\times \left(\begin{array}{l} |\Gamma(\frac{1}{4} + \frac{i\zeta}{2p})|^2 E^{(0)}_{-1/2+i\zeta/p}(e^{-i\pi/4}\sqrt{2p}\,\xi'') E^{(0)}_{-1/2+i\zeta/p}(e^{i\pi/4}\sqrt{2p}\,\eta'') \\ |\Gamma(\frac{3}{4} - \frac{i\zeta}{2p})|^2 E^{(1)}_{-1/2+i\zeta/p}(e^{-i\pi/4}\sqrt{2p}\,\xi'') E^{(1)}_{-1/2+i\zeta/p}(e^{i\pi/4}\sqrt{2p}\,\eta'') \end{array} \right)$$

$$\times \left(\begin{array}{l} |\Gamma(\frac{1}{4} + \frac{i\zeta}{2p})|^2 E^{(0)}_{-1/2-i\zeta/p}(e^{i\pi/4}\sqrt{2p}\,\xi') E^{(0)}_{-1/2-i\zeta/p}(e^{-i\pi/4}\sqrt{2p}\,\eta') \\ |\Gamma(\frac{3}{4} - \frac{i\zeta}{2p})|^2 E^{(1)}_{-1/2-i\zeta/p}(e^{i\pi/4}\sqrt{2p}\,\xi') E^{(1)}_{-1/2-i\zeta/p}(e^{-i\pi/4}\sqrt{2p}\,\eta') \end{array} \right) . \tag{4.4}$$

IV. Parabolic 2, $\xi \in \mathbb{R}, \eta > 0$:

$$\int_{\xi(t')=\xi'}^{\xi(t'')=\xi''} \mathcal{D}\xi(t) \int_{\eta(t')=\eta'}^{\eta(t'')=\eta''} \mathcal{D}\eta(t)(\eta^2 - \xi^2) \exp\left[\frac{im}{2\hbar} \int_{t'}^{t''} (\eta^2 - \xi^2)(\dot{\xi}^2 - \dot{\eta}^2) dt \right]$$

$$= \int_{\mathbb{R}} d\zeta \int_{\mathbb{R}} \frac{dp}{32\pi^4} e^{-i\hbar p^2 T/2m}$$

$$\times \left(\begin{array}{l} |\Gamma(\frac{1}{4} + \frac{i\zeta}{2p})|^2 E^{(0)}_{-1/2+i\zeta/p}(\sqrt{2p}\,\xi'') E^{(0)}_{-1/2+i\zeta/p}(\sqrt{2p}\,\eta'') \\ |\Gamma(\frac{3}{4} + \frac{i\zeta}{2p})|^2 E^{(1)}_{-1/2+i\zeta/p}(\sqrt{2p}\,\xi'') E^{(1)}_{-1/2+i\zeta/p}(\sqrt{2p}\,\eta'') \end{array} \right)$$

$$\times \left(\begin{array}{l} |\Gamma(\frac{1}{4} + \frac{i\zeta}{2p})|^2 E^{(0)}_{-1/2-i\zeta/p}(\sqrt{2p}\,\xi') E^{(0)}_{-1/2-i\zeta/p}(\sqrt{2p}\,\eta') \\ |\Gamma(\frac{3}{4} + \frac{i\zeta}{2p})|^2 E^{(1)}_{-1/2-i\zeta/p}(\sqrt{2p}\,\xi') E^{(1)}_{-1/2-i\zeta/p}(\sqrt{2p}\,\eta') \end{array} \right) . \tag{4.5}$$

V. Parabolic 3, $\xi \in \mathbb{R}, \eta > 0$:

$$\int_{\xi(t')=\xi'}^{\xi(t'')=\xi''} \mathcal{D}\xi(t) \int_{\eta(t')=\eta'}^{\eta(t'')=\eta''} \mathcal{D}\eta(t)(\xi - \eta) \exp\left[\frac{im}{2\hbar} \int_{t'}^{t''} (\xi - \eta)(\dot{\xi}^2 - \dot{\eta}^2) dt \right]$$

$$= 16 \int_0^\infty \frac{dp}{p^{1/3}} \int_{\mathbb{R}} d\zeta \, e^{-i\hbar p^2 T/2m}$$

$$\times \mathrm{Ai}\left[-\left(\xi' + \sqrt{2m}\frac{\zeta}{p^2} \right) p^{2/3} \right] \mathrm{Ai}\left[-\left(\xi'' + \sqrt{2m}\frac{\zeta}{p^2} \right) p^{2/3} \right]$$

$$\times \mathrm{Ai}\left[-\left(\eta' + \sqrt{2m}\frac{\zeta}{p^2} \right) p^{2/3} \right] \mathrm{Ai}\left[-\left(\eta'' + \sqrt{2m}\frac{\zeta}{p^2} \right) p^{2/3} \right] . \tag{4.6}$$

VI. Elliptic 1, $a \in \mathbb{R}, b > 0$:

$$\int_{a(t')=a'}^{a(t'')=a''} \mathcal{D}a(t) \int_{b(t')=b'}^{b(t'')=b''} \mathcal{D}b(t) d^2(\sinh^2 a - \sinh^2 b)$$

$$\times \exp\left[\frac{im}{2\hbar} d^2 \int_{t'}^{t''} (\sinh^2 a - \sinh^2 b)(\dot{a}^2 - \dot{b}^2) dt \right]$$

$$= \frac{1}{8\pi} \int_0^\infty p dp \int_{\mathbb{R}} dk \, e^{-\pi k} e^{-i\hbar p^2 T/2m} \mathrm{Me}_{ik}(b''; \tfrac{p^2 d^2}{4}) \mathrm{Me}^*_{ik}(b'; \tfrac{p^2 d^2}{4}) M^{(3)}_{ik}(a''; \tfrac{pd}{2}) M^{(3)\,*}_{ik}(a'; \tfrac{pd}{2}). \tag{4.7}$$

VII. Elliptic 2, $a \in \mathbb{R}, b > 0$:

$$\int_{a(t')=a'}^{a(t'')=a''} \mathcal{D}a(t) \int_{b(t')=b'}^{b(t'')=b''} \mathcal{D}b(t) d^2 (\sinh^2 a + \cosh^2 b)$$

$$\times \exp\left[\frac{im}{2\hbar} d^2 \int_{t'}^{t''} (\sinh^2 a + \cosh^2 b)(\dot{a}^2 - \dot{b}^2)dt\right]$$

$$= \frac{1}{8\pi} \int_0^\infty pdp \int_{\mathbb{R}} dk \, e^{-\pi k} e^{-i\hbar p^2 T/2m}$$

$$\times \mathrm{Me}_{ik}(b'' - i\tfrac{\pi}{2}; \tfrac{p^2 d^2}{4}) \mathrm{Me}_{ik}^*(b' - i\tfrac{\pi}{2}; \tfrac{p^2 d^2}{4}) M_{ik}^{(3)}(a''; \tfrac{pd}{2}) M_{ik}^{(3)\,*}(a'; \tfrac{pd}{2}) \quad . \tag{4.8}$$

VIII. Hyperbolic 1, $y_1, y_2 \in \mathbb{R}$:

$$\int_{y_1(t')=y_1'}^{y_1(t'')=y_1''} \mathcal{D}y_1(t) \int_{y_2(t')=y_2'}^{y_2(t'')=y_2''} \mathcal{D}y_2(t) \frac{d^2}{8} (\sinh y_1 - \sinh y_2)$$

$$\times \exp\left[\frac{imd^2}{16\hbar} \int_{t'}^{t''} (\sinh y_1 - \sinh y_2)(\dot{y}_1^2 - \dot{y}_2^2)dt\right]$$

$$= \frac{1}{32\pi} \int_0^\infty pdp \int_{\mathbb{R}} dk \, e^{-\pi k} e^{-i\hbar p^2 T/2m} \mathrm{Me}_{ik}(\tfrac{y_1''}{2} - i\tfrac{\pi}{4}; i\tfrac{p^2 d^2}{4}) \mathrm{Me}_{ik}^*(\tfrac{y_1'}{2} - i\tfrac{\pi}{4}; i\tfrac{p^2 d^2}{4})$$

$$\times M_{ik}^{(3)}(\tfrac{y_2''}{2} - i\tfrac{\pi}{4}; \sqrt{i}\tfrac{pd}{2}) M_{ik}^{(3)\,*}(\tfrac{y_2'}{2} - i\tfrac{\pi}{4}; \sqrt{i}\tfrac{pd}{2}) \quad . \tag{4.9}$$

IX. Hyperbolic 2, $y_1, y_2 \in \mathbb{R}$:

$$\int_{y_1(t')=y_1'}^{y_1(t'')=y_1''} \mathcal{D}y_1(t) \int_{y_1(t')=y_1'}^{y_1(t'')=y_1''} \mathcal{D}y_1(t)(e^{2y_1} + e^{2y_2}) \exp\left[\frac{im}{2\hbar} \int_{t'}^{t''} (e^{2y_1} + e^{2y_2})(\dot{y}_1^2 - \dot{y}_2^2)dt\right]$$

$$= \frac{2}{\pi^4} \int_0^\infty dk \, k \sinh \pi k \int_0^\infty dp \, p K_{ik}(e^{y_1''}p) K_{ik}(e^{y_1'}p) K_{ik}(-ie^{y_2'}p) K_{ik}(ie^{y_2''}p) \, e^{-i\hbar p^2 T/2m}. \tag{4.10}$$

X. Hyperbolic 3, $y_1, y_2 \in \mathbb{R}$:

$$\int_{y_1(t')=y_1'}^{y_1(t'')=y_1''} \mathcal{D}y_1(t) \int_{y_1(t')=y_1'}^{y_1(t'')=y_1''} \mathcal{D}y_1(t)(e^{2y_1} - e^{2y_2}) \exp\left[\frac{im}{2\hbar} \int_{t'}^{t''} (e^{2y_1} - e^{2y_2})(\dot{y}_1^2 - \dot{y}_2^2)dt\right]$$

$$= \frac{1}{\pi^2} \int_0^\infty pdp \, e^{-i\hbar p^2 T/2m} \left\{ \int_0^\infty \frac{kdk}{2\sinh \pi k} \right.$$

$$\times \left[J_{ik}(p\, e^{y_2''}) + J_{-ik}(p\, e^{y_2''})\right]^* \left[J_{ik}(p\, e^{y_2''}) + J_{-ik}(p\, e^{y_2''})\right] K_{ik}(ip\, e^{y_1''}) K_{ik}(-ip\, e^{y_1''})$$

$$\left. + \sum_{n \in \mathbb{N}} [2(2n + \alpha)] J_{2n+\alpha}(p\, e^{y_2''}) J_{2n+\alpha}(p\, e^{y_2''}) K_{2n+\alpha}(-ip\, e^{y_1''}) K_{2n+\alpha}(ip\, e^{y_1''}) \right\}. \tag{4.11}$$

Let us discuss the path integral solutions (4.1-4.11) in some detail.

Cartesian Coordinates.

This is the defining coordinate system. The path integrations (4.1, 4.2) in pseudo-cartesian coordinates are obvious and can be made in a straightforward way. Note that the spectrum of the Hamiltonian is taken in the physical domain $p_0^2 - p_1^2 = p^2 \geq 0$ in the momentum representation. This is due to the fact that we have taken into account the real Lie algebra $E(1,1)$ in the pseudo-Euclidean plane, c.f. [405, p.39].

Polar Coordinates.

Together with reflections and rotations the polar coordinate system covers all the pseudo-Euclidean plane except the lines $v_0 = \pm v_1$. The operations R (reflections) and I (inversions) act on the pseudo-Euclidean plane as $R(v_0, v_1) = (-v_0, v_1)$ and $I(v_0, v_1) = (v_1, v_0)$, respectively. In order to evaluate the path integration in pseudo-polar coordinates we consider the expansion ([66], [187, p.804], [249], in the following we omit the prefix "pseudo" if not explicitly stated otherwise)

$$e^{-z\cosh\alpha} = \sqrt{\frac{2}{\pi z}}(z\sinh\alpha)^{\frac{3-D}{2}}\int_0^\infty \left|\frac{\Gamma(ik + \frac{D-2}{2})}{\Gamma(ik)}\right|^2 \mathcal{P}_{ik-\frac{1}{2}}^{\frac{3-D}{2}}(\cosh\alpha)K_{ik}(z)dk \ , \quad (4.12)$$

which for $D = 2$ takes on the form

$$e^{-z\cosh\alpha} = \frac{1}{\pi}\int_{\mathbb{R}} dk\, e^{ik\alpha}K_{ik}(z) \ . \quad (4.13)$$

$K_\nu(z)$ is a MacDonald modified Bessel function, and $\mathcal{P}_\mu^\nu(z)$ is an associated Legendre function. We obtain the following expansion of the short-time kernel of the path integral in $\mathbb{R}^{(1,1)}$ in terms of the polar coordinates matrix elements

$$\frac{m}{2\pi\epsilon\hbar}\exp\left(\frac{im}{2\epsilon\hbar}|v'' - v'|^2\right)$$
$$= \frac{m}{2\pi\epsilon\hbar}\exp\left[\frac{im}{2\epsilon\hbar}(\varrho'^2 + \varrho''^2)\right]\frac{1}{\pi}\int_{\mathbb{R}} dk\, e^{ik(\tau''-\tau')}K_{ik}\left(\frac{m\varrho'\varrho''}{i\epsilon\hbar}\right) \ , \quad (4.14)$$

on the one hand, and on the other

$$\frac{m}{2\pi\epsilon\hbar}\exp\left(\frac{im}{2\epsilon\hbar}|v'' - v'|^2\right)$$
$$= \int_{\mathbb{R}^{(1,1)}}\frac{dp}{4\pi^2}\exp\left[-\frac{i\epsilon\hbar}{2m}p^2 + ip\cdot(v'' - v')\right]$$
$$= \frac{1}{4\pi^2}\int_0^\infty dp\, p\int_{\mathbb{R}} d\alpha\exp\left\{ip[\varrho''\cosh(\tau'' - \alpha) - \varrho'\cosh(\tau' - \alpha)] - \frac{i\epsilon\hbar p^2}{2m}\right\}$$
$$= \frac{1}{4\pi^2}\int_0^\infty dp\, p\int_{\mathbb{R}} d\alpha\frac{1}{\pi^2}\int_{\mathbb{R}} dk_1\int_{\mathbb{R}} dk_2\, e^{ik_1(\tau''-\alpha)-ik_2(\tau'-\alpha)-i\epsilon\hbar p^2/2m}K_{ik_1}(-ip\varrho'')K_{ik_2}(ip\varrho')$$
$$= \frac{1}{2\pi^3}\int_{\mathbb{R}} dk\, e^{ik(\tau''-\tau')}\int_0^\infty p\, dp K_{ik}(-ip\varrho'')K_{ik}(ip\varrho')e^{-i\epsilon\hbar p^2/2m} \ . \quad (4.15)$$

Exploiting the orthonormality relation (c.f. according to [385])

$$\frac{p}{\pi^2} \int_0^\infty r dr\, K_{ik}(-ipr) K_{ik}(ip'r) = \delta(p - p') \ , \qquad (4.16)$$

in each short-time path integration, then gives by means of our group matrix-element path integration [67] the result of (4.3). Separating the τ-path integration gives us as a by-result the *conjectural* path integral identity which is valid in the sense of distributions

$$\int\limits_{r(t')=r'}^{r(t'')=r''} \mathcal{D}r(t) \exp\left[\frac{i}{\hbar}\int_{t'}^{t''}\left(\frac{m}{2}\dot{r}^2 + \frac{k^2 + \frac{1}{4}}{2mr^2}\right)dt\right]$$

$$``=" \sqrt{r'r''} \int_0^\infty \frac{pdp}{\pi^2} K_{ik}(-ipr'') K_{ik}(ipr')\mathrm{e}^{-i\hbar p^2 T/2m} \ . \qquad (4.17)$$

Note that from the equivalence of the expansions (4.14) and (4.15) of the short-time kernel, we can derive the following integral representation ($\alpha, \beta, \gamma \in \mathbb{R}$) which is valid in the sense of distributions

$$\int_0^\infty dp\, p K_{ik}(i\alpha p) K_{ik}(-i\beta p)\, \mathrm{e}^{-i\gamma p^2} ``=" \frac{\pi}{2\gamma} \exp\left(-\frac{\alpha^2 + \beta^2}{4i\gamma}\right) K_{ik}\left(\frac{\alpha\beta}{2i\gamma}\right) \ . \qquad (4.18)$$

Let us shortly demonstrate the orthonormality (4.16) of the radial wavefunctions. We have

$$\Psi_p(r) = \frac{\sqrt{p}}{\pi} K_{ik}(-ipr) = \frac{\sqrt{p}}{2}\mathrm{e}^{\pi k/2} H_{-ik}^{(1)}(pr) \ . \qquad (4.19)$$

Using the integral formula for the product of two Bessel functions w_ν, W_μ [385, p.87]

$$\int_\epsilon^R \left[(\beta^2 - \alpha^2)r + \frac{\nu^2 - \mu^2}{r}\right] w_\nu(\alpha r) W_\mu(\beta r) dr$$

$$= \alpha r W_\mu(\beta r) w_{\nu-1}(\alpha r) - \beta r W_{\mu-1}(\beta r) w_\nu(\alpha r) + (\mu - \nu)W_\mu(\beta r) w_\nu(\alpha r)\bigg|_\epsilon^R , \qquad (4.20)$$

for $\nu = \mu = -ik$, $\alpha = p, \beta = p'$, and the asymptotic expansions for $r \to 0$ and $r \to \infty$ [1, pp.104, 108]

$$iH_\nu^{(1,2)}(r) \propto \pm\frac{1}{\pi}\Gamma(\nu)\left(\frac{r}{2}\right)^{-\nu} , \qquad H_\nu^{(1,2)}(r) \propto \sqrt{\frac{2}{\pi r}}\mathrm{e}^{\pm i(r-\nu\pi/2-\pi/4)} , \qquad (4.21)$$

and the distributional relation $\lim_{N\to\infty} \mathrm{e}^{iNx}/x = i\pi\delta(x)$, we obtain (4.16).

Parabolic 1 and 2 Coordinates.

The parabolic 1 and 2 systems are in comparison to the polar system regular at the origin and can therefore be treated without the difficulties which arise from the indefinite metric. All three coordinate systems cover all the pseudo-Euclidean plane, where in the case of the parabolic 3 system it is necessary to take into account reflection. After a time transformation we obtain, e.g., in the parabolic I system

$$
\int_{\xi(t')=\xi'}^{\xi(t'')=\xi''} \mathcal{D}\xi(t) \int_{\eta(t')=\eta'}^{\eta(t'')=\eta''} \mathcal{D}\eta(t)(\xi^2 - \eta^2)\exp\left[\frac{im}{2\hbar}\int_{t'}^{t''}(\xi^2 - \eta^2)(\dot{\xi}^2 - \dot{\eta}^2)dt\right]
$$

$$
= \int_{\mathbb{R}} \frac{dE}{2\pi\hbar}\, e^{-iET/\hbar}\int_0^\infty ds'' \int_{\xi(0)=\xi'}^{\xi(s'')=\xi''} \mathcal{D}\xi(s) \int_{\eta(0)=\eta'}^{\eta(s'')=\eta''} \mathcal{D}\eta(s)
$$

$$
\times \exp\left\{\frac{i}{\hbar}\int_0^{s''}\left[\frac{m}{2}(\dot{\xi}^2 - \dot{\eta}^2) + E(\xi^2 - \eta^2)\right]ds\right\}, \tag{4.22}
$$

thus yielding two repelling harmonic oscillators where each solution is given by

$$
\int_{x(t')=x'}^{x(t'')=x''} \mathcal{D}x(t)\exp\left[\frac{im}{2\hbar}\int_{t'}^{t''}(\dot{x}^2 + \omega^2 x^2)dt\right]
$$

$$
= \left(\frac{m\omega}{2\pi i\hbar\sinh\omega T}\right)^{1/2}\exp\left\{-\frac{m\omega}{2i\hbar}\left[(x'^2 + x''^2)\coth\omega T - 2\frac{x'x''}{\sinh\omega T}\right]\right\}. \tag{4.23}
$$

Each propagator is now analyzed by means of the expansion

$$
\frac{1}{\sqrt{2\pi\sin\alpha}}\exp\left[-(x+y)\cot\alpha\right]\exp\left(\frac{2\sqrt{xy}}{\sin\alpha}\right)
$$

$$
= \frac{1}{(2\pi)^2}\int_{\mathbb{R}}dp\, e^{-2\alpha p + \pi p}
$$

$$
\times\left[\left|\Gamma\left(\frac{1}{4}-ip\right)\right|^2 E^{(0)}_{-\frac{1}{2}+2ip}\left(e^{-i\pi/4}2\sqrt{x}\right)E^{(0)}_{-\frac{1}{2}-2ip}\left(e^{i\pi/4}2\sqrt{y}\right)\right.
$$

$$
\left. +\left|\Gamma\left(\frac{3}{4}-ip\right)\right|^2 E^{(1)}_{-\frac{1}{2}+2ip}\left(e^{-i\pi/4}2\sqrt{x}\right)E^{(1)}_{-\frac{1}{2}-2ip}\left(e^{i\pi/4}2\sqrt{y}\right)\right]. \tag{4.24}
$$

The $E^{(0)}_\nu(z)$ and $E^{(1)}_\nu(z)$ are even and odd parabolic cylinder functions [88], respectively. The final result is obtained by introducing a Coulomb coupling α in the s'' integration, performing a momentum variable transformation $(p_\xi, p_\eta) \to (\frac{1}{2p}(\frac{1}{a} + \zeta), -\frac{1}{2p}(\frac{1}{a} - \zeta))$ with Jacobean $J = 1/2ap^3$ ($a = \hbar^2/m\alpha$ is the Bohr radius) with the new variables (p, ζ) and finally setting $\alpha = 0$, i.e., $a = \infty$. ζ is the parabolic separation constant. In particular, if we abbreviate the product of the square brackets in the double expansion by $f(p_\xi, p_\eta)$ we get ($\omega = \sqrt{-2E/m}$)

$$
\frac{1}{2}\int_{\mathbb{R}}\frac{dE}{2\pi i}\, e^{-iET/\hbar}\frac{i}{\hbar}\int_0^\infty ds''\, e^{2i\alpha s''/\hbar}\frac{m\omega}{(4\pi^2)^2 i\hbar}\int dp_\xi \int dp_\eta\, e^{-2\omega s''(p_\xi - p_\eta)}f(p_\xi, p_\eta)\bigg|_{\alpha=0}
$$

$$= \frac{1}{2} \int_{\mathbb{R}} \frac{dE}{2\pi i} e^{-iET/\hbar} \frac{m\omega}{\hbar^2 (4\pi^2)^2} \int dp_\xi \int dp_\eta \frac{f(p_\xi, p_\eta)}{2\omega(p_\xi - p_\eta) - 2i\alpha/\hbar} \Big|_{\alpha=0}$$

$$= \int_{\mathbb{R}} \frac{dE}{2\pi i} e^{-iET/\hbar} \frac{m\omega}{2\alpha\hbar^2 (4\pi^2)^2} \int_0^\infty \frac{dp}{p^3} \int_{\mathbb{R}} d\zeta \frac{f(p_\xi, p_\eta)}{\omega/ap - i\alpha/\hbar} \Big|_{\alpha=0}$$

$$= \int_0^\infty dp \int_{\mathbb{R}} d\zeta \, e^{-i\hbar T p^2 / 2m} f\left(\frac{\zeta}{2p}, \frac{\zeta}{2p}\right) . \tag{4.25}$$

Thus we obtain (4.4) by properly taking into account that the entire wavefunctions are either complete symmetric or complete anti-symmetric.

In the second parabolic system we must take care of different phase factors in the arguments of the even and odd parabolic cylinder functions $E_\nu^{(0)}$ and $E_\nu^{(1)}$. Hence we obtain the path integral solutions (4.4, 4.5).

Parabolic 3 Coordinates.

The parabolic 3 systems does not differ from the former ones significantly. Instead of harmonic oscillators we deal with linear potentials after a time-transformation. Together with the path integral solution of the linear potential (Feynman and Hibbs [164] and Schulman [460]) we expand the kernel by using the identity

$$\left(\frac{m}{2\pi i\hbar T}\right)^{1/2} \exp\left[\frac{i}{\hbar}\left(\frac{m}{2}\frac{(x''-x')^2}{T} - \frac{kT}{2}(x'+x'') - \frac{k^2 T^3}{24m}\right)\right]$$

$$= \frac{1}{4\pi^2\hbar^2}\left(\frac{2m\hbar}{\sqrt{k}}\right)^{2/3} \int_{\mathbb{R}} dt \int_{\mathbb{R}} ds \int_{\mathbb{R}} dE$$

$$\times \exp\left[i\frac{t^3 - s^3}{3} + \frac{i}{\hbar}\left(\frac{2m\hbar}{k^2}\right)^{1/3} E(t-s) - \frac{i}{\hbar}\left(\frac{2mk}{\hbar^2}\right)^{1/3}(x't - x''s) - \frac{i}{\hbar}ET\right] . \tag{4.26}$$

We observe that

$$\frac{1}{2\pi\hbar}\left(\frac{2m\hbar}{\sqrt{k}}\right)^{1/3} \int_{\mathbb{R}} dt \exp\left[\frac{i}{\hbar}\frac{t^3}{3} - \frac{i}{\hbar}\left(\frac{E}{k} - x\right)\left(\frac{2mk}{\hbar^2}\right)^{1/3} t\right]$$

$$= \left(\frac{2m}{\hbar^2\sqrt{k}}\right)^{1/3} \text{Ai}\left[\left(x - \frac{E}{k}\right)\left(\frac{2mk}{\hbar^2}\right)^{1/3}\right] \tag{4.27}$$

$$= \sqrt{\frac{2m}{3\pi^2\hbar^2}\left(x - \frac{E}{k}\right)} K_{1/3}\left[\frac{2}{3}\left(x - \frac{E}{k}\right)^{3/2}\sqrt{\frac{2mk}{\hbar^2}}\right] , \tag{4.28}$$

and we can display the emerging wavefunctions in alternative ways. Ai denotes the Airy function [1]. In order to use the same procedure as in the two previous parabolic systems, we transform the two p-integrations emerging from the double expansion in the following way

$$\int dp_\xi \int dp_\eta \, e^{-i(p_\xi - p_\eta)s''/\hbar} \mapsto 4\omega^2\hbar^2 \int dp_1 \int dp_2 \, e^{-2\omega s''(p_1 - p_2)} , \tag{4.29}$$

with $\omega = \sqrt{-2E/m} = -i\hbar p/m$. Taking into account a proper consideration of the factors i we obtain in the argument of the Airy functions

$$\frac{p_\xi}{k} \mapsto -\frac{2im\omega\zeta}{p^3} = -\sqrt{2m}\frac{\zeta}{p^2} \ . \tag{4.30}$$

ζ is the parabolic separation constant. Putting everything together we obtain the path integral solution (4.6).

Elliptic 1 Coordinates.

The elliptic 1 system parameterizes all the pseudo-Euclidean plane. In order to discuss the elliptic 1, 2 and hyperbolic 1 systems we have to study a generalization of (5.9) for a hyperbolic metric. Using the theory of [399, p.185] we find ($\mathrm{Me}_\nu(z)$ and $M_\nu^{(3)}(z)$ are Mathieu functions)

$$\exp\left[ip(v_0\cosh\tau - v_1\sinh\tau)\right] = \frac{1}{2}\int_{\mathbb{R}} dk\, e^{-\pi k/2}\mathrm{Me}_{ik}(b;\tfrac{p^2d^2}{4})\mathrm{Me}_{ik}(\tau;\tfrac{p^2d^2}{4})M_{ik}^{(3)}(a;\tfrac{pd}{2}) \ , \tag{4.31}$$

which has the correct limit for $d \to 0$. $\mathrm{Me}_\nu(z; d^2) \propto e^{\nu z}$ and $M_\nu^{(3)}(z/d; d) \propto H_\nu^{(1)}(z)$ ($d \to 0$), yield the wavefunctions of the polar system. Using the orthonormality relation ($h = pd/2$)

$$\frac{1}{2\pi}\int_{\mathbb{R}} d\tau \mathrm{Me}_{ik}(\tau; h^2)\mathrm{Me}_{ik'}^*(\tau; h^2) = \delta(k' - k) \ , \tag{4.32}$$

we obtain for the short-time kernel

$$\frac{m}{2\pi\epsilon\hbar}\exp\left(\frac{im}{2\epsilon\hbar}|\mathbf{v}'' - \mathbf{v}'|^2\right)$$

$$= \int_{\mathbb{R}^{(1,1)}} \frac{d\mathbf{p}}{4\pi^2}\exp\left[-\frac{i\epsilon\hbar}{2m}\mathbf{p}^2 + i\mathbf{p}\cdot(\mathbf{v}'' - \mathbf{v}')\right]$$

$$= \frac{1}{4\pi^2}\int_0^\infty pdp\int_{\mathbb{R}} d\alpha\exp\left\{ip\left[(v_0'' - v_0')\cosh\alpha - (v_1'' - v_1')\sinh\alpha\right] - \frac{i\epsilon\hbar p^2}{2m}\right\}$$

$$= \frac{1}{16\pi^2}\int_0^\infty pdp\int_{\mathbb{R}} d\tau\int_{\mathbb{R}} dk\int_{\mathbb{R}} dk'\, e^{-\pi(k+k')/2}e^{-i\epsilon\hbar p^2/2m}$$

$$\times\mathrm{Me}_{ik}(b''; \tfrac{p^2d^2}{4})\mathrm{Me}_{ik'}^*(b'; \tfrac{p^2d^2}{4})M_{ik}^{(3)}(a''; \tfrac{pd}{2})M_{ik'}^{(3)*}(a'; \tfrac{pd}{2})\mathrm{Me}_{ik}(\tau; \tfrac{p^2d^2}{4})\mathrm{Me}_{ik'}^*(\tau; \tfrac{p^2d^2}{4})$$

$$= \frac{1}{8\pi}\int_0^\infty pdp\int_{\mathbb{R}} dk\, e^{-\pi k}e^{-i\epsilon\hbar p^2/2m}\mathrm{Me}_{ik}(b''; \tfrac{p^2d^2}{4})\mathrm{Me}_{ik}^*(b'; \tfrac{p^2d^2}{4})M_{ik}^{(3)}(a''; \tfrac{pd}{2})M_{ik}^{(3)*}(a'; \tfrac{pd}{2}) . \tag{4.33}$$

Exploiting the orthonormality relation

$$d^2\int_{\mathbb{R}} da\int_0^\infty db(\sinh^2 a - \sinh^2 b)\mathrm{Me}_{ik}(b; \tfrac{p^2d^2}{4})\mathrm{Me}_{ik'}^*(b; \tfrac{p'^2d^2}{4})M_{ik}^{(3)}(a; \tfrac{pd}{2})M_{ik'}^{(3)*}(a; \tfrac{p'd}{2})$$

$$= \frac{8\pi}{p\,e^{-\pi k}}\delta(k' - k)\delta(p' - p) \ , \tag{4.34}$$

in each step of a short-time path integration gives the result (4.7).

Elliptic 2 Coordinates.

In comparison to the elliptic 1 system the second elliptic system does not cover the entire pseudo-Euclidean plane and cannot be made so by adding reflections and rotations. For the path integration in the case of the elliptic 2 system, we can proceed similarly as for the elliptic 1 system. We just make a shift in the variable b, $b \mapsto b - i\pi/2$, and the result of (4.8) follows.

Hyperbolic 1 Coordinates.

In the following the first two hyperbolic systems cover the whole pseudo-Euclidean plane, where in the second a reflection and rotation are required. In order to treat the path integral on $\mathbb{R}^{(1,1)}$ in this coordinate system we perform the coordinate transformation $x_i = \frac{1}{2}(y_i + i\frac{\pi}{2})$, and abbreviate $\tilde{d}^2 = id^2$ $(i = 1, 2)$. This gives

$$
\int_{y_1(t')=y_1'}^{y_1(t'')=y_1''} \mathcal{D}y_1(t) \int_{y_2(t')=y_2'}^{y_2(t'')=y_2''} \mathcal{D}y_2(t) \frac{d^2}{8}(\sinh y_1 - \sinh y_2)
$$

$$
\times \exp\left[\frac{im}{16\hbar}d^2 \int_{t'}^{t''} (\sinh y_1 - \sinh y_2)(\dot{y}_1^2 - \dot{y}_2^2)dt\right]
$$

$$
= \frac{1}{4} \int_{x_1(t')=x_1'}^{x_1(t'')=x_1''} \mathcal{D}x_1(t) \int_{x_2(t')=x_2'}^{x_2(t'')=x_2''} \mathcal{D}x_2(t)\tilde{d}^2(\sinh^2 x_1 - \sinh^2 x_2)
$$

$$
\times \exp\left[\frac{im}{2\hbar}\tilde{d}^2 \int_{t'}^{t''} (\sinh^2 x_1 - \sinh^2 x_2)(\dot{x}_1^2 - \dot{x}_2^2)dt\right] , \qquad (4.35)
$$

and the result (4.9) follows by using the path integral solution (4.8) and re-inserting (y_1, y_2).

Hyperbolic 2 Coordinates.

In the hyperbolic 2 system we must consider due to the indefinite metric a proper combination of the polar coordinates in $\mathbb{R}^{(1,1)}$ and the well-known Liouville potential problem [247]. Performing a two-dimensional time-transformation and inserting for the y_1-path integration the Liouville potential problem [247] (c.f. the Lebedev transformation, e.g., [385, p.398])

$$
\int_{y(t')=y'}^{y(t'')=y''} \mathcal{D}y(t) \exp\left[\frac{i}{\hbar} \int_{t'}^{t''} \left(\frac{m}{2}\dot{y}^2 - \frac{\hbar^2\kappa^2}{2m}e^{2y}\right)dt\right]
$$

$$
= \frac{2}{\pi^2} \int_0^\infty dp\, p \sinh \pi p\, e^{-i\hbar p^2 T/2m} K_{ip}^*(e^{y'}\kappa)K_{ip}(e^{y''}\kappa) , \qquad (4.36)
$$

and the y_2-path integration being transformed to radial $\mathbb{R}^{(1,1)}$ coordinates by an additional space-time transformation we obtain ($\lambda = \sqrt{-2m\epsilon}/\hbar$, $q = e^{2y_2}$)

$$
\int_{y_1(t')=y_1'}^{y_1(t'')=y_1''} \mathcal{D}y_1(t) \int_{y_1(t')=y_1'}^{y_1(t'')=y_1''} \mathcal{D}y_1(t)(e^{2y_1} + e^{2y_2}) \exp\left[\frac{im}{2\hbar}\int_{t'}^{t''}(e^{2y_1} + e^{2y_2})(\dot{y}_1^2 - \dot{y}_2^2)dt\right]
$$

$$
= \int_{\mathbb{R}}\frac{dE}{2\pi\hbar}e^{-iET/\hbar}\int_0^\infty ds''
$$

$$
\times \int_{y_1(0)=y_1'}^{y_1(s'')=y_1''}\mathcal{D}y_1(s) \int_{y_2(0)=y_2'}^{y_2(s'')=y_2''}\mathcal{D}y_2(s)\exp\left\{\frac{i}{\hbar}\int_0^{s''}\left[\frac{m}{2}(\dot{y}_1^2 - \dot{y}_2^2) + E(e^{2y_1} + e^{2y_2})\right]ds\right\}
$$

$$
= \int_{\mathbb{R}}\frac{dE}{2\pi\hbar}e^{-iET/\hbar}\int_0^\infty ds'' \int_{y_1(0)=y_1'}^{y_1(s'')=y_1''}\mathcal{D}y_1(s)\exp\left[\frac{i}{\hbar}\int_0^{s''}\left(\frac{m}{2}\dot{y}_1^2 + Ee^{2y_1}\right)ds\right]
$$

$$
\times \int_{\mathbb{R}}\frac{d\epsilon}{2\pi\hbar}e^{i\epsilon s''/\hbar}\int_0^{-\infty}d\sigma''\frac{e^{-i\sigma''E/\hbar}}{\sqrt{q'q''}}
$$

$$
\times \int_{q(0)=q'}^{q(\sigma'')=q''}\mathcal{D}q(\sigma)\exp\left[\frac{i}{\hbar}\int_0^{\sigma''}\left(\frac{m}{2}\dot{q}^2 + \hbar^2\frac{2m\epsilon/\hbar^2 + \frac{1}{4}}{2mq^2}\right)d\sigma\right]
$$

$$
= \int_{\mathbb{R}}\frac{dE}{2\pi\hbar}e^{-iET/\hbar}\int_0^\infty dp\int_0^{-\infty}d\sigma''\exp\left[-\frac{i}{\hbar}\sigma''\left(E + \frac{\hbar^2 p^2}{2m}\right)\right]
$$

$$
\times \int_{\mathbb{R}}\frac{d\epsilon}{2\pi\hbar}e^{-i\epsilon\tau/\hbar}\int_0^\infty dk\int_0^\infty ds''\exp\left[\frac{i}{\hbar}s''\left(\epsilon - \frac{\hbar^2 k^2}{2m}\right)\right]\bigg|_{\kappa=0}
$$

$$
\times \frac{2}{\pi^4}k\sinh\pi k K_{ik}^*(e^{y_1'}\sqrt{-2mE}/\hbar)K_{ik}(e^{y_1''}\sqrt{-2mE}/\hbar)pK_\lambda(ip\,e^{y_2'})K_\lambda(-ip\,e^{y_2''})\ .
$$

$$
(4.37)
$$

The various integrations are now analyzed as follows: The y_1 path integration is for Liouville quantum mechanics with evolution parameter s'', the q (i.e., the transformed y_2) path integration a singular radial potential with evolution parameter $-\sigma''$ (backwards in time). The $\int d\sigma$-integrations yield a simple pole $E + \hbar^2 p^2/2m = 0$, which together with the $\int dE$-integrations allows the interpretation of a Green's function expansion where the wavefunctions are given by taking the residua at the poles or cuts. Similarly we analyse the $\int d\epsilon ds''$-integrations. The only difference being that we must finally set the evolution parameter $\tau = 0$. The $\int dk dp$-integrations are left as they stand. Considering the two residua emerging from the two contributions gives the propagator depending on the evolution parameter T (forwards in time), and thus (4.10).

Table 4.1: Coordinates in Two-Dimensional Pseudo-Euclidean Space

Coordinate System	Coordinates	Path Integral Solution
I. Cartesian	$v_0 = v_0'$ $v_1 = v_1'$	(4.1, 4.2)
II. Polar	$v_0 = \varrho \cosh \tau$ $v_1 = \varrho \sinh \tau$	(4.3)
III. Parabolic 1	$v_0 = \frac{1}{2}(\xi^2 + \eta^2)$ $v_1 = \xi\eta$	(4.4)
IV. Parabolic 2	$v_0 = \xi\eta$ $v_1 = \frac{1}{2}(\xi^2 + \eta^2)$	(4.5)
V. Parabolic 3	$v_0 = \frac{1}{2}(\eta - \xi)^2 + (\xi + \eta)$ $v_1 = \frac{1}{2}(\eta - \xi)^2 - (\xi + \eta)$	(4.6)
VI. Elliptic 1	$v_0 = d \cosh a \cosh b$ $v_1 = d \sinh a \sinh b$	(4.7)
VII. Elliptic 2	$v_0 = d \sinh a \cosh b$ $v_1 = d \cosh a \sinh b$	(4.8)
VIII. Hyperbolic 1	$v_0 = \frac{d}{2}\left(\cosh \frac{y_1 - y_2}{2} + \sinh \frac{y_1 + y_2}{2}\right)$ $v_1 = \frac{d}{2}\left(\cosh \frac{y_1 - y_2}{2} - \sinh \frac{y_1 + y_2}{2}\right)$	(4.9)
IX. Hyperbolic 2	$v_0 = \sinh(y_1 - y_2) + \frac{1}{2}e^{y_1 + y_2}$ $v_1 = \sinh(y_1 - y_2) - \frac{1}{2}e^{y_1 + y_2}$	(4.10)
X. Hyperbolic 3	$v_0 = \cosh(y_1 - y_2) + \frac{1}{2}e^{y_1 + y_2}$ $v_1 = \cosh(y_1 - y_2) - \frac{1}{2}e^{y_1 + y_2}$	(4.11)

Hyperbolic 3 Coordinates.

The parameterization of the hyperbolic 3 system does not cover the whole pseudo-Euclidean plane, only the domains $v_0 + v_1 > 0$, $v_0 - v_1 > 0$. For the path integration in the case of the hyperbolic 3 system we proceed in a similar way as before. Performing a time-transformation yields

$$
\int\limits_{y_1(t')=y_1'}^{y_1(t'')=y_1''} \mathcal{D}y_1(t) \int\limits_{y_1(t')=y_1'}^{y_1(t'')=y_1''} \mathcal{D}y_1(t) (e^{2y_1} - e^{2y_2}) \exp\left[\frac{im}{2\hbar}\int_{t'}^{t''}(e^{2y_1} - e^{2y_2})(\dot{y}_1^2 - \dot{y}_2^2)dt\right]
$$

$$
= \int_{\mathbb{R}} \frac{dE}{2\pi\hbar} e^{-iET/\hbar} \int_0^\infty ds'' \int\limits_{y_1(0)=y_1'}^{y_1(s'')=y_1''} \mathcal{D}y_1(s) \int\limits_{y_2(0)=y_2'}^{y_2(s'')=y_2''} \mathcal{D}y_2(s)
$$

$$\times \exp\left\{\frac{\mathrm{i}}{\hbar}\int_0^{s''}\left[\frac{m}{2}(\dot{y}_1^2 - \dot{y}_2^2) + E(\mathrm{e}^{2y_1} - \mathrm{e}^{2y_2})\right]ds\right\} \ . \tag{4.38}$$

For the y_1 path integration we must use the path integral solution coming from the pseudo-polar coordinates, i.e. (4.17). However, due to the different sign in the coupling of the two systems we must insert in the short-time kernel of the y_2-path integrations the path integral identity from the inverted Liouville problem. Inserting for the y_1-path integration the results of the usual Liouville path integration gives by a similar Green's function analysis as before the result (4.11).

The Coordinate Systems in the Pseudo-Euclidean Plane.

We summarize our results of the path integration in two-dimensional pseudo-Euclidean space including an enumeration of the coordinate systems according to Kalnins [300] and Miller [405]. Note that the coordinate systems of the IR^2 and the two-dimensional pseudo-Euclidean space cover all coordinate systems of the complex Helmholtz, respectively Schrödinger equation [405, p.62]. In the parametric coordinate systems d is a positive parameter.

4.2   Three-Dimensional Pseudo-Euclidean Space

We have the path integral representations in $\mathrm{IR}^{(2,1)}$ ($\mathbf{a}\cdot\mathbf{b} = a_0b_0 - a_1b_1 - a_2b_2$, the important spherical coordinate system XIII. is stated separately)

I. Cartesian and General Form of the Propagator , $(v_0, v_1, v_2) = \mathbf{v}\in\mathrm{IR}^{(2,1)}$:

$$\int_{\mathbf{v}(t')=\mathbf{v}'}^{\mathbf{v}(t'')=\mathbf{v}''}\mathcal{D}\mathbf{v}(t)\exp\left(\frac{\mathrm{i}m}{2\hbar}\int_{t'}^{t''}\dot{\mathbf{v}}^2 dt\right) = \sqrt{\mathrm{i}}\left(\frac{m}{2\pi\hbar T}\right)^{3/2}\exp\left(\frac{\mathrm{i}m}{2\hbar T}|\mathbf{v}'' - \mathbf{v}'|^2\right) \tag{4.39}$$

$$= \int_{\mathrm{IR}^{(2,1)}}\frac{d\mathbf{p}}{8\pi^3}\exp\left[-\frac{\mathrm{i}\hbar T}{2m}\mathbf{p}^2 + \mathrm{i}\mathbf{p}\cdot(\mathbf{v}'' - \mathbf{v}')\right] \ . \tag{4.40}$$

II.–X. Cylindrical $\mathrm{IR}^{(1,1)}$, $(V_0, V_1) = \mathbf{V}\in\mathrm{IR}^{(1,1)}, v\in\mathrm{IR}$:

$$\int_{\mathbf{V}(t')=\mathbf{V}'}^{\mathbf{V}(t'')=\mathbf{V}''}\mathcal{D}\mathbf{V}(t)\int_{v(t')=v'}^{v(t'')=v''}\mathcal{D}v(t)\exp\left[\frac{\mathrm{i}}{\hbar}\int_{t'}^{t''}\left(\mathcal{L}_{\mathrm{IR}^{(1,1)}}(\mathbf{V},\dot{\mathbf{V}}) - \frac{m}{2}\dot{v}^2\right)dt\right]$$

$$= \left(\frac{\mathrm{i}m}{2\pi\hbar T}\right)^{1/2}\exp\left(\frac{m}{2\mathrm{i}\hbar T}(v''-v')^2\right)\cdot K^{\mathrm{IR}^{(1,1)}}(\mathbf{V}'', \mathbf{V}'; T) \ . \tag{4.41}$$

XI.–XIII. Cylindrical IR^2, $(x, y) = \mathbf{x}\in\mathrm{IR}^2, v\in\mathrm{IR}$:

$$\int_{\mathbf{x}(t')=\mathbf{x}'}^{\mathbf{x}(t'')=\mathbf{x}''}\mathcal{D}\mathbf{x}(t)\int_{v(t')=v'}^{v(t'')=v''}\mathcal{D}v(t)\exp\left[\frac{\mathrm{i}}{\hbar}\int_{t'}^{t''}\left(\frac{m}{2}\dot{v}^2 - \mathcal{L}_{\mathrm{IR}^2}(\mathbf{x},\dot{\mathbf{x}})\right)dt\right]$$

$$= \left(\frac{m}{2\pi i\hbar T}\right)^{1/2} \exp\left(-\frac{m}{2i\hbar T}(v'' - v')^2\right) \cdot K^{\mathbb{R}^2}(\mathbf{x}'', \mathbf{x}''; -T) \ . \tag{4.42}$$

XIV. Spherical, $\tau \in \mathbb{R}, \varphi \in [0, 2\pi), r > 0$:

$$\int_{r(t')=r'}^{r(t'')=r''} \mathcal{D}r(t)r^2 \int_{\tau(t')=\tau'}^{\tau(t'')=\tau''} \mathcal{D}\tau(t)\sinh\tau \int_{\varphi(t')=\varphi'}^{\varphi(t'')=\varphi''} \mathcal{D}\varphi(t)$$

$$\times \exp\left\{\frac{i}{\hbar}\int_{t'}^{t''}\left[\frac{m}{2}\left(\dot{r}^2 - r^2(\dot{\tau}^2 + \sinh^2\tau\dot{\varphi}^2)\right) + \frac{1}{8mr^2}\left(1 - \frac{1}{\sinh^2\tau}\right)\right]dt\right\}$$

$$= \frac{1}{\sqrt{r'r''}}\sum_{\nu\in\mathbb{Z}}\frac{e^{i\nu(\varphi''-\varphi')}}{2\pi}\int_0^\infty \frac{dk}{\pi}k\sinh\pi k|\Gamma(\tfrac{1}{2}+ik+\nu)|^2\mathcal{P}_{ik-1/2}^{-\nu}(\cosh\tau'')\mathcal{P}_{ik-1/2}^{-\nu}(\cosh\tau')$$

$$\times \int_0^\infty \frac{pdp}{\pi^2}K_{ik}(-ipr'')K_{ik}(ipr')\,e^{-i\hbar p^2 T/2m} \ . \tag{4.43}$$

XV.–XXII. Spherical $\Lambda^{(2)}$, $\mathbf{u} \in \Lambda^{(2)}, r > 0$:

$$\int_{r(t')=r'}^{r(t'')=r''} \mathcal{D}r(t)r^2 \int_{\mathbf{u}(t')=\mathbf{u}'}^{\mathbf{u}(t'')=\mathbf{u}''} \frac{\mathcal{D}\mathbf{u}(t)}{u_0}\exp\left\{\frac{i}{\hbar}\int_{t'}^{t''}\left[\frac{m}{2}\left(\dot{r}^2 + r^2\dot{\mathbf{u}}^2\right) - \frac{1}{r^2}\Delta V_{\Lambda^{(2)}}(\mathbf{u})\right]dt\right\}$$

$$= \frac{1}{\sqrt{r'r''}}\int d\omega \int_0^\infty dk\,\Psi_{\omega,k}^{\Lambda^{(2)}}(\mathbf{u}'')\Psi_{\omega,k}^{\Lambda^{(2)}*}(\mathbf{u}')\int_0^\infty \frac{pdp}{\pi^2}K_{ik}(-ipr'')K_{ik}(ipr')\,e^{-i\hbar p^2 T/2m}. \tag{4.44}$$

XXIII. Parabolic 1, $\xi, \eta > 0, \varphi \in [0, 2\pi)$:

$$\int_{\xi(t')=\xi'}^{\xi(t'')=\xi''} \mathcal{D}\xi(t) \int_{\eta(t')=\eta'}^{\eta(t'')=\eta''} \mathcal{D}\eta(t)(\xi^2-\eta^2)\xi\eta \int_{\varphi(t')=\varphi'}^{\varphi(t'')=\varphi''} \mathcal{D}\varphi(t)$$

$$\times \exp\left\{\frac{i}{\hbar}\int_{t'}^{t''}\left[\frac{m}{2}\left((\xi^2-\eta^2)(\dot{\xi}^2-\dot{\eta}^2) - \xi^2\eta^2\dot{\varphi}^2\right) - \frac{\hbar^2}{8m\xi^2\eta^2}\right]dt\right\}$$

$$= \sum_{\nu\in\mathbb{Z}}\frac{e^{i\nu(\varphi''-\varphi')}}{2\pi}\int_{\mathbb{R}}d\zeta\int_0^\infty \frac{dp}{p}\frac{|\Gamma(\frac{1+|\nu|}{2}+i\zeta/p)|^4}{4\pi^2\xi'\xi''\eta'\eta''\Gamma^4(1+|\nu|)}e^{-i\hbar p^2 T/2m}$$

$$\times M_{-i\zeta/2p,|\nu|/2}(-ip\xi'')M_{i\zeta/2p,|\nu|/2}(ip\xi')M_{-i\zeta/2p,|\nu|/2}(ip\eta'')M_{i\zeta/2p,|\nu|/2}(-ip\eta') \ . \tag{4.45}$$

XXIV. Parabolic 2, $\xi, \eta > 0, \tau \in \mathbb{R}$:

$$\int_{\xi(t')=\xi'}^{\xi(t'')=\xi''} \mathcal{D}\xi(t) \int_{\eta(t')=\eta'}^{\eta(t'')=\eta''} \mathcal{D}\eta(t)(\eta^2-\xi^2)\xi\eta \int_{\tau(t')=\tau'}^{\tau(t'')=\tau''} \mathcal{D}\tau(t)$$

$$\times \exp\left\{\frac{i}{\hbar}\int_{t'}^{t''}\left[\frac{m}{2}\left((\eta^2-\xi^2)(\dot{\xi}^2-\dot{\eta}^2) - \xi^2\eta^2\dot{\tau}^2\right) + \frac{\hbar^2}{8m\xi^2\eta^2}\right]dt\right\} \tag{4.46}$$

$$= \int_{\mathbb{R}}\frac{dk}{2\pi}e^{ik(\tau''-\tau')}\int_{\mathbb{R}}d\zeta\int_0^\infty \frac{dp}{p}\frac{|\Gamma[\frac{1}{2}(1+ik)+i\zeta/p]|^4}{16\pi^2\xi'\xi''\eta'\eta''\sinh^2\pi k\,\Gamma^4(1+ik)}e^{-i\hbar p^2 T/2m}$$

$$\times \left[M_{i\zeta/2p,ik/2}(ip\xi'') - M_{i\zeta/2p,-ik/2}(-ip\xi'')\right]\left[M_{-i\zeta/2p,ik/2}(-ip\xi') - M_{-i\zeta/2p,-ik/2}(ip\xi')\right]$$

$$\times \left[M_{-\mathrm{i}\zeta/2p,\mathrm{i}k/2}(\mathrm{i}p\eta'') - M_{-\mathrm{i}\zeta/2p,-\mathrm{i}k/2}(-\mathrm{i}p\eta'') \right] \left[M_{\mathrm{i}\zeta/2p,\mathrm{i}k/2}(-\mathrm{i}p\eta') - M_{\mathrm{i}\zeta/2p,-\mathrm{i}k/2}(\mathrm{i}p\eta') \right].$$
(4.47)

XXV. Prolate-Spheroidal 1, $\xi, \eta > 0, \varphi \in [0, 2\pi)$:

$$\int_{\xi(t')=\xi'}^{\xi(t'')=\xi''} \mathcal{D}\xi(t) \int_{\eta(t')=\eta'}^{\eta(t'')=\eta''} \mathcal{D}\eta(t)(\sinh^2 \xi - \sinh^2 \eta) d^3 \sinh \xi \sinh \eta \int_{\varphi(t')=\varphi'}^{\varphi(t'')=\varphi''} \mathcal{D}\varphi(t)$$

$$\times \exp\left\{ \frac{\mathrm{i}}{\hbar} \int_{t'}^{t''} \left[\frac{m}{2}\Big(d^2(\sinh^2 \xi - \sinh^2 \eta)(\dot\eta^2 - \dot\xi^2) - \sinh^2 \xi \sinh^2 \eta\, \dot\varphi^2 \Big) \right. \right.$$

$$\left. \left. - \frac{\hbar^2}{8md^2 \sinh^2 \xi \sinh^2 \eta} \right] dt \right\}$$

$$= \sum_{\nu\in\mathbb{Z}} \frac{e^{\mathrm{i}\nu(\varphi''-\varphi')}}{2\pi} \frac{1}{\pi} \int_0^\infty dk\, k \sinh \pi k |\Gamma(\tfrac{1}{2} + \mathrm{i}k + \nu)|^2 \int_0^\infty \frac{p^2 dp}{2\pi} e^{-\pi k} e^{-\mathrm{i}\hbar p^2 T/2m}$$

$$\times \mathrm{Ps}_{\mathrm{i}k-1/2}^{-\nu}(\cosh \eta''; p^2 d^2) \mathrm{Ps}_{\mathrm{i}k-1/2}^{-\nu\,*}(\cosh \eta'; p^2 d^2)$$

$$\times S_{\mathrm{i}k-1/2}^{\nu\,(3)}(\cosh \xi''; pd) S_{\mathrm{i}k-1/2}^{\nu\,(3)\,*}(\cosh \xi'; pd) \;.$$
(4.48)

XXVI. Prolate-Spheroidal 2, $\xi \in \mathbb{R}, \eta > 0, \varphi \in [0, 2\pi)$:

$$\int_{\eta(t')=\eta'}^{\eta(t'')=\eta''} \mathcal{D}\eta(t) \int_{\varphi(t')=\varphi'}^{\varphi(t'')=\varphi''} \mathcal{D}\varphi(t) d^3(\sinh^2 \eta + \sin^2 \varphi) \sinh \eta \sin \varphi \int_{\xi(t')=\xi'}^{\xi(t'')=\xi''} \mathcal{D}\xi(t)$$

$$\times \exp\left\{ \frac{\mathrm{i}}{\hbar} \int_{t'}^{t''} \left[\frac{m}{2} d^2 \Big(-(\sinh^2 \eta + \sin^2 \varphi)(\dot\eta^2 + \dot\varphi^2) + \sinh^2 \eta \sin^2 \varphi\, \dot\xi^2 \Big) \right. \right.$$

$$\left. \left. - \frac{\hbar^2}{8md^2 \sinh^2 \eta \sin^2 \varphi} \right] dt \right\}$$

$$= \int_{\mathbb{R}} \frac{d\lambda}{2\pi} e^{\mathrm{i}\lambda(\xi''-\xi')} \int_0^\infty dp \int_0^\infty \frac{dk}{\cosh \pi(\lambda - k)} \frac{k \sinh \pi k}{\cosh^2 \pi k + \sinh^2 \pi\lambda} e^{-\mathrm{i}\hbar p^2 T/2m}$$

$$\times \sum_{\epsilon=\pm1} \mathrm{ps}_{\mathrm{i}k-1/2}^{\mathrm{i}\lambda}(\epsilon \cos \varphi''; p^2 d^2) \mathrm{ps}_{\mathrm{i}k-1/2}^{\mathrm{i}\lambda\,*}(\epsilon \cos \varphi'; p^2 d^2)$$

$$\times S_{\mathrm{i}k-1/2}^{\mathrm{i}\lambda\,(1)}(\cosh \eta''; pd) S_{\mathrm{i}k-1/2}^{\mathrm{i}\lambda\,(1)\,*}(\cosh \eta'; pd) \;.$$
(4.49)

XXVII. Prolate-Spheroidal 3, $\xi, \eta, \tau \in \mathbb{R}$:

$$\int_{\xi(t')=\xi'}^{\xi(t'')=\xi''} \mathcal{D}\xi(t) \int_{\eta(t')=\eta'}^{\eta(t'')=\eta''} \mathcal{D}\eta(t) d^3(\cosh^2 \xi - \cosh^2 \eta) \cosh \xi \cosh \eta \int_{\tau(t')=\tau'}^{\tau(t'')=\tau''} \mathcal{D}\tau(t)$$

$$\times \exp\left\{ \frac{\mathrm{i}}{\hbar} \int_{t'}^{t''} \left[\frac{m}{2}\Big(d^2(\cosh^2 \xi - \cosh^2 \eta)(\dot\xi^2 - \dot\eta^2) - \cosh^2 \xi \cosh^2\eta\, \dot\tau^2 \Big) \right. \right.$$

$$\left. \left. + \frac{\hbar^2}{8md^2 \cosh^2 \xi \cosh^2 \eta} \right] dt \right\}$$

$$
= \sum_{\nu \in \mathbb{Z}} \frac{e^{i\nu(\varphi'' - \varphi')}}{2\pi} \frac{1}{\pi} \int_0^\infty dk\, k \sinh \pi k |\Gamma(\tfrac{1}{2} + ik + \nu)|^2 \int_0^\infty \frac{p^2 dp}{2\pi} e^{-\pi k} e^{-i\hbar p^2 T/2m}
$$
$$
\times \mathrm{Psi}_{ik-1/2}^{-\nu}(\sinh \eta''; p^2 d^2)\mathrm{Psi}_{ik-1/2}^{-\nu\,*}(\sinh \eta'; p^2 d^2)
$$
$$
\times \mathrm{Si}_{ik-1/2}^{\nu\,(3)}(\sinh \xi''; pd)\mathrm{Si}_{ik-1/2}^{\nu\,(3)\,*}(\sinh \xi'; pd)\ . \tag{4.50}
$$

<u>XXVIII. Oblate-Spheroidal 1</u>, $\xi \in \mathbb{R}, \eta > 0, \varphi \in [0, 2\pi)$:

$$
\int_{\xi(t')=\xi'}^{\xi(t'')=\xi''} \mathcal{D}\xi(t) \int_{\eta(t')=\eta'}^{\eta(t'')=\eta''} \mathcal{D}\eta(t) d^3 (\cosh^2 \xi + \sinh^2 \eta) \sinh \eta \cosh \xi \int_{\varphi(t')=\varphi'}^{\varphi(t'')=\varphi''} \mathcal{D}\varphi(t)
$$
$$
\times \exp\left\{\frac{i}{\hbar}\int_{t'}^{t''}\left[\frac{m}{2}\Big(d^2(\sinh^2 \eta + \cosh^2 \xi)(\dot{\eta}^2 - \dot{\xi}^2) - \sinh^2 \eta \cosh^2 \xi\, \dot{\varphi}^2\Big) \right.\right.
$$
$$
\left.\left. - \frac{\hbar^2}{8md^2 \sinh^2 \eta \cosh^2 \xi}\right]dt\right\}
$$
$$
= \sum_{\nu \in \mathbb{Z}} \frac{e^{i\nu(\varphi'' - \varphi')}}{2\pi} \frac{1}{\pi} \int_0^\infty dk\, k \sinh \pi k |\Gamma(\tfrac{1}{2} + ik + \nu)|^2 \int_0^\infty \frac{p^2 dp}{2\pi} e^{-\pi k} e^{-i\hbar p^2 T/2m}
$$
$$
\times \mathrm{Psi}_{ik-1/2}^{-\nu}(\cosh \eta''; p^2 d^2)\mathrm{Psi}_{ik-1/2}^{-\nu\,*}(\cosh \eta'; p^2 d^2)
$$
$$
\times \mathrm{Si}_{ik-1/2}^{\nu\,(3)}(\cosh \xi''; pd)\mathrm{Si}_{ik-1/2}^{\nu\,(3)\,*}(\cosh \xi'; pd)\ . \tag{4.51}
$$

<u>XXIX. Oblate-Spheroidal 2</u>, $\xi > 0, \eta \in \mathbb{R}, \varphi \in [0, 2\pi)$:

$$
\int_{\eta(t')=\eta'}^{\eta(t'')=\eta''} \mathcal{D}\eta(t) \int_{\varphi(t')=\varphi'}^{\varphi(t'')=\varphi''} \mathcal{D}\varphi(t) d^3(\cos^2 \varphi - \cosh^2 \eta)\cosh \eta \cos \varphi \int_{\xi(t')=\xi'}^{\xi(t'')=\xi''} \mathcal{D}\xi(t)
$$
$$
\times \exp\left\{\frac{i}{\hbar}\int_{t'}^{t''}\left[\frac{m}{2}\Big(d^2(\cos^2 \varphi - \cosh^2 \eta)(\dot{\eta}^2 + \dot{\varphi}^2) + \cosh^2 \eta \cos^2 \varphi\, \dot{\xi}^2\Big)\right.\right.
$$
$$
\left.\left. - \frac{\hbar^2}{8md^2 \cosh^2 \eta \cos^2 \varphi}\right]dt\right\}
$$
$$
= \int_{\mathbb{R}} \frac{d\lambda}{2\pi} e^{i\lambda(\xi'' - \xi')}\lambda \sinh \pi \lambda \int_0^\infty dp \int_0^\infty \frac{dk\, k \sinh \pi k}{(\cosh^2 \pi k + \sinh^2 \pi \lambda)^2} e^{-i\hbar p^2 T/2m}
$$
$$
\times \sum_{\epsilon, \epsilon' = \pm 1} \mathrm{ps}_{i\lambda-1/2}^{ik}(\epsilon \sin \varphi''; p^2 d^2)\mathrm{ps}_{i\lambda-1/2}^{ik\,*}(\epsilon \sin \varphi'; p^2 d^2)
$$
$$
\times S_{i\lambda-1/2}^{ik\,(1)}(\epsilon' \tanh \eta''; pd)S_{i\lambda-1/2}^{ik\,(1)\,*}(\epsilon' \tanh \eta'; pd)\ . \tag{4.52}
$$

<u>XXX. Oblate-Spheroidal 3</u>, $\xi > 0, \eta \in \mathbb{R}, \tau \in \mathbb{R}$:

$$
\int_{\eta(t')=\eta'}^{\eta(t'')=\eta''} \mathcal{D}\eta(t) \int_{\xi(t')=\xi'}^{\xi(t'')=\xi''} \mathcal{D}\xi(t) d^3(\sinh^2 \xi + \cosh^2 \eta)\sinh \xi \cosh \eta \int_{\tau(t')=\tau'}^{\tau(t'')=\tau''} \mathcal{D}\tau(t)
$$
$$
\times \exp\left\{\frac{i}{\hbar}\int_{t'}^{t''}\left[\frac{m}{2}\Big(d^2(\sinh^2 \xi + \cosh^2 \eta)(\dot{\xi}^2 - \dot{\eta}^2) - \sinh^2 \xi \cosh^2 \eta\, \dot{\tau}^2\Big)\right.\right.
$$

$$\left. - \frac{\hbar^2}{8md^2 \cosh^2 \eta \sinh^2 \xi} \right] dt \right\}$$

$$= \sum_{\nu \in \mathbb{Z}} \frac{\mathrm{e}^{\mathrm{i}\nu(\varphi'' - \varphi')}}{2\pi} \frac{1}{\pi} \int_0^\infty dk \, k \sinh \pi k |\Gamma(\tfrac{1}{2} + \mathrm{i}k + \nu)|^2 \int_0^\infty \frac{p^2 dp}{2\pi} \mathrm{e}^{-\pi k} \mathrm{e}^{-\mathrm{i}\hbar p^2 T/2m}$$

$$\times \mathrm{Ps}_{\mathrm{i}k-1/2}^{-\nu}(\sinh \eta''; p^2 d^2) \mathrm{Ps}_{\mathrm{i}k-1/2}^{-\nu *}(\sinh \eta'; p^2 d^2) S_{\mathrm{i}k-1/2}^{\nu\,(3)}(\sinh \xi''; pd) S_{\mathrm{i}k-1/2}^{\nu\,(3)*}(\sinh \xi'; pd).$$

$$(4.53)$$

<u>XXXI. Hyperbolic 1, $y_{1,2} \in \mathbb{R}, z \in \mathbb{R}$:</u>

$$\int_{y_1(t')=y_1'}^{y_1(t'')=y_1''} \mathcal{D}y_1(t) \int_{y_2(t')=y_2'}^{y_2(t'')=y_2''} \mathcal{D}y_2(t) (\mathrm{e}^{2y_1} + \mathrm{e}^{2y_2}) \mathrm{e}^{y_1+y_2} \int_{z(t')=z'}^{z(t'')=z''} \mathcal{D}z(t)$$

$$\times \exp\left\{ \frac{\mathrm{i}m}{2\hbar} \int_{t'}^{t''} \left[(\mathrm{e}^{2y_1} + \mathrm{e}^{2y_2})(\dot{y}_1^2 - \dot{y}_2^2) - \mathrm{e}^{2(y_1+y_2)}\dot{z}^2 \right] dt \right\}$$

$$= \mathrm{e}^{-(y_1'+y_1''+y_2'+y_2'')/2} \int_{\mathbb{R}} d\lambda \frac{\mathrm{e}^{\mathrm{i}\lambda(z''-z')}}{2\pi} \int_{\mathbb{R}} dk \, \mathrm{e}^{-\pi k} \int_0^\infty \frac{pdp}{8\pi} \mathrm{e}^{-\mathrm{i}\hbar p^2 T/2m}$$

$$\times \mathrm{Me}_{\mathrm{i}k}\left(y_2'' + \frac{1}{2}\ln\frac{p}{|\lambda|} + \frac{\pi}{4}; 4p|\lambda| \right) \mathrm{Me}_{\mathrm{i}k}^*\left(y_2' + \frac{1}{2}\ln\frac{p}{|\lambda|} + \frac{\pi}{4}; 4p|\lambda| \right)$$

$$\times M_{\mathrm{i}k}^{(3)}\left(y_1'' + \frac{1}{2}\ln\frac{p}{|\lambda|}; 2\sqrt{p|\lambda|} \right) M_{\mathrm{i}k}^{(3)*}\left(y_1' + \frac{1}{2}\ln\frac{p}{|\lambda|}; 2\sqrt{p|\lambda|} \right) \ . \quad (4.54)$$

<u>XXXII. Hyperbolic 2, $y_{1,2} \in \mathbb{R}, \tau \in \mathbb{R}$:</u>

$$\int_{y_1(t')=y_1'}^{y_1(t'')=y_1''} \mathcal{D}y_1(t) \int_{y_2(t')=y_2'}^{y_2(t'')=y_2''} \mathcal{D}y_2(t) (\mathrm{e}^{2y_2} - \mathrm{e}^{2y_1}) \mathrm{e}^{y_1+y_2} \int_{\tau(t')=\tau'}^{\tau(t'')=\tau''} \mathcal{D}\tau(t)$$

$$\times \exp\left\{ \frac{\mathrm{i}m}{2\hbar} \int_{t'}^{t''} \left[(\mathrm{e}^{2y_2} - \mathrm{e}^{2y_1})(\dot{y}_1^2 - \dot{y}_2^2) - \mathrm{e}^{2(y_1+y_2)}\dot{\tau}^2 \right] dt \right\}$$

$$= \mathrm{e}^{-(y_1'+y_1''+y_2'+y_2'')/2} \int_{\mathbb{R}} d\lambda \frac{\mathrm{e}^{\mathrm{i}\lambda(\tau''-\tau')}}{2\pi} \int_{\mathbb{R}} dk \, \mathrm{e}^{-\pi k} \int_0^\infty \frac{pdp}{8\pi} \mathrm{e}^{-\mathrm{i}\hbar p^2 T/2m}$$

$$\times \mathrm{Me}_{\mathrm{i}k}\left(y_2'' + \frac{1}{2}\ln\frac{p}{|\lambda|} + \frac{\mathrm{i}\pi}{4}; 4\mathrm{i}|\lambda|p \right) \mathrm{Me}_{\mathrm{i}k}^*\left(y_2' + \frac{1}{2}\ln\frac{p}{|\lambda|} + \frac{\mathrm{i}\pi}{4}; 4\mathrm{i}|\lambda|p \right)$$

$$\times M_{\mathrm{i}k}^{(3)}\left(y_1'' + \frac{1}{2}\ln\frac{p}{|\lambda|} + \frac{\mathrm{i}\pi}{4}; 2\sqrt{\mathrm{i}p|\lambda|} \right) M_{\mathrm{i}k}^{(3)*}\left(y_1' + \frac{1}{2}\ln\frac{p}{|\lambda|} + \frac{\mathrm{i}\pi}{4}; 2\sqrt{\mathrm{i}p|\lambda|} \right). \quad (4.55)$$

<u>Systems XXXIII.–LIV.:</u>

$$\int_{\varrho(t')=\varrho'}^{\varrho(t'')=\varrho''} \mathcal{D}\boldsymbol{\varrho}(t) \sqrt{g(\boldsymbol{\varrho})} \exp\left\{ \frac{\mathrm{i}}{\hbar} \int_{t'}^{t''} \left[\frac{m}{2} \sum_{\substack{ijk \text{ cyclic} \\ k \neq i \neq j \neq k}} \frac{(\varrho_i - \varrho_j)(\varrho_i - \varrho_k)}{P(\varrho_i)} \dot{\varrho}_i^2 - \Delta V_{PF}(\boldsymbol{\varrho}) \right] dt \right\} \ .$$

$$(4.56)$$

Let us discuss the various path integral representations and their solutions.

The Systems I.–XXII.

The path integral solutions for the first 22 coordinate systems can be written down in a straightforward way. The path integral solution for the cartesian system also gives the general form of the propagator. The solutions of the cylindrical systems are of type II and III$_1$ and are constructed according to Chapter 3. In the spherical systems the three-dimensional version of (4.12) has been used and the quantum numbers (ω, k) stand for any of the set of the nine inequivalent sets of observables on the two-dimensional pseudosphere. The system XIV. (called spherical) is the spherical coordinate system corresponding to its Euclidean space counterpart.

The Parabolic Systems.

The path integral solutions in terms of the parabolic coordinates follow by analytic continuation of the path integral solution in parabolic coordinates in $\mathrm{I\!R}^3$, c.f. (5.21). In the case of the parabolic 1 coordinates we must consider $(\omega^2 = -2E/m)$

$$
\int_{\xi(t')=\xi'}^{\xi(t'')=\xi''} \mathcal{D}\xi(t) \int_{\eta(t')=\eta'}^{\eta(t'')=\eta''} \mathcal{D}\eta(t)(\xi^2-\eta^2)\xi\eta \int_{\varphi(t')=\varphi'}^{\varphi(t'')=\varphi''} \mathcal{D}\varphi(t)
$$

$$
\times \exp\left\{\frac{i}{\hbar}\int_{t'}^{t''}\left[\frac{m}{2}\left((\xi^2-\eta^2)(\dot\xi^2-\dot\eta^2)-\xi^2\eta^2\dot\varphi^2\right)-\frac{\hbar^2}{8m\xi^2\eta^2}\right]dt\right\}
$$

$$
= (\xi'\xi''\eta'\eta'')^{-1/2}\sum_{\nu\in\mathbb{Z}}\frac{e^{i\nu(\varphi''-\varphi')}}{2\pi}\int_{\mathrm{I\!R}}\frac{dE}{2\pi\hbar}e^{-iET/\hbar}\int_0^\infty ds''\int_{\xi(0)=\xi'}^{\xi(s'')=\xi''}\mathcal{D}\xi(s)\int_{\eta(0)=\eta'}^{\eta(s'')=\eta''}\mathcal{D}\eta(s)
$$

$$
\times \exp\left\{\frac{i}{\hbar}\int_0^{s''}\left[\frac{m}{2}(\dot\xi^2+\dot\eta^2)+E(\xi^2-\eta^2)-\frac{\hbar^2}{2m}\left(\frac{\nu^2-\frac14}{\xi^2}-\frac{\nu^2-\frac14}{\eta^2}\right)\right]ds\right\}
$$

$$
= \sum_{\nu\in\mathbb{Z}}\frac{e^{i\nu(\varphi''-\varphi')}}{2\pi}\int_{\mathrm{I\!R}}\frac{dE}{2\pi\hbar}e^{-iET/\hbar}\int_0^\infty ds''\left(\frac{m\omega}{\hbar\sin\omega s''}\right)^2
$$

$$
\times \exp\left[-\frac{m\omega}{2i\hbar}\left(\xi'^2+\xi''^2-\eta'^2-\eta''^2\right)\right]I_\nu\left(\frac{im\omega\eta'\eta''}{\hbar\sin\omega s''}\right)I_\nu\left(\frac{m\omega\xi'\xi''}{i\hbar\sin\omega s''}\right)\,. \tag{4.57}
$$

In the next step one must apply the expansion ([154, p.414], [187, p.884])

$$
\frac{1}{\sin\alpha}\exp\left[-(x+y)\cot\alpha\right]I_{2\mu}\left(\frac{2\sqrt{xy}}{\sin\alpha}\right)
$$

$$
= \frac{1}{2\pi\sqrt{xy}}\int_{\mathrm{I\!R}}\frac{\Gamma(\frac12+\mu+ip)\Gamma(\frac12+\mu-ip)}{\Gamma^2(1+2\mu)}e^{-2\alpha p+\pi p}M_{+ip,\mu}(-2ix)M_{-ip,\mu}(+2iy)dp\,.
$$

$$
\tag{4.58}
$$

The $M_{\lambda,\mu}(z), W_{\lambda,\mu}(z)$ are Whittaker functions. The remaining s''-integration is best performed [223] by considering a momentum variable transformation $(p_\xi, p_\eta) \to ((1/a+$

$\zeta)/2p, -(1/a-\zeta)/2p)$ similarly as in the parabolic systems in $\mathbb{R}^{(1,1)}$. ζ is the parabolic separation constant.

In the parabolic 2 system one proceeds similarly. I propose the path integral identity

$$
\int\limits_{\varrho(t')=\varrho'}^{\varrho(t'')=\varrho''} \mathcal{D}\varrho(t) \exp\left\{ \frac{i}{\hbar} \int_{t'}^{t''} \left[\frac{m}{2}(\dot{\varrho}^2 - \omega^2\varrho^2) + \hbar^2 \frac{k^2 + \frac{1}{4}}{2m\varrho^2} \right] dt \right\}
$$

$$
\text{„} = \text{”} \frac{m\omega\sqrt{\varrho'\varrho''}}{\pi\hbar\sin\omega T} \exp\left[-\frac{m\omega}{2i\hbar}(\varrho'^2 + \varrho''^2)\cot\omega T \right] K_{ik}\left(\frac{im\omega\varrho'\varrho''}{\hbar\sin\omega T} \right) , \qquad (4.59)
$$

which is derived by using the path integral solution of the harmonic oscillator-like potential in (pseudo-) cartesian coordinates, rewriting the solution into spherical coordinates and expanding it by means of (4.13). As the path integral identity (4.17) this path integral can only be seen as valid in the sense of distributions. Proceeding with the same steps as in (4.57), one can use (4.60) again by rewriting the MacDonald functions into modified Bessel functions I_{ik}, and the result of (4.47) follows. Note that this path integral identity is *conjectural* in the sense that its validity can only be shown in the distributional sense, c.f. the discussion of the polar coordinate system on the pseudo-Euclidean plane. The path integral solution (4.47) is therefore also only conjectural.

The Spheroidal Systems.

In order to discuss the spheroidal systems, let us consider first the coordinate system XXV. We must extend the expansion (4.31) to three dimensions with the restriction that for $d \to 0$ we get back the spherical system. The proper spheroidal functions consequently are $\mathrm{Ps}_\nu^\mu(z;\gamma^2)$ and $S_{ik-1/2}^{\mu\,(3)}(z/\gamma;\gamma)$. They have the asymptotic behaviour $\mathrm{Ps}_\nu^\mu(z;\gamma^2) \propto \mathcal{P}_\nu^\mu(z)$ and $S_{ik-1/2}^{\mu\,(3)}(z/\gamma;\gamma) \propto \sqrt{\pi/2z}\,H_{ik}^{(1)}(z)$ ($\gamma = pd$, $d \to 0$), giving the spherical wavefunctions. From these considerations we propose by using the theory of spheroidal functions [399] the following interbasis expansion

$$
\exp\left[ipd(\cosh\xi\cosh\eta\cosh\alpha - \sinh\xi\sinh\eta\sinh\alpha\cos\varphi) \right]
$$

$$
= \frac{1}{\pi^{3/2}} \sum_{n\in\mathbb{Z}} \int_0^\infty dk\, k\sinh\pi k |\Gamma(\tfrac{1}{2}+ik+n)|^2\, e^{-\pi k/2} e^{in\varphi}
$$

$$
\times \mathrm{Ps}_{ik-1/2}^{-n}(\cosh\eta; p^2d^2)\mathrm{Ps}_{ik-1/2}^{-n}(\cosh\alpha; p^2d^2)S_{ik-1/2}^{n\,(3)}(\cosh\xi; pd) . \qquad (4.60)
$$

The short-time kernel in cartesian coordinates is then expanded by means of (4.60) along the lines of (4.15), together with the orthonormality relation

$$
\int_0^{2\pi} d\psi \int_0^\infty \sinh\alpha d\alpha \mathrm{Ps}_{ik-1/2}^{-n}(\cosh\alpha)\mathrm{Ps}_{ik'-1/2}^{-n'}(\cosh\alpha)\, e^{i\psi(n-n')}
$$

$$
= \frac{2\pi^2}{k\sinh\pi k} |\Gamma(\tfrac{1}{2}+ik+n)|^{-2}\delta_{nn'}\delta(k-k') , \qquad (4.61)
$$

and one obtains for the short-time kernel

$$
\mathrm{i}\left(\frac{m}{2\pi\epsilon\hbar}\right)^{3/2}\exp\left(\frac{\mathrm{i}m}{2\epsilon\hbar}|\mathbf{v}''-\mathbf{v}'|^2\right)
$$

$$
=\sum_{n\in\mathbb{Z}}\frac{\mathrm{e}^{\mathrm{i}n(\varphi''-\varphi')}}{2\pi}\frac{1}{\pi}\int_0^\infty dk\,k\sinh\pi k|\Gamma(\tfrac{1}{2}+\mathrm{i}k+n)|^2\int_0^\infty\frac{p^2dp}{2\pi}\mathrm{e}^{-\pi k}\mathrm{e}^{-\mathrm{i}\epsilon\hbar p^2/2m}
$$

$$
\times\mathrm{Ps}^{-n}_{\mathrm{i}k-1/2}(\cosh\eta'';p^2d^2)\mathrm{Ps}^{-n\,*}_{\mathrm{i}k-1/2}(\cosh\eta';p^2d^2)
$$

$$
\times S^{n\,(3)}_{\mathrm{i}k-1/2}(\cosh\xi'';pd)S^{n\,(3)\,*}_{\mathrm{i}k-1/2}(\cosh\xi';pd)\ . \tag{4.62}
$$

Exploiting the orthonormality relation

$$
d^3\int_0^\infty d\xi\int_0^\infty d\eta(\sinh^2\xi-\sinh^2\eta)\sinh\xi\sinh\eta\int_0^{2\pi}d\varphi\,\mathrm{e}^{\mathrm{i}\varphi(n-n')}
$$

$$
\times\mathrm{Ps}^{-n}_{\mathrm{i}k-1/2}(\cosh\eta;p^2d^2)\mathrm{Ps}^{-n'\,*}_{\mathrm{i}k'-1/2}(\cosh\eta;p'^2d^2)S^{n\,(3)}_{\mathrm{i}k-1/2}(\cosh\xi;pd)S^{n'\,(3)\,*}_{\mathrm{i}k'-1/2}(\cosh\xi;p'd)
$$

$$
=\frac{\pi}{k\sinh\pi k}|\Gamma(\tfrac{1}{2}+\mathrm{i}k+n)|^{-2}\frac{2\pi}{p^2}\mathrm{e}^{\pi k}\delta_{nn'}\delta(k-k')\delta(p-p')\ , \tag{4.63}
$$

the group path integration can be performed, and we obtain the result of the path integral in spheroidal coordinates. The path integral representation (4.48) then gives the path integral identity

$$
\int_{\xi(t')=\xi'}^{\xi(t'')=\xi''}\mathcal{D}\xi(t)\int_{\eta(t')=\eta'}^{\eta(t'')=\eta''}\mathcal{D}\eta(t)d^2(\sinh^2\xi-\sinh^2\eta)
$$

$$
\times\exp\left\{\frac{\mathrm{i}}{\hbar}\int_{t'}^{t''}\left[\frac{m}{2}d^2(\sinh^2\xi-\sinh^2\eta)(\dot\xi^2-\dot\eta^2)+\frac{\hbar^2}{2md^2}\frac{\lambda^2-\frac{1}{4}}{\sinh^2\xi\sinh^2\eta}\right]dt\right\}
$$

$$
=(d^2\sinh\xi'\sinh\xi''\sinh\eta'\sinh\eta'')^{-1/2}
$$

$$
\times\frac{1}{\pi}\int_0^\infty dk\,k\sinh\pi k|\Gamma(\tfrac{1}{2}+\mathrm{i}k+\lambda)|^2\int_0^\infty\frac{p^2dp}{2\pi}\mathrm{e}^{-\pi k}\mathrm{e}^{-\mathrm{i}\hbar p^2 T/2m}
$$

$$
\times\mathrm{Ps}^{-\lambda}_{\mathrm{i}k-1/2}(\cosh\eta'';p^2d^2)\mathrm{Ps}^{-\lambda\,*}_{\mathrm{i}k-1/2}(\cosh\eta';p^2d^2)S^{\lambda\,(3)}_{\mathrm{i}k-1/2}(\cosh\xi'';pd)S^{\lambda\,(3)\,*}_{\mathrm{i}k-1/2}(\cosh\xi';pd)\ . \tag{4.64}
$$

The system XXVIII. is in analogy to three-dimensional Euclidean space an analytic continuation to oblate-spheroidal coordinates. Here one obtains similarly as before the path integral identity

$$
\int_{\xi(t')=\xi'}^{\xi(t'')=\xi''}\mathcal{D}\xi(t)\int_{\eta(t')=\eta'}^{\eta(t'')=\eta''}\mathcal{D}\eta(t)d^2(\cosh^2\xi+\sinh^2\eta)
$$

$$
\times\exp\left\{\frac{\mathrm{i}}{\hbar}\int_{t'}^{t''}\left[\frac{m}{2}d^2(\cosh^2\xi+\sinh^2\eta)(\dot\xi^2-\dot\eta^2)+\frac{\hbar^2}{2md^2}\frac{\lambda^2-\frac{1}{4}}{\cosh^2\xi\sinh^2\eta}\right]dt\right\}
$$

$$
=(d^2\cosh\xi'\cosh\xi''\sinh\eta'\sinh\eta'')^{-1/2}
$$

$$\times \frac{1}{\pi} \int_0^\infty dk\, k \sinh \pi k |\Gamma(\tfrac{1}{2} + ik + \lambda)|^2 \int_0^\infty \frac{p^2 dp}{2\pi} e^{-\pi k} e^{-i\hbar p^2 T/2m}$$
$$\times \mathrm{Ps}_{ik-1/2}^{-\lambda}(\cosh \eta''; p^2 d^2) \mathrm{Ps}_{ik-1/2}^{-\lambda\,*}(\cosh \eta'; p^2 d^2)$$
$$\times S_{ik-1/2}^{\lambda\,(3)}(\sinh \xi''; pd) S_{ik-1/2}^{\lambda\,(3)\,*}(\sinh \xi'; pd) \ . \tag{4.65}$$

The systems XXVI., XXVII., respectively XXIX. and XXX. turn out to be related to prolate-spheroidal and oblate-spheroidal coordinate path integrals. However, simple applications of the path integral identities (5.31, 5.34) are not possible because the parameter λ is purely imaginary. We consider the path integral for the prolate spheroidal 2 system first and we construct its path integral solution in a heuristic way. Separating off the ξ-path integration and a time-transformation yields

$$\int_{\eta(t')=\eta'}^{\eta(t'')=\eta''} \mathcal{D}\eta(t) \int_{\varphi(t')=\varphi'}^{\varphi(t'')=\varphi''} \mathcal{D}\varphi(t) d^3 (\sinh^2 \eta + \sin^2 \varphi) \sinh \eta \sin \varphi \int_{\xi(t')=\xi'}^{\xi(t'')=\xi''} \mathcal{D}\xi(t)$$

$$\times \exp\left\{ \frac{i}{\hbar} \int_{t'}^{t''} \left[\frac{m}{2} d^2 \left(-(\sinh^2 \eta + \sin^2 \varphi)(\dot\eta^2 + \dot\varphi^2) + \sinh^2 \eta \sin^2 \varphi\, \dot\xi^2 \right) \right. \right.$$
$$\left. \left. - \frac{\hbar^2}{8md^2 \sinh^2 \eta \sin^2 \varphi} \right] dt \right\}$$

$$= (d^2 \sinh \eta' \sinh \eta'' \sin \varphi' \sin \varphi'')^{-1/2} \int_{\mathbb{R}} \frac{d\lambda}{2\pi} e^{i\lambda(\xi'' - \xi')} \int_{\mathbb{R}} \frac{dE}{2\pi\hbar} e^{-iET/\hbar}$$

$$\times \int_0^\infty ds'' e^{-i\hbar s''(\lambda^2 - \frac{1}{4})/2m} \int_{\eta(0)=\eta'}^{\eta(s'')=\eta''} \mathcal{D}\eta(s) \int_{\varphi(0)=\varphi'}^{\varphi(s'')=\varphi''} \mathcal{D}\varphi(s) \frac{\sinh^2 \eta + \sin^2 \varphi}{\sinh^2 \eta \sin^2 \varphi}$$

$$\times \exp\left\{ -\frac{i}{\hbar} \int_0^{s''} \left[\frac{m}{2} \frac{\sinh^2 \eta + \sin^2 \varphi}{\sinh^2 \eta \sin^2 \varphi} (\dot\eta^2 + \dot\varphi^2) - d^2 E \sinh^2 \eta \sinh^2 \varphi \right] ds \right\} \ . \tag{4.66}$$

For $E = 0$ this is a path integral which is analogous to the path integral for the hyperbolic-parabolic coordinate system on $\Lambda^{(2)}$, c.f. (7.8) in Chapter 7. We must look therefore for the appropriate spheroidal wavefunctions which have for $E = 0$ a solution according to (7.8). We find

$$\mathrm{ps}_\nu^\mu(x; 0) = P_\nu^\mu(x) \ , \qquad (|x| \le 1) \ , \qquad S_\nu^{\mu\,(1)}(z; 0) = \mathcal{P}_\nu^\mu(z) \ , \qquad |z| \ge 1 \ . \tag{4.67}$$

Therefore the appropriate spheroidal wavefunctions are in our case $\Psi_1(\varphi) \propto \mathrm{ps}_{ik-1/2}^{i\lambda}$ $(\cos \varphi; p^2 d^2)$ and $\Psi_2(\eta) \propto S_{ik-1/2}^{i\lambda\,(1)}(\cosh \eta; pd)$, respectively. Note that the ds''-integration gives $\mu = i\lambda$. Putting everything together yields the result of (4.49). The case of the oblate spheroidal 2 system is treated similarly to (4.52). Here the $E = 0$ corresponding path integral is the path integral of the elliptic-parabolic coordinate system in $\Lambda^{(2)}$, c.f. (7.7) in Chapter 7. Note that this kind of heuristic construction may not be completely satisfactory; however, it is legitimate as a first step towards a better understanding of a particular path integral. A similar heuristic approach was performed by Schulman [459] in order to construct a path integral for spin.

The path integral solutions of the systems XXVII. and XXX. can be finally obtained by observing that a shift of variables $(\xi, \eta) \to (\xi + \frac{i\pi}{2}, \eta + \frac{i\pi}{2})$ transforms them into the systems XXV. and XXVIII., respectively. An appropriate change of variables in (4.48) and (4.50) then yields the path integral representations (4.51) and (4.53).

The Hyperbolic Systems.

For the systems XXXI. and XXXII. we use the results from the two-dimensional Minkowski-space in hyperbolic 1 and 2 coordinates. For instance, for the system XXXI. we have ($E = p_E^2 \hbar^2 / 2m$, $z_i = y_i + \ln \sqrt{p/|\lambda|}$)

$$
\int_{y_1(t')=y_1'}^{y_1(t'')=y_1''} \mathcal{D}y_1(t) \int_{y_2(t')=y_2'}^{y_2(t'')=y_2''} \mathcal{D}y_2(t) (e^{2y_1} + e^{2y_2}) e^{y_1+y_2} \int_{z(t')=z'}^{z(t'')=z''} \mathcal{D}z(t)
$$

$$
\times \exp\left\{ \frac{im}{2\hbar} \int_{t'}^{t''} \left[(e^{2y_1} + e^{2y_2})(\dot{y}_1^2 - \dot{y}_2^2) - e^{2(y_1+y_2)} \dot{z}^2 \right] dt \right\}
$$

$$
= e^{-(y_1'+y_1''+y_2'+y_2'')/2}
$$

$$
\times \int_{\mathbb{R}} d\lambda \frac{e^{i\lambda(z''-z')}}{2\pi} \int_{\mathbb{R}} \frac{dE}{2\pi\hbar} e^{-iET/\hbar} \int_0^\infty ds'' \int_{y_1(0)=y_1'}^{y_1(s'')=y_1''} \mathcal{D}y_1(s) \int_{y_2(0)=y_2'}^{y_2(s'')=y_2''} \mathcal{D}y_2(s)
$$

$$
\times \exp\left\{ \frac{i}{\hbar} \int_0^{s''} \left[\frac{m}{2}(\dot{y}_1^2 - \dot{y}_2^2) + E(e^{2y_1} + e^{2y_2}) + \frac{\hbar^2 \lambda^2}{2m} (e^{-2y_1} + e^{-2y_2}) \right] ds \right\}
$$

$$
= e^{-(y_1'+y_1''+y_2'+y_2'')/2} \int_{\mathbb{R}} d\lambda \frac{e^{i\lambda(z''-z')}}{2\pi}
$$

$$
\times \int_{\mathbb{R}} \frac{dE}{2\pi\hbar} e^{-iET/\hbar} \int_0^\infty ds'' \int_{z_1(0)=z_1'}^{z_1(s'')=z_1''} \mathcal{D}z_1(s) \int_{z_2(0)=z_2'}^{z_2(s'')=z_2''} \mathcal{D}z_2(s)
$$

$$
\times \exp\left\{ \frac{i}{\hbar} \int_0^{s''} \left[\frac{m}{2}(\dot{z}_1^2 - \dot{z}_2^2) + \frac{2\hbar^2 \lambda p_E}{m}(\cosh^2 z_1 + \sinh^2 z_2) \right] ds \right\} . \tag{4.68}
$$

This path integral is now of the form of the elliptic 2 system of the pseudo-Euclidean plane. In order that we can apply its solution in the present case we must consider an inverse time-transformation. The propagator and the energy-dependent Green's function are related by [285]

$$
K(x'', x'; T) = \int_{\mathbb{R}} \frac{dE}{2\pi i} e^{-iET/\hbar} G(x'', x'; E) , \tag{4.69}
$$

$$
G(x'', x'; E) = \frac{i}{\hbar} \int_{\mathbb{R}} dT \, e^{iET/\hbar} K(x'', x'; T) \Theta(T) , \tag{4.70}
$$

where x stands for some coordinate. A time transformation gives us the transformation formula

$$
G(x'', x'; E) = \frac{i}{\hbar} \int_{\mathbb{R}} ds'' \, e^{iEs''/\hbar} \hat{K}(x'', x'; s'') , \tag{4.71}
$$

$$\hat{K}(x'', x'; s'') = e^{-iEs''/\hbar}\tilde{K}(x'', x'; s'')\Theta(s'') \ , \tag{4.72}$$

from which follows the inverse time-transformation formula for the propagator $\tilde{K}(s'')$

$$\tilde{K}(x'', x'; s'')\Theta(s'') = e^{iEs''/\hbar}\int_{\mathbb{R}}\frac{dE'}{2\pi i}e^{-iE's''/\hbar}G(x'', x'; E') \ . \tag{4.73}$$

Let us now denote the propagator in (4.68) by $(p_E = \sqrt{2mE}/\hbar)$

$$\tilde{K}_{p_E}(z_1'', z_1', z_2'', z_2'; s'') = \int\limits_{z_1(0)=z_1'}^{z_1(s'')=z_1''}\mathcal{D}z_1(s)\int\limits_{z_2(0)=z_2'}^{z_2(s'')=z_2''}\mathcal{D}z_2(s)$$

$$\times \exp\left\{\frac{i}{\hbar}\int_0^{s''}\left[\frac{m}{2}(\dot{z}_1^2 - \dot{z}_2^2) + \frac{2\hbar^2\lambda p_E}{m}(\cosh^2 z_1 + \sinh^2 z_2)\right]ds\right\} \ , \tag{4.74}$$

with corresponding Green's function $G_{p_E}(E')$. We then obtain

$$\int_{\mathbb{R}}\frac{dE}{2\pi\hbar}e^{-iET/\hbar}\int_{\mathbb{R}}ds''\,e^{-iEs''/\hbar}K_{p_E}(z_1'', z_1', z_2'', z_2'; s'')\Theta(s'')$$

$$= \int_{\mathbb{R}}\frac{dE}{2\pi\hbar}e^{-iET/\hbar}\int_{\mathbb{R}}ds''\Theta(s'')\,e^{-iEs''/\hbar}\int_{\mathbb{R}}\frac{dE'}{2\pi i}e^{-iE's''/\hbar}G_{p_E}(z_1'', z_1', z_2'', z_2'; E')$$

$$= \int_{\mathbb{R}}\frac{dE}{2\pi\hbar}e^{-iET/\hbar}\int_{\mathbb{R}}dE'\delta(E - E')G_{p_E}(z_1'', z_1', z_2'', z_2'; E')$$

$$= K_{p_E}(z_1'', z_1', z_2'', z_2'; T)\Theta(T)\Big|_{E=\hbar^2p^2/2m} \ . \tag{4.75}$$

Therefore we can apply the path integral representation (4.8) by analytic continuation in the parameters, and arrive at the path integral solution (4.54). The path integral representation (4.55) is obtained in an analogous way.

Systems XXXIII.–LIV.

Path integral solutions in terms of the parametric coordinates XXXIII.–LIV. are not known, let alone the fact that a solution in terms of these coordinates for the corresponding Minkowski-Laplacian is not known at all. In table 4.2 I list the definition of these coordinate systems, and give also the explicit expression for the quantity $P(\varrho)$ as used in (4.56). The corresponding quantum potentials ΔV can be constructed from (2.11).

Coordinate Systems on Pseudo-Euclidean Space.

We summarize our results of the path integration in three-dimensional pseudo-Euclidean space including an enumeration of the coordinate systems according to [300] and Miller [404].

Table 4.2: Coordinates in Three-Dimensional Pseudo-Euclidean Space

$(V_0, V_1 \in \mathbb{R}^{(1,1)}; \ x, y \in \mathbb{R}^2; \ \mathrm{u} = (u_0, u_1, u_2) \equiv \boldsymbol{\tau} \in \Lambda^{(2)}. \ d > 0.)$

Coordinate System	Coordinates	Path Integral Solution
I. Cartesian	$v_0 = v_0'$, $v_1 = v_1'$, $v_2 = v_2'$	(4.39, 4.40)
II.–X. Cylindrical $\mathbb{R}^{(1,1)}$	$v_0 = V_0$, $v_1 = V_1$, $v_2 = v_2'$	(4.41)
XI.–XIII. Cylindrical $\mathbb{R}^2$	$v_0 = v$, $v_1 = x$, $v_2 = y$	(4.42)
XIV.–XXII. Spherical $\Lambda^{(2)}$	$v_0 = r u_0$, $v_1 = r u_1$, $v_2 = r u_2$	(4.44)
XXIII. Parabolic 1	$v_0 = \frac{1}{2}(\xi^2 + \eta^2)$ $v_1 = \xi \eta \cos\varphi$, $v_2 = \xi \eta \sin\varphi$	(4.45)
XXIV. Parabolic 2	$v_0 = \xi \eta \cosh\tau$, $v_1 = \xi \eta \sinh\tau$ $v_2 = \frac{1}{2}(\xi^2 + \eta^2)$	(4.47)
XXV. Prolate-Spheroidal 1	$v_0 = d \cosh\xi \cosh\eta$ $v_1 = d \sinh\xi \sinh\eta \cos\varphi$ $v_2 = d \sinh\xi \sinh\eta \sin\varphi$	(4.48)
XXVI. Prolate-Spheroidal 2	$v_0 = d \sinh\eta \sin\varphi \sinh\xi$ $v_1 = d \sinh\eta \sin\varphi \cosh\xi$ $v_2 = d \cosh\eta \cos\varphi$	(4.49)
XXVII. Prolate-Spheroidal 3	$v_0 = d \cosh\xi \cosh\eta \cosh\tau$ $v_1 = d \cosh\xi \cosh\eta \sinh\tau$ $v_2 = d \sinh\xi \sinh\eta$	(4.50)
XXVIII. Oblate-Spheroidal 1	$v_0 = d \sinh\xi \cosh\eta$ $v_1 = d \cosh\xi \sinh\eta \cos\varphi$ $v_2 = d \cosh\xi \sinh\eta \sin\varphi$	(4.51)
XXIX. Oblate-Spheroidal 2	$v_0 = d \cosh\eta \cos\varphi \sinh\xi$ $v_1 = d \cosh\eta \cos\varphi \cosh\xi$ $v_2 = d \sinh\eta \sin\varphi$	(4.52)
XXX. Oblate-Spheroidal 3	$v_0 = d \sinh\xi \cosh\eta \cosh\tau$ $v_1 = d \sinh\xi \cosh\eta \sinh\tau$ $v_2 = d \cosh\xi \sinh\eta$	(4.53)
XXXI. Hyperbolic 1	$v_0 = \sinh(y_1 - y_2) + \frac{1}{2}e^{y_1+y_2}(1 + z^2)$ $v_1 = \sinh(y_1 - y_2) + \frac{1}{2}e^{y_1+y_2}(z^2 - 1)$ $v_2 = z e^{y_1+y_2}$	(4.54)
XXXII. Hyperbolic 2	$v_0 = \cosh(y_1 - y_2) + \frac{1}{2}e^{y_1+y_2}(\tau^2 + 1)$ $v_1 = \cosh(y_1 - y_2) + \frac{1}{2}e^{y_1+y_2}(\tau^2 - 1)$ $v_2 = \tau e^{y_1+y_2}$	(4.55)

Table 4.2 (cont.): Coordinates in Three-Dimensional Pseudo-Euclidean Space

Coordinate System	Coordinates	Related Systems	$P(\varrho)$
XXXIII.–XXXVII. Ellipsoidal , $\varrho_3 > 1 > \varrho_2 > 0,\ \varrho_1 > a$	$v_0^2 = \dfrac{(\varrho_1 - a)(\varrho_2 - a)(\varrho_3 - a)}{a(1 - a)}$ $v_1^2 = \dfrac{\varrho_1 \varrho_2 \varrho_3}{a}$ $v_2^2 = \dfrac{(\varrho_1 - 1)(\varrho_2 - 1)(\varrho_3 - 1)}{(1 - a)}$	(iv_0, iv_1, iv_2) (v_2, iv_1, v_0) (iv_2, v_1, iv_0) (iv_1, v_0, iv_2)	$4(\varrho - a)$ $\times (\varrho - 1)\varrho$
XXXVIII., XXXIX. Hyperboloidal, $\varrho_{1,2,3} > 0$	$(v_2 + iv_0)^2$ $= \dfrac{2(\varrho_1 - a)(\varrho_2 - a)(\varrho_3 - a)}{a(1 - b)}$ $v_1^2 = \dfrac{\varrho_1 \varrho_2 \varrho_3}{ab}$	(iv_0, iv_1, iv_2) $\varrho_{1,2,3} < 0$	$(\varrho - a)$ $\times (\varrho - b)\varrho$
XL.–XLIII. $\varrho_{1,2,3} > 1$	$(v_0 - v_1)^2 = \varrho_1 \varrho_2 \varrho_3$ $v_0^2 - v_1^2 = -(\varrho_1 \varrho_3 + \varrho_2 \varrho_3 + \varrho_1 \varrho_2)$ $\qquad + \varrho_1 \varrho_2 \varrho_3$ $v_2^2 = (\varrho_1 - 1)(\varrho_2 - 1)(\varrho_3 - 1)$	(iv_0, iv_1, iv_2) (v_1, v_0, iv_2) (v_1, v_0, iv_2)	$(\varrho - 1)\varrho^2$
XLIV.–XLVI. $\varrho_1 > 1 > 0 > \varrho_{2,3}$	$(v_0 - v_2)^2 = \varrho_1 \varrho_2 \varrho_3$ $v_0^2 - v_1^2 = -\varrho_1 \varrho_2$ $v_2^2 = (\varrho_1 - 1)(\varrho_2 - 1)(\varrho_3 - 1)$	(iv_0, iv_1, iv_2) (v_2, iv_1, v_0)	$(\varrho - 1)\varrho^2$
XLVII., XLVIII. $\varrho_{1,2} > 0 > \varrho_3$	$(v_0 - v_1)^2 = \varrho_1 \varrho_2 \varrho_3$ $2v_2(v_0 - v_1) = \varrho_1 \varrho_2 + \varrho_1 \varrho_3 + \varrho_2 \varrho_3$ $v_1^2 + v_2^2 - v_0^2 = \varrho_1 + \varrho_2 + \varrho_3$	(iv_0, iv_1, iv_2)	ϱ^2
IL., L. $\varrho_{1,2} > 1 > \varrho_3 > 0$	$v_0 = \frac{1}{2}(\varrho_1 + \varrho_2 + \varrho_3)$ $v_1^2 = \varrho_1 \varrho_2 \varrho_3$ $v_2^2 = -(\varrho_1 - 1)(\varrho_2 - 1)(\varrho_3 - 1)$	(v_2, iv_1, v_0)	$\varrho(\varrho - 1)$
LI., LII.	$(v_0 - v_1)^2 = \varrho_1 \varrho_2 \varrho_3$ $v_0^2 - v_1^2 = \varrho_1 \varrho_2 + \varrho_1 \varrho_3 + \varrho_2 \varrho_3$ $v_2 = \frac{1}{2}(\varrho_1 + \varrho_2 + \varrho_3)$	(v_0, iv_1, v_2)	ϱ^2
LIII., LIV.	$2(v_0 - v_2) = \varrho_1 \varrho_2 + \varrho_1 \varrho_3 + \varrho_2 \varrho_3$ $\qquad - \frac{1}{2}(\varrho_1^2 + \varrho_2^2 + \varrho_3^2)$ $2(v_2 - v_0) = \varrho_1 + \varrho_2 + \varrho_3$ $v_1^2 = -\varrho_1 \varrho_2 \varrho_3$	(v_2, iv_1, v_0)	ϱ

Chapter 5

Path Integrals in Euclidean Spaces

5.1 Two-Dimensional Euclidean Space

We have the path integral representations in two-dimensional Euclidean space [223] (see also [67])

I. Cartesian and General Form of the Propagator , $(x, y) = \mathbf{x} \in \mathrm{I\!R}^2$:

$$\int_{\mathbf{x}(t')=\mathbf{x}'}^{\mathbf{x}(t'')=\mathbf{x}''} \mathcal{D}\mathbf{x}(t) \exp\left(\frac{\mathrm{i}m}{2\hbar}\int_{t'}^{t''}\dot{\mathbf{x}}^2 dt\right) = \frac{m}{2\pi\mathrm{i}\hbar T}\exp\left(\frac{\mathrm{i}m}{2\hbar T}|\mathbf{x}''-\mathbf{x}''|^2\right) \tag{5.1}$$

$$= \int_{\mathrm{I\!R}^2}\frac{d\mathbf{p}}{4\pi^2}\exp\left[-\frac{\mathrm{i}\hbar T}{2m}\mathbf{p}^2 + \mathrm{i}\mathbf{p}\cdot(\mathbf{x}''-\mathbf{x}')\right] . \tag{5.2}$$

II. Polar, $\varrho > 0, \varphi \in [0, 2\pi)$:

$$\int_{\varrho(t')=\varrho'}^{\varrho(t'')=\varrho''} \mathcal{D}\varrho(t)\varrho \int_{\varphi(t')=\varphi'}^{\varphi(t'')=\varphi''} \mathcal{D}\varphi(t)\exp\left\{\frac{\mathrm{i}}{\hbar}\int_{t'}^{t''}\left[\frac{m}{2}\left(\dot\varrho^2 + \varrho^2\dot\varphi^2\right) + \frac{\hbar^2}{8m\varrho^2}\right]dt\right\}$$

$$= \frac{m}{2\pi\mathrm{i}\hbar T}\exp\left[\frac{\mathrm{i}m}{2\hbar T}(\varrho'^2 + \varrho''^2)\right]\sum_{\nu\in\mathbb{Z}}\mathrm{e}^{\mathrm{i}\nu(\varphi''-\varphi')}I_\nu\left(\frac{m\varrho'\varrho''}{\mathrm{i}\hbar T}\right) \tag{5.3}$$

$$= \sum_{\nu\in\mathbb{Z}}\mathrm{e}^{\mathrm{i}\nu(\varphi''-\varphi')}\int_0^\infty dp\, p J_\nu(p\varrho')J_\nu(p\varrho'')\,\mathrm{e}^{-\mathrm{i}\hbar T p^2/2m} \tag{5.4}$$

$$= \frac{m}{2\pi\mathrm{i}\hbar T}\exp\left[\frac{\mathrm{i}m}{2\hbar T}\left(\varrho'^2 + \varrho''^2 - 2\varrho'\varrho''\cos(\varphi''-\varphi')\right)\right] . \tag{5.5}$$

III. Elliptic, $\mu > 0, \nu \in [-\pi, \pi)$:

$$\int_{\mu(t')=\mu'}^{\mu(t'')=\mu''} \mathcal{D}\mu(t) \int_{\nu(t')=\nu'}^{\nu(t'')=\nu''} \mathcal{D}\nu(t)d^2(\sinh^2\mu + \sin^2\nu)$$

$$\times \exp\left[\frac{\mathrm{i}m}{2\hbar}d^2\int_{t'}^{t''}(\sinh^2\mu + \sin^2\nu)(\dot\mu^2 + \dot\nu^2)dt\right]$$

$$= \frac{1}{2\pi} \sum_{n \in \mathbb{Z}} \int_0^\infty p\,dp\, \mathrm{me}_n^*(\nu'; \tfrac{d^2 p^2}{4}) \mathrm{me}_n(\nu''; \tfrac{d^2 p^2}{4}) M_n^{(1)*}(\mu'; \tfrac{dp}{2}) M_n^{(1)}(\mu''; \tfrac{dp}{2})\, \mathrm{e}^{-\mathrm{i}\hbar p^2 T/2m}.$$

(5.6)

IV. Parabolic, $\xi \in \mathbb{R}, \eta > 0$:

$$\int\limits_{\xi(t')=\xi'}^{\xi(t'')=\xi''} \mathcal{D}\xi(t) \int\limits_{\eta(t')=\eta'}^{\eta(t'')=\eta''} \mathcal{D}\eta(t)(\xi^2 + \eta^2) \exp\left[\frac{\mathrm{i}m}{2\hbar} \int_{t'}^{t''} (\xi^2 + \eta^2)(\dot\xi^2 + \dot\eta^2)dt\right]$$

$$= \int_{\mathbb{R}} d\zeta \int_{\mathbb{R}} \frac{dp}{32\pi^4} \mathrm{e}^{-\mathrm{i}\hbar p^2 T/2m}$$

$$\times \begin{pmatrix} |\Gamma(\tfrac{1}{4} + \tfrac{\mathrm{i}\zeta}{2p})|^2 E_{-1/2+\mathrm{i}\zeta/p}^{(0)}(\mathrm{e}^{-\mathrm{i}\pi/4}\sqrt{2p}\,\xi'') E_{-1/2-\mathrm{i}\zeta/p}^{(0)}(\mathrm{e}^{-\mathrm{i}\pi/4}\sqrt{2p}\,\eta'') \\ |\Gamma(\tfrac{3}{4} + \tfrac{\mathrm{i}\zeta}{2p})|^2 E_{-1/2+\mathrm{i}\zeta/p}^{(1)}(\mathrm{e}^{\mathrm{i}\pi/4}\sqrt{2p}\,\xi'') E_{-1/2-\mathrm{i}\zeta/p}^{(1)}(\mathrm{e}^{\mathrm{i}\pi/4}\sqrt{2p}\,\eta'') \end{pmatrix}$$

$$\times \begin{pmatrix} |\Gamma(\tfrac{1}{4} + \tfrac{\mathrm{i}\zeta}{2p})|^2 E_{-1/2-\mathrm{i}\zeta/p}^{(0)}(\mathrm{e}^{\mathrm{i}\pi/4}\sqrt{2p}\,\xi') E_{-1/2+\mathrm{i}\zeta/p}^{(0)}(\mathrm{e}^{\mathrm{i}\pi/4}\sqrt{2p}\,\eta') \\ |\Gamma(\tfrac{3}{4} + \tfrac{\mathrm{i}\zeta}{2p})|^2 E_{-1/2-\mathrm{i}\zeta/p}^{(1)}(\mathrm{e}^{-\mathrm{i}\pi/4}\sqrt{2p}\,\xi') E_{-1/2+\mathrm{i}\zeta/p}^{(1)}(\mathrm{e}^{-\mathrm{i}\pi/4}\sqrt{2p}\,\eta') \end{pmatrix} .$$

(5.7)

General Form of the Green's Function:

$$= \int_{\mathbb{R}} \frac{dE}{2\pi\mathrm{i}} \mathrm{e}^{-\mathrm{i}ET/\hbar} \frac{m}{\pi\hbar^2} K_0\left(\frac{|\mathbf{x}'' - \mathbf{x}''|}{\hbar}\sqrt{-2mE}\right) .$$

(5.8)

Elliptic Coordinates.

Because all but one path integration have been discussed already in [223] we do not need to do this once again. We can present the path integral solution of the elliptic system now explicitly. In [223] we have only stated the corresponding path integral solution by a heuristic construction. By means of the interbasis expansion of plane waves into Mathieu functions it is possible to evaluate the corresponding path integral. We consider the expansion for an arbitrary α ($h = pd/2$, [399, p.185])

$$\exp\left[\mathrm{i}p(x\cos\alpha + y\sin\alpha)\right] = 2\sum_{n=0}^\infty \mathrm{i}^n \mathrm{ce}_n(\alpha; h^2) M_n^{(1)}(\mu; h)\mathrm{ce}_n(\nu; h^2)$$

$$+ 2\sum_{n=1}^\infty \mathrm{i}^{-n} \mathrm{se}_n(\alpha; h^2) M_{-n}^{(1)}(\mu; h)\mathrm{se}_n(\nu; h^2) .$$

(5.9)

$\mathrm{me}_n, \mathrm{Me}_n^{(1)}$ are periodic and non-periodic Mathieu functions, and $\mathrm{ce}_n, \mathrm{se}_n, \mathrm{Mc}_n^{(1)}, \mathrm{Ms}_n^{(1)}$ the corresponding even and odd Mathieu functions, respectively. Insertion into the short-time kernel yields

$$\frac{m}{2\pi\mathrm{i}\epsilon\hbar} \exp\left(\frac{\mathrm{i}m}{2\epsilon\hbar}(\mathbf{x}'' - \mathbf{x}')^2\right)$$

$$= \frac{1}{4\pi^2} \int_{\mathbb{R}} dp_x \int_{\mathbb{R}} dp_y\, \mathrm{e}^{\mathrm{i}p_x(x''-x')+\mathrm{i}p_y(y''-y')-\mathrm{i}\hbar(p_x^2+p_y^2)\epsilon/2m}$$

$$= \frac{1}{\pi^2} \int_0^\infty p\,dp\, \mathrm{e}^{-\mathrm{i}\epsilon\hbar p^2/2m} \int_{-\pi}^\pi d\alpha$$

$$\times \left[\sum_{n=0}^{\infty} i^n \mathrm{ce}_n(\alpha; h^2) M_n^{(1)}(\mu''; h) \mathrm{ce}_n(\nu''; h^2) + \sum_{n=1}^{\infty} i^{-n} \mathrm{se}_n(\alpha; h^2) M_{-n}^{(1)}(\mu''; h) \mathrm{se}_n(\nu''; h^2)\right]$$

$$\times \left[\sum_{l=0}^{\infty} i^l \mathrm{ce}_l(\alpha; h^2) M_l^{(1)}(\mu'; h) \mathrm{ce}_l(\nu'; h^2) + \sum_{l=1}^{\infty} i^{-l} \mathrm{se}_l(\alpha; h^2) M_{-l}^{(1)}(\mu'; h) \mathrm{se}_l(\nu'; h^2)\right]^*$$

$$= \frac{1}{\pi^2} \int_0^\infty p\,dp\, e^{-i\epsilon\hbar p^2/2m} \left[\sum_{n=0}^{\infty} \mathrm{Mc}_n^{(1)}(\mu''; h) \mathrm{Mc}_n^{(1)*}(\mu'; h) \mathrm{ce}_n(\nu''; h^2) \mathrm{ce}_n^*(\nu'; h^2)\right.$$

$$\left. + \sum_{n=1}^{\infty} \mathrm{Ms}_n^{(1)}(\mu''; h) \mathrm{Ms}_n^{(1)*}(\mu'; h) \mathrm{ce}_n(\nu''; h^2) \mathrm{ce}_n^*(\nu'; h^2)\right]$$

$$= \frac{1}{2\pi} \sum_{n\in\mathbb{Z}} \int_0^\infty p\,dp\, e^{-i\epsilon\hbar p^2/2m} \mathrm{me}_n^*(\nu'; \tfrac{d^2p^2}{4}) \mathrm{me}_n(\nu''; \tfrac{d^2p^2}{4}) M_n^{(1)*}(\mu'; \tfrac{dp}{2}) M_n^{(1)}(\mu''; \tfrac{dp}{2}) \ ,$$

$$(5.10)$$

by means of the relations $\mathrm{ce}_n(z; h^2) = \mathrm{me}_n(z; h^2)/\sqrt{2}$, $M_n^{(1)}(z) = \mathrm{Mc}_n^{(1)}(z; h)$ ($n = 0, 1, \dots$) and $\mathrm{se}_n(z; h^2) = i \cdot \mathrm{me}_n(z; h^2)/\sqrt{2}$, $M_{-n}^{(1)}(z) = (-1)^{-n} \mathrm{Ms}_n^{(1)}(z; h)$ ($n = 1, 2, \dots$). Use has been made of the orthonormality relations ([399, p.114], [412, p.200])

$$\frac{1}{\pi} \int_{-\pi}^{\pi} \mathrm{ce}_n(\vartheta) \mathrm{ce}_l^*(\vartheta) d\vartheta = \frac{1}{\pi} \int_{-\pi}^{\pi} \mathrm{se}_n(\vartheta) \mathrm{se}_l^*(\vartheta) d\vartheta = \frac{1}{2\pi} \int_{-\pi}^{\pi} \mathrm{me}_n(\vartheta) \mathrm{me}_l^*(\vartheta) d\vartheta = \delta_{nl} \ ,$$

$$(5.11)$$

$$\frac{1}{\pi} \int_{-\pi}^{\pi} \mathrm{ce}_n(\vartheta) \mathrm{se}_l^*(\vartheta) d\vartheta = 0 \ . \tag{5.12}$$

The orthonormality relation has the form ($h = pd/2$)

$$d^2 \int_0^\infty d\mu \int_{-\pi}^{\pi} d\nu (\sinh^2\mu + \sin^2\nu) M_n^{(1)}(\mu; h) M_l^{(1)*}(\mu; h') \mathrm{me}_n(\nu; h^2) \mathrm{me}_l^*(\nu; h'^2)$$

$$= \frac{2\pi}{p} \delta_{nl} \delta(p - p') \ , \tag{5.13}$$

and can be derived by the use of the expansion [399, p.183] $h = pd/2$, $h' = p'd/2$ (ϱ, φ corresponding polar coordinates in $\mathbb{R}^2$)

$$\mathrm{me}_s(\nu; h^2) M_s^{(1)}(\mu; h) = \sum_{r\in\mathbb{Z}} (-1)^r c_{2r}^s J_{2r+s}(p\varrho) e^{-i(2r+s)\varphi} \ , \qquad \sum_{r\in\mathbb{Z}} |c_{2r}^s|^2 = 1 \ . \tag{5.14}$$

The path integration is then performed by expanding the short-time kernel by means of the expansion (5.10) and the orthonormality relation (5.13) (similarly as in the case of the path integration on the pseudo-Euclidean plane). Equation (5.6) provides a path integral identity for elliptic coordinates.

Coordinate Systems in Two-Dimensional Euclidean Space.

The table 5.1 summarizes our knowledge of path integration in $\mathbb{R}^2$, including an enumeration of the coordinate systems [416].

Table 5.1: Coordinates in Two-Dimensional Euclidean Space

Coordinate System	Coordinates	Path Integral Solution
I. Cartesian	$x = x'$ $y = y'$	(5.1, 5.2)
II. Polar	$x = \varrho \cos \varphi$ $y = \varrho \sin \varphi$	(5.3–5.5)
III. Elliptic	$x = d \cosh \mu \cos \nu$ $y = d \sinh \mu \sin \nu$	(5.6)
IV. Parabolic	$x = \frac{1}{2}(\eta^2 - \xi^2)$ $y = \xi \eta$	(5.7)

5.2   Three-Dimensional Euclidean Space

We have the path integral representations in three-dimensional Euclidean space [67, 223]

I. Cartesian and General Form of the Propagator , $(x, y, z) = \mathbf{x} \in \mathbb{R}^3$:

$$\int_{\mathbf{x}(t')=\mathbf{x}'}^{\mathbf{x}(t'')=\mathbf{x}''} \mathcal{D}\mathbf{x}(t) \exp\left(\frac{im}{2\hbar} \int_{t'}^{t''} \dot{\mathbf{x}}^2 dt \right) = \left(\frac{m}{2\pi i \hbar T} \right)^{3/2} \exp\left(\frac{im}{2\hbar T} |\mathbf{x}'' - \mathbf{x}''|^2 \right) \tag{5.15}$$

$$= \int_{\mathbb{R}^3} \frac{d\mathbf{p}}{(2\pi)^3} \exp\left[-\frac{i\hbar T}{2m}\mathbf{p}^2 + i\mathbf{p} \cdot (\mathbf{x}'' - \mathbf{x}') \right] . \tag{5.16}$$

II.–IV. Circular $\mathbb{R}^2$ (without I.), $(x, y) = \mathbf{x} \in \mathbb{R}^2, z \in \mathbb{R}$:

$$\int_{\mathbf{x}(t')=\mathbf{x}'}^{\mathbf{x}(t'')=\mathbf{x}''} \mathcal{D}\mathbf{x}(t) \int_{z(t')=z'}^{z(t'')=z''} \mathcal{D}z(t) \exp\left[\frac{im}{2\hbar} \int_{t'}^{t''} (\dot{\mathbf{x}}^2 + \dot{z}^2) dt \right]$$

$$= \left(\frac{m}{2\pi i \hbar T} \right)^{1/2} \exp\left(\frac{im}{2\hbar T} |z'' - z'|^2 \right) \cdot K^{\mathbb{R}^2}(\mathbf{x}'', \mathbf{x}'; T) . \tag{5.17}$$

V. Sphero-Conical, $r > 0, \alpha \in [-K, K], \beta \in [-2K', 2K']$:

$$\int_{r(t')=r'}^{r(t'')=r''} \mathcal{D}r(t) r^2 \int_{\alpha(t')=\alpha'}^{\alpha(t'')=\alpha''} \mathcal{D}\alpha(t) \int_{\beta(t')=\beta'}^{\beta(t'')=\beta''} \mathcal{D}\beta(t) (k^2 \mathrm{cn}^2 \alpha + k'^2 \mathrm{cn}^2 \beta)$$

$$\times \exp\left[\frac{im}{2\hbar} \int_{t'}^{t''} \left(\dot{r}^2 + r^2 (k^2 \mathrm{cn}^2 \alpha + k'^2 \mathrm{cn}^2 \beta)(\dot{\alpha}^2 + \dot{\beta}^2) \right) dt \right]$$

$$= \frac{m}{i\hbar T \sqrt{r' r''}} \exp\left[\frac{im}{2\hbar T} (r'^2 + r''^2) \right]$$

$$\times \sum_{l=0}^{\infty} \sum_{\lambda} \sum_{p,q=\pm} \Lambda_{l,h}^{p}(\alpha'') \Lambda_{l,h}^{p\,*}(\alpha') \Lambda_{l,\bar{h}}^{q}(\beta'') \Lambda_{l,\bar{h}}^{q\,*}(\beta') I_{l+1/2}\left(\frac{mr'r''}{i\hbar T}\right) . \tag{5.18}$$

VI. Spherical, $r > 0, \vartheta \in (0, \pi), \varphi \in [0, 2\pi)$:

$$\int_{r(t')=r'}^{r(t'')=r''} \mathcal{D}r(t)r^2 \int_{\vartheta(t')=\vartheta'}^{\vartheta(t'')=\vartheta''} \mathcal{D}\vartheta(t)\sin\vartheta \int_{\varphi(t')=\varphi'}^{\varphi(t'')=\varphi''} \mathcal{D}\varphi(t)$$

$$\times \exp\left\{\frac{i}{\hbar}\int_{t'}^{t''}\left[\frac{m}{2}\left(\dot{r}^2 + r^2\dot{\vartheta}^2 + r^2\sin^2\vartheta\dot{\varphi}^2\right) + \frac{\hbar^2}{8mr^2}\left(1 + \frac{1}{\sin^2\vartheta}\right)\right]dt\right\}$$

$$= \frac{m}{4\pi i\hbar T\sqrt{r'r''}} \exp\left[-\frac{m}{2i\hbar T}(r'^2 + r''^2)\right] \sum_{l=0}^{\infty}(2l+1)P_l(\cos\psi_{S^{(2)}})I_{l+1/2}\left(\frac{mr'r''}{i\hbar T}\right) \tag{5.19}$$

$$= \frac{m}{4\pi i\hbar T\sqrt{r'r''}} \sum_{l=0}^{\infty}\sum_{n=-l}^{l} Y_l^n(\vartheta'', \varphi'') Y_l^{n\,*}(\vartheta', \varphi') \int_0^{\infty} pdp J_{l+1/2}(pr'')J_{l+1/2}(pr')\,e^{-i\hbar p^2 T/2m}. \tag{5.20}$$

VII. Parabolic, $\xi, \eta > 0, \varphi \in [0, 2\pi)$:

$$\int_{\xi(t')=\xi'}^{\xi(t'')=\xi''} \mathcal{D}\xi(t) \int_{\eta(t')=\eta'}^{\eta(t'')=\eta''} \mathcal{D}\eta(t)(\xi^2 + \eta^2)\xi\eta \int_{\varphi(t')=\varphi'}^{\varphi(t'')=\varphi''} \mathcal{D}\varphi(t)$$

$$\times \exp\left\{\frac{i}{\hbar}\int_{t'}^{t''}\left[\frac{m}{2}\left((\xi^2 + \eta^2)(\dot{\xi}^2 + \dot{\eta}^2) + \xi^2\eta^2\dot{\varphi}^2\right) + \frac{\hbar^2}{8m\xi^2\eta^2}\right]dt\right\}$$

$$= \sum_{n\in\mathbb{Z}} \frac{e^{in(\varphi''-\varphi')}}{2\pi} \int_{\mathbb{R}} d\zeta \int_0^{\infty} \frac{dp}{p} \frac{|\Gamma(\frac{1+|n|}{2} + \frac{i\zeta}{2p})|^4}{4\pi^2\xi'\xi''\eta'\eta''\Gamma^4(1 + |n|)} e^{-i\hbar p^2 T/2m}$$

$$\times M_{-i\zeta/2p,|n|/2}(-ip\xi''^2)M_{i\zeta/2p,|n|/2}(ip\xi'^2)M_{i\zeta/2p,|n|/2}(-ip\eta''^2)M_{-i\zeta/2p,|n|/2}(ip\eta'^2). \tag{5.21}$$

VIII. Prolate-Spheroidal, $\mu > 0, \nu \in (0, \pi), \varphi \in [0, 2\pi)$:

$$\int_{\mu(t')=\mu'}^{\mu(t'')=\mu''} \mathcal{D}\mu(t) \int_{\nu(t')=\nu'}^{\nu(t'')=\nu''} \mathcal{D}\nu(t)d^3(\sinh^2\mu + \sin^2\nu)\sinh\mu\sin\nu \int_{\varphi(t')=\varphi'}^{\varphi(t'')=\varphi''} \mathcal{D}\varphi(t)$$

$$\times \exp\left\{\frac{i}{\hbar}\int_{t'}^{t''}\left[\frac{m}{2}d^2\left((\sinh^2\mu + \sin^2\nu)(\dot{\mu}^2 + \dot{\nu}^2) + \sinh^2\mu\sin^2\nu\dot{\varphi}^2\right)\right.\right.$$

$$\left.\left. + \frac{\hbar^2}{8md^2\sinh^2\mu\sin^2\nu}\right]dt\right\}$$

$$= \sum_{l=0}^{\infty}\sum_{n=-l}^{l} e^{in(\varphi''-\varphi')}\frac{2l+1}{2\pi^2}\frac{(l-n)!}{(l+n)!} \int_0^{\infty} p^2 dp\,e^{-i\hbar p^2 T/2m}$$

$$\times \mathrm{ps}_l^{n\,*}(\cos\nu'; p^2 d^2)\mathrm{ps}_l^n(\cos\nu''; p^2 d^2)S_l^{n\,(1)\,*}(\cosh\mu'; pd)S_l^{n\,(1)}(\cosh\mu''; pd) . \tag{5.22}$$

IX. Oblate-Spheroidal , $\xi > 0, \nu \in (0, \pi), \varphi \in [0, 2\pi)$:

$$\int_{\xi(t')=\xi'}^{\xi(t'')=\xi''} \mathcal{D}\xi(t) \int_{\nu(t')=\nu'}^{\nu(t'')=\nu''} \mathcal{D}\nu(t) d^3 (\cosh^2 \xi - \sin^2 \nu) \sinh \xi \sin \nu \int_{\varphi(t')=\varphi'}^{\varphi(t'')=\varphi''} \mathcal{D}\varphi(t)$$

$$\times \exp \left\{ \frac{i}{\hbar} \int_{t'}^{t''} \left[\frac{m}{2} d^2 \Big((\cosh^2 \xi - \sin^2 \nu)(\dot{\xi}^2 + \dot{\nu}^2) + \cosh^2 \xi \sin^2 \nu \dot{\varphi}^2 \Big) \right. \right.$$

$$\left. \left. + \frac{\hbar^2}{8md^2 \cosh^2 \xi \sin^2 \nu} \right] dt \right\}$$

$$= \sum_{l=0}^{\infty} \sum_{n=-l}^{l} e^{in(\varphi'' - \varphi')} \frac{2l+1}{2\pi^2} \frac{(l-n)!}{(l+n)!} \int_0^{\infty} p^2 dp \, e^{-i\hbar p^2 T/2m}$$

$$\times \mathrm{psi}_l^{n*}(\cos \nu'; p^2 d^2) \mathrm{psi}_l^n(\cos \nu''; p^2 d^2) \mathrm{Si}_l^{n\,(1)*}(\cosh \xi'; pd) \mathrm{Si}_l^{n\,(1)}(\cosh \xi''; pd) \ . \quad (5.23)$$

X. Ellipsoidal, $\alpha \in [iK', K + iK'], \beta \in [K, K + 2iK'], \gamma \in [0, 4K]$:

$$\int_{\alpha(t')=\alpha'}^{\alpha(t'')=\alpha''} \mathcal{D}_{MP}\alpha(t) \int_{\beta(t')=\beta'}^{\beta(t'')=\beta''} \mathcal{D}_{MP}\beta(t) \int_{\gamma(t')=\gamma'}^{\gamma(t'')=\gamma''} \mathcal{D}_{MP}\gamma(t)$$

$$\times \left(\frac{a^2 - b^2}{\sqrt{a^2 - c^2}} \right)^3 \sqrt{(\mathrm{sn}^2\alpha - \mathrm{sn}^2\beta)(\mathrm{sn}^2\beta - \mathrm{sn}^2\gamma)(\mathrm{sn}^2\alpha - \mathrm{sn}^2\gamma)}$$

$$\times \exp \left\{ \frac{i}{\hbar} \int_{t'}^{t''} \left[\frac{m}{2} \frac{(a^2 - b^2)^2}{a^2 - c^2} \Big((\mathrm{sn}^2\alpha - \mathrm{sn}^2\gamma)(\mathrm{sn}^2\alpha - \mathrm{sn}^2\beta)\dot{\alpha}^2 \right. \right.$$

$$+ (\mathrm{sn}^2\beta - \mathrm{sn}^2\alpha)(\mathrm{sn}^2\beta - \mathrm{sn}^2\gamma)\dot{\beta}^2$$

$$\left. \left. + (\mathrm{sn}^2\gamma - \mathrm{sn}^2\beta)(\mathrm{sn}^2\gamma - \mathrm{sn}^2\alpha)\dot{\gamma}^2 \Big) - \sum_{i=\alpha,\beta,\gamma} \Delta V_i(\alpha, \beta, \gamma) \right] dt \right\}$$

$$= \sum_{n,\nu} \int_0^{\infty} dp \, e^{-i\hbar p^2 T/2m} \mathrm{El}_{n,p}^{\nu}(\alpha'', \beta'', \gamma'') \mathrm{El}_{n,p}^{\nu*}(\alpha', \beta', \gamma') \ . \quad (5.24)$$

XI. Paraboloidal, $\alpha, \gamma > 0, \beta \in (0, \pi)$:

$$\int_{\alpha(t')=\alpha'}^{\alpha(t'')=\alpha''} \mathcal{D}_{MP}\alpha(t) \int_{\beta(t')=\beta'}^{\beta(t'')=\beta''} \mathcal{D}_{MP}\beta(t) \int_{\gamma(t')=\gamma'}^{\gamma(t'')=\gamma''} \mathcal{D}_{MP}\gamma(t)$$

$$\times 8d^3 (\cosh^2 \alpha + \sinh^2 \gamma)(\sinh^2 \alpha + \sin^2 \beta)(\cos^2 \beta + \sinh^2 \gamma)$$

$$\times \exp \left\{ \frac{i}{\hbar} \int_{t'}^{t''} \left[\frac{m}{2} 4d^2 \Big((\cosh^2 \alpha + \sinh^2 \gamma)(\sinh^2 \alpha + \sin^2 \beta)\dot{\alpha}^2 \right. \right.$$

$$+ (\cos^2 \beta + \sinh^2 \gamma)(\sin^2 \beta + \sinh^2 \alpha)\dot{\beta}^2$$

$$\left. \left. + (\sinh^2 \gamma + \cos^2 \beta)(\sinh^2 \gamma + \cosh^2 \alpha)\dot{\gamma}^2 \Big) - \sum_{i=\alpha,\beta,\gamma} \Delta V_i(\alpha, \beta, \gamma) \right] dt \right\}$$

$$= \int_{\mathbb{R}} d\lambda \sum_n \int_0^\infty dp\, e^{-i\hbar p^2 T/2m} \left[\mathrm{gc}_n\left(i\alpha''; 2dp; \frac{\lambda}{2p}\right) \mathrm{gc}_n\left(\beta''; 2dp; \frac{\lambda}{2p}\right) \mathrm{gc}_n\left(i\gamma'' + \frac{\pi}{2}; 2dp; \frac{\lambda}{2p}\right) \right.$$

$$\times \mathrm{gc}_n^*\left(i\alpha'; 2dp; \frac{\lambda}{2p}\right) \mathrm{gc}_n^*\left(\beta'; 2dp; \frac{\lambda}{2p}\right) \mathrm{gc}_n^*\left(i\gamma' + \frac{\pi}{2}; 2dp; \frac{\lambda}{2p}\right)$$

$$+ \mathrm{gs}_n\left(i\alpha''; 2dp; \frac{\lambda}{2p}\right) \mathrm{gs}_n\left(\beta''; 2dp; \frac{\lambda}{2p}\right) \mathrm{gs}_n\left(i\gamma'' + \frac{\pi}{2}; 2dp; \frac{\lambda}{2p}\right)$$

$$\left. \times \mathrm{gs}_n^*\left(i\alpha'; 2dp; \frac{\lambda}{2p}\right) \mathrm{gs}_n^*\left(\beta'; 2dp; \frac{\lambda}{2p}\right) \mathrm{gs}_n^*\left(i\gamma' + \frac{\pi}{2}; 2dp; \frac{\lambda}{2p}\right) \right]. \quad (5.25)$$

General Form of the Green's Function:
$$= \int_{\mathbb{R}} \frac{dE}{2\pi\hbar} e^{-iET/\hbar} \frac{m}{4\pi\hbar^2 |\mathbf{x}'' - \mathbf{x}'|} \exp\left(-\frac{|\mathbf{x}'' - \mathbf{x}'|}{\hbar} \sqrt{-2mE} \right). \quad (5.26)$$

The Two Spheroidal Systems.

All but the last five path integrations have been discussed already in [223] and we do not need to do this once again. Obviously, the path integral solutions of the circular systems follow from the corresponding two-dimensional cases. The sphero-conical system will be discussed in the next Chapter.

Similarly as in two-dimensional Euclidean space, we can solve the path integrals in the two spheroidal systems. In the case of prolate-spheroidal coordinates we have to consider the expansion of the short-time kernel according to [399, p.315]

$$\exp\left[ipd(\sinh\mu \sin\nu \sin\vartheta \cos\varphi + \cosh\mu \cos\nu \cos\vartheta) \right]$$

$$= \sum_{l=0}^\infty \sum_{n=-l}^l (2l+1) i^{l+2n} S_l^{n\,(1)}(\cosh\mu; pd) \mathrm{ps}_l^n(\cos\nu; p^2 d^2) \mathrm{ps}_l^{-n}(\cos\vartheta; p^2 d^2)\, e^{in\varphi}. \quad (5.27)$$

$S_l^{n\,(1)}, \mathrm{ps}_l^n$ are prolate spheroidal wavefunctions. The short-time kernel is expanded by means of (5.27) yielding

$$\left(\frac{m}{2\pi i\epsilon\hbar} \right)^{3/2} \exp\left(\frac{im}{2\epsilon\hbar} |\mathbf{x}'' - \mathbf{x}''|^2 \right)$$

$$= \sum_{l=0}^\infty \sum_{n=-l}^l e^{in(\varphi''-\varphi')} \frac{2l+1}{2\pi^2} \frac{(l-n)!}{(l+n)!} \int_0^\infty p^2 dp\, e^{-i\epsilon\hbar p^2/2m}$$

$$\times \mathrm{ps}_l^{n\,*}(\cos\nu'; p^2 d^2) \mathrm{ps}_l^n(\cos\nu''; p^2 d^2) S_l^{n\,(1)\,*}(\cosh\mu'; pd) S_l^{n\,(1)}(\cosh\mu''; pd), \quad (5.28)$$

and in the group path integration one makes use of the orthonormality relation

$$d^3 \int_0^\infty d\mu \int_0^\pi d\nu (\sinh^2\mu + \sin^2\nu) \int_0^{2\pi} d\varphi$$

$$\times S_l^{n\,(1)}(\cosh\mu; pd) S_{l'}^{n'\,(1)}(\cosh\mu; p'd) \mathrm{ps}_l^n(\cos\nu; p^2 d^2) \mathrm{ps}_{l'}^{n'}(\cos\nu; p'^2 d^2)\, e^{i\varphi(n-n')}$$

$$= \frac{2\pi^2}{2l+1} \frac{(l+n)!}{(l-n)!} \frac{1}{p^2} \delta_{nn'} \delta_{ll'} \delta(p - p'), \quad (5.29)$$

together with [399, p.286]

$$\int_0^\pi \sin\vartheta d\vartheta \int_0^{2\pi} d\varphi \, \mathrm{ps}_l^{-n}(\cos\vartheta; p^2d^2)\mathrm{ps}_{l'}^{-n'\,*}(\cos\vartheta; p^2d^2)e^{i\varphi(n-n')} = \frac{4\pi}{2l!+!1}\frac{(l-n)!}{(l+n)!}\delta_{nn'}\delta_{ll'}\,,$$
$$(5.30)$$

to evaluate the path integral. The path integral (5.22) gives us as a by-result the identity

$$\int_{\mu(t')=\mu'}^{\mu(t'')=\mu''} \mathcal{D}\mu(t) \int_{\nu(t')=\nu'}^{\nu(t'')=\nu''} \mathcal{D}\nu(t)d^2(\sinh^2\mu + \sin^2\nu)$$

$$\times \exp\left\{\frac{i}{\hbar}\int_{t'}^{t''}\left[\frac{m}{2}d^2(\sinh^2\mu+\sin^2\nu)(\dot\mu^2+\dot\nu^2) - \frac{\hbar^2}{2md^2}\frac{\lambda^2-1/4}{\sinh^2\mu\sin^2\nu}\right]dt\right\}$$

$$= d\sqrt{\sin\nu'\sin\nu''\sinh\mu'\sinh\mu''}\sum_{l=0}^{\infty}\frac{2l+1}{\pi}\frac{\Gamma(l-\lambda+1)}{\Gamma(l+\lambda+1)}\int_0^\infty p^2dp\, e^{-i\hbar p^2 T/2m}$$

$$\times \mathrm{ps}_l^{\lambda\,*}(\cos\nu'; p^2d^2)\mathrm{ps}_l^{\lambda}(\cos\nu''; p^2d^2)S_l^{\lambda\,(1)\,*}(\cosh\mu'; pd)S_l^{\lambda\,(1)}(\cosh\mu''; pd)\ .\ (5.31)$$

Analogously we have in the oblate-spheroidal case the expansion

$$\exp\left[ipd(\cosh\xi\sin\nu\sin\vartheta\cos\varphi + \sinh\xi\cos\nu\cos\vartheta)\right]$$

$$= \sum_{l=0}^{\infty}\sum_{n=-l}^{l}(2l+1)i^{l+2n}\mathrm{Si}_l^{n\,(1)}(\cosh\xi; pd)\mathrm{psi}_l^{n}(\cos\nu; p^2d^2)\mathrm{psi}_l^{-n}(\cos\vartheta; p^2d^2)\,e^{in\varphi}\quad (5.32)$$

$(\mathrm{Si}_l^{n\,(1)}, \mathrm{psi}_l^n$ are oblate spheroidal wavefunctions), the orthonormality relation

$$d^3\int_0^\infty d\xi \int_0^\pi d\nu(\cosh^2\xi - \sin^2\nu)\int_0^{2\pi}d\varphi$$

$$\times \mathrm{Si}_l^{n\,(1)}(\cosh\xi; pd)\mathrm{Si}_{l'}^{n'\,(1)}(\cosh\xi; p'd)\mathrm{psi}_l^n(\cos\nu; p^2d^2)\mathrm{psi}_{l'}^{n'}(\cos\nu; p'^2d^2)\,e^{i\varphi(n-n')}$$

$$= \frac{2\pi^2}{2l+1}\frac{(l+n)!}{(l-n)!}\frac{1}{p^2}\delta_{nn'}\delta_{ll'}\delta(p-p')\ ,\qquad\qquad (5.33)$$

and as a by-result the path integral identity

$$\int_{\xi(t')=\xi'}^{\xi(t'')=\xi''} \mathcal{D}\xi(t) \int_{\nu(t')=\nu'}^{\nu(t'')=\nu''} \mathcal{D}\nu(t)d^2(\cosh^2\xi - \sin^2\nu)$$

$$\times \exp\left\{\frac{i}{\hbar}\int_{t'}^{t''}\left[\frac{m}{2}d^2(\cosh^2\xi-\sin^2\nu)(\dot\mu^2+\dot\nu^2) - \frac{\hbar^2}{2md^2}\frac{\lambda^2-1/4}{\cosh^2\mu\sin^2\nu}\right]dt\right\}$$

$$= d\sqrt{\sin\nu'\sin\nu''\cosh\xi'\cosh\xi''}\sum_{l=0}^{\infty}\frac{2l+1}{\pi}\frac{\Gamma(l-\lambda+1)}{\Gamma(l+\lambda+1)}\int_0^\infty p^2dp\, e^{-i\hbar p^2 T/2m}$$

$$\times \mathrm{psi}_l^{\lambda\,*}(\cos\nu'; p^2d^2)\mathrm{psi}_l^{\lambda}(\cos\nu''; p^2d^2)\mathrm{Si}_l^{\lambda\,(1)\,*}(\cosh\xi'; pd)\mathrm{Si}_l^{\lambda\,(1)}(\cosh\xi''; pd)\ .\ (5.34)$$

The Green's function for the kernel in prolate-spheroidal coordinates has the form (where we abbreviate $\gamma = d\sqrt{2mE}/\hbar$, [399, p.312])

$$G(\mu'', \mu', \nu'', \nu', \varphi'', \varphi'; E) = \frac{m}{2\pi\hbar}\sqrt{\frac{2mE}{\hbar^2}} \sum_{l=0}^{\infty} \sum_{n=-l}^{l} (2l+1)(-1)^n$$

$$\times S_l^{n\,(3)}(\cosh \mu_>; \gamma) S_l^{n\,(1)}(\cosh \mu_<; \gamma) \mathrm{ps}_l^n(\cos \nu''; \gamma^2) \mathrm{ps}_l^{-n}(\cos \nu'; \gamma^2)\, e^{in(\varphi''-\varphi')} \ . \quad (5.35)$$

The case of oblate-spheroidal coordinates is, of course, similar, and is obtained by the replacement $\mu \to -i\xi$, $h \to ih$ and $\nu \to \nu$.

The Ellipsoidal and Paraboloidal Systems.

In the expansions in the case of ellipsoidal and paraboloidal coordinates we have adopted the notation of [9, 11, 405] for the ellipsoidal coordinates, and [10, 500, 501] for the paraboloidal coordinates. In both cases the quantum potential in the mid-point lattice description [248] has the form

$$\Delta V_{MP}(\alpha, \beta, \gamma) = \Delta V_\alpha(\alpha, \beta, \gamma) + \Delta V_\beta(\alpha, \beta, \gamma) + \Delta V_\gamma(\alpha, \beta, \gamma) \ , \quad (5.36)$$

where, e.g., ΔV_α is given by

$$\Delta V_\alpha(\alpha, \beta, \gamma) = \frac{\hbar^2}{8m} \frac{AB(A_{,\alpha\alpha}B + AB_{,\alpha\alpha}) - (A_{,\alpha}^2 B^2 + A^2 B_{,\alpha}^2)}{h^2 A^3 B^3} = \frac{\hbar^2}{8m} \frac{\partial_\alpha^2 \ln(AB)}{h^2 AB} \ , \quad (5.37)$$

with $h^2 = 4d^2$ in the paraboloidal case and $h^2 = (a^2 - b^2)^2/(a^2 - c^2)$ in the ellipsoidal case, respectively. By cyclic permutation we have for $\Delta V_\beta = \Delta V_\alpha(A \to B, B \to C)$ and $\Delta V_\gamma = \Delta V_\beta(B \to C, C \to A)$, together with

$$
\begin{aligned}
A_X &= \mathrm{sn}^2\alpha - \mathrm{sn}^2\beta & A_{XI} &= \cosh^2\alpha + \sinh^2\gamma \\
B_X &= \mathrm{sn}^2\beta - \mathrm{sn}^2\gamma & B_{XI} &= \sinh^2\alpha + \sin^2\beta \\
C_X &= \mathrm{sn}^2\alpha - \mathrm{sn}^2\gamma \ , & C_{XI} &= \cos^2\beta + \sinh^2\gamma \ .
\end{aligned}
\quad (5.38)
$$

This gives

$$\Delta V_{MP}(\alpha, \beta, \gamma) = \frac{\hbar^2}{8mh^2}\left(\frac{\partial_\alpha^2 \ln(AB)}{AB} + \frac{\partial_\beta^2 \ln(BC)}{BC} + \frac{\partial_\gamma^2 \ln(AC)}{AC}\right) \ . \quad (5.39)$$

However, the mid-point prescription is not very appropriate to apply the separation formula (2.39). The product-form prescription is better suited for this consideration, and the corresponding quantum potential for, e.g., the paraboloidal system has the form

$$\Delta V_{PF} = \frac{\hbar^2}{32md^2}\left[\frac{1 - 1/\sinh^2\alpha}{(\cosh^2\alpha + \sinh^2\gamma)(\sinh^2\alpha + \sin^2\beta)}\right.$$

$$\left. - \frac{1 + 1/\sin^2\beta}{(\cos^2\beta + \sinh^2\gamma)(\sinh^2\alpha + \sin^2\beta)} + \frac{1 + 1/\cosh^2\gamma}{(\cosh^2\alpha + \sinh^2\gamma)(\sinh^2\gamma + \cos^2\beta)}\right] \ .$$

$$(5.40)$$

Table 5.2: Coordinates in Three-Dimensional Euclidean Space

Coordinate System	Coordinates	Path Integral Solution
I. Cartesian	$x = x'$ $y = y'$ $z = z'$	(5.15, 5.16)
II. Circular Polar	$x = r \cos \varphi$ $y = r \sin \varphi$ $z = z'$	(5.17)
III. Circular Elliptic	$x = d \cosh \mu \cos \nu$ $y = d \sinh \mu \sin \nu$ $z = z'$	(5.17)
IV. Circular Parabolic	$x = \frac{1}{2}(\eta^2 - \xi^2)$ $y = \xi\eta$ $z = z'$	(5.17)
V. Sphero-Conical	$x = r\mathrm{sn}(\alpha, k)\mathrm{dn}(\beta, k')$ $y = r\mathrm{cn}(\alpha, k)\mathrm{cn}(\beta, k')$ $z = r\mathrm{dn}(\alpha, k)\mathrm{sn}(\beta, k')$	(5.18)
VI. Spherical	$x = r \sin \vartheta \cos \varphi$ $y = r \sin \vartheta \sin \varphi$ $z = r \cos \vartheta$	(5.19, 5.20)
VII. Parabolic	$x = \xi\eta \cos \varphi$ $y = \xi\eta \sin \varphi$ $z = \frac{1}{2}(\eta^2 - \xi^2)$	(5.21)
VIII. Prolate Spheroidal	$x = d \sinh \mu \sin \nu \cos \varphi$ $y = d \sinh \mu \sin \nu \sin \varphi$ $z = d \cosh \mu \cos \nu$	(5.22)
IX. Oblate Spheroidal	$x = d \cosh \mu \sin \nu \sin \varphi$ $y = d \cosh \mu \sin \nu \sin \varphi$ $z = d \sinh \mu \cos \nu$	(5.23)
X. Ellipsoidal	$x = k^2\sqrt{a^2 - c^2}\mathrm{sn}\alpha\mathrm{sn}\beta\mathrm{sn}\gamma$ $y = -(k^2/k')\sqrt{a^2 - c^2}\mathrm{cn}\alpha\mathrm{cn}\beta\mathrm{cn}\gamma$ $z = (\mathrm{i}/k')\sqrt{a^2 - c^2}\mathrm{dn}\alpha\mathrm{dn}\beta\mathrm{dn}\gamma$	(5.24)
XI. Paraboloidal	$x = 2d \cosh \alpha \cos \beta \sinh \gamma$ $y = 2d \sinh \alpha \sin \beta \cosh \gamma$ $z = d(\cosh^2 \alpha + \cos^2 \beta - \cosh^2 \gamma)$	(5.25)

Concerning the ellipsoidal wavefunctions, interbasis expansions between the conical and the ellipsoidal bases are discussed in [11, p.247] (see also [9, 387, 410]). However, these kinds of expansion are on a very formal level, and the whole theory seems rather poorly developed. Saying that a path integration is possible by means of an interbasis expansion and exploiting the unitarity of the coefficients requires a very pragmatic point of view: One knows that an interbasis exists, the coefficients satisfy a three-term recurrence relation, and the wavefunctions and the coefficients can be classified according to their parity properties. The spectral expansions are done according to the eigenvalues of the corresponding operators characterizing the ellipsoidal and paraboloidal coordinate systems, respectively, therefore corresponding in each case to a set of inequivalent observables. Hence, the problem is mathematically defined and has a solution. This solution may be quite a formidable one if one tries a, say numerical, investigation. Having the mathematical theory, the interbasis expansion from an appropriate coordinate space representation in the propagator can be performed, and we are done. Of course, the same line of reasoning applies to the case of the paraboloidal system. Details of this procedure, including a proper classification of the wavefunctions, and the corresponding interbasis expansion coefficients will be given elsewhere.

Coordinate Systems in Three-Dimensional Space.

Table 5.2 summarizes our knowledge of path integration in $\mathbb{R}^3$, including an enumeration of the coordinate systems [412, 416]. In the spheroidal coordinate systems and in the paraboloidal system d is a positive parameter. In ellipsoidal coordinates $k^2 = (a^2 - b^2)/(a^2 - c^2), k'^2 = (b^2 - c^2)/(a^2 - c^2), 0 < k, k' < 1 = 1$, where $a^2 = -a_1^2, b^2 = -a_2^2, c^2 = -a_3^2$, and I have adopted the notation of [153, 416]. For the paraboloidal system c.f. [9, 11, 405].

Chapter 6

Path Integrals on Spheres

6.1 The Two-Dimensional Sphere

We have the path integral representations on the two-dimensional unit-sphere $S^{(2)}$ [223] (see also [66, 68, 94, 248, 459])

I. Spherical and General Form of the Propagator, $\vartheta \in (0, \pi), \varphi \in [0, 2\pi)$:

$$
\int_{\vartheta(t')=\vartheta'}^{\vartheta(t'')=\vartheta''} \mathcal{D}\vartheta(t) \sin\vartheta \int_{\varphi(t')=\varphi'}^{\varphi(t'')=\varphi''} \mathcal{D}\varphi(t)
$$

$$
\times \exp\left\{\frac{i}{\hbar} \int_{t'}^{t''} \left[\frac{m}{2}\left(\dot\vartheta^2 + r^2\sin^2\vartheta\dot\varphi^2\right) + \frac{\hbar^2}{8m}\left(1 + \frac{1}{\sin^2\vartheta}\right)\right]dt\right\}
$$

$$
= \sum_{l=0}^{\infty} \frac{2l+1}{4\pi} P_l(\cos\psi_{S^{(2)}}) e^{-i\hbar T l(l+1)/2m} \tag{6.1}
$$

$$
= \sum_{l=0}^{\infty} \sum_{n=-l}^{l} e^{-i\hbar T l(l+1)/2m} Y_l^m(\vartheta'', \varphi'') Y_l^{m*}(\vartheta', \varphi') \ . \tag{6.2}
$$

II. Elliptic, $\alpha \in [-K, K], \beta \in [-2K', 2K']$:

$$
\int_{\alpha(t')=\alpha'}^{\alpha(t'')=\alpha''} \mathcal{D}\alpha(t) \int_{\beta(t')=\beta'}^{\beta(t'')=\beta''} \mathcal{D}\beta(t)(k^2\mathrm{cn}^2\alpha + k'^2\mathrm{cn}^2\beta)
$$

$$
\times \exp\left[\frac{im}{2\hbar}\int_{t'}^{t''}(k^2\mathrm{cn}^2\alpha + k'^2\mathrm{cn}^2\beta)(\dot\alpha^2 + \dot\beta^2)dt\right]
$$

$$
= \sum_{l=0}^{\infty} \sum_{\lambda} \sum_{p,q=\pm} \Lambda_{l,h}^p(\alpha'') \Lambda_{l,h}^{p*}(\alpha') \Lambda_{l,\bar h}^q(\beta'') \Lambda_{l,\bar h}^{q*}(\beta') \, e^{-i\hbar T l(l+1)/2m} \ . \tag{6.3}
$$

General Form of Green's Function:

$$
= \int_{\mathbb{R}} \frac{dE}{2\pi i} e^{-iET/\hbar} \frac{m}{2\hbar^2} \frac{P_{-1/2+\sqrt{2mE/\hbar^2+1/4}}(-\cos\psi_{S^{(2)}})}{\sin\left[\pi(1/2-\sqrt{2mE/\hbar^2+1/4}\,)\right]} \; .
\tag{6.4}
$$

Spherical Coordinates.

Because the spherical system has been already extensively discussed in the literature, e.g., [12, 66, 68, 144, 223, 248, 439, 478], we do not need to do this once again. The Kepler problem has been discussed in [45], and more general potentials in [242].

Elliptic Coordinates.

The elliptic coordinate system reads in algebraic form as follows ($a_1 \leq \varrho_1 \leq a_2 \leq \varrho_2 \leq a_3$)

$$
s_1^2 = R^2 \frac{(\varrho_1 - a_1)(\varrho_2 - a_1)}{(a_2 - a_1)(a_3 - a_1)}, \quad s_2^2 = R^2 \frac{(\varrho_1 - a_2)(\varrho_2 - a_2)}{(a_3 - a_2)(a_1 - a_2)}, \quad s_3^2 = R^2 \frac{(\varrho_1 - a_3)(\varrho_2 - a_3)}{(a_1 - a_3)(a_2 - a_3)} \; .
\tag{6.5}
$$

If we put $\varrho_1 = a_1 + (a_2 - a_1)\mathrm{sn}^2(\alpha, k)$ and $\varrho_2 = a_2 + (a_3 - a_2)\mathrm{cn}^2(\beta, k')$, where $\mathrm{sn}(\alpha, k), \mathrm{cn}(\alpha, k)$ and $\mathrm{dn}(\alpha, k)$ are the Jacobi elliptic functions with modulus k, we obtain for the coordinates **s** on the sphere

$$
\begin{aligned}
s_1 &= R\mathrm{sn}(\alpha, k)\mathrm{dn}(\beta, k') \; , & -K \leq \alpha \leq K \; , \\
s_2 &= R\mathrm{cn}(\alpha, k)\mathrm{cn}(\beta, k') \; , & -2K' \leq \beta \leq 2K' \; , \\
s_3 &= R\mathrm{dn}(\alpha, k)\mathrm{sn}(\beta, k') \; ,
\end{aligned}
\tag{6.6}
$$

where

$$
k^2 = \frac{a_2 - a_1}{a_3 - a_1} = \sin^2 f \; , \qquad k'^2 = \frac{a_3 - a_2}{a_3 - a_1} = \cos^2 f \; , \qquad k^2 + k'^2 = 1 \; .
\tag{6.7}
$$

$K = K(k) = \frac{\pi}{2}\,{}_2F_1(\frac{1}{2}, \frac{1}{2}; 1; k^2)$ and $K' = K(k')$ are complete elliptic integrals, and $2f$ is the interfocus distance on the upper semisphere of the ellipses on the sphere. Note the relations $\mathrm{cn}^2\alpha + \mathrm{sn}^2\alpha = 1$ and $\mathrm{dn}^2\alpha = 1 - k^2\mathrm{sn}^2\alpha$. In the following we omit the moduli k and k' of the Jacobi elliptic functions if it is obvious that the variable α goes with k and β goes with k'. For the periodic Lamé polynomials $\Lambda_{ln}^p(z)$ we have adopted the notation of [432] (compare for alternative notations [153, p.64], [333]). These ellipsoidal harmonics on $S^{(2)}$ satisfy the orthonormality relation

$$
\int_{-K}^{K} d\alpha \int_{-2K'}^{2K'} d\beta (k^2\mathrm{cn}^2\alpha + k'^2\mathrm{cn}^2\beta)\Lambda_{l',\tilde{h}}^{p'}(\alpha)\Lambda_{l,h}^{p*}(\alpha)\Lambda_{l',\tilde{h}'}^{q'}(\beta)\Lambda_{l,h'}^{q*}(\beta) = \delta_{ll'}\delta_{qq'}\delta_{pp'}\delta_{h\tilde{h}} \; .
\tag{6.8}
$$

The quantity $\cos\psi_{S^{(2)}}$ is defined as, e.g., in the spherical system

$$
\cos\psi_{S^{(2)}} = \cos\vartheta'' \cos\vartheta' + \sin\vartheta'' \sin\vartheta' \cos(\varphi'' - \varphi') \; ,
\tag{6.9}
$$

and describes the invariant distance on $S^{(2)}$. The interbasis expansion between the spherical coordinates ϑ, φ and the conical coordinates α, β has been established by Lukàcs [373, 374], Möglich [410], MacFayden and Winternitz [381], Miller [405], and Patera and Winternitz [432]. In particular we rely on [432]. In [432] it is shown that the wavefunctions of the spherical basis $|lm\rangle$ can be expanded into the wavefunctions of the elliptical basis $|l\lambda\rangle$ and vice versa according to

$$|lm\rangle = \sum_\lambda X_{\lambda,m}^{l*}|l\lambda\rangle , \qquad |l\lambda\rangle = \sum_m X_{\lambda,m}^l|lm\rangle . \qquad (6.10)$$

Here λ is the eigenvalue of the operator $E = -4(L_1^2 + k'^2 L_2^2)$ which commutes with the Hamiltonian, i.e., the Legendre operator on $S^{(2)}$ $(0 < k'^2 \leq 1)$. For each $l \in \mathbb{N}_0$ there are $2l + 1$ eigenvalues λ. $(X_{\lambda,m}^l)$ is a $(2l + 1) \times (2l + 1)$ unitary matrix. The interbasis expansion coefficients $X_{\lambda,m}^l$ satisfy a three-term recurrence relation reading

$$X_{\lambda,m-2}^l A_{m-2} + X_{\lambda,m}^l(B_m - \lambda) + X_{\lambda,m+2}^l A_m = 0 , \qquad (6.11)$$

where $(k^2 = 1 - k'^2)$

$$A_m = k^2\sqrt{(l-m)(l+m+1)(l-m-1)(l+m+2)} , \qquad (6.12)$$
$$B_m = 2(1 + k'^2)[l(l+1) - m^2] .$$

Furthermore one has the properties $A_{-m} = A_{m-2}$ and $B_{-m} = B_m$. There seems to be no closed expression of these coefficients in terms of Lamé polynomials $\propto \Lambda_{lh'}^p$ as it is the case in the elliptic system in $\mathbb{R}^2$.

As it turns out, the basis $|l\lambda\rangle$ can be classified by its parity with respect to reflections. One finds

$$\begin{array}{llllll} P: x \to -x, & y \to -y, & z \to -z; & P|l,m\rangle & = & (-1)^l|l,m\rangle, \\ Z: x \to x, & y \to y, & z \to -z; & Z|l,m\rangle & = & (-1)^{l-m}|l,m\rangle, \\ X: x \to -x, & y \to y, & z \to z; & X|l,m\rangle & = & |l,-m\rangle, \\ Y: x \to x, & y \to -y, & z \to z; & XY|l,m\rangle & = & (-1)^m|l,-m\rangle . \end{array} \right\} \qquad (6.13)$$

Applying these operators to the eigen-functions of the elliptic basis yields $(p, q = \pm 1)$

$$\begin{array}{llll} P|l\lambda pq\rangle & = & (-1)^l q|l\lambda pq\rangle , & X|l\lambda pq\rangle & = & p|l\lambda pq\rangle , \\ Z|l\lambda pq\rangle & = & (-1)^l|l\lambda pq\rangle , & Y|l\lambda pq\rangle & = & pq|l\lambda pq\rangle . \end{array} \qquad (6.14)$$

Hence one obtains four operators defining the $O(3)$ basis functions in the non-subgroup basis

$$\begin{array}{llll} \Delta|lmpq\rangle & = & l(l+1)|l\lambda pq\rangle , & X|l\lambda pq\rangle & = & p|l\lambda pq\rangle , \\ E|l\lambda pq\rangle & = & \lambda|l\lambda pq\rangle , & XY|l\lambda pq\rangle & = & q|l\lambda pq\rangle . \end{array} \qquad (6.15)$$

The basis $|l\lambda pq\rangle$ can now be identified with a product of two Lamé polynomials according to

$$\langle \alpha\beta|l\lambda pq\rangle = \Lambda_{lh}^p(\alpha)\Lambda_{l\tilde{h}}^q(\beta) , \qquad h + \tilde{h} = l(l+1) , \qquad (6.16)$$

where Λ_{lh}^p is a Lamé polynomial satisfying $(k^2 = 1 - k'^2)$

$$\left.\begin{array}{l} \dfrac{\mathrm{d}^2\Lambda_{lh}^p}{\mathrm{d}\alpha^2} + [h - l(l+1)k^2\mathrm{sn}^2(\alpha,k)]\Lambda_{lh}^p = 0 \ , \\[2mm] \Lambda_{lh}^p(-\alpha) = p\Lambda_{lh}(\alpha) \ , \qquad h = -\dfrac{\lambda}{4} + l(l+1) \ . \end{array}\right\} \qquad (6.17)$$

The function $\Lambda_{lh'}^q(\beta)$ satisfies the same equation with $\alpha \to \beta, k \to k', h \to \tilde{h} = \lambda/4$ and $p \to q$. The functions (6.10) are called ellipsoidal harmonics. They can be identified with the wavefunctions of the asymmetric top [360, 373, 376, 473, 525].

The unitarity of the interbasis coefficients implies

$$\sum_n X_{\lambda,n}^{pq,l} X_{\lambda',n}^{p'q',l'} = \delta_{\lambda,\lambda'}\delta_{p,p'}\delta_{q,q'} \ . \qquad (6.18)$$

Therefore we obtain in each short-time kernel

$$\sum_{l=0}^{\infty}\sum_{n=-l}^{l} \mathrm{e}^{-i\epsilon\hbar l(l+1)/2m} Y_l^m(\vartheta'',\varphi'') Y_l^{m*}(\vartheta',\varphi')$$

$$= \sum_{l=0}^{\infty} \mathrm{e}^{-i\epsilon\hbar l(l+1)/2m} \sum_{\lambda}\sum_{p,q=\pm}\sum_{\lambda'}\sum_{p',q'=\pm} \Lambda_{l',h'}^{p'}(\alpha'')\Lambda_{l',\tilde{h}'}^{q'}(\beta'')\Lambda_{l,\tilde{h}}^{q*}(\beta')\Lambda_{l,h}^{p*}(\alpha') \sum_n X_{\lambda,n}^{pq,l} X_{\lambda',n}^{p'q',l'}$$

$$= \sum_{l=0}^{\infty}\sum_{\lambda}\sum_{p,q=\pm} \Lambda_{l,h}^{p}(\alpha'')\Lambda_{l,h}^{p*}(\alpha')\Lambda_{l,\tilde{h}}^{q}(\beta'')\Lambda_{l,\tilde{h}}^{q*}(\beta') \, \mathrm{e}^{-i\epsilon\hbar l(l+1)/2m} \ . \qquad (6.19)$$

This shows the expansion

$$P_l(\cos\psi_{S^{(2)}}) = \frac{4\pi}{2l+1}\sum_{\lambda}\sum_{p,q=\pm} \Lambda_{l,h}^{p}(\alpha'')\Lambda_{l,h}^{p*}(\alpha')\Lambda_{l,\tilde{h}}^{q}(\beta'')\Lambda_{l,\tilde{h}}^{q*}(\beta') \ , \qquad (6.20)$$

which is the analogue of the usual expansion of $P_l(\cos\psi_{S^{(2)}})$ in terms of the spherical harmonics in polar coordinates. Performing the group path integration yields the result (6.3).

Let us mention another parameterization of elliptic coordinates on the sphere. We set [373] $\varrho_1 = a_1 + (a_2 - a_1)\cos^2\mu$ and $\varrho_2 = a_2 - (a_3 - a_2)\cos^2\nu$ $(0 \le \mu \le 2\pi, 0 \le \nu \le \pi)$, and obtain

$$s_1 = \sqrt{1 - k'^2\cos^2\nu}\,\cos\mu \ , \quad s_2 = \sin\mu\sin\nu \ , \quad s_3 = \sqrt{1 - k^2\cos^2\mu}\,\cos\nu \ . \quad (6.21)$$

We then have for instance

$$\frac{ds^2}{d^2t} = (k^2\sin^2\mu + k'^2\sin^2\nu)\left(\frac{\dot{\mu}^2}{1 - k^2\cos^2\mu} + \frac{\dot{\nu}^2}{1 - k'^2\cos^2\nu}\right) \ . \qquad (6.22)$$

Lukač [373] has shown that for this parameterization it is possible to prove the expansion

$$P_l(\cos\psi_{S^{(2)}}) = \frac{4\pi}{2l+1}\sum_{m=-l}^{l} E_{l,m}(\mu'',\nu'')E_{l,m}^*(\mu',\nu') \ , \qquad (6.23)$$

with $E_{l,m}(\mu,\nu)$ the spherical harmonics on $S^{(2)}$ in elliptic coordinates in the (μ,ν) parameterization which is the analogue of (6.20). The parameterization (6.21) has the advantage that the spherical harmonics can be stated explicitly, c.f. [373] for a small table of these wavefunctions. The corresponding path integral has the form

$$
\int_{\mu(t')=\mu'}^{\mu(t'')=\mu''} \mathcal{D}\mu(t) \int_{\nu(t')=\nu'}^{\nu(t'')=\nu''} \mathcal{D}\nu(t) \frac{k^2\sin^2\mu + k'^2\sin^2\nu}{\sqrt{(1-k^2\cos^2\mu)(1-k'^2\cos^2\nu)}}
$$

$$
\times \exp\left[\frac{im}{2\hbar}\int_{t'}^{t''}(k^2\sin^2\mu + k'^2\sin^2\nu)\left(\frac{\dot{\mu}^2}{1-k^2\cos^2\mu} + \frac{\dot{\nu}^2}{1-k'^2\cos^2\nu}\right)dt\right]
$$

$$
= \sum_{l=0}^{\infty}\sum_{m=-l}^{l} e^{-i\hbar Tl(l+1)/2m} E_{l,m}(\mu'',\nu'') E_{l,m}^*(\mu',\nu') \ . \tag{6.24}
$$

Table 6.1: Coordinates on the Two-Dimensional Sphere

Coordinate System	Coordinates	Path Integral Solution
I. Spherical	$s_1 = \cos\vartheta$	(6.1, 6.2)
	$s_2 = \sin\vartheta\cos\varphi$	
	$s_3 = \sin\vartheta\sin\varphi$	
II. Elliptic	$s_1 = \mathrm{sn}(\alpha,k)\mathrm{dn}(\beta,k')$	(6.3, 6.24)
	$s_2 = \mathrm{cn}(\alpha,k)\mathrm{cn}(\beta,k')$	
	$s_3 = \mathrm{dn}(\alpha,k)\mathrm{sn}(\beta,k')$	

Coordinate Systems on the Two-Dimensional Sphere.

In the table 6.1 we summarize the results on path integration on $S^{(2)}$, including an enumeration of the coordinate systems according to [425]. A rotated elliptic system is given by

$$
s_1 = \frac{1}{a_3-a_1}\left(\sqrt{(\varrho_1-a_3)(\varrho_2-a_3)} - \sqrt{(\varrho_1-a_1)(\varrho_2-a_1)}\right)
$$

$$
= k'\mathrm{sn}(\alpha,k)\mathrm{dn}(\beta,k') + k\,\mathrm{dn}(\alpha,k)\mathrm{sn}(\beta,k') \ , \tag{6.25}
$$

$$
s_2 = \sqrt{\frac{(\varrho_1-a_2)(\varrho_2-a_2)}{(a_3-a_2)(a_2-a_1)}} = \mathrm{cn}(\alpha,k)\mathrm{cn}(\beta,k') \ , \tag{6.26}
$$

$$
s_3 = \frac{1}{a_3-a_1}\left(\sqrt{\frac{a_3-a_2}{a_2-a_1}}(\varrho_1-a_1)(\varrho_2-a_1) - \sqrt{\frac{a_2-a_1}{a_3-a_2}}(\varrho_1-a_3)(\varrho_2-a_3)\right)
$$

$$
= k'\mathrm{dn}(\alpha,k)\mathrm{sn}(\beta,k') - k\,\mathrm{sn}(\alpha,k)\mathrm{dn}(\beta,k) \ . \tag{6.27}
$$

In the notation, I have chosen $\varrho_1 = a_1 + (a_2 - a_1)\text{sn}^2(\alpha, k)$ and $\varrho_2 = a_2 + (a_3 - a_2)\text{cn}^2(\beta, k')$. Let us note that in the flat space limit the special case of the elliptic system with $k = k' = \frac{1}{2}$ goes over into the cartesian system, and the corresponding rotated system into the parabolic system in $\mathbb{R}^2$ [292]. The usual elliptic system has as its limit the elliptic system, and the rotated system the elliptic II system in two-dimensional Euclidean space, respectively.

6.2   The Three-Dimensional Sphere

We have the path integral representations on the three-dimensional sphere $S^{(3)}$ [223] (for the spectral expansion of the spherical system c.f. [223, 240], see also [66, 68, 94])

I. Cylindrical, $\vartheta \in (0, \pi/2), \varphi_{1,2} \in [0, 2\pi)$:

$$
\int_{\vartheta(t')=\vartheta'}^{\vartheta(t'')=\vartheta''} \mathcal{D}\vartheta(t) \sin\vartheta \cos\vartheta \int_{\varphi_1(t')=\varphi_1'}^{\varphi_1(t'')=\varphi_1''} \mathcal{D}\varphi_1(t) \int_{\varphi_2(t')=\varphi_2'}^{\varphi_2(t'')=\varphi_2''} \mathcal{D}\varphi_2(t)
$$

$$
\times \exp\left\{ \frac{i}{\hbar} \int_{t'}^{t''} \left[\frac{m}{2} \left(\dot\vartheta^2 + \cos^2\vartheta \dot\varphi_1^2 + \sin^2\vartheta \dot\varphi_2^2 \right) + \frac{\hbar^2}{8m} \left(4 + \frac{1}{\cos^2\vartheta} + \frac{1}{\sin^2\vartheta} \right) \right] dt \right\}
$$

$$
= \sum_{J=0}^{\infty} \sum_{k_1, k_2 \in \mathbb{Z}} \frac{e^{i[k_1(\varphi_1'' - \varphi_1') + k_2(\varphi_2'' - \varphi_2')]}}{4\pi^2} \frac{2(J+1) \left(\frac{J - |m_1| - |m_2|}{2} \right)! \left(\frac{J + |m_1| + |m_2|}{2} \right)!}{\left(\frac{J - |m_1| + |m_2|}{2} \right)! \left(\frac{J + |m_1| - |m_2|}{2} \right)!}
$$

$$
\times (\sin\vartheta' \sin\vartheta'')^{|m_1|} (\cos\vartheta' \cos\vartheta'')^{|m_2|}
$$

$$
\times P_{(J - |m_1| - |m_2|)/2}^{(|m_1|, |m_2|)}(\cos 2\vartheta') P_{(J - |m_1| - |m_2|)/2}^{(|m_1|, |m_2|)}(\cos 2\vartheta'') \, e^{-i\hbar T J(J+2)/2m} \quad . \tag{6.28}
$$

II. Sphero-Elliptic, $\chi \in (0, \pi), \alpha \in [-K, K], \beta \in [-2K', 2K']$:

$$
\int_{\chi(t')=\chi'}^{\chi(t'')=\chi''} \mathcal{D}\chi(t) \sin^2\chi \int_{\alpha(t')=\alpha'}^{\alpha(t'')=\alpha''} \mathcal{D}\alpha(t) \int_{\beta(t')=\beta'}^{\beta(t'')=\beta''} \mathcal{D}\beta(t) (k^2 \text{cn}^2\alpha + k'^2 \text{cn}^2\beta)
$$

$$
\times \exp\left\{ \frac{im}{2\hbar} \int_{t'}^{t''} \left[\dot\chi^2 + \sin^2\chi (k^2 \text{cn}^2\alpha + k'^2 \text{cn}^2\beta)(\dot\alpha^2 + \dot\beta^2) \right] dt + \frac{i\hbar T}{2m} \right\}
$$

$$
= \sum_{J=0}^{\infty} \sum_{l=-J}^{J} \sum_{\lambda} \sum_{p,q=\pm} \Lambda_{l,h}^p(\alpha'') \Lambda_{l,h}^{p*}(\alpha') \Lambda_{l,\tilde{h}}^q(\beta'') \Lambda_{l,\tilde{h}}^{q*}(\beta')
$$

$$
\times (J+1) \frac{(l+J+1)!}{|J-l|!} \, e^{-i\hbar T J(J+2)/2m} P_{J+1/2}^{-l-1/2}(\sin\chi'') P_{J+1/2}^{-l-1/2}(\sin\chi') \quad . \tag{6.29}
$$

III. Spherical, $\chi, \vartheta \in (0, \pi), \varphi \in [0, 2\pi)$:

$$\int_{\chi(t')=\chi'}^{\chi(t'')=\chi''} \mathcal{D}\chi(t) \sin^2\chi \int_{\vartheta(t')=\vartheta'}^{\vartheta(t'')=\vartheta''} \mathcal{D}\vartheta(t) \sin\vartheta \int_{\varphi(t')=\varphi'}^{\varphi(t'')=\varphi''} \mathcal{D}\varphi(t)$$

$$\times \exp\left\{ \frac{i}{\hbar} \int_{t'}^{t''} \left[\frac{m}{2}\left(\dot\chi^2 + \sin^2\chi(\dot\vartheta^2 + \sin^2\vartheta\dot\varphi^2) \right) + \frac{\hbar^2}{8m}\left(4 + \frac{1}{\sin^2\chi} + \frac{1}{\sin^2\chi\sin^2\vartheta} \right) \right] dt \right\}$$

$$= \frac{1}{2\pi^2} \sum_{J=0}^{\infty} (J+1) C_J^1(\cos\psi_{S^{(3)}}) \exp\left[-\frac{i\hbar T}{2m} J(J+2) \right] \tag{6.30}$$

$$= \sum_{J=0}^{\infty} \sum_{m_1, m_2} \Psi_{J,m_1,m_2}(\chi'', \vartheta'', \varphi'') \Psi_{J,m_1,m_2}^*(\chi', \vartheta', \varphi') \exp\left[-\frac{i\hbar T}{2m} J(J+2) \right] , \tag{6.31}$$

$$\Psi_{J,m_1,m_2}(\chi, \vartheta, \varphi) = N^{-1/2} e^{im_1} (\sin\chi)^{m_1} C_{J-m_1}^{m_1+2}(\cos\chi)(\sin\vartheta)^{m_2} C_{m_1-m_2}^{m_2+3/2}(\cos\vartheta) , \tag{6.32}$$

$$N = \frac{2\pi^3 2^{-(1+2m_1+2m_2)}}{(J+1)(m_1+\frac{3}{2})(J-m_1)!(m_1-m_2)!)} \cdot \frac{\Gamma(J+m_1+2)\Gamma(m_1+m_2+1)}{\Gamma^2(m_1+1)\Gamma^2(m_2+\frac{3}{2})} . \tag{6.33}$$

IV. Oblate Elliptic, $\alpha \in [0, 2K], \beta \in [-K', K'], \varphi \in [0, 2\pi), a \geq 0$:

$$\int_{\alpha(t')=\alpha'}^{\alpha(t'')=\alpha''} \mathcal{D}\alpha(t) \int_{\beta(t')=\beta'}^{\beta(t'')=\beta''} \mathcal{D}\beta(t)(k^2\mathrm{cn}^2\alpha + k'^2\mathrm{cn}^2\beta)\mathrm{sn}\alpha\,\mathrm{dn}\beta \int_{\varphi(t')=\varphi'}^{\varphi(t'')=\varphi''} \mathcal{D}\varphi(t)$$

$$\times \exp\left\{ \frac{i}{\hbar} \int_{t'}^{t''} \left[\frac{m}{2}\left((k^2\mathrm{cn}^2\alpha + k'^2\mathrm{cn}^2\beta)(\dot\alpha^2 + \dot\beta^2) + \mathrm{sn}^2\alpha\,\mathrm{dn}^2\beta\dot\varphi^2 \right) \right.\right.$$

$$\left.\left. + \frac{\hbar^2}{8m}\frac{1}{k^2\mathrm{cn}^2\alpha + k'^2\mathrm{cn}^2\beta}\left(\frac{\mathrm{cn}^2\beta\mathrm{dn}^2\beta}{\mathrm{sn}^2\beta} + k^4\frac{\mathrm{sn}^2\alpha\mathrm{cn}^2\alpha}{\mathrm{dn}^2\alpha} \right) \right] dt + \frac{i\hbar T}{2m} \right\}$$

$$= \frac{1}{2\pi} \sum_{J=0}^{\infty} \sum_{r,p=\pm 1} \sum_{qk_2} e^{ik_2(\varphi''-\varphi')} e^{-i\hbar T J(J+2)/2m}$$

$$\times \psi_{1,Jqk_2}^{(r,p)}(\alpha''; a) \psi_{1,Jqk_2}^{(r,p)*}(\alpha'; a) \psi_{2,Jqk_2}^{(r,p)}(\beta''; a) \psi_{2,Jqk_2}^{(r,p)*}(\beta'; a) . \tag{6.34}$$

V. Prolate Elliptic, $\alpha \in [-K, K], \beta \in [-K', K'], \varphi \in [0, 2\pi), a \in [-1, 0]$:

$$\int_{\alpha(t')=\alpha'}^{\alpha(t'')=\alpha''} \mathcal{D}\alpha(t) \int_{\beta(t')=\beta'}^{\beta(t'')=\beta''} \mathcal{D}\beta(t)(k^2\mathrm{cn}^2\alpha + k'^2\mathrm{cn}^2\beta)\mathrm{cn}\alpha\,\mathrm{cn}\beta \int_{\varphi(t')=\varphi'}^{\varphi(t'')=\varphi''} \mathcal{D}\varphi(t)$$

$$\times \exp\left\{ \frac{i}{\hbar} \int_{t'}^{t''} \left[\frac{m}{2}\left((k^2\mathrm{cn}^2\alpha + k'^2\mathrm{cn}^2\beta)(\dot\alpha^2 + \dot\beta^2) + \mathrm{cn}^2\alpha\,\mathrm{cn}^2\beta\dot\varphi^2 \right) \right.\right.$$

$$\left.\left. + \frac{\hbar^2}{8m}\frac{1}{k^2\mathrm{cn}^2\alpha + k'^2\mathrm{cn}^2\beta}\left(\frac{\mathrm{sn}^2\beta\mathrm{dn}^2\beta}{\mathrm{cn}^2\beta} + k^4\frac{\mathrm{sn}^2\alpha\mathrm{dn}^2\alpha}{\mathrm{cn}^2\alpha} \right) \right] dt + \frac{i\hbar T}{2m} \right\}$$

$$= \frac{1}{2\pi} \sum_{J=0}^{\infty} \sum_{r,p=\pm 1} \sum_{qk_2} \mathrm{e}^{\mathrm{i}k_2(\varphi''-\varphi')} \mathrm{e}^{-2\mathrm{i}\hbar T J(J+2)/2m}$$

$$\times \psi_{1,Jqk_2}^{(r,p)}(\alpha'';a) \psi_{1,Jqk_2}^{(r,p)\,*}(\alpha';a) \psi_{2,Jqk_2}^{(r,p)}(\beta'';a) \psi_{2,Jqk_2}^{(r,p)\,*}(\beta';a) \;. \tag{6.35}$$

VI. Ellipsoidal: $d < \varrho_3 < c < \varrho_2 < b < \varrho_1 < a$:

$$\int_{\varrho_1(t')=\varrho_1'}^{\varrho_1(t'')=\varrho_1''} \mathcal{D}\varrho_1(t) \int_{\varrho_2(t')=\varrho_2'}^{\varrho_2(t'')=\varrho_2''} \mathcal{D}\varrho_2(t) \int_{\varrho_3(t')=\varrho_3'}^{\varrho_3(t'')=\varrho_3''} \mathcal{D}\varrho_3(t) \frac{(\varrho_2-\varrho_1)(\varrho_3-\varrho_2)(\varrho_3-\varrho_1)}{8\sqrt{P(\varrho_1)P(\varrho_2)P(\varrho_3)}}$$

$$\times \exp\left\{ \frac{\mathrm{i}}{\hbar} \int_{t'}^{t''} \left[\frac{m}{2} \sum_{i=1}^{3} g_{\varrho_i \varrho_i} \dot{\varrho}_i^2 - \Delta V_{PF}(\boldsymbol{\varrho}) \right] dt \right\}$$

$$= \sum_{2s=0}^{\infty} \sum_{\lambda,\mu} \mathrm{e}^{-2\mathrm{i}\hbar T s(s+1)/m} \Psi_{s,\lambda,\mu}^{*}(\varrho_1',\varrho_2',\varrho_3') \Psi_{s,\lambda,\mu}(\varrho_1'',\varrho_2'',\varrho_3'') \;. \tag{6.36}$$

General Form of the Propagator:

$$= \frac{\mathrm{e}^{\mathrm{i}\hbar T/2m}}{4\pi^2} \frac{\mathrm{d}}{\mathrm{d}\cos\psi_{S^{(3)}}} \Theta_3 \left(\frac{\psi_{S^{(3)}}}{2} \middle| -\frac{\hbar T}{2\pi m} \right) \;. \tag{6.37}$$

General Form of the Green's Function:

$$= \int_{\mathbb{R}} \frac{dE}{2\pi} \mathrm{e}^{-\mathrm{i}ET/\hbar} \frac{m}{2\pi\hbar^2} \frac{\sin[(\pi-\psi_{S^{(3)}})(\gamma+1/2)]}{\sin[\pi(\gamma+1/2)]\sin\psi_{S^{(3)}}} \;. \tag{6.38}$$

The quantity $\cos\psi_{S^{(3)}}$ is in spherical coordinates given by ($\gamma = -1/2 + \sqrt{2mE/\hbar^2 + 1}$)

$$\cos\psi_{S^{(3)}}(\mathbf{q}'',\mathbf{q}') = \cos\vartheta_1' \cos\vartheta_1'' + \sin\vartheta_1' \sin\vartheta_1'' \Big(\cos\vartheta_2' \cos\vartheta_2'' + \sin\vartheta_2' \sin\vartheta_2'' \cos(\varphi''-\varphi') \Big) \;. \tag{6.39}$$

Cylindrical, Sphero-Elliptic and Spherical Coordinates.

The path integral solutions (6.28-6.31) follow straightforward from the consideration of the various subgroup path integration, i.e., from the path integration on $S^{(2)}$. They have been discussed in [223, 248] and are obtained by means of the path integral solutions of the Pöschl–Teller path integral and the path integral on the D-dimensional sphere $S^{(D-1)}$ [248]. Equation (6.29) follows from the corresponding case of the two-dimensional sphere.

Oblate- and Prolate Elliptic Coordinates.

The two ellipso-cylindrical systems can be treated by a group path integration together with an interbasis expansion from the cylindrical or spherical basis to the ellipso-cylindrical ones. For instance, one has [240] ($r,p = \pm 1$)

$$\Psi_{J,k_1,k_2}^{(r,p)}(\vartheta,\varphi_1,\varphi_2) = \sum_q T_{Jqk_2}^{(r,p)\,*} \Psi_{Jqk_2}^{(r,p)}(\alpha,\beta,\varphi;a) \;, \tag{6.40}$$

where the $a \geq$ corresponds to the oblate elliptic, and $a \in [-1, 0]$ to the prolate elliptic system, respectively, and $\Psi(\alpha, \beta, \varphi; a) = \psi_1(\alpha; a)\psi_2(\beta; a)e^{ik_2\varphi}/\sqrt{2\pi}$. For the associated Lamé polynomials ψ_{i,Jqk_2}, $i = 1, 2$, we adopt the notations of [240, 333]. The relevant quantum numbers have the following meaning: The functions $\psi_{i,Jqk_2}^{(r,p)}(z)$ are called associated Lamé polynomials and satisfy the normalized associated Lamé equation. For the principal quantum number we have $l \in \mathbb{N}_0$. $(r, p) = \pm 1$ denotes one of the four parity classes of solutions of dimension $(J+1)^2$, i.e., the multiplicity of the degeneracy of the level J, one for each class of the corresponding recurrence relations as given in [240, 333] and the parity classes from the periodic Lamé functions Λ_{lh}^p from the spherical harmonics on the sphere can be applied. These expansions have been considered in [240, 333] together with three-term recurrence relations for the interbasis coefficients. They can be determined by taking into account that a basis in O(4) is related in a unique way to the cylindrical and spherical bases on $S^{(3)}$ by using the properties of the elliptic operator Λ on $S^{(3)}$ with eigenvalue q [240, 333]. Due to the unitarity of these coefficients the path integration is then performed by inserting in each short-time kernel in the cylindrical system first the expansions (6.40), exploiting the unitarity, and thus yielding the results (6.34, 6.35), respectively.

Ellipsoidal Coordinates.

The metric tensor in ellipsoidal coordinates has the form

$$(g_{ab}) = -\frac{1}{4}\text{diag}\left(\frac{(\varrho_1-\varrho_2)(\varrho_1-\varrho_3)}{P(\varrho_1)}, \frac{(\varrho_2-\varrho_3)(\varrho_2-\varrho_1)}{P(\varrho_2)}, \frac{(\varrho_3-\varrho_1)(\varrho_3-\varrho_2)}{P(\varrho_3)}\right), \quad (6.41)$$

and $P(\varrho) = (\varrho - a)(\varrho - b)(\varrho - c)(\varrho - d)$. The corresponding quantum potential ΔV can be constructed from (2.11). In the following we have adopted the notation of Karayan et al. [240]. (An alternative approach is due to Harnad and Winternitz [262, 263] and Kuznetsov et al. [306, 356, 365, 366]. However, there seems to be no obvious way to construct interbasis expansion between the various coordinate systems from this approach.) The quantum numbers are the eigenvalues of the operators which characterize the ellipsoidal system on the sphere, thus giving a complete set of observables for ellipsoidal coordinates on the sphere [374, 333, 375, 356].

Coordinate Systems on the Three-Dimensional Sphere.

In order to illustrate the various coordinate systems, let us discuss and visualize the coordinate systems on the three-dimensional sphere. Each coordinate system is described by a pair of operators, which are actually the angular momentum operators:

$$I_{kl} = s_k\frac{\partial}{\partial s_l} - s_l\frac{\partial}{\partial s_k}, \qquad k, l = 1, 2, 3, 4, \qquad k \neq l, \qquad I_{kl} = -I_{lk} \quad (6.42)$$

which obey the commutation relations

$$[I_{kl}, I_{st}] = \delta ls I_{kt} - \delta ks I_{lt} - \delta lt I_{ks} + \delta kt I_{ls}. \quad (6.43)$$

Table 6.2: Coordinates on the Three-Dimensional Sphere

Coordinate System	Coordinates	Path Integral Solution
I. Cylindrical	$s_1 = \cos\vartheta \cos\varphi_1$ $s_2 = \cos\vartheta \sin\varphi_1$ $s_3 = \sin\vartheta \cos\varphi_2$ $s_4 = \sin\vartheta \sin\varphi_2$	(6.28)
II. Sphero-Elliptic	$s_1 = \sin\chi \mathrm{sn}(\alpha,k)\mathrm{dn}(\beta,k')$ $s_2 = \sin\chi \mathrm{cn}(\alpha,k)\mathrm{cn}(\beta,k')$ $s_3 = \sin\chi \mathrm{dn}(\alpha,k)\mathrm{sn}(\beta,k')$ $s_4 = \cos\chi$	(6.29)
III. Spherical	$s_1 = \sin\chi \sin\vartheta \cos\varphi$ $s_2 = \sin\chi \sin\vartheta \sin\varphi$ $s_3 = \sin\chi \cos\vartheta$ $s_4 = \cos\chi$	(6.30, 6.31)
IV. Oblate Elliptic	$s_1 = \mathrm{sn}(\alpha,k)\mathrm{dn}(\beta,k')\cos\varphi$ $s_2 = \mathrm{sn}(\alpha,k)\mathrm{dn}(\beta,k')\sin\varphi$ $s_3 = \mathrm{cn}(\alpha,k)\mathrm{cn}(\beta,k')$ $s_4 = \mathrm{dn}(\alpha,k)\mathrm{sn}(\beta,k')$	(6.34)
V. Prolate Elliptic	$s_1 = \mathrm{cn}(\alpha,k)\mathrm{cn}(\beta,k')\cos\varphi$ $s_2 = \mathrm{cn}(\alpha,k)\mathrm{cn}(\beta,k')\sin\varphi$ $s_3 = \mathrm{sn}(\alpha,k)\mathrm{dn}(\beta,k')$ $s_4 = \mathrm{dn}(\alpha,k)\mathrm{sn}(\beta,k')$	(6.35)
VI. Ellipsoidal	$s_1^2 = \dfrac{(\varrho_1-d)(\varrho_2-d)(\varrho_3-d)}{(a-d)(b-d)(c-d)}$ $s_2^2 = \dfrac{(\varrho_1-c)(\varrho_2-c)(\varrho_3-c)}{(a-c)(b-c)(d-c)}$ $s_3^2 = \dfrac{(\varrho_1-a)(\varrho_2-a)(\varrho_3-a)}{(d-a)(c-a)(b-a)}$ $s_4^2 = \dfrac{(\varrho_1-b)(\varrho_2-b)(\varrho_3-b)}{(d-b)(c-b)(a-b)}$	(6.36)

Spherical Coordinates:

The spherical systems is the most known system and is defined as

$$
\begin{aligned}
s_1 &= R\sin\chi \sin\theta \cos\phi \ , \quad 0 < \theta, \chi < \pi \ , \\
s_2 &= R\sin\chi \sin\theta \sin\phi \ , \quad 0 \le \phi < 2\pi \ , \\
s_3 &= R\sin\chi \cos\theta \ . \\
s_4 &= R\cos\chi \ .
\end{aligned}
\tag{6.44}
$$

The spherical system is defined by the operators $L_1 = I_{23}^2, L_2 = I_{14}^2$. It is characterized by $\varphi =$ const. which is a family of coaxial cones (axis OY), $\vartheta =$ const. which is a family of planes with axis OY, $\chi =$ const. which is a family of spheres with center at 0.

Sphero-Conical Coordinates:

On the two-dimensional sphere elliptic coordinates are defined in the algebraic form as $(a_1 \leq \rho_1 \leq a_2 \leq \rho_2 \leq a_3)$

$$s_1^2 = R^2 \frac{(\rho_1 - a_1)(\rho_2 - a_1)}{(a_2 - a_1)(a_3 - a_1)}, \quad s_2^2 = R^2 \frac{(\rho_1 - a_2)(\rho_2 - a_2)}{(a_3 - a_2)(a_1 - a_2)}, \quad s_3^2 = R^2 \frac{(\rho_1 - a_3)(\rho_2 - a_3)}{(a_1 - a_3)(a_2 - a_3)} .$$

$$(6.45)$$

If we put $\rho_1 = a_1 + (a_2 - a_1)\text{sn}^2(\mu, k)$ and $\rho_2 = a_2 + (a_3 - a_2)\text{cn}^2(\nu, k')$, where $\text{sn}(\mu, k), \text{cn}(\mu, k)$ and $\text{dn}(\mu, k)$ are the Jacobi elliptic functions with modulus k, we obtain an alternative description of the elliptic coordinates on the sphere

$$\begin{aligned}
s_1 &= R\text{sn}(\mu, k)\text{dn}(\nu, k') , & -K \leq \mu \leq K , \\
s_2 &= R\text{cn}(\mu, k)\text{cn}(\nu, k') , & -2K' \leq \nu \leq 2K' , \\
s_3 &= R\text{dn}(\mu, k)\text{sn}(\nu, k') .
\end{aligned}$$

$$(6.46)$$

The three-dimensional analogue is consequentely:

$$\begin{aligned}
s_1 &= R\sin\chi\,\text{sn}(\mu, k)\text{dn}(\nu, k') , & -K \leq \mu \leq K , \\
s_2 &= R\sin\chi\,\text{cn}(\mu, k)\text{cn}(\nu, k') , & -2K' \leq \nu \leq 2K' , \\
s_3 &= R\sin\chi\,\text{dn}(\mu, k)\text{sn}(\nu, k') , & 0 \leq \chi \leq \pi , \\
s_4 &= R\cos\chi
\end{aligned}$$

$$(6.47)$$

$(0 < \chi < \pi)$. The pair of operators is given by $L_1 = I_{12}^2 + I_{13}^2 + I_{23}^2, L_2 = I_{23}^2 + k^2 I_{13}^2$, with k the modulus of the elliptic functions. The sphero-conical coordinates are characterized by $\varrho_{1,2}$ which are two families of ellipses and hyperbolas that are orthogonal to each other.

Cylindrical Coordinates:

The cylindrical system is defined as

$$\begin{aligned}
s_1 &= \sin\theta\cos\phi_1 , & 0 < \theta < \pi/2 , \\
s_2 &= \sin\theta\sin\phi_1 , & 0 \leq \phi_{1,2} < 2\pi , \\
s_3 &= \cos\theta\cos\phi_2 , \\
s_4 &= \cos\theta\sin\phi_2 .
\end{aligned}$$

$$(6.48)$$

The spherical system is defined by the operators $L_1 = I_{12}^2 + I_{13}^2 + I_{23}^2, L_2 = I_{23}^2$. It is characterized by $\vartheta =$ const. which are surfaces obtained from rotation of a family of Clifford surfaces, $\varphi_1 =$ const. which is a family of planes along the x-axis, and $\varphi_2 =$ const. which is a family of planes perpendicular to the x-axis.

Table 6.3: Description of Coordinate Systems on Spheres

System	Coordinates	Description of System
The Two-Dimensional Sphere:		
I. Spherical	$s_1 = \cos\vartheta$ $s_2 = \sin\vartheta\cos\varphi$ $s_3 = \sin\vartheta\sin\varphi$	$\vartheta = $ const. are concentric circles with centre 0 $\varphi = $ const. are a bunch of stright lines passing through 0
II. Elliptic	$s_1 = \text{sn}(\alpha,k)\text{dn}(\beta,k')$ $s_2 = \text{cn}(\alpha,k)\text{cn}(\beta,k')$ $s_3 = \text{dn}(\alpha,k)\text{sn}(\beta,k')$	Algebraic: $\dfrac{s_1^2}{\varrho_i - b} + \dfrac{s_2^2}{\varrho_i - a} + \dfrac{s_3^2}{\varrho_i - c} = 0$. $\varrho_{1,2}$ are two families of ellipses and hyperbolas that are orthogonal to each other $(c < \varrho_2 < b < \varrho_1 < a)$
The Three-Dimensional Sphere:		
I. Cylindrical	$s_1 = \cos\vartheta\cos\varphi_1$ $s_2 = \cos\vartheta\sin\varphi_1$ $s_3 = \sin\vartheta\cos\varphi_2$ $s_4 = \sin\vartheta\sin\varphi_2$	$\vartheta = $ const. are surfaces obtained from rotation of a family of Clifford surfaces $\varphi_1, \varphi_2 = $ const. is a family of planes along/perpendicular the x-axis
II. Sphero- Elliptic	$s_1 = \sin\chi\text{sn}(\alpha,k)\text{dn}(\beta,k')$ $s_2 = \sin\chi\text{cn}(\alpha,k)\text{cn}(\beta,k')$ $s_3 = \sin\chi\text{dn}(\alpha,k)\text{sn}(\beta,k')$ $s_4 = \cos\chi$	$\chi = $ const. is a family of spheres with center at 0 The lines $\varrho_{1,2}$ are two families of ellipses and hyperbolas that are orthogonal to each other $(c < \varrho_2 < b < \varrho_1 < a)$ (see system II. Elliptic)
III. Spherical	$s_1 = \sin\chi\sin\vartheta\cos\varphi$ $s_2 = \sin\chi\sin\vartheta\sin\varphi$ $s_3 = \sin\chi\cos\vartheta$ $s_4 = \cos\chi$	$\varphi = $ const. is a family of coaxial cones (axis OY) $\vartheta = $ const. is a family of planes with axis OY $\chi = $ const. is a family of spheres with center at 0
IV. Oblate Elliptic	$s_1 = \text{sn}(\alpha,k)\text{dn}(\beta,k')\cos\varphi$ $s_2 = \text{sn}(\alpha,k)\text{dn}(\beta,k')\sin\varphi$ $s_3 = \text{cn}(\alpha,k)\text{cn}(\beta,k')$ $s_4 = \text{dn}(\alpha,k)\text{sn}(\beta,k')$	$\dfrac{s_1^2+s_2^2}{\varrho_i-b} + \dfrac{s_3^2}{\varrho_i-a} + \dfrac{s_4^2}{\varrho_i-c} = 0$. $(c<\varrho_2<b<\varrho_1<a)$ $\varrho_1 = $ const. is a family of confocal oblate ellipsoids $\varrho_2 = $ const. is a family of confocal hyperboloids $\varphi = $ const. is a family of planes with axis OY
V. Prolate Elliptic	$s_1 = \text{cn}(\alpha,k)\text{cn}(\beta,k')\cos\varphi$ $s_2 = \text{cn}(\alpha,k)\text{cn}(\beta,k')\sin\varphi$ $s_3 = \text{sn}(\alpha,k)\text{dn}(\beta,k')$ $s_4 = \text{dn}(\alpha,k)\text{sn}(\beta,k')$	$\dfrac{s_1^2+s_2^2}{\varrho_i-b} + \dfrac{s_3^2}{\varrho_i-a} + \dfrac{s_4^2}{\varrho_i-c} = 0$. $(c<\varrho_2<b<\varrho_1<a)$ $\varrho_1 = $ const. is a family of confocal prolate ellipsoids $\varrho_2 = $ const. is a family of confocal hyperboloids $\varphi = $ const. is a family of planes with axis X
VI. Ellipsoidal	$s_1^2 = \dfrac{(\varrho_1-d)(\varrho_2-d)(\varrho_3-d)}{(a-d)(b-d)(c-d)}$ $s_2^2 = \dfrac{(\varrho_1-c)(\varrho_2-c)(\varrho_3-c)}{(a-c)(b-c)(d-c)}$ $s_3^2 = \dfrac{(\varrho_1-a)(\varrho_2-a)(\varrho_3-a)}{(d-a)(c-a)(b-a)}$ $s_4^2 = \dfrac{(\varrho_1-b)(\varrho_2-b)(\varrho_3-b)}{(d-b)(c-b)(a-b)}$	$(d < \varrho_3 < c < \varrho_2 < b < \varrho_1 < a)$ $\varrho_{1,2,3} = $ const. are three families of confocal ellipsoids

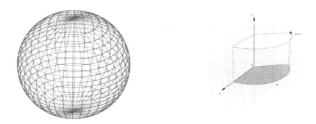

Figure 6.1: (a, b) Wireframe of a (two-dimensional) sphere (left) and the depiction of cylindrical coordinates (right) [521].

In Figure 6.1 the spherical and cylindrical coordinates are illustrated by the wireframe of a sphere and by the superposition of the relevant surfaces for the cylinder, respectively. In Figure 6.2 a double-cone and the superposition of the relevant surfaces of the sphero-conical coordinates are depicted. They are most conveniently visualized by the projection of the terminus of the angular momentum vector which traces out such a cone of elliptic cross section about the z-axis.

Elliptic Cylindrical Coordinates 1 (Oblate Elliptic):

The first elliptical cylindrical coordinate system in algebraic form is given by ($a_1 \leq \rho_1 \leq a_2 \leq \rho_2 \leq a_3$)

$$\left.\begin{aligned}
s_1^2 &= R^2 \frac{(\rho_1 - a_1)(\rho_2 - a_1)}{(a_2 - a_1)(a_3 - a_1)} \cos^2\phi \ , \\
s_2^2 &= R^2 \frac{(\rho_1 - a_1)(\rho_2 - a_1)}{(a_2 - a_1)(a_3 - a_1)} \sin^2\phi \ , \\
s_3^2 &= R^2 \frac{(\rho_1 - a_2)(\rho_2 - a_2)}{(a_3 - a_2)(a_1 - a_2)} \ , \\
s_4^2 &= R^2 \frac{(\rho_1 - a_3)(\rho_2 - a_3)}{(a_1 - a_3)(a_2 - a_3)} \ .
\end{aligned}\right\} \tag{6.49}$$

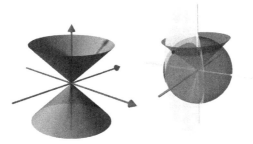

Figure 6.2: (a, b) Depicting a double-cone (left) and a conical coordinate system (right) [521].

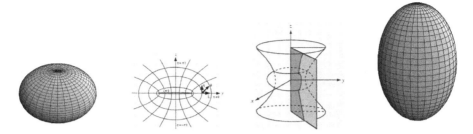

Figure 6.3: (a–d) An oblate spheroid, wireframe of an oblate spheroid, the respective surfaces, and a prolate spheroid (from left to right) [521].

In terms of the Jacobi elliptic functions we have $(-K \leq \mu \leq K, -2K' \leq \nu \leq 2K', 0 \leq \phi < 2\pi)$

$$\left.\begin{array}{l} s_1 = R\operatorname{sn}(\mu, k)\operatorname{dn}(\nu, k')\cos\phi \ , \\ s_2 = R\operatorname{sn}(\mu, k)\operatorname{dn}(\nu, k')\sin\phi \ , \\ s_3 = R\operatorname{cn}(\mu, k)\operatorname{cn}(\nu, k') \ , \\ s_4 = R\operatorname{dn}(\mu, k)\operatorname{sn}(\nu, k') \ . \end{array}\right\} \tag{6.50}$$

For $R \to \infty$ the oblate spheroidal coordinate system in $\mathbb{R}^3$ is recovered. The pair of operators reads $L_1 = I_{12}^2 + I_{13}^2 + k^2 I_{14}^2$, $L_2 = I_{23}^2$. It is characterized by $\varrho_1 = $ const. which is a family of confocal oblate ellipsoids, $\varrho_2 = $ const. which is a family of confocal hyperboloids, and $\varphi = $ const. which is a family of planes with axis OY.

Elliptic Cylindrical Coordinates 2 (Prolate Elliptic):

The second elliptical cylindrical coordinate system in algebraic form is given by $(a_1 \leq \rho_1 \leq a_2 \leq \rho_2 \leq a_3)$

$$\left.\begin{array}{l} s_1^2 = R^2 \dfrac{(\rho_1 - a_2)(\rho_2 - a_2)}{(a_3 - a_2)(a_1 - a_2)}\cos^2\phi \ , \\[2mm] s_2^2 = R^2 \dfrac{(\rho_1 - a_2)(\rho_2 - a_2)}{(a_3 - a_2)(a_1 - a_2)}\sin^2\phi \ , \\[2mm] s_3^2 = R^2 \dfrac{(\rho_1 - a_1)(\rho_2 - a_1)}{(a_2 - a_1)(a_3 - a_1)} \ , \\[2mm] s_4^2 = R^2 \dfrac{(\rho_1 - a_3)(\rho_2 - a_3)}{(a_2 - a_3)(a_1 - a_3)} \ . \end{array}\right\} \tag{6.51}$$

In terms of the Jacobi elliptic functions we have $(-K \leq \mu \leq K, -2K' \leq \nu \leq 2K', 0 \leq \phi < 2\pi)$

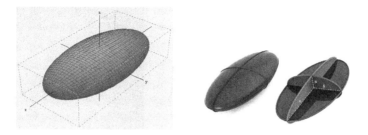

Figure 6.4: An ellipsoid with wireframe. Tri-axial ellipsoid with distinct semi-axes lengths $c > b > a$ [521].

$$\left.\begin{aligned}
s_1 &= R\mathrm{cn}(\mu,k)\mathrm{cn}(\nu,k')\cos\phi ~, \\
s_2 &= R\mathrm{cn}(\mu,k)\mathrm{cn}(\nu,k')\sin\phi ~, \\
s_3 &= R\mathrm{sn}(\mu,k)\mathrm{dn}(\nu,k') ~, \\
s_4 &= R\mathrm{dn}(\mu,k)\mathrm{sn}(\nu,k') ~.
\end{aligned}\right\} \tag{6.52}$$

For $R \to \infty$ the prolate spheroidal coordinate system in $\mathbb{R}^3$ is recovered. The pair of operators reads $L_1 = I_{12}^2 + k^2 I_{13}^2 + I_{14}^2$, $L_2 = I_{23}^2$. It is characterized by $\varrho_1 = $ const. which is a family of confocal prolate ellipsoids, $\varrho_2 = $ const. which is a family of confocal hyperboloids, and $\varphi = $ const. which is a family of planes with axis X.

In Figure 6.3 an oblate an prolate ellipsoide is depicted, including a wireframe of an oblate spheroid.

Ellipsoidal Coordinates:

Let us consider the coordinate system defined by

$$\frac{s_1^2}{\rho_i - a_1} + \frac{s_2^2}{\rho_i - a_2} + \frac{s_3^2}{\rho_i - a_3} + \frac{s_4^2}{\rho_i - a_4} = 0 ~, \qquad (i = 1,2,3) ~,$$
$$\text{and } s_1^2 + s_2^2 + s_3^2 + s_4^2 = R^2 ~. \tag{6.53}$$

Explicitly $(a_1 < \rho_1 < a_2 < \rho_2 < a_3 < \rho_3 < a_4)$:

$$\left.\begin{aligned}
s_1^2 &= R^2 \frac{(\rho_1 - a_1)(\rho_2 - a_1)(\rho_3 - a_1)}{(a_2 - a_1)(a_3 - a_1)(a_4 - a_1)} ~, \\
s_2^2 &= R^2 \frac{(\rho_1 - a_2)(\rho_2 - a_2)(\rho_3 - a_2)}{(a_1 - a_2)(a_3 - a_2)(a_4 - a_2)} ~, \\
s_3^2 &= R^2 \frac{(\rho_1 - a_3)(\rho_2 - a_3)(\rho_3 - a_3)}{(a_1 - a_3)(a_2 - a_3)(a_4 - a_3)} ~, \\
s_4^2 &= R^2 \frac{(\rho_1 - a_4)(\rho_2 - a_4)(\rho_3 - a_4)}{(a_1 - a_4)(a_2 - a_4)(a_3 - a_4)} ~.
\end{aligned}\right\} \tag{6.54}$$

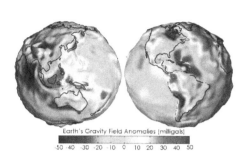

Figure 6.5: (a, b) Earth and Jupiter as ellipsoids. Note that the Earth is only approximated by an ellipsoid, also called goeid, whereas Jupiter is almost a perfect ellipsoid due to its fast rotation period (ca. 10h). [419]

Unfortunately, the ellipsoidal system is a two-parametric system and to write it in terms of elliptic functions is difficult and cumbersome because we need two moduli. The set of commuting operators is given by

$$
\left.
\begin{aligned}
\mathcal{L}_1 &= abI_{12}^2 + aI_{13}^2 + bI_{14}^2 \ , \\
\mathcal{L}_2 &= (a+b)I_{12}^2 + (a+1)I_{13}^2 + (b+1)I_{14}^2 + aI_{32}^2 + bI_{42}^2 + I_{43}^2 \ .
\end{aligned}
\right\} \quad (6.55)
$$

The ellipsoidal system is characterized by $\varrho_{1,2,3} = $ const. which are three families of confocal ellipsoids. In Figure 6.4 an ellipsoid is depicted.

In the table 6.2 we summarize the results on path integration on $S^{(3)}$, including an enumeration of the coordinate systems according to Kalnins et al. [333], Olevski [425], and Winternitz et al. [525].

To close this Chapter I have depicted in figure 6.5 the planets Earth and Jupiter as examples for ellipsoids. In fact, the Earth is only approximated by an ellipsoid and is actually rather distorted due to anomalies in its crust. However, the planet Jupiter is almost a perfect rotational ellipsoid, it consists of gases which smoothly adopt due the fast rotation (ca. 10h) into the perfect shape of an ellipsoid.

Chapter 7

Path Integrals on Hyperboloids

7.1 The Two-Dimensional Pseudosphere

We have the following path integral representations on the pseudosphere $\Lambda^{(2)}$ [223] (including those which have not been treated in [223], compare also [66, 94, 202, 210, 247, 249, 361])

I. Spherical, $\tau > 0$, $\varphi \in [0, 2\pi)$ (Pseudosphere, Poincaré Disc):

$$
\int_{\tau(t')=\tau'}^{\tau(t'')=\tau''} \mathcal{D}\tau(t) \sinh\tau \int_{\varphi(t')=\varphi'}^{\varphi(t'')=\varphi''} \mathcal{D}\varphi(t)
$$

$$
\times \exp\left\{ \frac{i}{\hbar}\int_{t'}^{t''}\left[\frac{m}{2}(\dot\tau^2 + \sinh^2\tau\,\dot\varphi^2) - \frac{\hbar^2}{8m}\left(1 - \frac{1}{\sinh^2\tau}\right)\right]dt \right\}
$$

$$
= \int_0^\infty dp \sum_{\nu\in\mathbb{Z}} \frac{e^{i\nu(\varphi''-\varphi')}}{2\pi^2}\exp\left[-\frac{i\hbar T}{2m}\left(p^2+\frac14\right)\right]
$$

$$
\times p\sinh\pi p\left|\Gamma\left(\frac12 + ip + \nu\right)\right|^2 \mathcal{P}_{ip-\frac12}^{-\nu}(\cosh\tau')\mathcal{P}_{ip-\frac12}^{-\nu}(\cosh\tau'') \ . \tag{7.1}
$$

II. Equidistant, $X \in \mathbb{R}$, $|Y| < \pi/2$ (Hyperbolic Strip):

$$
\int_{X(t')=X'}^{X(t'')=X''} \mathcal{D}X(t) \int_{Y(t')=Y'}^{Y(t'')=Y''} \frac{\mathcal{D}Y(t)}{\cos^2 Y}\exp\left(\frac{im}{2\hbar}\int_{t'}^{t''}\frac{\dot X^2 + \dot Y^2}{\cos^2 Y}dt\right)
$$

$$
= \frac{1}{2\pi}\sqrt{\cos Y'\cos Y''}\int_{\mathbb{R}} dk \int_{\mathbb{R}} \frac{dp\,p\sinh\pi p}{\cosh^2\pi k + \sinh^2\pi p}
$$

$$
\times \exp\left[-\frac{i\hbar T}{2m}\left(p^2+\frac14\right)\right]e^{ik(X''-X')}P_{ik-1/2}^{ip}(\sin Y'')P_{ik-1/2}^{-ip}(\sin Y') \ . \tag{7.2}
$$

III. Horicyclic , $x \in \mathbb{R}$, $y > 0$ (Poincaré Upper Half-Plane):

$$
\int_{x(t')=x'}^{x(t'')=x''} \mathcal{D}x(t) \int_{y(t')=y'}^{y(t'')=y''} \frac{\mathcal{D}y(t)}{y^2} \exp\left(\frac{im}{2\hbar} \int_{t'}^{t''} \frac{\dot{x}^2 + \dot{y}^2}{y^2} dt\right)
$$
$$
= \frac{\sqrt{y'y''}}{\pi^3} \int_{\mathbb{R}} dk\, e^{ik(x''-x')} \int_0^\infty dp\, p \sinh \pi p K_{ip}(|k|y')\, K_{ip}(|k|y'')\, e^{-i\hbar T(p^2+\frac{1}{4})/2m} \quad . \tag{7.3}
$$

IV. Elliptic, $\nu \in [K, K + 2iK']$, $\eta \in [iK', iK' + 2K]$:

$$
\int_{\nu(t')=\nu'}^{\nu(t'')=\nu''} \mathcal{D}\nu(t) \int_{\eta(t')=\eta'}^{\eta(t'')=\eta''} \mathcal{D}\eta(t) k^2 (\text{sn}^2\nu - \text{sn}^2\eta) \exp\left[\frac{im}{2\hbar} k^2 \int_{t'}^{t''} (\text{sn}^2\nu - \text{sn}^2\eta)(\dot{\nu}^2 - \dot{\eta}^2) dt\right]
$$
$$
= \sum_{\lambda=0}^\infty \int_0^\infty dp\, e^{-i\hbar T(p^2+\frac{1}{4})/2m} \Big[\text{Ec}_{ip-1/2}^\lambda(\nu'') \text{Ec}_{ip-1/2}^{\lambda*}(\nu') \text{Ec}_{ip-1/2}^\lambda(\eta'') \text{Ec}_{ip-1/2}^{\lambda*}(\eta')
$$
$$
+ \text{Es}_{ip-1/2}^\lambda(\nu'') \text{Es}_{ip-1/2}^{\lambda*}(\nu') \text{Es}_{ip-1/2}^\lambda(\eta'') \text{Es}_{ip-1/2}^{\lambda*}(\eta') \Big] \quad . \tag{7.4}
$$

V. Hyperbolic, $\alpha \in [iK', iK' + 2K]$, $\beta \in [0, 2iK']$:

$$
\int_{\alpha(t')=\alpha'}^{\alpha(t'')=\alpha''} \mathcal{D}\alpha(t) \int_{\beta(t')=\beta'}^{\beta(t'')=\beta''} \mathcal{D}\beta(t) k^2 (\text{sn}^2\alpha - \text{sn}^2\beta) \exp\left[\frac{im}{2\hbar} k^2 \int_{t'}^{t''} (\text{sn}^2\alpha - \text{sn}^2\beta)(\dot{\alpha}^2 - \dot{\beta}^2) dt\right]
$$
$$
= \sum_{\lambda=0}^\infty \int_0^\infty dp\, e^{-i\hbar T(p^2+\frac{1}{4})/2m} \Big[\text{Fc}_{ip-1/2}^\lambda(\alpha'') \text{Fc}_{ip-1/2}^{\lambda*}(\alpha') \text{Fc}_{ip-1/2}^\lambda(\beta'') \text{Fc}_{ip-1/2}^{\lambda*}(\beta')
$$
$$
+ \text{Fs}_{ip-1/2}^\lambda(\alpha'') \text{Fs}_{ip-1/2}^{\lambda*}(\alpha') \text{Fs}_{ip-1/2}^\lambda(\beta'') \text{Fs}_{ip-1/2}^{\lambda*}(\beta') \Big] \quad . \tag{7.5}
$$

VI. Semi-Hyperbolic, $\mu_{1,2} > 0$, $P(\mu) = \mu(1 + \mu^2)$:

$$
\int_{\mu_1(t')=\mu_1'}^{\mu_1(t'')=\mu_1''} \mathcal{D}\mu_1(t) \int_{\mu_1(t')=\mu_1'}^{\mu_1(t'')=\mu_1''} \mathcal{D}\mu_1(t) \frac{\mu_1 + \mu_2}{4\sqrt{P(\mu_1)P(\mu_2)}}
$$
$$
\times \exp\left\{ \frac{i}{\hbar} \int_{t'}^{t''} \left[\frac{m}{2} \frac{\mu_1+\mu_2}{4} \left(\frac{\dot{\mu}_1^2}{P(\mu_1)} - \frac{\dot{\mu}_2^2}{P(\mu_2)} \right) - \frac{\hbar^2/2m}{\mu_1+\mu_2} \sum_{i=1,2} \left(P''(\mu_i) - \frac{3P'^2(\mu_i)}{4P(\mu_i)} \right) \right] dt \right\}
$$
$$
= \sum_{\lambda=0}^\infty \int_0^\infty dp\, e^{-i\hbar T(p^2+\frac{1}{4})/2m} \Big[K_{ip-1/2}^\lambda(\mu_1'') K_{ip-1/2}^{\lambda*}(\mu_1') K_{ip-1/2}^\lambda(\mu_2'') K_{ip-1/2}^{\lambda*}(\mu_2')
$$
$$
+ M_{ip-1/2}^\lambda(\mu_1'') M_{ip-1/2}^{\lambda*}(\mu_1') M_{ip-1/2}^\lambda(\mu_2'') M_{ip-1/2}^{\lambda*}(\mu_2') \Big] \quad . \tag{7.6}
$$

VII. Elliptic-Parabolic , $a > 0$, $|\vartheta| < \pi/2$:

$$
\int_{a(t')=a'}^{a(t'')=a''} \mathcal{D}a(t) \int_{\vartheta(t')=\vartheta'}^{\vartheta(t'')=\vartheta''} \mathcal{D}\vartheta(t) \frac{\cosh^2 a - \cos^2 \vartheta}{\cos^2 \vartheta \cosh^2 a} \exp\left[\frac{im}{2\hbar} \int_{t'}^{t''} \frac{\cosh^2 a - \cos^2 \vartheta}{\cos^2 \vartheta \cosh^2 a} (\dot{a}^2 + \dot{\vartheta}^2) dt\right]
$$
$$
= \sqrt{\cos \vartheta' \cos \vartheta''} \int_0^\infty dp\, p \sinh \pi p \int_0^\infty \frac{dk\, k \sinh \pi k}{(\cosh^2 \pi k + \sinh^2 \pi p)^2} e^{-i\hbar T(p^2+\frac{1}{4})/2m}
$$

$$\times \sum_{\epsilon,\epsilon'=\pm1} P_{ip-1/2}^{ik}(\epsilon\tanh a'')P_{ip-1/2}^{-ik}(\epsilon\tanh a')P_{ik-1/2}^{ip}(\epsilon'\sin\vartheta'')P_{ik-1/2}^{-ip}(\epsilon'\sin\vartheta'). \quad (7.7)$$

VIII. Hyperbolic-Parabolic, $b > 0$, $0 < \vartheta < \pi$:

$$\int_{b(t')=b'}^{b(t'')=b''} \mathcal{D}b(t) \int_{\vartheta(t')=\vartheta'}^{\vartheta(t'')=\vartheta''} \mathcal{D}\vartheta(t) \frac{\sinh^2 b + \sin^2\vartheta}{\sin^2\vartheta\sinh^2 b} \exp\left[\frac{im}{2\hbar}\int_{t'}^{t''}\frac{\sinh^2 b+\sin^2\vartheta}{\sin^2\vartheta\sinh^2 b}(\dot b^2+\dot\vartheta^2)dt\right]$$

$$= \sqrt{\sin\vartheta'\sin\vartheta''\sinh b'\sinh b''}\int_0^\infty dp\int_0^\infty \frac{dk\,k\sinh\pi k}{\cosh^2\pi k+\sinh^2\pi p}\frac{e^{-i\hbar T(p^2+\frac14)/2m}}{\cosh\pi(p-k)}$$

$$\times \sum_{\epsilon=\pm1} P_{ik-1/2}^{ip}(\epsilon\cos\vartheta'')P_{ik-1/2}^{-ip}(\epsilon\cos\vartheta')P_{ik-1/2}^{ip}(\cosh b'')P_{ik-1/2}^{-ip}(\cosh b') . \quad (7.8)$$

IX. Semi-Circular-Parabolic, $\xi,\eta > 0$:

$$\int_{\xi(t')=\xi'}^{\xi(t'')=\xi''} \mathcal{D}\xi(t) \int_{\eta(t')=\eta'}^{\eta(t'')=\eta''} \mathcal{D}\eta(t) \frac{\xi^2+\eta^2}{\xi^2\eta^2}\exp\left(\frac{im}{2\hbar}\int_{t'}^{t''}\frac{\xi^2+\eta^2}{\xi^2\eta^2}(\dot\xi^2+\dot\eta^2)dt\right)$$

$$= \frac{\sqrt{\xi'\xi''\eta'\eta''}}{4\pi^2}\int_0^\infty kdk\int_0^\infty dp\,p\sinh^2\pi p\exp\left[-\frac{i\hbar T}{2m}\left(p^2+\frac14\right)\right]$$

$$\times\left[H_{-ip}^{(1)}(k\eta')H_{ip}^{(1)}(k\eta'')K_{ip}(k\xi')K_{ip}(k\xi'')+K_{ip}(k\eta')K_{ip}(k\eta'')H_{-ip}^{(1)}(k\xi')H_{ip}^{(1)}(k\xi'')\right]. \quad (7.9)$$

General Expression for the Propagator ($\cosh d$ = invariant distance on $\Lambda^{(2)}$ [269]):

$$= \frac{1}{2\pi}\int_0^\infty pdp\tanh\pi p\,\mathcal{P}_{ip-1/2}\Big(\cosh d_{\Lambda^{(2)}}(u'',u')\Big)e^{-i\hbar T(p^2+\frac14)/2m} . \quad (7.10)$$

General Form for the Green's Function:

$$= \int_{\mathrm{IR}}\frac{dE}{2\pi i}e^{-iET/\hbar}\frac{m}{\pi\hbar^2}\mathcal{Q}_{-1/2-i\sqrt{2mE/\hbar^2-\frac14}}\Big(\cosh d_{\Lambda^{(2)}}(u'',u')\Big) . \quad (7.11)$$

Spherical, Equidistant and Horicyclic Coordinates.

These three coordinate systems have been discussed thoroughly in [66] (polar system), [202, 223] (all three systems, note the different coordinate realizations), and [247, 361] (Poincaré upper half-plane). Magnetic fields have been taken into account in [197, 200], and the Kepler problem was treated in [46, 201]. This will not be repeated here. $\epsilon = \pm1$ denotes that both contributions (linearly independent solutions) must be taken into account.

Elliptic, Hyperbolic and Semi-Hyperbolic Coordinates.

For (7.4) we can argue in a similar way as for (6.10). We consider the interbasis expansions with $|p\nu>$ the spherical basis

$$|p\nu> = \sum_{\lambda_{IV}} X_{\lambda_{IV}}^{\nu *}|p\lambda_{IV}> \quad (7.12)$$

$$= \sum_{\lambda_V} Y_{\lambda_V}^{\nu *} |p\lambda_V> \tag{7.13}$$

$$= \sum_{\lambda_{VI}} Z_{\lambda_{VI}}^{\nu *} |p\lambda_{VI}> \tag{7.14}$$

where the expansion coefficients $X_\lambda^\nu, Y_\lambda^\nu, Z_\lambda^l$ satisfy three-term recurrence relations similarly as in (6.11) for the elliptic system on $S^{(2)}$. We obtain the following expansions of the short-time kernel on $\Lambda^{(2)}$ for the elliptic system

$$\frac{m}{2\pi i \epsilon \hbar} \exp\left[-\frac{i}{\hbar} \frac{m}{\epsilon} \left(1 - \cosh d_{\Lambda^{(2)}}^{j,j-1} \right) \right]$$

$$= \frac{1}{2\pi} \int_0^\infty dp\, p \tanh \pi p P_{ip-1/2}(\cosh d_{\Lambda^{(2)}}^{j,j-1})\, e^{-i\epsilon \hbar(p^2 + \frac{1}{4})/2m}$$

$$= \int_0^\infty dp \sum_{\nu \in \mathbb{Z}} \frac{e^{i\nu(\varphi'' - \varphi')}}{2\pi^2} p \sinh \pi p |\Gamma(\tfrac{1}{2} + ip + \nu)|^2$$

$$\times e^{-i\epsilon \hbar(p^2 + \frac{1}{4})/2m} \mathcal{P}_{ip-\frac{1}{2}}^{-\nu}(\cosh \tau') \mathcal{P}_{ip-\frac{1}{2}}^{-\nu}(\cosh \tau'')$$

$$= \sum_{\lambda=0}^\infty \int_0^\infty dp\, e^{-i\epsilon \hbar(p^2 + \frac{1}{4})/2m} \Big[\mathrm{Ec}_{ip-1/2}^\lambda(\nu'') \mathrm{Ec}_{ip-1/2}^{\lambda *}(\nu') \mathrm{Ec}_{ip-1/2}^\lambda(\eta'') \mathrm{Ec}_{ip-1/2}^{\lambda *}(\eta')$$

$$+ \mathrm{Es}_{ip-1/2}^\lambda(\nu'') \mathrm{Es}_{ip-1/2}^{\lambda *}(\nu') \mathrm{Es}_{ip-1/2}^\lambda(\eta'') \mathrm{Es}_{ip-1/2}^{\lambda *}(\eta') \Big] . \tag{7.15}$$

The two other systems are treated in an analogous way. The functions $\mathrm{Ec}_{ip-1/2}^n, \mathrm{Es}_{ip-1/2}^n$ are periodic Lamé-functions. Similarly as for the sphere $S^{(2)}$ we have four classes of functions [309] $(n, \in \mathbb{N}_0, l = ip - \frac{1}{2})$

Elliptic Function	Period
$\mathrm{Ec}_l^{2n}(z; k^2)$	$2K$
$\mathrm{Ec}_l^{2n+1}(z; k^2)$	$4K$
$\mathrm{Es}_l^{2n+1}(z; k^2)$	$2K$
$\mathrm{Es}_l^{2n+2}(z; k^2)$	$4K$

The functions $\mathrm{Fc}_{ip-1/2}^n, \mathrm{Fs}_{ip-1/2}^n$, and $K_{ip-1/2}^n$ and $W_{ip-1/2}^n$ are Lamé-Wangerian functions. All these functions have been normalized according to

$$k^2 \int_K^{K+2iK'} d\nu \int_{iK}^{iK+2K'} d\eta (\mathrm{sn}^2 \nu - \mathrm{sn}^2 \eta)$$

$$\times \left(\begin{array}{c} \mathrm{Ec}_{ip-1/2}^n(\nu) \mathrm{Ec}_{ip'-1/2}^{n' *}(\nu) \mathrm{Ec}_{ip-1/2}^n(\eta) \mathrm{Ec}_{ip'-1/2}^{n' *}(\eta) \\ \mathrm{Es}_{ip-1/2}^n(\nu) \mathrm{Es}_{ip'-1/2}^{n' *}(\nu) \mathrm{Es}_{ip-1/2}^n(\eta) \mathrm{Es}_{ip'-1/2'}^{n' *}(\eta) \end{array} \right) = \delta_{nn'} \delta(p' - p) , \tag{7.16}$$

$$k^2 \int_{iK'}^{iK'+2K} d\alpha \int_{-iK'}^{iK'} d\beta (\mathrm{sn}^2 \alpha - \mathrm{sn}^2 \beta)$$

$$\times \left(\begin{array}{c} \mathrm{Fc}_{ip-1/2}^n(\alpha) \mathrm{Fc}_{ip'-1/2}^{n' *}(\alpha) \mathrm{Fc}_{ip-1/2}^n(\beta) \mathrm{Fc}_{ip'-1/2}^{n' *}(\beta) \\ \mathrm{Fs}_{ip-1/2}^n(\alpha) \mathrm{Fs}_{ip'-1/2}^{n' *}(\alpha) \mathrm{Fs}_{ip-1/2}^n(\beta) \mathrm{Fs}_{ip'-1/2'}^{n' *}(\beta) \end{array} \right) = \delta_{nn'} \delta(p' - p) , \tag{7.17}$$

$$\frac{1}{4}\int_0^\infty d\mu_1 \int_0^\infty d\mu_2 \frac{\mu_1 + \mu_2}{\sqrt{P(\mu_1)P(\mu_2)}}$$

$$\times \left(\begin{array}{c} K^n_{ip-1/2}(\mu_1)K^{n'\,*}_{ip'-1/2}(\mu_1)K^n_{ip-1/2}(\mu_2)K^{n'\,*}_{ip'-1/2}(\mu_2) \\ W^n_{ip-1/2}(\mu_1)W^{n'\,*}_{ip'-1/2}(\mu_1)W^n_{ip-1/2}(\mu_2)W^{n'\,*}_{ip'-1/2'}(\mu_2) \end{array} \right) = \delta_{nn'}\delta(p'-p) \; . \quad (7.18)$$

The former functions are analytic continuations in the parameter $l \mapsto ip - 1/2$ of the corresponding elliptical wavefunctions on the sphere $S^{(2)}$ (we have adopted the nomenclature of [153, p.74] and of Kalnins and Miller [309]). The numbers k, k' are, as usual, the moduli of the elliptic functions with $k^2 + k'^2 = 1$, and the elliptic functions have in all cases the same modulus, i.e., $\mathrm{sn}\nu = \mathrm{sn}(\nu, k), \mathrm{sn}\eta = \mathrm{sn}(\eta, k)$, etc. $\cosh d_{\Lambda^{(2)}}$ denotes the invariant distance on $\Lambda^{(2)}$ which is in the spherical system given by

$$\cosh d_{\Lambda^{(2)}}(\mathrm{u}'', \mathrm{u}') = \cosh \tau'' \cosh \tau' - \sinh \tau'' \sinh \tau' \cos(\varphi'' - \varphi') \; . \quad (7.19)$$

In the semi-hyperbolic system we have abbreviated $P(\mu) = \mu(1 + \mu^2)$. Note that we have used in (7.6) the midpoint prescription [248] which gives in the present case a simpler quantum potential as for the product ordering.

Elliptic-, Hyperbolic- and Semi-Circular Parabolic Coordinates.

Let us add some remarks concerning the solutions of the elliptic-parabolic, hyperbolic-parabolic and semi-circular-parabolic coordinate systems. In all three one performs a two-dimensional time transformation yielding potential problems in terms of a symmetric Pöschl–Teller ($\propto 1/\cos^2 \vartheta$) and a symmetric Rosen–Morse potential ($\propto 1/\cosh^2 a$), a symmetric Pöschl–Teller ($\propto 1/\sin^2 \vartheta$) and a hyperbolic centrifugal potential ($\propto 1/\sinh^2 b$), and $\propto 1/\xi^2$, $\propto 1/\eta^2$ potentials, respectively, with the energy E as the coupling. We obtain for instance in the elliptic-parabolic system

$$\int_{a(t')=a'}^{a(t'')=a''} \mathcal{D}a(t) \int_{\vartheta(t')=\vartheta'}^{\vartheta(t'')=\vartheta''} \mathcal{D}\vartheta(t) \frac{\cosh^2 a - \cos^2 \vartheta}{\cos^2 \vartheta \cosh^2 a} \exp\left(\frac{im}{2\hbar}\int_{t'}^{t''} \frac{\cosh^2 a - \cos^2 \vartheta}{\cos^2 \vartheta \cosh^2 a}(\dot{a}^2 + \dot{\vartheta}^2)dt \right)$$

$$= \int_{\mathbb{R}} \frac{dE}{2\pi\hbar} e^{-iET/\hbar} \int_0^\infty ds'' \int_{a(0)=a'}^{a(s'')=a''} \mathcal{D}a(s) \int_{\vartheta(0)=\vartheta'}^{\vartheta(s'')=\vartheta''} \mathcal{D}\vartheta(s)$$

$$\times \exp\left\{ \frac{i}{\hbar}\int_0^{s''} \left[\frac{m}{2}(\dot{a}^2 + \dot{\vartheta}^2) + \frac{E}{\cos^2 \vartheta} - \frac{E}{\cosh^2 a}\right]ds \right\}$$

$$= \int_{\mathbb{R}} \frac{dE}{2\pi\hbar} e^{-iET/\hbar} \int_0^\infty ds'' \int_{\mathbb{R}} \frac{dE'}{2\pi i} e^{-iE's''/\hbar}$$

$$\times \frac{m}{\hbar^2}\sqrt{\cos\vartheta'\cos\vartheta''}\,\Gamma(\lambda - L_{E'})\Gamma(L_{E'} - \lambda + 1)P^{-\lambda}_{L_{E'}}(-\sin\vartheta_<)P^{-\lambda}_{L_{E'}}(\sin\vartheta_>)$$

$$\times \sum_{\epsilon=\pm1}\int_0^\infty \frac{dk\, k \sinh \pi k}{\cos^2 \pi\lambda + \sinh^2 \pi k} P^{ik}_{\lambda-1/2}(\epsilon \tanh a'')P^{-ik}_{\lambda-1/2}(\epsilon \tanh a')\, e^{-i\hbar k^2 s''/2m} \; , \quad (7.20)$$

with $\lambda = \sqrt{1/4 - 2mE/\hbar^2}$, $L_E = -\frac{1}{2} + \sqrt{2mE}/\hbar$. Here we have written the kernel $K(s'')$ as

$$
\begin{aligned}
&K(a'', a', \vartheta'', \vartheta'; s'') \\
&= K_a(a'', a'; s'') \cdot K_\vartheta(\vartheta'', \vartheta'; s'') \\
&= \frac{1}{2} K_a(a'', a'; s'') \cdot \int_{\mathbb{R}} \frac{dE'}{2\pi i} e^{-iE's''/\hbar} G_\vartheta(\vartheta'', \vartheta'; E') \\
&\quad + \frac{1}{2} K_\vartheta(\vartheta'', \vartheta'; s'') \cdot \int_{\mathbb{R}} \frac{dE'}{2\pi i} e^{-iE's''/\hbar} G_a(a'', a'; E') \qquad (7.21)
\end{aligned}
$$

and, of course, both contributions must be taken into account. In the present case, however, both terms turn out to be equivalent. The Green's function expression is evaluated by means of the relation [353], [385, p.170]

$$
\begin{aligned}
P_\nu^{-\mu}(-y) &= \frac{\Gamma(\nu - \mu + 1)}{\Gamma(\nu + \mu + 1)} \left[P_\nu^\mu(-y) \cos \pi\mu - \frac{2}{\pi} Q_\nu^\mu(-y) \sin \pi\mu \right] \\
&= \frac{\Gamma(\nu - \mu + 1)}{\Gamma(\nu + \mu + 1)} \frac{\sin \pi\mu P_\nu^\mu(y) + \sin \pi\nu P_\nu^\mu(-y)}{\sin \pi(\nu + \mu)} . \qquad (7.22)
\end{aligned}
$$

Thus we obtain

$$
\begin{aligned}
\Psi_{kp}(\vartheta'') \Psi_{kp}^*(\vartheta') &\propto \frac{1}{i\pi} \Big[\Gamma(\tfrac{1}{2} + ik + ip) \Gamma(\tfrac{1}{2} - ik - ip) P_{ik-1/2}^{ip}(-\sin \vartheta'') P_{ik-1/2}^{ip}(\sin \vartheta') \\
&\quad - \Gamma(\tfrac{1}{2} + ik - ip) \Gamma(\tfrac{1}{2} - ik + ip) P_{ik-1/2}^{-ip}(-\sin \vartheta'') P_{ik-1/2}^{-ip}(\sin \vartheta') \Big] \\
&= \frac{p \sinh \pi p}{\cosh^2 \pi k + \sinh^2 \pi p} \sum_{\epsilon = \pm 1} P_{ik-1/2}^{ip}(\epsilon \sin \vartheta'') P_{ik-1/2}^{-ip}(\epsilon \sin \vartheta') . \qquad (7.23)
\end{aligned}
$$

Using the same method of Green's function analysis as in the Chapter 4, c.f. (4.37), we apply in the elliptic parabolic case the path integral solutions for the symmetric Pöschl–Teller and symmetric Rosen–Morse potential, and in the hyperbolic parabolic case the solution of the symmetric Pöschl–Teller and the hyperbolic centrifugal potential, respectively.

In the semi-circular parabolic system one uses on the one hand the path integral solution of the radial potential which gives together with the integral [187, p.709]

$$
\int_0^\infty e^{-ax} J_\nu(\beta x) J_\nu(\gamma x) dx = \frac{1}{\pi \sqrt{\beta\gamma}} Q_{\nu-1/2}\left(\frac{a^2 + \beta^2 + \gamma^2}{2\beta\gamma} \right) , \qquad (7.24)
$$

and the invariant distance

$$
\cosh d_{\Lambda^{(2)}}(u'', u') = \frac{(\xi'^2 + \eta'^2)^2 + (\xi''^2 + \eta''^2)^2 - 2(\xi'^2 - \eta'^2)(\xi''^2 - \eta''^2)}{8\xi'\xi''\eta'\eta''} , \qquad (7.25)
$$

explicitly the Green's function on $\Lambda^{(2)}$. On the other hand, we obtain by using the path integral solution for the radial $1/r^2$-potential ($\lambda = \sqrt{1/4 - 2mE/\hbar}$)

$$
\int\limits_{\xi(t')=\xi'}^{\xi(t'')=\xi''} \mathcal{D}\xi(t) \int\limits_{\eta(t')=\eta'}^{\eta(t'')=\eta''} \mathcal{D}\eta(t) \frac{\xi^2 + \eta^2}{\xi^2\eta^2} \exp\left(\frac{im}{2\hbar} \int_{t'}^{t''} \frac{\xi^2 + \eta^2}{\xi^2\eta^2} (\dot\xi^2 + \dot\eta^2) dt \right)
$$

$$
= \int_{\mathbb{R}} \frac{dE}{2\pi\hbar} e^{-iET/\hbar} \int_0^\infty ds'' \int_{\xi(0)=\xi'}^{\xi(s'')=\xi''} \mathcal{D}\xi(s) \int_{\eta(0)=\eta'}^{\eta(s'')=\eta''} \mathcal{D}\eta(s)
$$

$$
\times \exp\left\{ \frac{i}{\hbar} \int_0^{s''} \left[\frac{m}{2}(\dot\xi^2 + \dot\eta^2) - \hbar^2 \frac{\lambda^2 - \frac14}{2m}\left(\frac{1}{\xi^2} + \frac{1}{\eta^2}\right) \right] ds \right\}
$$

$$
= \frac{m^2}{i\hbar^3} \sqrt{\xi'\xi''\eta'\eta''} \int_{\mathbb{R}} \frac{dE}{2\pi\hbar} e^{-iET/\hbar} \int_0^\infty \frac{ds''}{s''} \int_{\mathbb{R}} \frac{dE'}{2\pi i} e^{-iE's''/\hbar}
$$

$$
\times I_\lambda\left(\sqrt{-2mE'}\,\frac{\xi_<}{\hbar}\right) K_\lambda\left(\sqrt{-2mE'}\,\frac{\xi_>}{\hbar}\right) \exp\left[-\frac{m}{2i\hbar s''}(\eta'^2 + \eta''^2) \right] I_\lambda\left(\frac{m\eta'\eta''}{i\hbar s''}\right)
$$

$$
= \frac{1}{2\pi^2} \sqrt{\xi'\xi''\eta'\eta''} \int_{\mathbb{R}} \frac{dE}{2\pi i} e^{-iET/\hbar} \int_0^\infty \frac{k\,dk}{\pi i} I_\lambda(-ik\xi_<) K_\lambda(-ik\xi_>)
$$

$$
\times \int_0^\infty \frac{dp\,p\sinh\pi p}{\frac{\hbar^2}{2m}(p^2 + \frac14) - E} K_{ip}(k\eta') K_{ip}(k\eta'') + (\xi \leftrightarrow \eta) \ . \tag{7.26}
$$

Here, for the η-dependent part one has used the dispersion relation [197]

$$
I_\lambda(z) = \frac{\hbar^2}{\pi^2 m} \int_0^\infty \frac{dp\,p\sinh\pi p}{\frac{\hbar^2}{2m}(p^2 + \frac14) - E} K_{ip}(z) \ , \tag{7.27}
$$

together with the integral representation [187, p.725]

$$
\int_0^\infty e^{-x/2 - (z^2 + w^2)/2x} K_\nu\left(\frac{zw}{x}\right) \frac{dx}{x} = 2K_\nu(z) K_\nu(w) \ . \tag{7.28}
$$

In order to analyze the ξ-dependent part, we first rewrite the Green's function according to

$$
I_\lambda(-ik\xi_<) K_\lambda(-ik\xi_>) = \frac{i\pi}{2} J_\lambda(k\xi_<) H_\lambda^{(1)}(k\xi_>) \ . \tag{7.29}
$$

The wavefunctions on the cut are then obtained by using

$$
\Psi_{k,p}(\eta'') \Psi_{k,p}^*(\eta') \ \propto \ \left[J_{-ip}(k\xi'') H_{-ip}^{(1)}(k\xi') - J_{ip}(k\xi'') H_{ip}^{(1)}(k\xi') \right]
$$

$$
= \ \sinh\pi p\, H_{ip}^{(1)}(k\xi'') H_{-ip}^{(1)}(k\xi') \tag{7.30}
$$

$(\lambda = -ip)$ and the definition of the Hankel-function, i.e., $H_\nu^{(1)}(z) = i[e^{-i\nu\pi} J_\nu(z) - J_{-\nu}(z)]/\sin\pi\nu$. Interchanging ξ and η and adding both contributions then gives the path integral representation (7.9).

Coordinate Systems on the Two-Dimensional Pseudosphere.

In table 7.1 we summarize the results on path integration on $\Lambda^{(2)}$, including an enumeration of the coordinate systems according to [309, 318, 425, 525]. The rotated elliptic system is given by [243]

Table 7.1: Coordinates on the Two-Dimensional Pseudosphere

Coordinate System	Coordinates	Path Integral Solution
I. Spherical	$u_0 = \cosh \tau$ $u_1 = \sinh \tau \cos \varphi$ $u_2 = \sinh \tau \sin \varphi$	(7.1)
II. Equidistant	$u_0 = \cosh \tau_1 \cosh \tau_2$ $u_1 = \cosh \tau_1 \sinh \tau_2$ $u_2 = \sinh \tau_1$	(7.2)
III. Horicyclic	$u_0 = (y^2 + x^2 + 1)/2y$ $u_1 = (y^2 + x^2 - 1)/2y$ $u_2 = x/y$	(7.3)
IV. Elliptic	$u_0 = k\,\mathrm{sn}(\nu, k)\mathrm{sn}(\eta, k)$ $u_1 = (k/k')\mathrm{cn}(\nu, k)\mathrm{cn}(\eta, k)$ $u_2 = (\mathrm{i}/k')\mathrm{dn}(\nu, k)\mathrm{dn}(\eta, k)$	(7.4)
V. Hyperbolic	$u_0 = (\mathrm{i}k/k')\mathrm{cn}(\alpha, k)\mathrm{cn}(\beta, k)$ $u_1 = (\mathrm{i}/k')\mathrm{dn}(\alpha, k)\mathrm{dn}(\beta, k)$ $u_2 = \mathrm{i}k\,\mathrm{sn}(\alpha, k)\mathrm{sn}(\beta, k)$	(7.5)
VI. Semi-Hyperbolic	$u_0 = \frac{1}{\sqrt{2}}(\sqrt{(1+\mu_1^2)(1+\mu_2^2)}+\mu_1\mu_2+1)^{1/2}$ $u_1 = \frac{1}{\sqrt{2}}(\sqrt{(1+\mu_1^2)(1+\mu_2^2)}-\mu_1\mu_2-1)^{1/2}$ $u_2 = \sqrt{\mu_1\mu_2}$	(7.6)
VII. Elliptic-Parabolic	$u_0 = \dfrac{\cosh^2 a + \cos^2 \vartheta}{2\cosh a \cos \vartheta}$ $u_1 = \dfrac{\sinh^2 a - \sin^2 \vartheta}{2\cosh a \cos \vartheta}$ $u_2 = \tan \vartheta \tanh a$	(7.7)
VIII. Hyperbolic-Parabolic	$u_0 = \dfrac{\cosh^2 b + \cos^2 \vartheta}{2\sinh b \sin \vartheta}$ $u_1 = \dfrac{\sinh^2 b - \sin^2 \vartheta}{2\sinh b \sin \vartheta}$ $u_2 = \cot \vartheta \coth b$	(7.8)
IX. Semi-Circular-Parabolic	$u_0 = \dfrac{(\xi^2 + \eta^2)^2 + 4}{8\xi\eta}$ $u_1 = \dfrac{(\xi^2 + \eta^2)^2 - 4}{8\xi\eta}$ $u_2 = \dfrac{\eta^2 - \xi^2}{2\xi\eta}$	(7.9)

$$u_0 = \frac{1}{a_2 - a_3}\left(\sqrt{(\varrho_1 - a_3)(\varrho_2 - a_3)} + \sqrt{(\varrho_1 - a_2)(\varrho_2 - a_2)}\right)$$

$$= \mathrm{sn}\nu\mathrm{sn}\eta + \mathrm{cn}\nu\mathrm{cn}\eta\ , \tag{7.31}$$

$$u_1 = \frac{1}{a_2 - a_3}\left(\sqrt{\frac{a_1 - a_2}{a_1 - a_3}(\varrho_1 - a_3)(\varrho_2 - a_3)} + \sqrt{\frac{a_1 - a_3}{a_1 - a_2}(\varrho_1 - a_2)(\varrho_2 - a_2)}\right)$$

$$= k'\mathrm{sn}\nu\mathrm{sn}\eta + \frac{1}{k'}\mathrm{cn}\nu\mathrm{cn}\eta\ , \tag{7.32}$$

$$u_2 = \sqrt{\frac{(\varrho_1 - a_1)(a_1 - \varrho_2)}{(a_1 - a_2)(a_1 - a_3)}} = \frac{\mathrm{i}}{k'}\mathrm{dn}\nu\mathrm{dn}\eta\ . \tag{7.33}$$

In the notation we have chosen $a_1 = 0, a_2 = -k'^2, a_3 = k^2$ and have rewritten u in terms of the Jacobi elliptic functions, $\varrho_1 = k^2\mathrm{cn}^2(\nu, k), \varrho_2 = k^2\mathrm{cn}^2(\eta, k)$.

7.2 The Three-Dimensional Pseudosphere

We have the following path integral representations on the pseudosphere $\Lambda^{(3)}$ [223] (including those which have not been treated in [223], compare also [66, 202, 210, 247, 249, 361])

I. Cylindrical, $\tau_{1,2} > 0, \varphi \in [0, 2\pi)$:

$$\int_{\tau_1(t')=\tau_1'}^{\tau_1(t'')=\tau_1''} \mathcal{D}\tau_1(t)\cosh\tau_1\sinh\tau_1 \int_{\tau_2(t')=\tau_2'}^{\tau_2(t'')=\tau_2''} \mathcal{D}\tau_2(t) \int_{\varphi(t')=\varphi'}^{\varphi(t'')=\varphi''} \mathcal{D}\varphi(t)$$

$$\times \exp\left\{\frac{\mathrm{i}}{\hbar}\int_{t'}^{t''}\left[\frac{m}{2}\left(\dot{\tau}_1^2 + \cosh^2\tau_1\dot{\tau}_2^2 + \sinh^2\tau_1\dot{\varphi}^2\right)\right.\right.$$

$$\left.\left. - \frac{\hbar^2}{8m}\left(4 + \frac{1}{\cosh^2\tau_1} - \frac{1}{\sinh^2\tau_1}\right)\right]dt\right\}$$

$$= \left(\sinh\tau_1'\sinh\tau_1''\cosh\tau_1'\cosh\tau_1''\right)^{-1/2}\sum_{\nu\in\mathbb{Z}}\frac{\mathrm{e}^{\mathrm{i}\nu(\varphi''-\varphi')}}{2\pi}\int_{\mathbb{R}}\frac{dp_1}{2\pi}\mathrm{e}^{\mathrm{i}p_1(\tau_2''-\tau_2')}$$

$$\times \int_0^\infty dp\,\exp\left[-\frac{\mathrm{i}\hbar T}{2m}(p^2 + 1)\right]\Psi_p^{(p_1,\nu)*}(\tau_1')\Psi_p^{(p_1,l)}(\tau_1'')\ , \tag{7.34}$$

$$\Psi_p^{(p_1,\nu)}(\tau_1) = N_p^{(p_1,\nu)}(\cosh\tau_1)^{\mathrm{i}p_1}(\tanh\tau_1)^{|\nu|-1/2}$$

$$\times {}_2F_1\left[\frac{1}{2}(|\nu| + \mathrm{i}p_1 - \mathrm{i}p + 1), \frac{1}{2}(|\nu| - \mathrm{i}p_1 - \mathrm{i}p - 1); |\nu| + 1; \tanh^2\tau_1\right]\ , \tag{7.35}$$

$$N_p^{(p_1,\nu)} = \frac{1}{|\nu|!}\sqrt{\frac{p\sinh\pi p}{2\pi^2}}\Gamma\left(\frac{1 + |\nu| + \mathrm{i}p_1 + \mathrm{i}p}{2}\right)\Gamma\left(\frac{1 + |\nu| + \mathrm{i}p_1 - \mathrm{i}p}{2}\right)\ . \tag{7.36}$$

II. Horicyclic, $y > 0, \mathbf{x} \in \mathbb{R}^2$:

$$\int_{y(t')=y'}^{y(t'')=y''} \frac{\mathcal{D}y(t)}{y^3} \int_{\mathbf{x}(t')=\mathbf{x}'}^{\mathbf{x}(t'')=\mathbf{x}''} \mathcal{D}\mathbf{x}(t) \exp\left[\frac{\mathrm{i}}{\hbar}\int_{t'}^{t''}\left(\frac{m}{2}\frac{\dot{\mathbf{x}}^2+\dot{y}^2}{y^2}-\frac{3\hbar^2}{8m}\right)dt\right]$$

$$= \int_{\mathbb{R}^2}\frac{d\mathbf{k}}{(2\pi)^2}\mathrm{e}^{\mathrm{i}\mathbf{k}(\mathbf{x}''-\mathbf{x}'')}\frac{2y'y''}{\pi^2}$$

$$\times \int_0^\infty dp\,p\sinh\pi p\exp\left[-\frac{\mathrm{i}\hbar T}{2m}(p^2+1)\right]K_{\mathrm{i}p}(|\mathbf{k}|y')K_{\mathrm{i}p}(|\mathbf{k}|y'') \ . \tag{7.37}$$

III. Sphero-Elliptic, $\tau > 0, \alpha \in [-K, K], \beta \in [-2K', 2K']$:

$$\int_{\tau(t')=\tau'}^{\tau(t'')=\tau''} \mathcal{D}\tau(t)\sinh^2\tau \int_{\alpha(t')=\alpha'}^{\alpha(t'')=\alpha''} \mathcal{D}\alpha(t) \int_{\beta(t')=\beta'}^{\beta(t'')=\beta''} \mathcal{D}\beta(t)(k^2\mathrm{cn}^2\alpha + k'^2\mathrm{cn}^2\beta)$$

$$\times \exp\left\{\frac{\mathrm{i}}{\hbar}\int_{t'}^{t''}\left[\frac{m}{2}\left(\dot{\tau}^2 + \sinh^2\tau(k^2\mathrm{cn}^2\alpha + k'^2\mathrm{cn}^2\beta)(\dot{\alpha}^2+\dot{\beta}^2)\right) - \frac{\hbar^2}{2m}\right]dt\right\}$$

$$= (\sinh\tau'\sinh\tau'')^{-1/2}\sum_{l=0}^\infty\sum_\lambda\sum_{p,q=\pm}\Lambda_{l,h}^p(\alpha'')\Lambda_{l,h}^{p*}(\alpha')\Lambda_{l,\tilde{h}}^q(\beta'')\Lambda_{l,\tilde{h}}^{q*}(\beta')\int_0^\infty dp\,\mathrm{e}^{-\mathrm{i}\hbar T(p^2+1)/2m}$$

$$\times\frac{p\sinh\pi p}{\pi}|\Gamma(\mathrm{i}p+k+1/2)|^2\mathcal{P}_{\mathrm{i}p-1/2}^{-l-1/2}(\cosh\tau')\mathcal{P}_{\mathrm{i}p-1/2}^{-l-1/2}(\cosh\tau'') \ . \tag{7.38}$$

IV.–IX., XI., XIII. Equidistant $\Lambda^{(2)}$, $\tau \in \mathbb{R}, \mathrm{u} \in \Lambda^{(2)}$:

$$\int_{\tau(t')=\tau'}^{\tau(t'')=\tau''} \mathcal{D}\tau(t)\cosh^2\tau \int_{\mathbf{u}(t')=\mathbf{u}'}^{\mathbf{u}(t'')=\mathbf{u}''} \frac{\mathcal{D}\mathbf{u}(t)}{u_0}$$

$$\times\exp\left\{\frac{\mathrm{i}}{\hbar}\int_{t'}^{t''}\left[\frac{m}{2}(\dot{\tau}^2 - \cosh^2\tau\dot{\mathrm{u}}^2) - \frac{\hbar^2}{2m} - \frac{\Delta V(\mathrm{u})}{\cosh^2\tau}\right]dt\right\}$$

$$= (\cosh\tau'\cosh\tau'')^{-1}\int d\lambda\int_0^\infty dk\Psi_{\lambda,k}^{\Lambda^{(2)}}(\mathrm{u}'')\Psi_{\lambda,k}^{\Lambda^{(2)}*}(\mathrm{u}')$$

$$\times\int_{\mathbb{R}}\frac{p\sinh\pi p\,dp}{\cosh^2\pi k + \sinh^2\pi p}P_{\mathrm{i}k-\frac{1}{2}}^{-\mathrm{i}p}(\tanh\tau')P_{\mathrm{i}k-\frac{1}{2}}^{\mathrm{i}p}(\tanh\tau'')\,\mathrm{e}^{-\mathrm{i}\hbar T(p^2+1)/2m} \ . \tag{7.39}$$

X. Spherical , $\tau > 0, \vartheta \in (0, \pi), \varphi \in [0, 2\pi)$:

$$\int_{\tau(t')=\tau'}^{\tau(t'')=\tau''} \mathcal{D}\tau(t)\sinh^2\tau \int_{\vartheta(t')=\vartheta'}^{\vartheta(t'')=\vartheta''} \mathcal{D}\vartheta(t)\sin\vartheta \int_{\varphi(t')=\varphi'}^{\varphi(t'')=\varphi''} \mathcal{D}\varphi(t)$$

$$\times\exp\left\{\frac{\mathrm{i}}{\hbar}\int_{t'}^{t''}\left[\frac{m}{2}\left(\dot{\tau}^2+\sinh^2\tau(\dot{\vartheta}^2+\sin^2\vartheta\dot{\varphi}^2)\right) - \frac{\hbar^2}{8m}\left(4-\frac{1}{\sinh^2\tau}\left(1+\frac{1}{\sin^2\vartheta}\right)\right)\right]dt\right\}$$

$$= (\sinh\tau'\sinh\tau'')^{-1/2}\sum_{l=0}^\infty\sum_{m=-l}^l Y_l^{m*}(\vartheta',\varphi')Y_l^m(\vartheta'',\varphi'')$$

$$\times \int_0^\infty dp \frac{p \sinh \pi p}{\pi} |\Gamma(ip+l+1)|^2 \mathcal{P}_{ip-1/2}^{-\frac{1}{2}-l}(\cosh \tau') \mathcal{P}_{ip-1/2}^{-\frac{1}{2}-l}(\cosh \tau'') \, \mathrm{e}^{-i\hbar T(p^2+1)/2m} \ .$$

$$(7.40)$$

XII. Equidistant , $\tau_{1,2,3} \in \mathbb{R}$:

$$\int_{\tau_1(t')=\tau_1'}^{\tau_1(t'')=\tau_1''} \mathcal{D}\tau_1(t) \cosh^2 \tau_1 \int_{\tau_2(t')=\tau_2'}^{\tau_2(t'')=\tau_2''} \mathcal{D}\tau_2(t) \cosh \tau_2 \int_{\tau_3(t')=\tau_3'}^{\tau_3(t'')=\tau_3''} \mathcal{D}\tau_3(t)$$

$$\times \exp\left\{ \frac{i}{\hbar} \int_{t'}^{t''} \left[\frac{m}{2}\left(\dot\tau_1^2 + \cosh^2 \tau_1 \dot\tau_2^2 + \cosh^2 \tau_1 \cosh^2 \tau_2 \dot\tau_3^2 \right) \right.\right.$$

$$\left.\left. - \frac{\hbar^2}{8m}\left(4 + \frac{1}{\cosh^2 \tau_1} + \frac{1}{\cosh^2 \tau_1 \cosh^2 \tau_2} \right) \right] dt \right\}$$

$$= \left(\cosh^2 \tau_1' \cosh^2 \tau_1'' \cosh \tau_2' \cosh \tau_2'' \right)^{-1/2} \int_{\mathbb{R}} \frac{dk}{2\pi} \mathrm{e}^{ik(\tau_3''-\tau_3')}$$

$$\times \int_{\mathbb{R}} \frac{p_1 \sinh \pi p_1 \, dp_1}{\cosh^2 \pi k + \sinh^2 \pi p_1} P_{ik-\frac{1}{2}}^{-ip_1}(\tanh \tau_2') P_{ik-\frac{1}{2}}^{ip_1}(\tanh \tau_2'')$$

$$\times \int_{\mathbb{R}} \frac{p \sinh \pi p \, dp}{\cosh^2 \pi p_1 + \sinh^2 \pi p} \mathrm{e}^{-i\hbar T(p^2+1)/2m} P_{ip_1-\frac{1}{2}}^{-ip}(\tanh \tau_3') P_{ip_1-\frac{1}{2}}^{ip}(\tanh \tau_3'') \ . \qquad (7.41)$$

XIV., XV., XVI. Horicyclic $\mathbb{R}^2$, (without II.), $y > 0, \mathbf{x} \in \mathbb{R}^2$:

$$\int_{y(t')=y'}^{y(t'')=y''} \frac{\mathcal{D}y(t)}{y^3} \int_{\mathbf{x}(t')=\mathbf{x}'}^{\mathbf{x}(t'')=\mathbf{x}''} \mathcal{D}\mathbf{x}(t) \exp\left[\frac{i}{\hbar} \int_{t'}^{t''} \left(\frac{m}{2}\frac{\dot{\mathbf{x}}^2 + \dot{y}^2}{y^2} - \frac{3\hbar^2}{8m} \right) dt \right]$$

$$= \int d\lambda \int_0^\infty dk \Psi_{\lambda,k}^{\mathbb{R}^2}(\mathbf{x}'') \Psi_{\lambda,k}^{\mathbb{R}^2 *}(\mathbf{x}')$$

$$\times \frac{2y'y''}{\pi^2} \int_0^\infty dp \, p \sinh \pi p \exp\left[-\frac{i\hbar T}{2m}(p^2+1) \right] K_{ip}(ky') K_{ip}(ky'') \ . \qquad (7.42)$$

XVII. Prolate Elliptic, $\alpha \in [K, K+2iK'], \beta \in [iK', iK'+2K], \varphi \in [0, 2\pi)$:

$$\int_{\alpha(t')=\alpha'}^{\alpha(t'')=\alpha''} \mathcal{D}\alpha(t) \int_{\beta(t')=\beta'}^{\beta(t'')=\beta''} \mathcal{D}\beta(t) \frac{k^3}{k'}(\mathrm{sn}^2\alpha - \mathrm{sn}^2\beta)\mathrm{cn}\alpha\mathrm{cn}\beta \int_{\varphi(t')=\varphi'}^{\varphi(t'')=\varphi''} \mathcal{D}\varphi(t)$$

$$\times \exp\left\{ \frac{i}{\hbar} \int_{t'}^{t''} \left[\frac{m}{2}\left(k^2(\mathrm{sn}^2\alpha - \mathrm{sn}^2\beta)(\dot\alpha^2 - \dot\beta^2) + \frac{k^2}{k'^2}\mathrm{cn}^2\alpha\mathrm{cn}^2\beta\dot\varphi^2 \right) \right.\right.$$

$$\left.\left. - \frac{\hbar^2}{2m}\left(1 - \frac{k'^2}{4k^2(\mathrm{sn}^2\alpha - \mathrm{sn}^2\beta)}\left(\frac{\mathrm{sn}^2\alpha\mathrm{dn}^2\alpha}{\mathrm{cn}^2\alpha} - \frac{\mathrm{sn}^2\beta\mathrm{dn}^2\beta}{\mathrm{cn}^2\beta} \right) \right) \right] dt \right\}$$

$$= \left(\frac{k^2}{k'^2}\mathrm{cn}\alpha'\mathrm{cn}\alpha''\mathrm{cn}\beta'\mathrm{cn}\beta'' \right)^{-1/2} \sum_{\nu\in\mathbb{Z}} \frac{\mathrm{e}^{i\nu(\varphi''-\varphi')}}{2\pi} \int_0^\infty dp \sum_n \sum_{\epsilon,\epsilon'=\pm 1} \mathrm{e}^{-i\hbar T(p^2+1)/2m}$$

$$\times L_{ip-1/2,\nu,n}^{\epsilon\epsilon'}(\alpha'') L_{ip-1/2,\nu,n}^{\epsilon\epsilon' *}(\alpha') L_{ip-1/2,\nu,n}^{\epsilon\epsilon'}(\beta'') L_{ip-1/2,\nu,n}^{\epsilon\epsilon' *}(\beta') \ . \qquad (7.43)$$

XVIII. Oblate Elliptic, $\alpha \in [K, K + 2iK'], \beta \in [iK', iK' + 2K], \varphi \in [0, 2\pi)$:

$$\int_{\alpha(t')=\alpha'}^{\alpha(t'')=\alpha''} \mathcal{D}\alpha(t) \int_{\beta(t')=\beta'}^{\beta(t'')=\beta''} \mathcal{D}\beta(t) \frac{k^2}{k'}(\mathrm{sn}^2\alpha - \mathrm{sn}^2\beta)\mathrm{dn}\alpha\mathrm{dn}\beta \int_{\varphi(t')=\varphi'}^{\varphi(t'')=\varphi''} \mathcal{D}\varphi(t)$$

$$\times \exp\left\{ \frac{i}{\hbar}\int_{t'}^{t''} \left[\frac{m}{2}\left(k^2(\mathrm{sn}^2\alpha - \mathrm{sn}^2\beta)(\dot\alpha^2 - \dot\beta^2) - \frac{1}{k'^2}\mathrm{dn}^2\alpha\mathrm{dn}^2\beta\dot\varphi^2 \right) \right.\right.$$

$$\left.\left. - \frac{\hbar^2}{2m}\left(1 - \frac{k'^2}{4k^2(\mathrm{sn}^2\alpha - \mathrm{sn}^2\beta)}\left(k'^4\frac{\mathrm{sn}^2\alpha\mathrm{cn}^2\alpha}{\mathrm{dn}^2\alpha} - k^4\frac{\mathrm{sn}^2\beta\mathrm{cn}^2\beta}{\mathrm{dn}^2\beta} \right) \right) \right]dt \right\}$$

$$= \sum_{\nu\in\mathbb{Z}} \frac{e^{i\nu(\varphi''-\varphi')}}{2\pi}(\mathrm{dn}\alpha'\mathrm{dn}\alpha''\mathrm{dn}\beta'\mathrm{dn}\beta'')^{|\nu|-1/2} \int_0^\infty dp \sum_n\sum_s e^{-i\hbar T(p^2+1)/2m}$$

$$\times K_{ip-1/2,\nu,n}^{Ps}(\alpha'')K_{ip-1/2,\nu,n}^{Ps*}(\alpha')K_{ip-1/2,\nu,n}^{Ps}(\beta'')K_{ip-1/2,\nu,n}^{Ps*}(\beta') \ . \tag{7.44}$$

XIX. Elliptic-Cylindrical, $\alpha \in [K, K + 2iK'], \beta \in [iK', iK' + 2K], \tau \in \mathbb{R}$:

$$\int_{\alpha(t')=\alpha'}^{\alpha(t'')=\alpha''} \mathcal{D}\alpha(t) \int_{\beta(t')=\beta'}^{\beta(t'')=\beta''} \mathcal{D}\beta(t)k^3(\mathrm{sn}^2\alpha - \mathrm{sn}^2\beta)\mathrm{sn}\alpha\mathrm{sn}\beta \int_{\tau(t')=\tau'}^{\tau(t'')=\tau''} \mathcal{D}\tau(t)$$

$$\times \exp\left\{ \frac{i}{\hbar}\int_{t'}^{t''} \left[\frac{m}{2}\left(k^2(\mathrm{sn}^2\alpha - \mathrm{sn}^2\beta)(\dot\alpha^2 - \dot\beta^2) - k^2\mathrm{sn}^2\alpha\mathrm{sn}^2\beta\dot\tau^2 \right) \right.\right.$$

$$\left.\left. - \frac{\hbar^2}{2m}\left(1 - \frac{1}{4k^2(\mathrm{sn}^2\alpha - \mathrm{sn}^2\beta)}\left(k^2\frac{\mathrm{cn}^2\alpha\mathrm{dn}^2\alpha}{\mathrm{sn}^2\alpha} - k'^2\frac{\mathrm{cn}^2\beta\mathrm{dn}^2\beta}{\mathrm{sn}^2\beta} \right) \right) \right]dt \right\} \ . \tag{7.45}$$

XX. Hyperbolic-Cylindrical 1 , $\alpha \in [iK', iK' + 2K], \beta \in [0, 2iK']$:

$$\int_{\alpha(t')=\alpha'}^{\alpha(t'')=\alpha''} \mathcal{D}\alpha(t) \int_{\beta(t')=\beta'}^{\beta(t'')=\beta''} \mathcal{D}\beta(t)\frac{k^3}{k'}(\mathrm{sn}^2\alpha - \mathrm{sn}^2\beta)\mathrm{sn}\alpha\mathrm{sn}\beta \int_{\tau(t')=\tau'}^{\tau(t'')=\tau''} \mathcal{D}\tau(t)$$

$$\times \exp\left\{ \frac{i}{\hbar}\int_{t'}^{t''} \left[\frac{m}{2}\left(k^2(\mathrm{sn}^2\alpha - \mathrm{sn}^2\beta)(\dot\alpha^2 - \dot\beta^2) + \frac{k^2}{k'^2}\mathrm{cn}^2\alpha\mathrm{cn}^2\beta\dot\tau^2 \right) \right.\right.$$

$$\left.\left. - \frac{\hbar^2}{2m}\left(1 - \frac{1}{4k^4(\mathrm{sn}^2\alpha - \mathrm{sn}^2\beta)}\left(k^2\frac{\mathrm{sn}^2\alpha\mathrm{dn}^2\alpha}{\mathrm{cn}^2\alpha} - k'^2\frac{\mathrm{sn}^2\beta\mathrm{dn}^2\beta}{\mathrm{cn}^2\beta} \right) \right) \right]dt \right\} \ . \tag{7.46}$$

XXI. Hyperbolic-Cylindrical 2 , $\alpha \in [iK', iK' + 2K], \beta \in [0, 2iK], \varphi \in [0, 2\pi)$:

$$\int_{\alpha(t')=\alpha'}^{\alpha(t'')=\alpha''} \mathcal{D}\alpha(t) \int_{\beta(t')=\beta'}^{\beta(t'')=\beta''} \mathcal{D}\beta(t)\frac{k^2}{k'}(\mathrm{sn}^2\alpha - \mathrm{sn}^2\beta)\mathrm{dn}\alpha\mathrm{dn}\beta \int_{\varphi(t')=\varphi'}^{\varphi(t'')=\varphi''} \mathcal{D}\varphi(t)$$

$$\times \exp\left\{ \frac{i}{\hbar}\int_{t'}^{t''} \left[\frac{m}{2}\left(k^2(\mathrm{sn}^2\alpha - \mathrm{sn}^2\beta)(\dot\alpha^2 - \dot\beta^2) - \frac{1}{k'^2}\mathrm{dn}^2\alpha\mathrm{dn}^2\beta\dot\varphi^2 \right) \right.\right.$$

$$\left.\left. - \frac{\hbar^2}{2m}\left(1 - \frac{k'^2}{4k^2(\mathrm{sn}^2\alpha - \mathrm{sn}^2\beta)}\left(k'^4\frac{\mathrm{sn}^2\alpha\mathrm{cn}^2\alpha}{\mathrm{dn}^2\alpha} - k^4\frac{\mathrm{sn}^2\beta\mathrm{cn}^2\beta}{\mathrm{dn}^2\beta} \right) \right) \right]dt \right\}$$

$$= \left(-\frac{1}{k'^2} \mathrm{dn}\alpha'\, \mathrm{dn}\alpha''\, \mathrm{dn}\beta'\, \mathrm{dn}\beta'' \right)^{-1} \sum_{\nu \in \mathbb{Z}} \int_0^\infty dp \sum_{n \in \mathbb{Z}} \mathrm{e}^{-\mathrm{i}\hbar T(p^2+1)/2m}$$

$$\times \frac{\mathrm{e}^{\mathrm{i}\nu(\varphi''-\varphi')}}{2\pi} W_{\mathrm{ip}-1/2,\nu}^n(\alpha'') W_{\mathrm{ip}-1/2,\nu}^{n\,*}(\alpha') W_{\mathrm{ip}-1/2,\nu}^n(\beta'') W_{\mathrm{ip}-1/2,\nu}^{n\,*}(\beta') \ . \tag{7.47}$$

XXII. Semi-Hyperbolic, $\tau \in \mathbb{R}, \mu_{1,2} > 0, \varphi \in [0,2\pi), P(\mu) = \mu(1+\mu^2)$:

$$\int_{\mu_1(t')=\mu_1'}^{\mu_1(t'')=\mu_1''} \mathcal{D}\mu_1(t) \int_{\mu_2(t')=\mu_2'}^{\mu_2(t'')=\mu_2''} \mathcal{D}\mu_2(t) \int_{\varphi(t')=\varphi'}^{\varphi(t'')=\varphi''} \mathcal{D}\varphi(t) \frac{(\mu_1+\mu_2)\mu_1\mu_2}{4\sqrt{P(\mu_1)P(\mu_2)}}$$

$$\times \exp\left\{ \frac{\mathrm{i}}{\hbar} \int_{t'}^{t''} \left[\frac{m}{2}\left(\frac{\mu_1+\mu_2}{4}\left(\frac{\dot\mu_1^2}{P(\mu_1)} - \frac{\dot\mu_2^2}{P(\mu_2)} \right) + \mu_1\mu_2\dot\varphi^2 \right) \right.\right.$$

$$- \frac{\hbar^2}{8m}\frac{4}{\mu_1+\mu_2}\left(P''(\mu_1) - P''(\mu_2) - \frac{3P'(\mu_1)}{4P(\mu_1)} - \frac{3P'(\mu_2)}{4P(\mu_2)} \right)$$

$$\left.\left. - \frac{\hbar^2}{8m(\mu_1+\mu_2)}\left(\frac{P(\mu_1)}{\mu_1^2} - \frac{P(\mu_2)}{\mu_2^2} \right) \right] dt \right\} \ . \tag{7.48}$$

XXIII. Elliptic-Parabolic 1 , $a > 0, |\vartheta| < \pi/2, \varrho \in \mathbb{R}$:

$$\int_{a(t')=a'}^{a(t'')=a''} \mathcal{D}a(t) \int_{\vartheta(t')=\vartheta'}^{\vartheta(t'')=\vartheta''} \mathcal{D}\vartheta(t) \frac{\cosh^2 a - \cos^2\vartheta}{\cosh^3 a \cos^3\vartheta} \int_{\varrho(t')=\varrho'}^{\varrho(t'')=\varrho''} \mathcal{D}\varrho(t)$$

$$\times \exp\left[\frac{\mathrm{i}m}{2\hbar} \int_{t'}^{t''} \frac{(\cosh^2 a - \cos^2\vartheta)(\dot a^2 + \dot\vartheta^2) + \dot\varrho^2}{\cosh^2 a \cos^2\vartheta} dt - \frac{3\mathrm{i}\hbar T}{8m} \right]$$

$$= \sqrt{\cosh a' \cosh a''} \cos\vartheta' \cos\vartheta'' \int_{\mathbb{R}} \frac{dk_\varrho}{2\pi} \mathrm{e}^{\mathrm{i}k_\varrho(\varrho''-\varrho')}$$

$$\times \int_0^\infty dp \sinh\pi p \int_0^\infty \frac{dk\, k \sinh\pi k}{(\cosh^2\pi k + \sinh^2\pi p)^2} \mathrm{e}^{-\mathrm{i}\hbar T(p^2+1)/2m}$$

$$\times \sum_{\epsilon,\epsilon'=\pm 1} S_{\mathrm{ip}-1/2}^{\mathrm{i}k\,(1)}(\epsilon \tanh a''; \mathrm{i}k_\varrho) S_{\mathrm{ip}-1/2}^{\mathrm{i}k\,(1)\,*}(\epsilon \tanh a'; \mathrm{i}k_\varrho)$$

$$\times \mathrm{ps}_{\mathrm{i}k-1/2}^{\mathrm{i}p}(\epsilon' \sin\vartheta''; -k_\varrho^2) \mathrm{ps}_{\mathrm{i}k-1/2}^{\mathrm{i}p\,*}(\epsilon' \sin\vartheta'; -k_\varrho^2) \ . \tag{7.49}$$

XXIV. Elliptic-Hyperbolic 1 , $b > 0, 0 < \vartheta < \pi, \varrho \in \mathbb{R}$:

$$\int_{b(t')=b'}^{b(t'')=b''} \mathcal{D}b(t) \int_{\vartheta(t')=\vartheta'}^{\vartheta(t'')=\vartheta''} \mathcal{D}\vartheta(t) \frac{\sinh^2 b + \sin^2\vartheta}{\sinh^3 b \sin^3\vartheta} \int_{\varrho(t')=\varrho'}^{\varrho(t'')=\varrho''} \mathcal{D}\varrho(t)$$

$$\times \exp\left[\frac{\mathrm{i}m}{2\hbar} \int_{t'}^{t''} \frac{(\sinh^2 b + \sin^2\vartheta)(\dot b^2 + \dot\vartheta^2) + \dot\varrho^2}{\sinh^2 b \sin^2\vartheta} dt - \frac{3\mathrm{i}\hbar T}{8m} \right]$$

$$= \sinh b' \sinh b'' \sin\vartheta' \sin\vartheta'' \int_{\mathbb{R}} \frac{dk_\varrho}{2\pi} \mathrm{e}^{\mathrm{i}k_\varrho(\varrho''-\varrho')}$$

$$\times \int_0^\infty dp \int_0^\infty \frac{dk\, k \sinh\pi k}{\cosh^2\pi k + \sinh^2\pi p} \frac{1}{\cosh\pi(p-k)} \mathrm{e}^{-\mathrm{i}\hbar T(p^2+1)/2m}$$

$$\times \sum_{\epsilon=\pm 1} S_{ik-1/2}^{ip(1)}(\cosh b''; ik_\varrho) S_{ik-1/2}^{ip(1)\,*}(\cosh b'; ik_\varrho)$$

$$\times \mathrm{ps}_{ik-1/2}^{ip}(\epsilon \cos \vartheta''; -k_\varrho^2)\mathrm{ps}_{ik-1/2}^{ip\,*}(\epsilon \cos \vartheta'; -k_\varrho^2) \ . \tag{7.50}$$

XXV. Elliptic-Parabolic 2 , $a > 0, |\vartheta| < \pi/2, \varphi \in [0, 2\pi)$:

$$\int_{a(t')=a'}^{a(t'')=a''} \mathcal{D}a(t) \int_{\vartheta(t')=\vartheta'}^{\vartheta(t'')=\vartheta''} \mathcal{D}\vartheta(t) \tanh a \tan \vartheta \frac{\cosh^2 a - \cos^2 \vartheta}{\cosh^2 a \cos^2 \vartheta} \int_{\varphi(t')=\varphi'}^{\varphi(t'')=\varphi''} \mathcal{D}\varphi(t)$$

$$\times \exp\left\{\frac{i}{\hbar}\int_{t'}^{t''}\left[\frac{m}{2}\frac{(\cosh^2 a - \cos^2 \vartheta)(\dot{a}^2 + \dot{\vartheta}^2) + \sinh^2 a \sin^2 \vartheta \dot{\varphi}^2}{\cosh^2 a \cos^2 \vartheta}\right.\right.$$

$$\left.\left.+\frac{\hbar^2}{8m}\frac{\cosh^2 a + \cos^2 \vartheta - 1}{\sinh^2 a \sin^2 \vartheta}\right]dt - \frac{i\hbar T}{2m}\right\}$$

$$= \sum_{\nu\in\mathbb{Z}} \frac{e^{i\nu(\varphi''-\varphi')}}{2\pi} \int_0^\infty dk \int_0^\infty dp \, \exp\left[-\frac{i\hbar T}{2m}(p^2+1)\right]$$

$$\times \Psi_k^{(\nu,ip)}(a'')\Psi_k^{(\nu,ip)\,*}(a')\Psi_k^{(\nu,ip)}(\vartheta'')\Psi_k^{(\nu,ip)\,*}(\vartheta') \ , \tag{7.51}$$

$$\Psi_k^{(\nu,ip)}(a) = \frac{\Gamma[\frac{1}{2}(1+|\nu|+ip+ik)]\Gamma[\frac{1}{2}(1+|\nu|+ip-ik)]}{\Gamma(1+|\nu|)}\sqrt{\frac{k \sinh \pi k}{2\pi^2}}$$

$$\times (\tanh a)^{|\nu|}(\cosh a)^{ik}\,_2F_1\left(\frac{1+|\nu|+ip+ik}{2}, \frac{1+|\nu|-ip+ik}{2}; |\nu|+1; \tanh^2 a\right) \ , \tag{7.52}$$

$$\Psi_k^{(\nu,ip)}(\vartheta) = \frac{\Gamma[\frac{1}{2}(1+|\nu|+ip+ik)]\Gamma[\frac{1}{2}(1+|\nu|+ip-ik)]}{\Gamma(1+|\nu|)}\sqrt{\frac{k \sinh \pi k}{2\pi^2}}$$

$$\times (\tan \vartheta)^{|\nu|}(\cos \vartheta)^{ip+|\nu|+1}\,_2F_1\left(\frac{1+|\nu|+ip+ik}{2}, \frac{1+|\nu|+ip-ik}{2}; |\nu|+1; -\sin^2 \vartheta\right). \tag{7.53}$$

XXVI. Elliptic-Hyperbolic 2 , $b > 0, 0 < \vartheta < \pi, \varphi \in [0, 2\pi)$:

$$\int_{b(t')=b'}^{b(t'')=b''} \mathcal{D}b(t) \int_{\vartheta(t')=\vartheta'}^{\vartheta(t'')=\vartheta''} \mathcal{D}\vartheta(t) \frac{\sinh^2 b + \sin^2 \vartheta}{\sinh^2 b \sin^2 \vartheta} \coth b \cot \vartheta \int_{\varphi(t')=\varphi'}^{\varphi(t'')=\varphi''} \mathcal{D}\varphi(t)$$

$$\times \exp\left\{\frac{i}{\hbar}\int_{t'}^{t''}\left[\frac{m}{2}\frac{(\sinh^2 b + \sin^2 \vartheta)(\dot{b}^2 + \dot{\vartheta}^2) + \cosh^2 b \cos^2 \vartheta \dot{\varphi}^2}{\sinh^2 b \sin^2 \vartheta}\right.\right.$$

$$\left.\left.-\frac{\hbar^2}{8m}\frac{\sin^2 \vartheta - \sinh^2 b - 1}{\cosh^2 b \cos^2 \vartheta}\right]dt - \frac{i\hbar T}{2m}\right\}$$

$$= \sum_{\nu\in\mathbb{Z}} \frac{e^{i\nu(\varphi''-\varphi')}}{2\pi} \int_0^\infty dk \int_0^\infty dp \, \Psi_k^{(\nu,ip)}(b'')\Psi_k^{(\nu,ip)\,*}(b')\Psi_k^{(\nu,ip)}(\vartheta'')\Psi_k^{(\nu,ip)\,*}(\vartheta') \, e^{-i\hbar T(p^2+1)/2m},$$

$$\tag{7.54}$$

$$\Psi_k^{(\nu,\mathrm{i}p)}(b) = \sqrt{\frac{k\sinh\pi k\sinh\pi p}{2\pi^3 p}}\,\Gamma[\tfrac{1}{2}(1+|\nu|+\mathrm{i}p+\mathrm{i}k)]\Gamma[\tfrac{1}{2}(1-|\nu|+\mathrm{i}p+\mathrm{i}k)]$$

$$\times(\tanh b)^{1+\mathrm{i}p}(\cosh b)^{1+\mathrm{i}k}{}_2F_1\left(\frac{1+|\nu|+\mathrm{i}p-\mathrm{i}k}{2},\frac{1-|\nu|+\mathrm{i}p-\mathrm{i}k}{2};1+\mathrm{i}p;\tanh^2 b\right),$$

$$(7.55)$$

$$\Psi_k^{(\nu,\mathrm{i}p)}(\vartheta) = \sqrt{\frac{k\sinh\pi k\sinh\pi p}{2\pi^3 p}}\,\Gamma[\tfrac{1}{2}(1+|\nu|+\mathrm{i}p+\mathrm{i}k)]\Gamma[\tfrac{1}{2}(1-|\nu|+\mathrm{i}p+\mathrm{i}k)]$$

$$\times(\cos\vartheta)^{|\nu|}(\sin\vartheta)^{\mathrm{i}p}{}_2F_1\left(\frac{1+|\nu|+\mathrm{i}p-\mathrm{i}k}{2},\frac{1+|\nu|+\mathrm{i}p+\mathrm{i}k}{2};1+\mathrm{i}p;-\sin^2\vartheta\right). \quad (7.56)$$

XXVII. Semi-Circular-Parabolic, $\xi,\eta > 0$, $\varrho \in \mathbb{R}$:

$$\int_{\xi(t')=\xi'}^{\xi(t'')=\xi''}\mathcal{D}\xi(t)\int_{\eta(t')=\eta'}^{\eta(t'')=\eta''}\mathcal{D}\eta(t)\frac{\xi^2+\eta^2}{\xi^3\eta^3}\int_{\varrho(t')=\varrho'}^{\varrho(t'')=\varrho''}\mathcal{D}\varrho(t)$$

$$\times\exp\left[\frac{\mathrm{i}m}{2\hbar}\int_{t'}^{t''}\frac{(\xi^2+\eta^2)(\dot{\xi}^2+\dot{\eta}^2)+\dot{\varrho}^2}{\xi^2\eta^2}dt-\frac{3\mathrm{i}\hbar T}{8m}\right]$$

$$=\int_{\mathbb{R}}\frac{dq}{2\pi}\frac{e^{\mathrm{i}q(\varrho''-\varrho')}}{\pi^2 q^2}\sum_{\pm}\int_0^\infty dp\,p(\sinh\pi p)^2\int_0^\infty dk\,k\Big|\Gamma[\tfrac{1}{2}(1\pm k^2/2q+\mathrm{i}p)]\Big|^4 e^{-\mathrm{i}\hbar T(p^2+1)/2m}$$

$$\times W_{\pm k^2/4q,\mathrm{i}k/2}(q\xi''^2)W_{\pm k^2/4q,\mathrm{i}k/2}(q\xi'^2)W_{\pm k^2/4q,\mathrm{i}p/2}(q\eta''^2)W_{\pm k^2/4q,\mathrm{i}p/2}(q\eta'^2)\ . \quad (7.57)$$

Systems XXVIII.–XXXIV.:

$$\int_{\varrho(t')=\varrho'}^{\varrho(t'')=\varrho''}\mathcal{D}\varrho(t)\sqrt{g}\exp\left\{\frac{\mathrm{i}}{\hbar}\int_{t'}^{t''}\left[\frac{m}{2}\sum_{i=1}^3 g_{\varrho_i\varrho_i}\dot{\varrho}_i^2-\Delta V_{PF}(\boldsymbol{\varrho})\right]dt\right\}\ . \quad (7.58)$$

General Expression for the Propagator ($\cosh d = $ invariant distance on $\Lambda^{(3)}$ [269]):

$$=\left(\frac{m}{2\pi\mathrm{i}\hbar T}\right)^{3/2}\frac{d_{\Lambda^{(3)}}(\mathrm{u}'',\mathrm{u}')}{\sinh d_{\Lambda^{(3)}}(\mathrm{u}'',\mathrm{u}')}\exp\left[\frac{\mathrm{i}m}{2\hbar T}d_{\Lambda^{(3)}}^2(\mathrm{u}'',\mathrm{u}')-\frac{\mathrm{i}\hbar T}{2m}\right]\ . \quad (7.59)$$

General Expression for the Green's Function:

$$=\int_{\mathbb{R}}\frac{dE}{2\pi\mathrm{i}}e^{-\mathrm{i}ET/\hbar}\frac{-m}{\pi^2\hbar^2\sinh d_{\Lambda^{(3)}}(\mathrm{u}'',\mathrm{u}')}\mathcal{Q}_{-1/2-\mathrm{i}\sqrt{2mE/\hbar^2-1}}^{1/2}\Big(\cosh d_{\Lambda^{(3)}}(\mathrm{u}'',\mathrm{u}')\Big)\ . \quad (7.60)$$

The metric tensor in ellipsoidal coordinates has the form

$$(g_{ab})=\frac{1}{4}\mathrm{diag}\left(\frac{(\varrho_1-\varrho_2)(\varrho_1-\varrho_3)}{P(\varrho_1)},\frac{(\varrho_2-\varrho_3)(\varrho_2-\varrho_1)}{P(\varrho_2)},\frac{(\varrho_3-\varrho_1)(\varrho_3-\varrho_2)}{P(\varrho_3)}\right)\ , \quad (7.61)$$

and $P(\varrho)=(\varrho-a)(\varrho-b)(\varrho-c)(\varrho-d)$. The corresponding quantum potentials ΔV can be constructed from (2.11).

The Systems I.–XVI.

The path integral solutions of the systems I.–XVI. follow either from the results of [223] or from the corresponding solutions of $\Lambda^{(2)}$. The numbers λ, k are the quantum numbers of the corresponding subsystems. The path integral for the cylindrical system is evaluated by separating off φ and τ_2, and then applying the path integral solution of the modified Pöschl–Teller potential. Only the continuous spectrum is taken into account. The horicyclic systems II., XIV.–XVI. are of the subgroup type $SO(3,1) \supset E(2)$. Therefore the path integral solutions in the Euclidean plane come into play. The system II. is usually called the "horicyclic system". The path integral representations corresponding to the systems III. and X. correspond to the subgroup chain $SO(3,1) \supset SO(3)$, and we can apply the two path integral solutions of the two-dimensional sphere. The nine equidistant coordinate systems correspond to the subgroup chain $SO(3,1) \supset SO(2,1)$ and the corresponding path integral solutions follow from the path integral solutions on the two-dimensional pseudosphere. The system XII. then is usually called the "equidistant system". The systems XXVIII.–XXXIV. will not be discussed because no solution for the Schrödinger equation is known, let alone a path integral approach.

The Elliptic- and Hyperbolic-Cylindrical Systems.

We are therefore left with the systems XVII.–XXVII. For the systems XIX., XX., XXII. and XXVIII.–XXXIV. we have not found any solution in the literature. In the semi-hyperbolic system we have set $P(\mu) = \mu(1+\mu^2)$. The (yet unknown) solutions of the cylindrical systems can in principle be determined similarly as the already known ones, there seems to be until now no attempt to solve the Schrödinger and the path integral problem for the remaining parametric coordinate systems.

In the systems XVII.–XXI. we adopt the notation of [318]. The functions $L^{\epsilon\epsilon'}_{ip-1/2,\nu,n}(z)$ and $K^{Ps}_{ip-1/2,\nu,n}(z)$, are called associated periodic Lamé functions of the first and second kind, respectively, and are labeled with their quantum numbers p, ν, n and their parity $\epsilon, \epsilon' = \pm 1, P = A, B, C, D$ (for the latter c.f. the sphere $S^{(3)}$) as defined by the corresponding recurrence relations. The functions $W^n_{ip-1/2,\nu}(z)$ are also associated Lamé functions. The corresponding interbasis expansion coefficient can be calculated from the known formulæ and overlap integrals similarly as for the case of the ellipso-cylindrical systems on the three-dimensional sphere. We have for the interbasis expansions of the cylindrical systems

$$|lpp_1 > = \sum_{n_{XVII}} X^{lp*}_{n_{XVII}} |pln_{XVII} > , \qquad (7.62)$$

$$= \sum_{n_{XVIII}} X^{lp*}_{n_{XVIII}} |pln_{XVIII} > , \qquad (7.63)$$

$$= \sum_{n_{XXI}} X^{lp*}_{n_{XXI}} |pln_{XXI} > . \qquad (7.64)$$

It is expected that this can also be done for the remaining cylindrical systems where I have not found an explicit discussion in the literature.

The Elliptic-Parabolic and -Hyperbolic Systems.

In the systems XXIII. and XXIV. one first separates the ϱ-path integration according to

$$K^{(XXIII)}(a'', a', \vartheta'', \vartheta', \varrho'', \varrho'; T)$$
$$= (\cosh a' \cosh a'' \cos \vartheta' \cos \vartheta'')^{1/2} e^{-i\hbar T/2m} \int_{\mathbb{R}} \frac{dk_\varrho}{2\pi} e^{ik_\varrho(\varrho''-\varrho')}$$

$$\times \int_{a(t')=a'}^{a(t'')=a''} \mathcal{D}a(t) \int_{\vartheta(t')=\vartheta'}^{\vartheta(t'')=\vartheta''} \mathcal{D}\vartheta(t) \frac{\cosh^2 a - \cos^2 \vartheta}{\cosh^2 a \cos^2 \vartheta}$$

$$\times \exp\left\{\frac{i}{\hbar} \int_{t'}^{t''} \left[\frac{m}{2} \frac{\cosh^2 a - \cos^2 \vartheta}{\cosh^2 a \cos^2 \vartheta}(\dot{a}^2 + \dot{\vartheta}^2) - \frac{\hbar^2 k_\varrho^2}{2m} \cosh^2 a \cos^2 \vartheta + \frac{\hbar^2}{8m}\right]dt\right\} . \quad (7.65)$$

We now observe that on the one hand the last path integral is for $k_\varrho = 0$ exactly the path integral for the elliptic parabolic coordinate system on $\Lambda^{(2)}$ (with shifted energy). On the other it yields after an additional time transformation the path integral for an oblate spheroidal coordinate systems, i.e.,

$$K^{(XXIII)}(a'', a', \vartheta'', \vartheta', \varrho'', \varrho'; T)$$
$$= (\cosh a' \cosh a'' \cos \vartheta' \cos \vartheta'')^{1/2} e^{-i\hbar T/2m} \int_{\mathbb{R}} \frac{dk_\varrho}{2\pi} e^{ik_\varrho(\varrho''-\varrho')}$$

$$\times \int_{\mathbb{R}} \frac{dE}{2\pi\hbar} e^{-iET/\hbar} \int_0^\infty ds'' \int_{a(0)=a'}^{a(s'')=a''} \mathcal{D}a(s) \int_{\vartheta(0)=\vartheta'}^{\vartheta(s'')=\vartheta''} \mathcal{D}\vartheta(s)$$

$$\times \exp\left\{\frac{i}{\hbar} \int_0^{s''} \left[\frac{m}{2}(\dot{a}^2 + \dot{\vartheta}^2) - \frac{\hbar^2 k_\varrho^2}{2m}(\cosh^2 a - \cos^2 \vartheta) - \frac{\hbar^2}{2m}\left(\frac{\lambda^2 - \frac{1}{4}}{\cos^2 \vartheta} - \frac{\lambda^2 - \frac{1}{4}}{\cosh^2 a}\right)\right]ds\right\},$$
$$(7.66)$$

where $\lambda = \sqrt{-2mE}/\hbar$. Because λ is for positive E purely imaginary we cannot apply the oblate spheroidal path integral identity (5.34) in a simple way. We must find a proper analytic continuation. Let us construct this analytic continuation heuristically. Since the (a, ϑ)-path integration in (7.65) corresponds for $k_\varrho = 0$ to the path integral (7.7) we look for those spheroidal wavefunctions which have for the parameter $k_\varrho = 0$ the limit of the wavefunctions of (7.7) and we find similarly as in Chapter 4

$$ps_\nu^\mu(x; 0) = P_\nu^\mu(x) , \qquad (|x| \leq 1) , \qquad S_\nu^{\mu(1)}(z; 0) = \mathcal{P}_\nu^\mu(z) , \qquad |z| \geq 1 . \quad (7.67)$$

Putting everything together yields (7.49). The case of elliptic hyperbolic 1 system (7.50) is done in an analogous way.

In the systems XXV. and XXVI. we get after the separation of the angular variable and a time transformation

$$K^{(XXV)}(a'', a', \vartheta'', \vartheta', \varphi'', \varphi'; T)$$

$$= (\coth a' \coth a'' \cot \vartheta' \cot \vartheta'')^{1/2} e^{-i\hbar T/2m} \sum_{\nu \in \mathbb{Z}} \frac{e^{i\nu(\varphi'' - \varphi')}}{2\pi}$$

$$\times \int_{\mathbb{R}} \frac{dE}{2\pi\hbar} e^{-iET/\hbar} \int_0^\infty ds'' \int_{a(0)=a'}^{a(s'')=a''} \mathcal{D}a(s) \int_{\vartheta(0)=\vartheta'}^{\vartheta(s'')=\vartheta''} \mathcal{D}\vartheta(s)$$

$$\times \exp\left\{ \frac{i}{\hbar} \int_0^{s''} \left[\frac{m}{2}(\dot{a}^2 + \dot{\vartheta}^2) - \frac{\hbar^2}{2m}\left(\frac{\nu^2 - \frac{1}{4}}{\sinh^2 a} - \frac{\lambda^2 - \frac{1}{4}}{\cosh^2 a} + \frac{\nu^2 - \frac{1}{4}}{\sin^2 \vartheta} + \frac{\lambda^2 - \frac{1}{4}}{\cos^2 \vartheta} \right) \right] ds \right\},$$

$$(7.68)$$

where $\lambda^2 = -2mE/\hbar^2$. One makes use of the path integral solution of the Pöschl–Teller and modified Pöschl–Teller path in terms of their Green's function representations according to [65, 66, 353], and performs a Green's function analysis similarly as in (4.37) to get (7.51).

The Semi-Circular Parabolic System.

In the semi-circular-parabolic system XXVII. one obtains after separating the φ-path integration and a time-transformation

$$K^{(XXVII)}(\xi'', \xi'', \eta'', \eta', \varphi'', \varphi'; T)$$

$$= (\xi'\xi''\eta'\eta'')^{1/2} e^{-i\hbar T/2m} \int_{\mathbb{R}} \frac{dq}{2\pi} e^{iq(\varrho'' - \varrho')}$$

$$\times \int_{\mathbb{R}} \frac{dE}{2\pi\hbar} e^{-iET/\hbar} \int_0^\infty ds'' \int_{\xi(0)=\xi'}^{\xi(s'')=\xi''} \mathcal{D}\xi(s) \int_{\eta(0)=\eta'}^{\eta(s'')=\eta''} \mathcal{D}\eta(s)$$

$$\times \exp\left\{ \frac{i}{\hbar} \int_0^{s''} \left[\frac{m}{2}\left((\dot{\xi}^2 + \dot{\eta}^2) - \omega^2(\xi^2 + \eta^2) \right) - \hbar^2 \frac{\lambda^2 - \frac{1}{4}}{2m}\left(\frac{1}{\xi^2} + \frac{1}{\eta^2} \right) \right] ds \right\}$$

$$= \frac{1}{2}\xi'\xi'' e^{-i\hbar T/2m} \int_{\mathbb{R}} \frac{dq}{2\pi} e^{iq(\varrho'' - \varrho')} \int_{\mathbb{R}} \frac{dE}{2\pi\hbar} e^{-iET/\hbar} \int_0^\infty ds'' \int_{\mathbb{R}} \frac{dE'}{2\pi i} e^{-iE's''/\hbar}$$

$$\times \frac{m\omega}{i\hbar \sin \omega s''} \exp\left[-\frac{m\omega}{2i\hbar}(\xi'^2 + \xi''^2) \cot \omega s'' \right] I_\lambda\left(\frac{m\omega \xi' \xi''}{i\hbar \sin \omega s''} \right)$$

$$\times \frac{\Gamma[\frac{1}{2}(1 + \lambda - E'/\hbar\omega)]}{\hbar\omega \Gamma(1 + \lambda)} W_{E'/2\hbar\omega, \lambda/2}\left(\frac{m\omega}{\hbar}\eta_>^2 \right) M_{E'/2\hbar\omega, \lambda/2}\left(\frac{m\omega}{\hbar}\eta_<^2 \right) + (\xi \leftrightarrow \eta), \quad (7.69)$$

where $\omega^2 = \hbar^2 q^2/m^2$, $\lambda^2 = -2mE/\hbar^2$; we must take into account a term with ξ and η interchanged. One uses the path integral solution of the radial harmonic oscillator [439] (and c.f. [248] for the functional weight formulation), where for the ξ-dependent part we expand the propagator by means of (7.27) and the integral representation ([88, p.85], [187, p.729])

$$W_{\chi,\frac{\mu}{2}}(a)W_{\chi,\frac{\mu}{2}}(b) = \frac{2\sqrt{ab}}{\Gamma(\frac{1+\mu}{2}-\chi)\Gamma(\frac{1-\mu}{2}-\chi)}\int_0^\infty e^{-\frac{a+b}{2}\cosh v}K_\mu(\sqrt{ab}\sinh v)\left(\coth\frac{v}{2}\right)^{2\chi}dv.$$
(7.70)

In the η-dependent part one uses the Green's function for the radial harmonic oscillator,

$$\frac{i}{\hbar}\int_0^\infty dT\, e^{iET/\hbar}\int_{r(t')=r'}^{r(t'')=r''}\mathcal{D}r(t)\mu_\lambda[r^2]\exp\left[\frac{im}{2\hbar}\int_{t'}^{t''}(\dot{r}^2-\omega^2r^2)dt\right]$$

$$= \frac{\Gamma[\frac{1}{2}(1+\lambda-E/\hbar\omega)]}{\hbar\omega\sqrt{r'r''}\,\Gamma(1+\lambda)}W_{E/2\hbar\omega,\lambda/2}\left(\frac{m\omega}{\hbar}r_>^2\right)M_{E/2\hbar\omega,\lambda/2}\left(\frac{m\omega}{\hbar}r_<^2\right),\qquad (7.71)$$

and the relation [187, p.1062]

$$W_{\lambda,\mu}(z) = \frac{\Gamma(-2\mu)}{\Gamma(\frac{1}{2}-\mu-\lambda)}M_{\lambda,\mu}(z) + \frac{\Gamma(2\mu)}{\Gamma(\frac{1}{2}+\mu-\lambda)}M_{\lambda,-\mu}(z).\qquad (7.72)$$

The final result (7.57) is then obtained by combining on the one hand together with $E' = k^2\hbar^2/2m$

$$\frac{m\omega}{\hbar^2}\int_0^\infty ds''\frac{ds''}{\sin\omega s''}\exp\left[-i\frac{E's''}{\hbar}-\frac{m\omega}{2i\hbar}(\xi'^2+\xi''^2)\cot\omega s''\right]I_\lambda\left(\frac{m\omega\xi'\xi''}{i\hbar\sin\omega s''}\right)$$

$$= \frac{1}{\pi^2 q}\int_0^\infty\frac{dp\,p\sinh\pi p}{p^2\hbar^2/2m-E}\left|\Gamma\left[\frac{1}{2}\left(1+ip-\frac{k^2}{2q}\right)\right]\right|^2 W_{-k^2/4q,ip/2}(q\xi'^2)W_{-k^2/4q,ip/2}(q\xi''^2),$$
(7.73)

and on the other

$$W_{k^2/4q,ip/2}(q\eta'^2)\left[\frac{\Gamma[\frac{1}{2}(1+ip-k^2/2q)]}{\hbar\omega\Gamma(1+ip)}M_{k^2/4q,ip/2}(q\eta''^2)\right.$$

$$\left.-\frac{\Gamma[\frac{1}{2}(1-ip-k^2/2q)]}{\hbar\omega\Gamma(1-ip)}M_{k^2/4q,-ip/2}(q\eta''^2)\right]$$

$$= \frac{im}{\pi\hbar^2 q}\sinh\pi p\left|\Gamma\left[\frac{1}{2}\left(1+ip-\frac{k^2}{2q}\right)\right]\right|^2 W_{-k^2/4q,ip/2}(q\eta'^2)W_{-k^2/4q,ip/2}(q\eta''^2).\qquad (7.74)$$

Coordinate Systems on the Three-Dimensional Pseudosphere.

In table 7.2 we summarize the results on path integration on $\Lambda^{(3)}$ including an enumeration of the coordinate systems according to [318, 425].

Table 7.2: Coordinates on the Three-Dimensional Pseudosphere

Coordinate System	Coordinates	Path Integral Solution
I. Cylindrical	$u_0 = \cosh\tau_1\cosh\tau_2$ $u_1 = \sinh\tau_1\cos\varphi$ $u_2 = \sinh\tau_1\sin\varphi$ $u_3 = \cosh\tau_1\sinh\tau_2$	(7.34)
II. Horicyclic	$u_0 = [y + (x_1^2 + x_2^2)/y + 1/y]/2$ $u_1 = x_1/y$ $u_2 = x_2/y$ $u_3 = [y + (x_1^2 + x_2^2)/y - 1/y]/2$	(7.37)
III. Sphero-Elliptic	$u_0 = \cosh\tau$ $u_1 = \sinh\tau\,\mathrm{sn}(\alpha,k)\mathrm{dn}(\beta,k')$ $u_2 = \sinh\tau\,\mathrm{cn}(\alpha,k)\mathrm{cn}(\beta,k')$ $u_3 = \sinh\tau\,\mathrm{dn}(\alpha,k)\mathrm{sn}(\beta,k')$	(7.38)
IV. Equidistant-Elliptic	$u_0 = k\,\mathrm{sn}(\nu,k)\mathrm{sn}(\eta,k)\cosh\tau$ $u_1 = (k/k')\mathrm{cn}(\nu,k)\mathrm{cn}(\eta,k)\cosh\tau$ $u_2 = (\mathrm{i}/k')\mathrm{dn}(\nu,k)\mathrm{dn}(\eta,k)\cosh\tau$ $u_3 = \sinh\tau$	(7.39)
V. Equidistant-Hyperbolic	$u_0 = \dfrac{\mathrm{i}k}{k'}\mathrm{cn}(\alpha,k)\mathrm{cn}(\beta,k)\cosh\tau$ $u_1 = \mathrm{i}k\,\mathrm{sn}(\alpha,k)\mathrm{sn}(\beta,k)\cosh\tau$ $u_2 = \dfrac{\mathrm{i}}{k'}\mathrm{dn}(\alpha,k)\mathrm{dn}(\beta,k)\cosh\tau$ $u_3 = \sinh\tau$	(7.39)
VI. Equidistant-Semi-Hyperbolic	$u_0 = \dfrac{1}{\sqrt{2}}\left(\sqrt{(1+\mu_1^2)(1+\mu_2^2)} + \mu_1\mu_2 + 1\right)^{1/2}\cosh\tau$ $u_2 = \dfrac{1}{\sqrt{2}}\left(\sqrt{(1+\mu_1^2)(1+\mu_2^2)} - \mu_1\mu_1 - 1\right)^{1/2}\cosh\tau$ $u_2 = \sqrt{\mu_1\mu_2}\,\cosh\tau$ $u_3 = \sinh\tau$	(7.39)
VII. Equidistant-Elliptic-Parabolic	$u_0 = \dfrac{\cosh^2 a + \cos^2\vartheta}{2\cosh a\cos\vartheta}\cosh\tau$ $u_1 = \dfrac{\sinh^2 a - \sin^2\vartheta}{2\cosh a\cos\vartheta}\cosh\tau$ $u_2 = \tan\vartheta\tanh a\cosh\tau$ $u_3 = \sinh\tau$	(7.39)

Table 7.2 (cont.) Coordinate System	Coordinates	Path Integral Solution
VIII. Equidistant- Hyperbolic-Parabolic	$u_0 = \dfrac{\cosh^2 b + \cos^2 \vartheta}{2 \sinh b \sin \vartheta} \cosh \tau$ $u_1 = \dfrac{\sinh^2 b - \sin^2 \vartheta}{2 \sinh b \sin \vartheta} \cosh \tau$ $u_2 = \cot \vartheta \coth b \cosh \tau$ $u_3 = \sinh \tau$	(7.39)
IX. Equidistant- Semi-Circular-Parabolic	$u_0 = \dfrac{(\xi^2 + \eta^2)^2 + 4}{8\xi\eta} \cosh \tau$ $u_1 = \dfrac{(\xi^2 + \eta^2)^2 - 4}{8\xi\eta} \cosh \tau$ $u_2 = \dfrac{\eta^2 - \xi^2}{2\xi\eta} \cosh \tau$ $u_3 = \sinh \tau$	(7.39)
X. Spherical	$u_0 = \cosh \tau$ $u_1 = \sinh \tau \sin \vartheta \cos \varphi$ $u_2 = \sinh \tau \sin \vartheta \sin \varphi$ $u_3 = \sinh \tau \cos \vartheta$	(7.40)
XI. Equidistant-Cylindrical	$u_0 = \cosh \tau_1 \cosh \tau_2$ $u_1 = \cosh \tau_1 \sinh \tau_2 \cos \varphi$ $u_2 = \cosh \tau_1 \sinh \tau_2 \sin \varphi$ $u_3 = \sinh \tau_1$	(7.39)
XII. Equidistant	$u_0 = \cosh \tau_1 \cosh \tau_2 \cosh \tau_3$ $u_1 = \cosh \tau_1 \cosh \tau_2 \sinh \tau_3$ $u_2 = \cosh \tau_1 \sinh \tau_2$ $u_3 = \sinh \tau_1$	(7.41)
XIII. Equidistant-Horicyclic	$u_0 = \cosh \tau(y + x^2/y + 1/y)/2$ $u_1 = \sinh \tau$ $u_2 = x \cosh \tau/y$ $u_3 = \cosh \tau(y + x^2/y - 1/y)/2$	(7.39)
XIV. Horicyclic-Cylindrical	$u_0 = (y + \varrho^2/y + 1/y)/2$ $u_1 = \varrho \cos \varphi/y$ $u_2 = \varrho \sin \varphi/y$ $u_3 = (y + \varrho^2/y - 1/y)/2$	(7.42)
XV. Horicyclic-Elliptic	$u_0 = [y + (\cosh^2 \mu - \sin^2 \nu)/y + 1/y]/2$ $u_1 = \cosh \mu \cos \nu/y$ $u_2 = \sinh \mu \sin \nu/y$ $u_3 = [y + (\cosh^2 \mu - \sin^2 \nu)/y - 1/y]/2$	(7.42)

Table 7.2 (cont.) Coordinate System	Coordinates	Path Integral Solution
XVI. Horicyclic-Parabolic	$u_0 = [y + (\xi^2 + \eta^2)^2/y + 1/y]/2$ $u_1 = (\eta^2 - \xi^2)/2y$ $u_2 = \xi\eta/y$ $u_3 = [y + (\xi^2 + \eta^2)^2/y - 1/y]/2$	(7.42)
XVII. Prolate Elliptic	$u_0 = k\,\mathrm{sn}(\nu, k)\mathrm{sn}(\eta, k)$ $u_1 = (k/k')\mathrm{cn}(\nu, k)\mathrm{cn}(\eta, k)\cos\varphi$ $u_2 = (k/k')\mathrm{cn}(\nu, k)\mathrm{cn}(\eta, k)\sin\varphi$ $u_3 = (\mathrm{i}/k')\mathrm{dn}(\nu, k)\mathrm{dn}(\eta, k)$	(7.43)
XVIII. Oblate Elliptic	$u_0 = k\,\mathrm{sn}(\nu, k)\mathrm{sn}(\eta, k)$ $u_1 = (\mathrm{i}/k')\mathrm{dn}(\nu, k)\mathrm{dn}(\eta, k)\cos\varphi$ $u_2 = (\mathrm{i}/k')\mathrm{dn}(\nu, k)\mathrm{dn}(\eta, k)\sin\varphi$ $u_3 = (k/k')\mathrm{cn}(\nu, k)\mathrm{cn}(\eta, k)$	(7.44)
XIX. Elliptic Cylindrical	$u_0 = k\,\mathrm{sn}(\nu, k)\mathrm{sn}(\eta, k)\cosh\tau$ $u_1 = k\,\mathrm{sn}(\nu, k)\mathrm{sn}(\eta, k)\sinh\tau$ $u_2 = (k/k')\mathrm{cn}(\nu, k)\mathrm{cn}(\eta, k)$ $u_3 = (\mathrm{i}/k')\mathrm{dn}(\nu, k)\mathrm{dn}(\eta, k)$	not known
XX. Hyperbolic-Cylindrical 1	$u_0 = (\mathrm{i}k/k')\mathrm{cn}(\alpha, k)\mathrm{cn}(\beta, k)\cosh\tau$ $u_1 = (\mathrm{i}k/k')\mathrm{cn}(\alpha, k)\mathrm{cn}(\beta, k)\sinh\tau$ $u_2 = (\mathrm{i}/k')\mathrm{dn}(\alpha, k)\mathrm{dn}(\beta, k)$ $u_3 = \mathrm{i}k\,\mathrm{sn}(\alpha, k)\mathrm{sn}(\beta, k)$	not known
XXI. Hyperbolic-Cylindrical 2	$u_0 = (\mathrm{i}k/k')\mathrm{cn}(\alpha, k)\mathrm{cn}(\beta, k)$ $u_1 = (\mathrm{i}k/k')\mathrm{dn}(\alpha, k)\mathrm{dn}(\beta, k)\cos\varphi$ $u_2 = (\mathrm{i}k/k')\mathrm{dn}(\alpha, k)\mathrm{dn}(\beta, k)\sin\varphi$ $u_3 = \mathrm{i}k\,\mathrm{sn}(\alpha, k)\mathrm{sn}(\beta, k)$	(7.47)
XXII. Semi-Hyperbolic	$u_0 = \frac{1}{\sqrt{2}}\left(\sqrt{(1+\mu_1^2)(1+\mu_2^2)} + \mu_1\mu_2 + 1\right)^{1/2}$ $u_1 = \sqrt{\mu_1\mu_2}\,\cos\varphi$ $u_2 = \sqrt{\mu_1\mu_2}\,\sin\varphi$ $u_3 = \frac{1}{\sqrt{2}}\left(\sqrt{(1+\mu_1^2)(1+\mu_2^2)} - \mu_1\mu_2 - 1\right)^{1/2}$	not known
XXIII. Elliptic-Parabolic 1	$u_0 = \dfrac{\cosh^2 a + \cos^2\vartheta + \varrho^2}{2\cosh a\cos\vartheta}$ $u_1 = \varrho/\cosh a\cos\vartheta$ $u_2 = \tanh a\tan\vartheta$ $u_3 = \dfrac{\cosh^2 a + \cos^2\vartheta - \varrho^2 - 2}{2\cosh a\cos\vartheta}$	(7.49)

Table 7.2 (cont.) Coordinate System	Coordinates	Path Integral Solution
XXIV. Hyperbolic-Parabolic 1	$u_0 = \dfrac{\sinh^2 b - \sin^2 \vartheta + \varrho^2 + 2}{2 \sinh b \sin \vartheta}$ $u_1 = \varrho / \sinh b \sin \vartheta$ $u_2 = \coth b \cot \vartheta$ $u_3 = \dfrac{\sinh^2 b - \sin^2 \vartheta - \varrho^2}{2 \sinh b \sin \vartheta}$	(7.50)
XXV. Elliptic-Parabolic 2	$u_0 = \dfrac{\cosh^2 a + \cos^2 \vartheta}{2 \cosh a \cos \vartheta}$ $u_1 = \tanh a \tan \vartheta \cos \varphi$ $u_2 = \tanh a \tan \vartheta \sin \varphi$ $u_3 = \dfrac{\sinh^2 a - \sin^2 \vartheta}{2 \cosh a \cos \vartheta}$	(7.51)
XXVI. Hyperbolic-Parabolic 2	$u_0 = \dfrac{\cos^2 \vartheta + \cosh^2 b}{2 \sinh b \sin \vartheta}$ $u_1 = \coth b \cot \vartheta \cos \varphi$ $u_2 = \coth b \cot \vartheta \sin \varphi$ $u_3 = \dfrac{\sinh^2 b - \sin^2 \vartheta}{2 \sinh b \sin \vartheta}$	(7.54)
XXVII. Semi-Circular-Parabolic	$u_0 = \dfrac{(\xi^2 - \eta^2)^2 + 4\varrho^2 + 4}{8 \xi \eta}$ $u_1 = (\xi^2 - \eta^2)/2\xi\eta$ $u_2 = \varrho/\xi\eta$ $u_3 = \dfrac{(\xi^2 - \eta^2)^2 + 4\varrho^2 - 4}{8 \xi \eta}$	(7.57)
XXVIII. Ellipsoidal $0 < 1 < \varrho_3 < b < \varrho_2 < a < \varrho_1$	$u_0^2 = \dfrac{\varrho_1 \varrho_2 \varrho_3}{ab}$ $u_1^2 = \dfrac{(\varrho_1 - 1)(\varrho_2 - 1)(\varrho_3 - 1)}{(a - 1)(b - 1)}$ $u_2^2 = -\dfrac{(\varrho_1 - b)(\varrho_2 - b)(\varrho_3 - b)}{(a - b)(b - 1)b}$ $u_3^2 = \dfrac{(\varrho_1 - a)(\varrho_2 - a)(\varrho_3 - a)}{(a - b)(a - 1)a}$	not known
XXIX. Hyperboloidal $\varrho_3 < 0 < 1 < b < \varrho_2 < a < \varrho_1$	$u_0^2 = -\dfrac{(\varrho_1 - 1)(\varrho_2 - 1)(\varrho_3 - 1)}{(a - 1)(b - 1)}$ $u_1^2 = -\dfrac{\varrho_1 \varrho_2 \varrho_3}{ab}$ $u_2^2 = -\dfrac{(\varrho_1 - b)(\varrho_2 - b)(\varrho_3 - b)}{(a - b)(b - 1)}$ $u_3^2 = \dfrac{(\varrho_1 - a)(\varrho_2 - a)(\varrho_3 - a)}{(a - b)(a - 1)a}$	not known

Table 7.2 (cont.) Coordinate System	Coordinates	Path Integral Solution
XXX. Paraboloidal $\varrho_3 < 0 < \varrho_2 < 1 < \varrho_1$ $a = b^* = \alpha + \mathrm{i}\beta, \alpha, \beta \in \mathbb{R}$	$(u_1 + \mathrm{i}u_0)^2 = 2\dfrac{(\varrho_1-1)(\varrho_2-1)(\varrho_3-1)}{(a-1)(b-1)b}$ $u_2^2 = \dfrac{(\varrho_1-1)(\varrho_2-1)(\varrho_3-1)}{(a-1)(b-1)}$ $u_3^2 = -\dfrac{\varrho_1\varrho_2\varrho_3}{ab}$	not known
XXXI.* $0 < \varrho_3 < 1 < \varrho_2 < a < \varrho_1$	$(u_1 + u_0)^2 = \dfrac{\varrho_1\varrho_2\varrho_3}{a}$ $(u_0^2 - u_1^2)$ $\quad = \dfrac{a(\varrho_1\varrho_2+\varrho_1\varrho_3+\varrho_2\varrho_3)-(a+1)\varrho_1\varrho_2\varrho_3}{a^2}$ $u_2^2 = -\dfrac{(\varrho_1-1)(\varrho_2-1)(\varrho_3-1)}{(a-1)}$ $u_3^2 = \dfrac{(\varrho_1-a)(\varrho_2-a)(\varrho_3-a)}{a^2(a-1)}$	not known
XXXII.* $-\varrho_3 < 0 < 1 < \varrho_2 < a < \varrho_1$	$(u_1 + u_0)^2 = -\dfrac{\varrho_1\varrho_2\varrho_3}{a}$ $(u_0^2 - u_1^2)$ $\quad = \dfrac{a(\varrho_1\varrho_2+\varrho_2\varrho_3+\varrho_1\varrho_3)-(a+1)\varrho_1\varrho_2\varrho_3}{a^2}$ $u_2^2 = -\dfrac{(\varrho_1-1)(\varrho_2-1)(\varrho_3-1)}{(a-1)}$ $u_3^2 = \dfrac{(\varrho_1-a)(\varrho_2-a)(\varrho_3-a)}{a^2(a-1)}$	not known
XXXIII.* $\varrho_3 < -1 < 0 < \varrho_2 < a < \varrho_1$	$(u_1 + u_0)^2 = -\dfrac{\varrho_1\varrho_2\varrho_3}{a}$ $(u_0^2 - u_1^2)$ $\quad = \dfrac{a(\varrho_1\varrho_3+\varrho_1\varrho_2+\varrho_2\varrho_3)-(a-1)\varrho_1\varrho_2\varrho_3}{a^2}$ $u_2^2 = \dfrac{(\varrho_1-a)(\varrho_2-a)(\varrho_3-a)}{a^2(a+1)}$ $u_3^2 = -\dfrac{(\varrho_1+1)(\varrho_2+1)(\varrho_3+1)}{(a+1)}$	not known
XXXIV.* $\varrho_3 < 0 < \varrho_2 < 1 < \varrho_1$	$(u_0 - u_1)^2 = -\varrho_1\varrho_2\varrho_3$ $2u_2(u_1 - u_0) = \varrho_1\varrho_2+\varrho_2\varrho_3+\varrho_1\varrho_3 - \varrho_1\varrho_2\varrho_3$ $u_1^2 + u_2^2 - u_0^2 = -\varrho_1\varrho_2\varrho_3$ $\quad + \varrho_1\varrho_2+\varrho_2\varrho_3+\varrho_1\varrho_3 - \varrho_1 - \varrho_2 - \varrho_3$ $u_3^2 = (\varrho_1 - 1)(\varrho_2 - 1)(\varrho_3 - 1)$	not known

* "In view of the absence of commonly agreed names of the various hyperboloids appearing here, we present only the equations of the respective families of surfaces in the cases XXXI.-XXXIV" [425].

Chapter 8

Path Integral on the Complex Sphere

8.1   The Two-Dimensional Complex Sphere

We have the following path integral representations on the two-dimensional complex sphere [223, 236, 242, 252]

I. Spherical, $\vartheta \in [0, \pi)$, $\varphi \in [0, 2\pi)$:

(The sphercial system exists on $S^{(2)}$ and $\Lambda^{(2)}$)

$$
\int_{\vartheta(t')=\vartheta'}^{\vartheta(t'')=\vartheta''} \mathcal{D}\vartheta(t) \sin\vartheta \int_{\varphi(t')=\varphi'}^{\varphi(t'')=\varphi''} \mathcal{D}\varphi(t)
$$

$$
\times \exp\left\{\frac{i}{\hbar}\int_{t'}^{t''}\left[\frac{m}{2}\left(\dot\vartheta^2 + \sin^2\vartheta\dot\varphi^2\right) + \frac{\hbar^2}{8m}\left(1 + \frac{1}{\sin^2\vartheta}\right)\right]dt\right\}
$$

$$
= \sum_{l=0}^{\infty}\sum_{n=-l}^{l} e^{-i\hbar Tl(l+1)/2m}Y_l^m(\vartheta'', \varphi'')Y_l^{m\,*}(\vartheta', \varphi') \ . \tag{8.1}
$$

II. Horocyclic, $\alpha \in [-K, K], \beta \in [-2K', 2K']$:

(The horocyclic system exists only on $\Lambda^{(2)}$)

$$
\int_{x(t')=x'}^{x(t'')=x''} \mathcal{D}x(t) \int_{y(t')=y'}^{y(t'')=y''} \mathcal{D}y(t)e^{ix}\exp\left\{\frac{i}{\hbar}\int_{t'}^{t''}\left[\frac{m}{2}(\dot x^2 + e^{2ix}\dot y^2) + \frac{\hbar^2}{8m}\right]dt\right\}
$$

$$
= \int_{\mathbb{R}} \frac{e^{ik_y(y''-y')}}{2\pi}e^{-\frac{i}{2}(x''-x')}\sum_{n_x\in\mathbb{N}_0}\tfrac{1}{2}H_{n_x+1/2}^{(1)}(|k_y|\,e^{-ix''})H_{n_x+1/2}^{(1)}(|k_y|\,e^{ix'})e^{-i\hbar Tl(l+1)/2m}.
$$

$$
\tag{8.2}
$$

III. Elliptic, $\alpha \in [-K, K], \beta \in [-2K', 2K']$:

(The elliptic system exists on $S^{(2)}$ and $\Lambda^{(2)}$)

$$\int_{\alpha(t')=\alpha'}^{\alpha(t'')=\alpha''} \mathcal{D}\alpha(t) \int_{\beta(t')=\beta'}^{\beta(t'')=\beta''} \mathcal{D}\beta(t)(k^2\mathrm{cn}^2\alpha + k'^2\mathrm{cn}^2\beta)$$

$$\times \exp\left[\frac{im}{2\hbar}\int_{t'}^{t''}(k^2\mathrm{cn}^2\alpha + k'^2\mathrm{cn}^2\beta)(\dot{\alpha}^2 + \dot{\beta}^2)dt\right]$$

$$= \sum_{l=0}^{\infty}\sum_{\lambda}\sum_{p,q=\pm}\Lambda_{l,\hbar}^p(\alpha'')\Lambda_{l,\hbar}^{p*}(\alpha')\Lambda_{l,\hbar}^q(\beta'')\Lambda_{l,\hbar}^{q*}(\beta')\,e^{-i\hbar Tl(l+1)/2m}\quad. \tag{8.3}$$

IV. Degenerate elliptic I, $\tau_1, \tau_2 \in \mathbb{R}$:

(The degenerate elliptic I system exists only on $\Lambda^{(2)}$)

$$\int_{\tau_1(t')=\tau_1'}^{\tau_1(t'')=\tau_1''} \mathcal{D}\tau_1(t) \int_{\tau_2(t')=\tau_2'}^{\tau_2(t'')=\tau_2''} \mathcal{D}\tau_2(t)\left(\frac{1}{\cosh^2\tau_1} - \frac{1}{\cosh^2\tau_2}\right)$$

$$\times \exp\left\{\frac{i}{\hbar}\int_{t'}^{t''}\left[\frac{m}{2}\left(\frac{1}{\cosh^2\tau_1} - \frac{1}{\cosh^2\tau_2}\right)(\dot{\tau}_1^2 - \dot{\tau}_2^2)\right]dt\right\}$$

$$= \sum_{\epsilon,\epsilon'=\pm 1}\int_0^{\infty}dp\,p\sinh\pi p\int_0^{\infty}\frac{dk\,k\sinh\pi k}{(\cosh^2\pi k + \sinh^2\pi p)^2}\exp\left[-\frac{i}{\hbar}\frac{\hbar^2(p^2+\frac{1}{4})}{2m}T\right]$$

$$v\qquad \times P_{ip-1/2}^{ik}(\epsilon\tanh\tau_1'')P_{ip-1/2}^{-ik}(\epsilon\tanh\tau_1')P_{ik-1/2}^{ip}(\epsilon'\tanh\tau_2'')P_{ik-1/2}^{-ip}(\epsilon'\tanh\tau_2') \tag{8.4}$$

V. Degenerate elliptic II, $\chi, \vartheta \in [0, \pi), \xi, \eta > 0$:

(The degenerate elliptic II system exists only on $\Lambda^{(3)}$)

$$\int_{\xi(t')=\xi'}^{\xi(t'')=\xi''} \mathcal{D}\xi(t) \int_{\eta(t')=\eta'}^{\eta(t'')=\eta''} \mathcal{D}\eta(t)\left(\frac{1}{\eta^2} - \frac{1}{\xi^2}\right)$$

$$\times \exp\left\{\frac{i}{\hbar}\int_{t'}^{t''}\left[\frac{m}{2}\left(\frac{1}{\eta^2} - \frac{1}{\xi^2}\right)(\dot{\xi}^2 - \dot{\eta}^2)\right]dt\right\}$$

$$= \frac{1}{4\pi^2}\sqrt{\mu'\mu''\nu'\nu''}\int dk\int_0^{\infty}dp\,p\sinh^2\pi p\exp\left[-\frac{i}{\hbar}\frac{\hbar^2(p^2+\frac{1}{4})}{2m}T\right]$$

$$\times H_p^{(1)}(\sqrt{k}\,\xi')H_p^{(1)*}(\sqrt{k}\,\xi'')H_p^{(1)}(\sqrt{k}\,\eta')H_p^{(1)*}(\sqrt{k}\,\eta'')\quad. \tag{8.5}$$

In the previous sections we have seen that on the two-dimensional sphere there are just two coordinate systems, spherical and elliptical; on the three-dimensional sphere there are six coordinate systems, cylindrical, sphero-elliptic. spherical, oblate elliptic, prolate elliptic and ellipsoidal. On the other hand, the corresponding number of coordinate systems on the two- and three-dimensional hyperboloid are nine, respectively 34 [425].

Table 8.1: Coordinates on the Two-Dimensional Complex Sphere

Coordinate System	Coordinates	Path Integral Solution
I. Spherical	$z_1 = \cos\vartheta$ $z_2 = \sin\vartheta\cos\varphi$ $z_3 = \sin\vartheta\sin\varphi$	(8.1)
II. Horocyclic	$z_1 = \frac{1}{2}\left[e^{-ix} + (1 - y^2)\,e^{ix}\right]$ $z_2 = y\,e^{ix}$ $z_3 = -\frac{i}{2}\left[e^{-ix} - (1 + y^2)\,e^{ix}\right]$	(8.2)
III. Elliptic	$z_1 = \mathrm{sn}(\alpha, k)\mathrm{dn}(\beta, k')$ $z_2 = \mathrm{cn}(\alpha, k)\mathrm{cn}(\beta, k')$ $z_3 = \mathrm{dn}(\alpha, k)\mathrm{sn}(\beta, k')$	(8.3)
IV. Degenerate elliptic I	$z_1 = \frac{1}{2}\left(\dfrac{\cosh\tau_2}{\cosh\tau_1} + \dfrac{\cosh\tau_1}{\cosh\tau_2}\right)$ $z_2 = \tanh\tau_1\tanh\tau_2$ $z_3 = \left[\dfrac{1}{\cosh\tau_1\cosh\tau_2}\right.$ $\left. -\frac{1}{2}\left(\dfrac{\cosh\tau_2}{\cosh\tau_1} + \dfrac{\cosh\tau_1}{\cosh\tau_2}\right)\right]$	(8.4)
V. Degenerate elliptic II	$z_1 = -\frac{i}{8\xi\eta}\left[(\xi^2 - \eta^2)^2 + 4\right]$ $z_2 = \frac{1}{2\xi\eta}(\xi^2 + \eta^2)^2$ $z_3 = \frac{1}{8\xi\eta}\left[-(\xi^2 - \eta^2)^2 + 4\right]$	(8.5)

Furthermore, on the two- and three-dimensional Euclidean and pseudo-Euclidean spaces the number of systems is 10 and 54, and 4 and 11, respectively. All the spaces listed above can have an Euclidean metric signature, or a Minkowskian metric signature, and they are all real. The situation changes, if we start to consider the corresponding complex spaces. On two-dimensional and three-dimensional complex Euclidean space there are 6 and 18, respectively, and on the two- and three-dimensional complex sphere there are 5 and 21 coordinate systems which separate the Schrödinger equation [84]. In comparison to the real two- and three-dimensional sphere there is a richer structure. This is not surprising because the two-dimensional flat space and the hyperboloids are contained as subgroup cases in the complex sphere.

In table 8.2 I have listed some properties of the coordinate systems on the three-dimensional complex sphere. The coordinate systems, which contain two-dimensional flat systems, i.e. the Euclidean plane (real and complex) are, (2), (5), (9)–(11) [405]. The coordinates systems which have the two-dimensional sphere (real and complex) as a subsystem are (3), (4), (6)–(8) [301, 316, 329, 405]. The coordinate systems which exist also on the real three-dimensional sphere are (1), (3), (6), (13) and (17) [333, 425].

Table 8.2: Coordinate Systems on S_{3C}

No. of System	Complexifcation of Real Space	Two-Dimensional Subsystem	Three-Dimensional Systems
No 1	$S^{(3)}, \Lambda^{(3)}, O(2,2)$	$S(2, \mathbb{R})$ Polar	Cylindrical
No 2	$\Lambda^{(3)}, O(2,2)$	$E(2, \mathbb{R})$ Cartesian	Horicyclic
No 3	$S^{(3)}, \Lambda^{(3)}, O(2,2)$	$S(2, \mathbb{R})$ Polar	Spherical
No 4	$\Lambda^{(3)}, O(2,2)$	$S(2, \mathbb{C})$ Horospherical	Horospherical
No 5	$\Lambda^{(3)}, O(2,2)$	$E(2, \mathbb{R})$ Polar	Horicyclic-polar
No 6	$S^{(3)}, \Lambda^{(3)}, O(2,2)$	$S(2, \mathbb{R})$ Conical	Sphero-elliptic
No 7	$\Lambda^{(3)}, O(2,2)$	$S(2, \mathbb{C})$ Degenerate elliptic I	Spherical-degenerate elliptic I
No 8	$\Lambda^{(3)}, O(2,2)$	$S(2, \mathbb{C})$ Degenerate elliptic II	Spherical-degenerate elliptic II
No 9	$\Lambda^{(3)}, O(2,2)$	$E(2, \mathbb{R})$ Elliptic	Horicyclic-elliptic
No 10	$O(2,2)$	$E(2, \mathbb{C})$ Hyperbolic	Horicyclic-hyperbolic
No 11	$\Lambda^{(3)}, O(2,2)$	$E(2, \mathbb{R})$ Parabolic	Horicyclic-parabolic I
No 12	$O(2,2)$	$E(2, \mathbb{C})$ Semi-parabolic	Horicyclic-parabolic II
No 13	$S^{(3)}, \Lambda^{(3)}, O(2,2)$	-	Elliptic-cylindrical
No 14	$\Lambda^{(3)}, O(2,2)$	-	Elliptic-parabolic
No 15	$\Lambda^{(3)}, O(2,2)$	-	Elliptic-hyperbolic
No 16	$O(2,2)$	-	Parabolic
No 17	$S^{(3)}, \Lambda^{(3)}, O(2,2)$	-	Ellipsoidal
No 18	$\Lambda^{(3)}, O(2,2)$	-	*System 18*
No 19	$O(2,2)$	-	*System 19*
No 20	$\Lambda^{(3)}, O(2,2)$	-	*System 20*
No 21	$O(2,2)$	-	*System 21*

According to [317], the complexification of the two elliptic cylindrical coordinate systems (i.e. spheroidal systems) on the $S(3, \mathbb{R})$-sphere and on the three-dimensional hyperboloid give just one coordinate system on $S(3, \mathbb{C})$, i.e. (13). In particular in [392, 329] coordinate systems on the two-dimensional complex sphere and corresponding superintegrable potentials, and in [329] coordinate systems on the two-dimensional (complex) plane and corresponding superintegrable potentials were discussed, including the corresponding interbases expansions. The goal of [329] was to extend the notion of superintegrable potentials of real spaces to the corresponding complexified spaces. The findings were such that there are in addition to the four coordinate systems on the real two-dimensional Euclidean plane three more coordinate systems and also three more superintegrable potentials. Similarly, in addition to the two coordinate systems on the real two-dimensional sphere there are three more coordinate systems on the complex sphere and also four more superintegrable potentials. This is not surprising because the complex plane contains not only the Euclidean plane but also the pseudo-Euclidean plane (10 coordinate systems [301, 300]) and the complex sphere contains not only the real sphere but also the two-dimensional hyperboloid (9 coordinate systems [301, 309, 425]).

On the other hand, it is also possible to complexify the 34 coordinate systems on the three-dimensional hyperboloid [310, 425]. This gives all 21 systems except (1), (12), (16), (19) and (21). These remaining system can be found be complexifying the coordinate systems on the real hyperboloid $SO(2, 2)$ [317].

In table 8.2 I have indicated which coordinate systems emerges from which real space by complexification; these spaces are indicated by $S^{(3)}$ (real sphere), $\Lambda^{(3)}$ (real hyperboloid), and the systems emerging from $O(2, 2)$ (real hyperboloid), respectively.

In the present section I have displayed the path integral representation for the two-dimensional complex sphere, and in the next one for the three-dimensional complex sphere. Many of these path integral representations remain on a formal level, because the complex sphere is an abstract space. If one wants to consider an actual path integral representation one must choose the actual real space which one wants to investigate and then interprete the corresponding path integral representation in terms of the real space, c.f. the chapters for the path integrals on spheres and pseudospheres, respectively.

Let us note that the complexification requires also the following consideration: The eigenvalues of the Hamiltonian on the three-dimensional complex sphere are denoted by $\propto -\sigma(\sigma + 2)$. On the real sphere this yields with $\sigma = J$ the eigenvalues $\propto J(J + 2)$ whereas on the real three-dimensional hyperboloid one has $\sigma = 1 + \mathrm{i}p$ and therefore the eigenvalues are $\propto (p^2 + 1)$. We must therefore look carefully which manifold we consider if we specify the coordinates including their ranges. Usually, an analytic continuation may be required, which is not performed, however.

We will denote in the following the quantum number of the Eigenvalues of the Hamiltonian by J, irrespective whether there is a discrete or a continuous spectrum. On the real sphere the spectrum is always discrete and on the hyperboloid (two-sheeted) continuous. The kernel in its wave-functions expansion and spectrum on the complex sphere will be displayed in the most cases in a discrete formulation. The corresponding restriction to the sphere or hyperboloid then will decide whether one can keep the discrete formulation as it is or one must analytically continue to the continuous spectrum. However, on the single-sheeted hyperboloid and on the $O(2, 2)$-hyperboloid one has in fact both: a discrete and a continuous part. The latter is not discussed in the sequel and is postponed to a future study. In the enumeration of the coordinate system we keep the convention of the corresponding systems on the sphere and hyperboloid from previous publications [425]. In some cases we note explicitly the correspondences to the sphere and hyperboloid cases to illustrate the examples.

In the following section I will enumerate the path integral representation on the three-dimensional complex sphere.

8.2   The Three-Dimensional Complex Sphere

We have the following path integral representations on the three-dimensional complex sphere [223, 236, 242, 252]

I. Cylindrical, $\vartheta \in [0, \pi/2)$, $\varphi_1, \varphi_2 \in [0, 2\pi)$:

(The cylindrical system exists on $S^{(3)}$, $\Lambda^{(3)}$, and $O(2,2)$)

$$
\int_{\vartheta(t')=\vartheta'}^{\vartheta(t'')=\vartheta''} \mathcal{D}\vartheta(t) \sin\vartheta \cos\vartheta \int_{\varphi_1(t')=\varphi_1'}^{\varphi_1(t'')=\varphi_1''} \mathcal{D}\varphi_1(t) \int_{\varphi_2(t')=\varphi_2'}^{\varphi_2(t'')=\varphi_2''} \mathcal{D}\varphi_2(t)
$$

$$
\times \exp\left\{ \frac{i}{\hbar}\int_{t'}^{t''}\left[\frac{m}{2}\left(\dot\vartheta^2 + \cos^2\vartheta\,\dot\varphi_1^2 + \sin^2\vartheta\,\dot\varphi_2^2 \right) + \frac{\hbar^2}{8m}\left(4 + \frac{1}{\cos^2\vartheta} + \frac{1}{\sin^2\vartheta} \right) \right] dt \right\}
$$

$$
= \sum_{J=0}^{\infty}\sum_{k_1,k_2\in\mathbb{Z}} \frac{e^{i[k_1(\varphi_1''-\varphi_1')+k_2(\varphi_2''-\varphi_2')]}}{4\pi^2}\frac{2(J+1)\left(\frac{J-|m_1|-|m_2|}{2}\right)!\left(\frac{J+|m_1|+|m_2|}{2}\right)!}{\left(\frac{J-|m_1|+|m_2|}{2}\right)!\left(\frac{J+|m_1|-|m_2|}{2}\right)!}
$$

$$
\times (\sin\vartheta'\sin\vartheta'')^{|m_1|}(\cos\vartheta'\cos\vartheta'')^{|m_2|}
$$

$$
\times P_{(J-|m_1|-|m_2|)/2}^{(|m_1|,|m_2|)}(\cos 2\vartheta')\, P_{(J-|m_1|-|m_2|)/2}^{(|m_1|,|m_2|)}(\cos 2\vartheta'')\, e^{-i\hbar TJ(J+2)/2m} \quad . \tag{8.6}
$$

II. Horicyclic, $x, y, z \in \mathbb{R}$:

(The horicyclic system exists on $\Lambda^{(3)}$ and $O(2,2)$)

$$
\int_{x(t')=x'}^{x(t'')=x''}\mathcal{D}x(t)\int_{y(t')=y'}^{y(t'')=y''}\mathcal{D}y(t)\int_{z(t')=z'}^{z(t'')=z''}\mathcal{D}z(t)e^{2ix}\exp\left\{\frac{i}{\hbar}\int_{t'}^{t''}\left[\frac{m}{2}\left(\dot x^2 + e^{2ix}(\dot y^2 + \dot z^2)\right)+\frac{\hbar^2}{2m}\right]dt\right\}
$$

$$
= e^{-i(x''-x')}\int_{\mathbb{R}}dk_y\int_{\mathbb{R}}dk_z\frac{e^{ik_y(y''-y')+ik_z(z''-z')}}{(2\pi)^2}
$$

$$
\times \sum_{J\in\mathbb{N}_0}\tfrac{1}{2}H_{J+1/2}^{(1)}\left(\sqrt{k_y^2+k_z^2}\,e^{-ix''}\right)H_{J+1/2}^{(1)}\left(\sqrt{k_y^2+k_z^2}\,e^{-ix'}\right)e^{-i\hbar TJ(J+2)/2m} \quad . \tag{8.7}
$$

III. Spherical, $\chi, \vartheta \in [0, \pi)$, $\varphi \in [0, 2\pi)$:

(The spherical system exists on $S^{(3)}$, $\Lambda^{(3)}$, and $O(2,2)$)

$$
= \int_{\chi(t')=\chi'}^{\chi(t'')=\chi''}\mathcal{D}\chi(t)\sin^2\chi\int_{\vartheta(t')=\vartheta'}^{\vartheta(t'')=\vartheta''}\mathcal{D}\vartheta(t)\sin\vartheta\int_{\varphi(t')=\varphi'}^{\varphi(t'')=\varphi''}\mathcal{D}\varphi(t)
$$

$$
\times \exp\left\{\frac{i}{\hbar}\int_{t'}^{t''}\left[\frac{m}{2}\left(\dot\chi^2+\sin^2\chi(\dot\vartheta^2+\sin^2\vartheta\,\dot\varphi^2)\right)+\frac{\hbar^2}{8m}\left(4+\frac{1}{\sin^2\chi}\left(1+\frac{1}{\sin^2\vartheta}\right)\right)\right]dt\right\}
$$

$$
= \frac{1}{2\pi^2}\sum_{J=0}^{\infty}(J+1)C_J^1(\cos\psi_{S^{(3)}})\exp\left[-\frac{i\hbar T}{2m}J(J+2)\right] \tag{8.8}
$$

$$= \sum_{J=0}^{\infty} \sum_{m_1,m_2} \Psi_{J,m_1,m_2}(\chi'',\vartheta'',\varphi'')\Psi^*_{J,m_1,m_2}(\chi',\vartheta',\varphi')\exp\left[-\frac{\mathrm{i}\hbar T}{2m}J(J+2)\right], \tag{8.9}$$

$$\Psi_{J,m_1,m_2}(\chi,\vartheta,\varphi) = N^{-1/2}\mathrm{e}^{\mathrm{i}m_1\varphi}(\sin\chi)^{m_1}C_{J-m_1}^{m_1+2}(\cos\chi)(\sin\vartheta)^{m_2}C_{m_1-m_2}^{m_2+3/2}(\cos\vartheta), \tag{8.10}$$

$$N = \frac{2\pi^3 2^{-(1+2m_1+2m_2)}}{(J+1)(m_1+3/2)(J-m_1)!(m_1-m_2)!}\frac{\Gamma(J+m_1+2)\Gamma(m_1+m_2+1)}{\Gamma^2(m_1+1)\Gamma^2(m_2+3/2)}.$$

IV. Horospherical, $x,y\in\mathbb{R}$, $\chi\in[0,\pi]$:

(The horospherical system exists on $\Lambda^{(3)}$ and $O(2,2)$)

$$\int_{\chi(t')=\chi'}^{\chi(t'')=\chi''}\mathcal{D}\chi(t)\int_{x(t')=x'}^{x(t'')=x''}\mathcal{D}x(t)\int_{y(t')=y'}^{y(t'')=y''}\mathcal{D}y(t)\mathrm{e}^{\mathrm{i}x}\sin^2\chi$$

$$\times\exp\left\{\frac{\mathrm{i}}{\hbar}\int_{t'}^{t''}\left[\frac{m}{2}\left(\dot\chi^2+\sin^2\chi(\dot x^2+\mathrm{e}^{2\mathrm{i}x}\dot y^2)\right)+\frac{\hbar^2}{8m}\left(4+\frac{1}{\sin^2\chi}\right)\right]dt\right\}$$

$$=(\sin\chi'\sin\chi'')^{-1/2}\mathrm{e}^{-\frac{1}{2}(x''-x')}\mathrm{e}^{\mathrm{i}\hbar T/2m}\int_{\mathbb{R}}\frac{\mathrm{e}^{\mathrm{i}k_y(y''-y')}}{2\pi}$$

$$\times\sum_{n_x\in\mathbb{N}_0}\tfrac{1}{2}H_{n_x+1/2}^{(1)}(|k_y|\,\mathrm{e}^{-\mathrm{i}x''})H_{n_x+1/2}^{(1)}(|k_y|\,\mathrm{e}^{\mathrm{i}x'})$$

$$\times\sum_{l\in\mathbb{N}_0}\left[(l+n_x+1)\frac{\Gamma(l+2n_x+2))}{l!}\right]P_{l+n_x+1/2}^{-n_x-1/2}(\cos\chi')P_{l+n_x+1/2}^{-n_x-1/2}(\cos\chi'')$$

$$\times\exp\left[-\frac{\mathrm{i}\,\hbar^2(l+n_x)(l+n_x+2)}{\hbar}T\right]. \tag{8.11}$$

V. Horicyclic-Polar, $x\in\mathbb{R}$, $\varrho>0$, $\varphi\in[0,2\pi)$:

(The horicyclic-polar system exists on $\Lambda^{(3)}$ and $O(2,2)$)

$$\int_{x(t')=x'}^{x(t'')=x''}\mathcal{D}x(t)\int_{\varrho(t')=\varrho'}^{\varrho(t'')=\varrho''}\mathcal{D}\varrho(t)\int_{\varphi(t')=\varphi'}^{\varphi(t'')=\varphi''}\mathcal{D}\varphi(t)\varrho\,\mathrm{e}^{2\mathrm{i}x}$$

$$\times\exp\left\{\frac{\mathrm{i}}{\hbar}\int_{t'}^{t''}\left[\frac{m}{2}\left(\dot x^2+\mathrm{e}^{2\mathrm{i}x}(\dot\varrho^2+\varrho^2\dot\varphi^2)\right)+\mathrm{e}^{-2\mathrm{i}x}\frac{\hbar^2}{8m\varrho^2}\right]dt\right\}$$

$$=\mathrm{e}^{-\mathrm{i}(x''-x')}\sum_{\nu\in\mathbb{Z}}\frac{\mathrm{e}^{\mathrm{i}\nu(\varphi''-\varphi')}}{2\pi}\int_0^{\infty}dk_\varrho k_\varrho J_\nu(k_\varrho\varrho')J_\nu(k_\varrho\varrho'')$$

$$\times\sum_{J\in\mathbb{N}_0}\tfrac{1}{2}H_{J+1/2}^{(1)}(|k_\varrho|\,\mathrm{e}^{-\mathrm{i}x''})H_{J+1/2}^{(1)}(|k_\varrho|\,\mathrm{e}^{\mathrm{i}x'})\exp\left[-\frac{\mathrm{i}\,\hbar^2 J(J+2)}{\hbar}T\right]. \tag{8.12}$$

VI. Sphero-Elliptic, $\chi\in[0,\pi)$, $-K\leq\alpha\leq K$, $-2K'\leq\beta\leq 2K'$:

(The sphero-elliptic system exists on $S^{(3)}$, $\Lambda^{(3)}$, and $O(2,2)$)

$$\int_{\chi(t')=\chi'}^{\chi(t'')=\chi''}\mathcal{D}\chi(t)\sin^2\chi\int_{\alpha(t')=\alpha'}^{\alpha(t'')=\alpha''}\mathcal{D}\alpha(t)\int_{\beta(t')=\beta'}^{\beta(t'')=\beta''}\mathcal{D}\beta(t)(k^2\mathrm{cn}^2\alpha+k'^2\mathrm{cn}^2\beta)$$

$$\times \exp\left\{\frac{im}{2\hbar}\int_{t'}^{t''}\left[\dot\chi^2 + \sin^2\chi(k^2\mathrm{cn}^2\alpha + k'^2\mathrm{cn}^2\beta)(\dot\alpha^2 + \dot\beta^2)\right]dt + \frac{i\hbar T}{2m}\right\}$$

$$= \sum_{J=0}^{\infty}\sum_{l=-J}^{J}\sum_{\lambda}\sum_{p,q=\pm}\Lambda_{l,h}^{p}(\alpha'')\Lambda_{l,h}^{p*}(\alpha')\Lambda_{l,\tilde h}^{q}(\beta'')\Lambda_{l,\tilde h}^{q*}(\beta')$$

$$\times (J+1)\frac{(l+J+1)!}{|J-l|!}\,P_{J+1/2}^{-l-1/2}(\sin\chi'')P_{J+1/2}^{-l-1/2}(\sin\chi')\,e^{-i\hbar T J(J+2)/2m}\ . \tag{8.13}$$

VII. Spherical-degenerate Elliptic I, $\chi, \vartheta \in [0,\pi)$, , $\tau_1, \tau_2 \in \mathbb{R}$:

(The spherical-degenerate elliptic I system exists on $\Lambda^{(3)}$ and $O(2,2)$)

$$\int_{\chi(t')=\chi'}^{\chi(t'')=\chi''}\mathcal{D}\chi(t)\int_{\tau_1(t')=\tau_1'}^{\tau_1(t'')=\tau_1''}\mathcal{D}\tau_1(t)\int_{\tau_2(t')=\tau_2'}^{\tau_2(t'')=\tau_2''}\mathcal{D}\tau_2(t)\sin^2\chi\left(\frac{1}{\cosh^2\tau_1}-\frac{1}{\cosh^2\tau_2}\right)$$

$$\times\exp\left\{\frac{i}{\hbar}\int_{t'}^{t''}\left[\frac{m}{2}\left(\dot\chi^2+\sin^2\chi\left(\frac{1}{\cosh^2\tau_1}-\frac{1}{\cosh^2\tau_2}\right)(\dot\tau_1^2-\dot\tau_2^2)\right)+\frac{\hbar^2}{8m}\left(4+\frac{1}{\sin^2\chi}\right)\right]dt\right\}$$

$$(\sin\chi'\sin\chi'')^{-1}e^{i\hbar T/2m}\int_0^\infty dp\,p\sinh\pi p\int_0^\infty\frac{dk\,k\sinh\pi k}{(\cosh^2\pi k+\sinh^2\pi p)^2}$$

$$\times\sum_{\epsilon,\epsilon'=\pm 1}P_{ip-1/2}^{ik}(\epsilon\tanh\tau_1'')P_{ip-1/2}^{-ik}(\epsilon\tanh\tau_1')P_{ik-1/2}^{ip}(\epsilon'\tanh\tau_2'')P_{ik-1/2}^{-ip}(\epsilon'\tanh\tau_2')$$

$$\times\sum_{J\in\mathbb{N}_0}\left[(J+p+\tfrac{1}{2})\frac{\Gamma(p+2J+1)}{J!}\right]P_{J+p+1/2}^{-J-1/2}(\cos\chi')P_{J+p+1/2}^{-J-1/2}(\cos\chi'')$$

$$\times\exp\left[-\frac{i}{\hbar}\frac{\hbar^2 J(J+2)}{2m}T\right]\ . \tag{8.14}$$

VIII. Spherical-degenerate Elliptic II, $\chi, \vartheta \in [0,\pi)$, $\xi, \eta > 0$:

(The spherical-degenerate elliptic II system exists on $\Lambda^{(3)}$ and $O(2,2)$)

$$\int_{\chi(t')=\chi'}^{\chi(t'')=\chi''}\mathcal{D}\chi(t)\int_{\xi(t')=\xi'}^{\xi(t'')=\xi''}\mathcal{D}\xi(t)\int_{\eta(t')=\eta'}^{\eta(t'')=\eta''}\mathcal{D}\eta(t)\sin^2\chi\left(\frac{1}{\eta^2}-\frac{1}{\xi^2}\right)$$

$$\times\exp\left\{\frac{i}{\hbar}\int_{t'}^{t''}\left[\frac{m}{2}\left(\dot\chi^2+\sin^2\chi\left(\frac{1}{\eta^2}-\frac{1}{\xi^2}\right)(\dot\xi^2-\dot\eta^2)\right)+\frac{\hbar^2}{8m}\left(4+\frac{1}{\sin^2\chi}\right)\right]dt\right\}$$

$$= \frac{1}{4\pi^2}\sqrt{\mu'\mu''\nu'\nu''}$$

$$\times\int dk\int_0^\infty dp\,p\sinh^2\pi p H_p^{(1)}(\sqrt{k}\,\xi')H_p^{(1)*}(\sqrt{k}\,\xi'')H_p^{(1)}(\sqrt{k}\,\eta')H_p^{(1)*}(\sqrt{k}\,\eta'')$$

$$\times\sum_{J\in\mathbb{N}_0}\left[(J+p+\tfrac{1}{2})\frac{\Gamma(p+2J+1)}{J!}\right]P_{J+p+1/2}^{-J-1/2}(\cos\chi')P_{J+p+1/2}^{-J-1/2}(\cos\chi'')$$

$$\times\exp\left[-\frac{i}{\hbar}\frac{\hbar^2 J(J+2)}{2m}T\right]\ . \tag{8.15}$$

IX. Horicyclic-Elliptic, $x\,,\tau_1,\tau_2\in\mathbb{R}$:

(The horicyclic-elliptic system exists on $\Lambda^{(3)}$ and $O(2,2)$)

$$\int_{x(t')=x'}^{x(t'')=x''}\mathcal{D}x(t)\int_{\tau_1(t')=\tau_1'}^{\tau_1(t'')=\tau_1''}\mathcal{D}\tau_1(t)\int_{\tau_2(t')=\tau_2'}^{\tau_2(t'')=\tau_2''}\mathcal{D}\tau_2(t)\mathrm{e}^{2\mathrm{i}x}(\cosh^2\tau_1-\cosh^2\tau_2)$$

$$\times\exp\left\{\frac{\mathrm{i}}{\hbar}\int_{t'}^{t''}\left[\frac{m}{2}\left(\dot{x}^2+\mathrm{e}^{2\mathrm{i}x}(\cosh^2\tau_1-\cosh^2\tau_2)(\dot{\tau}_1^2-\dot{\tau}_2^2)\right)+\frac{\hbar^2}{2m}\right]dt\right\}$$

$$=\frac{1}{8\pi}\int_0^\infty pdp\int_{\mathbb{R}}dk\,\mathrm{e}^{-\pi k}\mathrm{Me}_{\mathrm{i}k}(\tau_2'';\tfrac{p^2}{4})\mathrm{Me}_{\mathrm{i}k}^*(\tau_2';\tfrac{p^2}{4})M_{\mathrm{i}k}^{(3)}(\tau_1'';\tfrac{p}{2})M_{\mathrm{i}k}^{(3)*}(\tau_1';\tfrac{p}{2})$$

$$\times\sum_{J\in\mathbb{N}_0}\tfrac{1}{2}H_{J+1/2}^{(1)}(|k_\varrho|\,\mathrm{e}^{-\mathrm{i}x''})H_{J+1/2}^{(1)}(|k_\varrho|\,\mathrm{e}^{\mathrm{i}x'})\exp\left[-\frac{\mathrm{i}}{\hbar}\frac{\hbar^2J(J+2)}{2m}T\right]. \qquad (8.16)$$

X. Horicyclic-Hyberbolic, $x,y,z\in\mathbb{R}$:

(The horicyclic-hyperbolic system exists only on $O(2,2)$)

$$\int_{x(t')=x'}^{x(t'')=x''}\mathcal{D}x(t)\int_{y(t')=y'}^{y(t'')=y''}\mathcal{D}y(t)\int_{z(t')=z'}^{z(t'')=z''}\mathcal{D}z(t)\mathrm{e}^{2\mathrm{i}x}(\mathrm{e}^{2y}+\mathrm{e}^{2z})$$

$$\times\exp\left\{\frac{\mathrm{i}}{\hbar}\int_{t'}^{t''}\left[\frac{m}{2}\left(\dot{x}^2+\mathrm{e}^{2\mathrm{i}x}(\mathrm{e}^{2y}+\mathrm{e}^{2z})(\dot{y}^2-\dot{z}^2)\right)+\frac{\hbar^2}{2m}\right]dt\right\}$$

$$=\frac{2}{\pi^4}\int_0^\infty dk\,k\sinh\pi k\int_0^\infty dp\,pK_{\mathrm{i}k}(p\,\mathrm{e}^{y''})K_{\mathrm{i}k}(p\,\mathrm{e}^{y'})K_{\mathrm{i}k}(-\mathrm{i}p\,\mathrm{e}^{z'})K_{\mathrm{i}k}(\mathrm{i}p\,\mathrm{e}^{z''})$$

$$\times\sum_{J\in\mathbb{N}_0}\tfrac{1}{2}H_{J+1/2}^{(1)}(p\,\mathrm{e}^{-\mathrm{i}x''})H_{J+1/2}^{(1)}(p\,\mathrm{e}^{\mathrm{i}x'})\exp\left[-\frac{\mathrm{i}}{\hbar}\frac{\hbar^2J(J+2)}{2m}T\right]. \qquad (8.17)$$

XI. Horicyclic-Parabolic I, $x\,,\xi\in\mathbb{R}$, $\eta>0$:

(The horicyclic-parabolic I system exists on $\Lambda^{(3)}$ and $O(2,2)$)

$$\int_{x(t')=x'}^{x(t'')=x''}\mathcal{D}x(t)\int_{\xi(t')=\xi'}^{\xi(t'')=\xi''}\mathcal{D}\xi(t)\int_{\eta(t')=\eta'}^{\eta(t'')=\eta''}\mathcal{D}\eta(t)\mathrm{e}^{2\mathrm{i}x}(\xi^2+\eta^2)$$

$$\times\exp\left\{\frac{\mathrm{i}}{\hbar}\int_{t'}^{t''}\left[\frac{m}{2}\left(\mathrm{d}x^2+\mathrm{e}^{2\mathrm{i}x}(\xi^2+\eta^2)(\mathrm{d}\xi^2+\mathrm{d}\eta^2)\right)+\frac{\hbar^2}{2m}\right]dt\right\}$$

$$=\int_{\mathbb{R}}d\zeta\int_{\mathbb{R}}\frac{dp}{32\pi^4}$$

$$\times\left\{\begin{array}{l}|\Gamma(\tfrac{1}{4}+\tfrac{\mathrm{i}\zeta}{2p})|^2E_{-1/2+\mathrm{i}\zeta/p}^{(0)}(\mathrm{e}^{-\mathrm{i}\pi/4}\sqrt{2p}\,\xi'')E_{-1/2-\mathrm{i}\zeta/p}^{(0)}(\mathrm{e}^{-\mathrm{i}\pi/4}\sqrt{2p}\,\eta'')\\ |\Gamma(\tfrac{3}{4}+\tfrac{\mathrm{i}\zeta}{2p})|^2E_{-1/2+\mathrm{i}\zeta/p}^{(1)}(\mathrm{e}^{\mathrm{i}\pi/4}\sqrt{2p}\,\xi'')E_{-1/2-\mathrm{i}\zeta/p}^{(1)}(\mathrm{e}^{\mathrm{i}\pi/4}\sqrt{2p}\,\eta'')\end{array}\right\}$$

$$\times\left\{\begin{array}{l}|\Gamma(\tfrac{1}{4}+\tfrac{\mathrm{i}\zeta}{2p})|^2E_{-1/2-\mathrm{i}\zeta/p}^{(0)}(\mathrm{e}^{\mathrm{i}\pi/4}\sqrt{2p}\,\xi')E_{-1/2+\mathrm{i}\zeta/p}^{(0)}(\mathrm{e}^{\mathrm{i}\pi/4}\sqrt{2p}\,\eta')\\ |\Gamma(\tfrac{3}{4}+\tfrac{\mathrm{i}\zeta}{2p})|^2E_{-1/2-\mathrm{i}\zeta/p}^{(1)}(\mathrm{e}^{-\mathrm{i}\pi/4}\sqrt{2p}\,\xi')E_{-1/2+\mathrm{i}\zeta/p}^{(1)}(\mathrm{e}^{-\mathrm{i}\pi/4}\sqrt{2p}\,\eta')\end{array}\right\}$$

$$\times \sum_{J \in \mathbb{N}_0} \tfrac{1}{2} H_{J+1/2}^{(1)}(p \, \mathrm{e}^{-\mathrm{i}x''}) H_{J+1/2}^{(1)}(p \, \mathrm{e}^{\mathrm{i}x'}) \exp\left[-\frac{\mathrm{i}}{\hbar} \frac{\hbar^2 J(J+2)}{2m} T \right] . \qquad (8.18)$$

XII. Horicyclic-Parabolic II, $x, \xi \in \mathbb{R}$, $\eta > 0$:

(The horicyclic-parabolic II system exists only on $O(2,2)$)

$$\int_{x(t')=x'}^{x(t'')=x''} \mathcal{D}x(t) \int_{\xi(t')=\xi'}^{\xi(t'')=\xi''} \mathcal{D}\xi(t) \int_{\eta(t')=\eta'}^{\eta(t'')=\eta''} \mathcal{D}\eta(t) 4\mathrm{e}^{2\mathrm{i}x}(\xi - \eta)$$

$$\times \exp\left\{ \frac{\mathrm{i}}{\hbar} \int_{t'}^{t''} \left[\frac{m}{2}\left(\dot{x}^2 + 4\mathrm{e}^{2\mathrm{i}x}(\xi - \eta)(\dot{\xi}^2 - \dot{\eta}^2) \right) + \frac{\hbar^2}{2m} \right] dt \right\}$$

$$= 16 \int_0^\infty \frac{dp}{p^{1/3}} \int_{\mathbb{R}} d\zeta \, \mathrm{Ai}\left[-\left(\xi' + \sqrt{2m}\frac{\zeta}{p^2} \right)p^{2/3} \right] \mathrm{Ai}\left[-\left(\xi'' + \sqrt{2m}\frac{\zeta}{p^2} \right)p^{2/3} \right]$$

$$\times \mathrm{Ai}\left[-\left(\eta' + \sqrt{2m}\frac{\zeta}{p^2} \right)p^{2/3} \right] \mathrm{Ai}\left[-\left(\eta'' + \sqrt{2m}\frac{\zeta}{p^2} \right)p^{2/3} \right]$$

$$\times \sum_{J \in \mathbb{N}_0} \tfrac{1}{2} H_{J+1/2}^{(1)}(p \, \mathrm{e}^{-\mathrm{i}x''}) H_{J+1/2}^{(1)}(p \, \mathrm{e}^{\mathrm{i}x'}) \exp\left[-\frac{\mathrm{i}}{\hbar} \frac{\hbar^2 J(J+2)}{2m} T \right] . \qquad (8.19)$$

XIII. Elliptic-Cylindrical, $\varphi \in [0, 2\pi)$, $-K \leq \alpha \leq K$, $-2K' \leq \beta \leq 2K'$:

(The elliptic-cylindrical system exists on $S^{(3)}$, $\Lambda^{(3)}$, and $O(2,2)$)

$$\int_{\alpha(t')=\alpha'}^{\alpha(t'')=\alpha''} \mathcal{D}\alpha(t) \int_{\beta(t')=\beta'}^{\beta(t'')=\beta''} \mathcal{D}\beta(t)(k^2\mathrm{cn}^2\alpha + k'^2\mathrm{cn}^2\beta)\mathrm{cn}\alpha\mathrm{cn}\beta \int_{\varphi(t')=\varphi'}^{\varphi(t'')=\varphi''} \mathcal{D}\varphi(t)$$

$$\times \exp\left\{ \frac{\mathrm{i}}{\hbar} \int_{t'}^{t''} \left[\frac{m}{2}\left((k^2\mathrm{cn}^2\alpha + k'^2\mathrm{cn}^2\beta)(\dot{\alpha}^2 + \dot{\beta}^2) + \mathrm{cn}^2\alpha\mathrm{cn}^2\beta\dot{\varphi}^2 \right) \right. \right.$$

$$\left. \left. + \frac{\hbar^2}{8m}\frac{1}{k^2\mathrm{cn}^2\alpha + k'^2\mathrm{cn}^2\beta}\left(\frac{\mathrm{sn}^2\beta\mathrm{dn}^2\beta}{\mathrm{cn}^2\beta} + k^4\frac{\mathrm{sn}^2\alpha\mathrm{dn}^2\alpha}{\mathrm{cn}^2\alpha} \right) \right] dt + \frac{\mathrm{i}\hbar T}{2m} \right\}$$

$$= \frac{1}{2\pi} \sum_{J=0}^\infty \sum_{r,p=\pm 1} \sum_{qk_2} \mathrm{e}^{\mathrm{i}k_2(\varphi''-\varphi')} \mathrm{e}^{-2\mathrm{i}\hbar TJ(J+2)/2m}$$

$$\times \psi_{1,Jqk_2}^{(r,p)}(\alpha''; a)\psi_{1,Jqk_2}^{(r,p)*}(\alpha'; a)\psi_{2,Jqk_2}^{(r,p)}(\beta''; a)\psi_{2,Jqk_2}^{(r,p)*}(\beta'; a) . \qquad (8.20)$$

XIV. Elliptic-Parabolic, $\tau_{1,2} > 0$, $\tau_3 \in \mathbb{R}$: ($\eta = \mathrm{i}\sqrt{k_{\tau_3}^2 - \frac{1}{4}}$, $\nu = 2mE/\hbar^2 + 1 \equiv p^2 + 1$, $\Psi_p^{(\mu,\nu)}(\omega)$: modified Pöschl–Teller functions)

(The elliptic-parabolic system exists on $\Lambda^{(3)}$ and $O(2,2)$)

$$\int_{\tau_1(t')=\tau_1'}^{\tau_1(t'')=\tau_1''} \mathcal{D}\tau_1(t) \int_{\tau_2(t')=\tau_2'}^{\tau_2(t'')=\tau_2''} \mathcal{D}\tau_2(t) \int_{\tau_3(t')=\tau_3'}^{\tau_3(t'')=\tau_3''} \mathcal{D}\tau_3(t)(\tanh^2\tau_1 - \tanh^2\tau_2)\tanh\tau_1\tanh\tau_2$$

$$\times \exp\left\{\frac{i}{\hbar}\int_{t'}^{t''}\left[\frac{m}{2}\left((\tanh^2\tau_1-\tanh^2\tau_2)(\dot\tau_1^2-\dot\tau_2^2)+\tanh^2\tau_1\tanh^2\tau_2\dot\tau_3^2\right)-\frac{3\hbar^2}{8m}\right]dt\right\}$$

$$=(\tanh\tau_1'\tanh\tau_1''\tanh\tau_2'\tanh\tau_2'')^{-1/2}$$

$$\times\int_{\mathbb{R}}dk\frac{e^{ik\tau_3(\tau_3''-\tau_3')}}{2\pi}\int_0^\infty dk\int_0^\infty dp\exp\left[-\frac{i\hbar T}{2m}(p^2+1)\right]$$

$$\times\Psi_k^{(\eta,\nu)}(\tau_2'')\Psi_k^{(\eta,\nu)\,*}(\tau_2')\Psi_k^{(\eta,\nu)}(\tau_1'')\Psi_k^{(\eta,\nu)\,*}(\tau_1') \ . \tag{8.21}$$

<u>XV. Elliptic-Hyperbolic, $\tau_{1,2,3}\in\mathbb{R}$,:</u>

(The elliptic-hyperbolic system exists on $\Lambda^{(3)}$ and $O(2,2)$)

$$\int_{\tau_1(t')=\tau_1'}^{\tau_1(t'')=\tau_1''}\mathcal{D}\tau_1(t)\int_{\tau_2(t')=\tau_2'}^{\tau_2(t'')=\tau_2''}\mathcal{D}\tau_2(t)\int_{\tau_3(t')=\tau_3'}^{\tau_3(t'')=\tau_3''}\mathcal{D}\tau_3(t)\left(\frac{1}{\cosh^2\tau_1}-\frac{1}{\cosh^2\tau_2}\right)\frac{1}{\cosh\tau_1\cosh\tau_2}$$

$$\times\exp\left\{\frac{i}{\hbar}\int_{t'}^{t''}\left[\frac{m}{2}\left(\frac{\cosh^2\tau_2-\cosh^2\tau_1}{\cosh^2\tau_1\cosh^2\tau_2}(\dot\tau_1^2-\dot\tau_2^2)+\frac{\dot\tau_3^2}{\cosh^2\tau_1\cosh^2\tau_2}\right)-\frac{3\hbar^2}{8m}\right]dt\right\}$$

(coordinate substitution $\vartheta=i\tau_2$ and $\varrho=\tau_3$)

$$=\sqrt{\cosh a'\cosh a''}\,\cos\vartheta'\cos\vartheta''\int_{\mathbb{R}}\frac{dk_\varrho}{2\pi}e^{ik_\varrho(\varrho''-\varrho')}$$

$$\times\int_0^\infty dp\,\sinh\pi p\int_0^\infty\frac{dk\,k\sinh\pi k}{(\cosh^2\pi k+\sinh^2\pi p)^2}\exp\left[-\frac{i\hbar T}{2m}(p^2+1)\right]$$

$$\times\sum_{\epsilon,\epsilon'=\pm1}S_{ip-1/2}^{ik\,(1)}(\epsilon\tanh a'';ik_\varrho)S_{ip-1/2}^{ik\,(1)\,*}(\epsilon\tanh a';ik_\varrho)$$

$$\times\mathrm{ps}_{ik-1/2}^{ip}(\epsilon'\sin\vartheta'';-k_\varrho^2)\mathrm{ps}_{ik-1/2}^{ip\,*}(\epsilon'\sin\vartheta';-k_\varrho^2) \ . \tag{8.22}$$

<u>XVI. Parabolic, , $\xi,\eta>0$, $\tau\in\mathbb{R}$,:</u>

(The parabolic system exists only on $O(2,2)$)

$$\int_{\xi(t')=\xi'}^{\xi(t'')=\xi''}\mathcal{D}\xi(t)\int_{\eta(t')=\eta'}^{\eta(t'')=\eta''}\mathcal{D}\eta(t)\int_{\tau(t')=\tau'}^{\tau(t'')=\tau''}\mathcal{D}\tau(t)\frac{\xi^2+\eta^2}{\xi^3\eta^3}$$

$$\times\exp\left\{\frac{i}{\hbar}\int_{t'}^{t''}\left[\frac{m}{2}\left(\frac{\xi^2+\eta^2}{\xi^2\eta^2}(\dot\xi^2+\dot\eta^2)+\frac{\dot\tau^2}{\xi^2\eta^2}\right)-\frac{3\hbar^2}{8m}\right]dt\right\}$$

$$=\int_{\mathbb{R}}dk\sum_{n_\xi,n_\eta}\Psi_{kn_\xi n_\eta}(\xi',\eta',x')\Psi_{kn_\xi n_\eta}^*(\xi'',\eta'',x'')e^{-2i\hbar TJ(J+2)/2m} \tag{8.23}$$

$$\Psi_{kn_\xi n_\eta}(\xi,\eta,\tau)=\frac{e^{ik\tau}}{\sqrt{2\pi}}\sqrt{\hbar|k|\frac{\xi n_\xi!}{\Gamma(n_\xi+\lambda+1)}\frac{\eta n_\eta!}{\Gamma(n_\eta+\lambda+1)}\left(\frac{m^2\omega^2}{\hbar^2}\xi\eta\right)^{J+1}}$$

$$\times L_{n_\xi}^{(J+1)}\left(\frac{m\omega}{\hbar}\xi^2\right)L_{n_\eta}^{(J+1)}\left(\frac{m\omega}{\hbar}\eta^2\right)\exp\left[-\frac{m\omega}{2\hbar}(\xi^2+\eta^2)\right] \ . \tag{8.24}$$

XVII. <u>Ellipsoidal</u>, $d < \varrho_3 < c < \varrho_2 < b < \varrho_1 < a$:

(The ellipsoidal system exists on $S^{(3)}, \Lambda^{(3)}$, and $O(2,2)$)

(The wave-functions are only known on the sphere)

$$
\int_{\varrho_1(t')=\varrho_1'}^{\varrho_1(t'')=\varrho_1''} \mathcal{D}\varrho_1(t) \int_{\varrho_2(t')=\varrho_2'}^{\varrho_2(t'')=\varrho_2''} \mathcal{D}\varrho_2(t) \int_{\varrho_3(t')=\varrho_3'}^{\varrho_3(t'')=\varrho_3''} \mathcal{D}\varrho_3(t) \frac{(\varrho_2 - \varrho_1)(\varrho_3 - \varrho_2)(\varrho_3 - \varrho_1)}{8\sqrt{P(\varrho_1)P(\varrho_2)P(\varrho_3)}}
$$

$$
\times \exp\left\{\frac{i}{\hbar}\int_{t'}^{t''}\left[\frac{m}{2}\sum_{i=1}^{3}g_{\varrho_i\varrho_i}\dot{\varrho}_i^2 - \Delta V_{PF}(\boldsymbol{\varrho})\right]dt\right\}
$$

$$
= \sum_{2s=0}^{\infty}\sum_{\lambda,\mu}e^{-2i\hbar Ts(s+1)/m}\Psi_{s,\lambda,\mu}^{*}(\varrho_1',\varrho_2',\varrho_3')\Psi_{s,\lambda,\mu}(\varrho_1'',\varrho_2'',\varrho_3'') \ . \tag{8.25}
$$

General Expressions for the Sphere $S^{(3)}$:

General Expression for the Propagator ($\cos\psi_{S^{(3)}}$ is the invariant distance):

$$
= \frac{e^{i\hbar T/2m}}{4\pi^2}\frac{d}{d\cos\psi_{S^{(3)}}}\Theta_3\left(\frac{\psi_{S^{(3)}}}{2}\bigg| - \frac{\hbar T}{2\pi m}\right) \ . \tag{8.26}
$$

General Form for the Green's Function ($\gamma = -1/2 + \sqrt{2mE/\hbar^2 + 1}$):

$$
= \int_{\mathbb{R}}\frac{dE}{2\pi i}e^{-iET/\hbar}\frac{m}{2\pi\hbar^2}\frac{\sin[(\pi - \psi_{S^{(3)}})(\gamma + 1/2)]}{\sin[\pi(\gamma + 1/2)]\sin\psi_{S^{(3)}}} \ . \tag{8.27}
$$

General Expressions for the Pseudosphere $\Lambda^{(3)}$:

General Expression for the Propagator ($\cosh d$ is the invariant distance on $\Lambda^{(3)}$):

$$
= \left(\frac{m}{2\pi i\hbar T}\right)^{3/2}\frac{d_{\Lambda^{(3)}}(\mathbf{u}'',\mathbf{u}')}{\sinh d_{\Lambda^{(3)}}(\mathbf{u}'',\mathbf{u}')}\exp\left[\frac{im}{2\hbar T}d_{\Lambda^{(3)}}^2(\mathbf{u}'',\mathbf{u}') - \frac{i\hbar T}{2m}\right] \ . \tag{8.28}
$$

General Expression for the Green's Function:

$$
= \int_{\mathbb{R}}\frac{dE}{2\pi i}e^{-iET/\hbar}\frac{-m}{\pi^2\hbar^2\sinh d_{\Lambda^{(3)}}(\mathbf{u}'',\mathbf{u}')}\mathcal{Q}_{-1/2-i\sqrt{2mE/\hbar^2-1}}^{1/2}\left(\cosh d_{\Lambda^{(3)}}(\mathbf{u}'',\mathbf{u}')\right) \ . \tag{8.29}
$$

8.3   Path Integral Evaluations on the Complex Sphere

8.3.1   Path Integral Representations on S_{3C}: Part I

System 1: Cylindrical.

The cylindrical system exists on $S^{(3)}, \Lambda^{(3)}$, and $O(2,2)$, c.f. (6.28) on $S^{(3)}$. The corresponding path integral representation on $\Lambda^{(3)}$ is given by (7.34) and can be interpreted as an analytic continuation from a pure discrete to a pure continuous spectrum.

System 2: Horicyclic.

The horicyclic system exists on $\Lambda^{(3)}$ and $O(2,2)$. In the path integral evaluation the variables y and z are Gaussian, and one obtains:

$$K^{(S_{3C})}(x'', x', y'', y', z'', z' : T)$$
$$= e^{-i(x''-x')} \int_{\mathbb{R}} dk_y \int_{\mathbb{R}} dk_z \frac{e^{ik_y(y''-y')+ik_z(z''-z')}}{(2\pi)^2}$$
$$\times \int_{x(t')=x'}^{x(t'')=x''} \mathcal{D}x(t) \exp\left\{\frac{i}{\hbar} \int_0^T \left[\frac{m}{2}\dot{x}^2 - \frac{\hbar^2}{2m}(k_y^2 + k_z^2)e^{-2ix}\right] dt\right\}. \quad (8.30)$$

The new feature is now the complex potential $V(x) = \frac{\hbar^2}{2m}(k_y^2 + k_z^2)e^{-2ix}$. The remaining path integral in the variable x we can solve this path integral by an analytic continuation of the Liouville path integral solution. In [95] it was shown that the proper continuation are Hankel functions $\propto 2^{-1/2}H^{(1)}_{n+1/2}(k\,e^{-ix})$, hence we obtain (8.7) and are done.

The principal new representation we have achieved herewith is the one for the complex Liouville potential, i.e.

$$\int_{x(t')=x'}^{x(t'')=x''} \mathcal{D}x(t) \exp\left[\frac{i}{\hbar}\int_0^T \left(\frac{m}{2}\dot{x}^2 - \frac{\hbar^2}{2m}k^2 e^{-2ix}\right) dt\right]$$
$$= \sum_{J\in\mathbb{N}_0} \frac{1}{2}H^{(1)}_{J+1/2}\left(k\,e^{-ix''}\right) H^{(1)}_{J+1/2}\left(k\,e^{-ix'}\right) \exp\left[-\frac{i}{\hbar}\frac{\hbar^2 J(J+2)}{2m}T\right]. \quad (8.31)$$

This path integral representation is very useful in evaluating several path integral representations involving other horicyclic coordinate systems.

System 3: Spherical.

The spherical system exists on $S^{(3)}$, $\Lambda^{(3)}$, and $O(2,2)$. I have stated the well-known representation on the real three-dimensional sphere (6.28). The corresponding path integral representation on $\Lambda^{(3)}$ is given by (7.40) and can be interpreted as an analytic continuation from a pure discrete to a pure continuous spectrum.

Systems 4 and 5: Horospherical and Horicyclic-Polar.

These two coordinate systems exist on $\Lambda^{(3)}$ and $O(2,2)$. In these two systems usual path integral techniques apply, i.e. the symmetric Pöschl–Teller potential (horospherical) and the $1/r^2$ potential (horicyclic-polar), respectively. The new feature is the already mentioned complex potential $V(x) = \propto e^{-2ix}$ in the path integral. Applying the solution from the horicyclic system gives the representations (8.11, 8.11).

System 6: Sphero-Elliptic.

The sphero-elliptic system exists on $S^{(3)}, \Lambda^{(3)}$, and $O(2,2)$. I have stated the well-known representation on the real three-dimensional sphere (6.29). The corresponding path integral representation on $\Lambda^{(3)}$ is given by (7.38) and can be interpreted as an analytic continuation from a pure discrete to a pure continuous spectrum.

System 7: Spherical-Degenerate Elliptic I.

The spherical-degenerate elliptic-I system exists on $\Lambda^{(3)}$ and $O(2,2)$. In the spherical-degenerate elliptic-I systems we have a combination of a path integration of the elliptic parabolic systems on the two-dimensional hyperboloid (7.7) and the spherical system. This gives the result (8.14).

System 8: Spherical-Degenerate Elliptic II.

The spherical-degenerate elliptic-II system exists on $\Lambda^{(3)}$ and $O(2,2)$. In the path integral for the spherical-degenerate elliptic-II one has to perform a space-time transformation to separate off the spherical path integration in the variable χ and in the variables ξ, η, respectively. The corresponding Green function we can write as follows:

$$
\begin{aligned}
\hat{G}(\xi'', \xi', \eta'', \eta'; \mathcal{E}) &= \int_0^\infty ds'' \hat{K}(\xi'', \xi', \eta'', \eta'; s'') \\
&= \frac{m^2}{\hbar^3} \sqrt{\xi'\xi''\eta'\eta''} \int_0^\infty \frac{ds''}{s''} \int \frac{d\mathcal{E}}{2\pi i} e^{i\mathcal{E}s''/\hbar} \\
&\quad \times \exp\left[\frac{m}{2i\hbar s''}(\xi'^2 + \xi''^2)\right] I_\lambda\left(\frac{im\xi'\xi''}{\hbar s''}\right) I_{\tilde{\lambda}}\left(\sqrt{-2m\mathcal{E}}\,\frac{\eta_<}{\hbar}\right) K_{\tilde{\lambda}}\left(\sqrt{-2m\mathcal{E}}\,\frac{\eta_>}{\hbar}\right) \quad (8.32) \\
&= \frac{1}{4\pi^2} \sqrt{\mu'\mu''\nu'\nu''} \int d\kappa \int_0^\infty \frac{dk\, k \sinh^2 \pi k}{\hbar^2 k^2/2m - \mathcal{E}} \\
&\quad \times H_{-i\tilde{p}}^{(1)}(\sqrt{\kappa}\,\xi') H_{-i\tilde{p}}^{(1)*}(\sqrt{\kappa}\,\xi'') H_{-i\tilde{p}}^{(1)}(\sqrt{\kappa}\,\eta') H_{-i\tilde{p}}^{(1)*}(\sqrt{\kappa}\,\eta'') \quad (8.33)
\end{aligned}
$$

($\mathcal{E} = (\lambda^2 - \frac{1}{4})/2m$). Combing the results yields the representation (8.15).

System 9: Horicyclic-Elliptic.

The horicyclic-elliptic system exists on $\Lambda^{(3)}$ and $O(2,2)$. The path integral in the variables $\tau_{1,2}$ has the form of the path integral representation of an elliptic coordinate system on the two-dimensional pseudo-Euclidean plane. Together with the x-path integration of the horicyclic-polar system this yields the final result (8.16). $\mathrm{Me}_\nu(z)$ and $\mathrm{Me}_\nu^{(3)}(z)$ are Mathieu-functions [399], which are typical for the quantum motion in elliptic coordinates in two dimensions.

System 10: Horicyclic-Hyperbolic.

The horicyclic-hyperbolic system exists on $\Lambda^{(3)}$ and $O(2,2)$. We start by considering the (y,z)-subpath integration. This two-dimensional sub-system corresponds to the second of the hyperbolic systems on the two-dimensional pseudo-Euclidean plane $E(1,1)$. Using the result of the x-path integration of the horicyclic-polar system we get finally (8.17).

System 11: Horicyclic-Parabolic I.

The horicyclic-parabolic-I system exists on $\Lambda^{(3)}$ and $O(2,2)$. For this path integral we exploit the path integral representation on the two-dimensional Euclidean plane in parabolic coordinates and the already mentioned complex potential $V(x) = \propto$ e^{-2ix}. The wave-functions on S_{3C} $E^{(0)}_{-1/2-i\zeta/p}(z)$ are of even parity, whereas the wave-functions $E^{(1)}_{-1/2-i\zeta/p}(z)$ are of odd parity. The wave-functions on S_{3C} for even and odd parity, respectively. Combing yields the representation (8.18).

System 12: Horicyclic-Parabolic II.

The horicyclic-parabolic-II system exists on $\Lambda^{(3)}$ and $O(2,2)$. The (ξ,η)-subpath integration corresponds to the path integration of the third parabolic system on the two-dimensional pseudo-Euclidean plane. Using this result and the horicyclic system we get the representation (8.19).

8.3.2   Path Integral Representations on S_{3C}: Part II

We now come the those coordinate systems which do not have a subgroup structure. There are nine of them, and we can find for six of these cases a path integral representation.

System 13: Elliptic Cylindrical.

The elliptic cylindrical system exists on $S^{(3)}$ (oblate and prolate spheroidal (Systems IV and V.), $\Lambda^{(3)}$ (Systems XVII. and XVIII – prolate and oblate elliptic), and $O(2,2)$. In [242] we have constructed a kernel for the prolate elliptic coordinate system on the sphere $S^{(3)}$. The corresponding representation can be derived by a group path integration together with an interbasis expansion from the cylindrical or spherical basis to the ellipso-cylindrical ones. For instance, one has [240] $(r,p=\pm 1)$

$$\Psi^{(r,p)}_{J,k_1,k_2}(\vartheta,\varphi_1,\varphi_2) = \sum_q T^{(r,p)\,*}_{Jqk_2}\Psi^{(r,p)}_{Jqk_2}(\alpha,\beta,\varphi;a)\ , \qquad (8.34)$$

where the $a \geq 0$ corresponds to the oblate elliptic case (which we do not discuss here), and $a \in [-1,0]$ to the prolate elliptic system, respectively. We have the factorization $\Psi(\alpha,\beta,\varphi;a) = \psi_1(\alpha;a)\psi_2(\beta;a)e^{ik_2\varphi}/\sqrt{2\pi}$. For the associated Lamé polynomials

ψ_{i,Jqk_2}, $i = 1, 2$, we adopt the notations of [240, 333]. The relevant quantum numbers have the following meaning: The functions $\psi_{i,Jqk_2}^{(r,p)}(z)$ are called *associated Lamé
polynomials* and satisfy the associated Lamé equation. We take them for normalized.
For the principal quantum number we have $l \in \mathbb{N}_0$. $(r, p) = \pm 1$ denotes one of the
four parity classes of solutions of dimension $(J + 1)^2$, i.e., the multiplicity of the degeneracy of the level J, one for each class of the corresponding recurrence relations
as given in [240, 333] and the parity classes from the periodic Lamé functions Λ_{lh}^p
from the spherical harmonics on the sphere can be applied. These expansions have
been considered in [240, 333] together with three-term recurrence relations for the
interbasis coefficients. They can be determined by taking into account that a basis
in O(4) is related in a unique way to the cylindrical and spherical bases on $S^{(3)}$ by
using the properties of the elliptic operator Λ on $S^{(3)}$ with eigenvalue q. Details can
be found in [240, 333]. Due to the unitarity of these coefficients the path integration
is then performed by inserting in each short-time kernel in the cylindrical system first
the expansions (8.34), second, exploiting the unitarity, and thus yielding the result
(6.35). This justifies our representation (8.20).

System 14: Elliptic-Parabolic.

The elliptic-parabolic system exists on $\Lambda^{(3)}$ and O(2, 2). The corresponding path
integral representation can be obtained by a combination of modified Pöschl–Teller
potential path integrals yielding (8.21).

System 15: Elliptic-Hyperbolic.

The elliptic-hyperbolic system exists on $\Lambda^{(3)}$ and O(2, 2). The path integral for this
coordinate system corresponds to the path integral for the first elliptic parabolic
system on the three-dimensional hyperboloid in its complexified form. Actually, one
must make use of the path integral identity (5.34) and the ps_μ^ν and $S_\mu^{\nu\,(1)}$ are spheroidal
wave-functions. Putting everything together gives (8.22) and we are done.

System 16: Parabolic.

The parabolic system exists on O(2, 2). In order to solve the path integral we must
perform a time-transformation yielding a path integral for a radial harmonic oscillator. We can use the discrete spectrum-wave-functions and obtain (8.23, 8.24). This
concludes the discussion.

Systems 17–21.

The path integrals for the remaining coordinate systems cannot be evaluated due to
the too complicated structure of the coordinates.
    The archived results for the complex sphere are very satisfactory. However, we
must always keep in mind that the complex sphere is an abstract space, which means
that the various path integral representations require an interpretation depending

Table 8.3: Coordinates on the Three-Dimensional Complex Sphere

Coordinate System	Coordinates	Path Integral Solution
I. Cylindrical	$z_1 = \sin\vartheta\cos\varphi$ $z_2 = \sin\vartheta\sin\varphi_1$ $z_3 = \cos\vartheta\cos\varphi_2$ $z_4 = \cos\vartheta\sin\varphi_2$	(8.6)
II. Horicyclic	$z_1 = \frac{1}{2}\left[e^{-ix} + (1+y^2+z^2)e^{ix}\right]$ $z_2 = iye^{ix}$ $z_3 = ize^{ix}$ $z_4 = \frac{i}{2}\left[e^{-ix} + (1+y^2+z^2)e^{ix}\right]$	(8.7)
III. Spherical	$z_1 = \sin\chi\cos\vartheta$ $z_2 = \sin\chi\sin\vartheta\cos\varphi$ $z_3 = \sin\chi\sin\vartheta\sin\varphi$ $z_4 = \cos\chi$	(8.8–8.10)
IV. Horospherical	$z_1 = \frac{1}{2}\left[e^{-ix} + (1-y^2)e^{ix}\right]\sin\chi$ $z_2 = ye^{ix}\sin\chi$ $z_3 = -\frac{i}{2}\left[e^{-ix} - (1+y^2)e^{ix}\right]\sin\chi$ $z_4 = \cos\chi$	(8.11)
V. Horicyclic-Polar	$z_1 = \frac{1}{2}\left[e^{-ix} + (1+\varrho^2)e^{ix}\right]$ $z_2 = i\rho e^{ix}\cos\varphi$ $z_3 = i\varrho e^{ix}\sin\varphi$ $z_4 = \frac{i}{2}\left[e^{-ix} - (1-\varrho^2)e^{ix}\right]$	(8.12)
VI. Sphero-Elliptic	$z_1 = \sin\chi\,\mathrm{sn}(\alpha,k)\mathrm{dn}(\beta,k')$ $z_2 = \sin\chi\,\mathrm{cn}(\alpha,k)\mathrm{cn}(\beta,k')$ $z_3 = \sin\chi\,\mathrm{dn}(\alpha,k)\mathrm{sn}(\beta,k')$ $z_4 = \cos\chi$	(8.13)
VII. Spherical-Degenerate Elliptic I	$z_1 = \frac{1}{2}\sin\chi\left(\frac{\cosh\tau_2}{\cosh\tau_1} + \frac{\cosh\tau_1}{\cosh\tau_2}\right)$ $z_2 = \sin\chi\tanh\tau_1\tanh\tau_2$ $z_3 = \sin\chi\left[\frac{1}{\cosh\tau_1\cosh\tau_2}\right.$ $\left. -\frac{1}{2}\left(\frac{\cosh\tau_2}{\cosh\tau_1} + \frac{\cosh\tau_1}{\cosh\tau_2}\right)\right]$ $z_4 = \cos\chi$	(8.14)

Table 8.3 (cont.) Coordinate System	Coordinates	Path Integral Solution
VIII. Spherical- Degenerate Elliptic II	$z_1 = -\frac{\mathrm{i}\sin\chi}{8\xi\eta}\Big[(\xi^2 - \eta^2)^2 + 4\Big]$ $z_2 = \frac{\sin\chi}{2\xi\eta}(\xi^2 + \eta^2)^2$ $z_3 = \frac{\sin\chi}{8\xi\eta}\Big[-(\xi^2 - \eta^2)^2 + 4\Big]$ $z_4 = \cos\chi - \{I_{12}, I_{13}\} + \mathrm{i}\{\hat{I}_{12}, I_{23}\}$	(8.15)
IX. Horicyclic- Elliptic	$z_1 = \frac{1}{2}\Big[e^{-\mathrm{i}x} + (1 + \cosh^2\tau_1 + \sinh^2\tau_2)e^{\mathrm{i}x}\Big]$ $z_2 = \mathrm{i}\cosh\tau_1\cosh\tau_2 e^{\mathrm{i}x}$ $z_3 = \sinh\tau_1\sinh\tau_2 e^{\mathrm{i}x}$ $z_4 = \frac{1}{2}\Big[e^{-\mathrm{i}x} + (-1 + \cosh^2\tau_1 + \sinh^2\tau_2)e^{\mathrm{i}x}\Big]$	(8.16)
X. Horicyclic- Hyperbolic	$z_1 = \frac{1}{2}\Big[e^{-\mathrm{i}x} + (1 + e^{2y} - e^{2z})e^{\mathrm{i}x}\Big]$ $z_2 = \frac{\mathrm{i}}{\sqrt{2}}\Big[\sinh(y - z) + e^{y+z}\Big]e^{\mathrm{i}x}$ $z_3 = \frac{1}{\sqrt{2}}\Big[\sinh(y - z) - e^{y+z}\Big]e^{\mathrm{i}x}$ $z_4 = \frac{\mathrm{i}}{2}\Big[e^{-\mathrm{i}x} + (-1 + e^{2y} - e^{2z})e^{\mathrm{i}x}\Big]e^{\mathrm{i}x}$	(8.17)
XI. Horicyclic- Parabolic I	$z_1 = \frac{1}{2}\Big[e^{-\mathrm{i}x} + (1 + \frac{1}{4}(\xi^2 + \eta^2)^2)e^{\mathrm{i}x}\Big]$ $z_2 = \frac{1}{2}(\xi^2 - \eta^2)e^{\mathrm{i}x}$ $z_3 = \mathrm{i}\xi\eta e^{\mathrm{i}x}$ $z_4 = \frac{\mathrm{i}}{2}\Big[e^{-\mathrm{i}x} + (-1 + \frac{1}{4}(\xi^2 + \eta^2)^2)e^{\mathrm{i}x}\Big]$	(8.18)
XII. Horicyclic- Parabolic II	$z_1 = \frac{1}{2}\Big\{e^{-\mathrm{i}x} + [1 + 2(\xi - \eta)^2(\xi + \eta)]e^{\mathrm{i}x}\Big\}$ $z_2 = \mathrm{i}[\frac{1}{2}(\xi - \eta)^2 + (\xi + \eta)]e^{\mathrm{i}x}$ $z_3 = [\frac{1}{2}(\xi - \eta)^2 - (\xi + \eta)]e^{\mathrm{i}x}$ $z_4 = \frac{1}{2}\Big\{e^{-\mathrm{i}x} + [-1 + 2(\xi - \eta)^2(\xi + \eta)]e^{\mathrm{i}x}\Big\}$	(8.19)
XIII. Elliptic- Cylindrical	$z_1 = k\,\mathrm{sn}(\alpha, k)\mathrm{sn}(\beta, k)$ $z_2 = -\mathrm{i}\frac{k}{k'}\mathrm{cn}(\alpha, k)\mathrm{cn}(\beta, k)\cos\varphi$ $z_3 = -\mathrm{i}\frac{k}{k'}\mathrm{cn}(\alpha, k)\mathrm{cn}(\beta, k)\sin\varphi$ $z_4 = \frac{1}{k'}\mathrm{dn}(\alpha, k)\mathrm{dn}(\beta, k)$	(8.20)
XIV. Elliptic- Parabolic	$z_1 = \frac{1}{2}\Big(\frac{\cosh\tau_1}{\cosh\tau_2} + \frac{\cosh\tau_2}{\cosh\tau_1}\Big)$ $z_2 = \tanh\tau_1\tanh\tau_2\cosh\tau_3$ $z_3 = -\mathrm{i}\tanh\tau_1\tanh\tau_2\sinh\tau_3$ $z_4 = \frac{-\mathrm{i}}{\cosh\tau_1\cosh\tau_2} + \frac{\mathrm{i}}{2}\Big(\frac{\cosh\tau_1}{\cosh\tau_2} + \frac{\cosh\tau_2}{\cosh\tau_1}\Big)$	(8.21)
XV. Elliptic- Hyperbolic	$z_1 = -\frac{1}{2}\Big(\frac{\cosh\tau_2}{\cosh\tau_1} + \frac{\cosh\tau_1}{\cosh\tau_2}\Big) - \frac{\tau_3^2}{2\cosh\tau_1\cosh\tau_2}$ $z_2 = \frac{\mathrm{i}\tau_3}{\cosh\tau_1\cosh\tau_2}$ $z_3 = \tanh\tau_1\tanh\tau_2$ $z_4 = \mathrm{i}\Big[\frac{2 - \tau_3^2}{2\cosh\tau_1\cosh\tau_2} - \frac{1}{2}\Big(\frac{\cosh\tau_1}{\cosh\tau_2} + \frac{\cosh\tau_2}{\cosh\tau_1}\Big)\Big]$	(8.22)

Table 8.3 (cont.) Coordinate System	Coordinates	Path Integral Solution
XVI. Parabolic	$z_1 = \dfrac{(\xi^2 + \eta^2)^2 + 4}{8\xi\eta} + \dfrac{\tau^2}{2\xi\eta}$ $z_2 = -\mathrm{i}\dfrac{\tau}{\xi\eta}$ $z_3 = -\dfrac{\mathrm{i}}{2}\left(\dfrac{\xi}{\eta} - \dfrac{\eta}{\xi}\right)$ $z_4 = \mathrm{i}\left(\dfrac{(\xi^2 + \eta^2)^2 - 4}{8\xi\eta} + \dfrac{\tau^2}{2\xi\eta}\right)$	(8.23, 8.24)
XVII. Ellipsoidal	$z_1^2 = -\dfrac{\varrho_1\varrho_2\varrho_3}{ab}$ $z_2^2 = \dfrac{(\varrho_1 - 1)(\varrho_2 - 1)(\varrho_3 - 1)}{(a - 1)(b - 1)}$ $z_3^2 = -\dfrac{(\varrho_1 - b)(\varrho_2 - b)(\varrho_3 - b)}{(a - b)(b - 1)b}$ $z_4^2 = \dfrac{(\varrho_1 - a)(\varrho_2 - a)(\varrho_3 - a)}{(a - b)(a - 1)a}$	(8.25)
XVIII.*	$(\mathrm{i}z_1 + z_2)^2 = \dfrac{\varrho_1\varrho_2\varrho_3}{a}$ $z_1^2 + z_2^2 = \dfrac{1}{a^2}\left[(a+1)\varrho_1\varrho_2\varrho_3 - a(\varrho_1\varrho_2 + \varrho_1\varrho_3 + \varrho_2\varrho_3)\right]$ $z_3^2 = \dfrac{(\varrho_1 - 1)(\varrho_2 - 1)(\varrho_3 - 1)}{1 - a}$ $z_4^2 = \dfrac{(\varrho_1 - a)(\varrho_2 - a)(\varrho_3 - a)}{a^2(a - 1)}$	(not known)
XIX.*	$(z_1 + \mathrm{i}z_2)^2 = -(\varrho_1 - 1)(\varrho_2 - 1)(\varrho_3 - 1)$ $z_1^2 + z_2^2 = 2\varrho_1\varrho_2\varrho_3 - (\varrho_1\varrho_3 + \varrho_2\varrho_3 + \varrho_1\varrho_2) + 1$ $(z_3 + \mathrm{i}z_4)^2 = -\varrho_1\varrho_2\varrho_3$ $z_3^2 + z_4^2 = \varrho_1\varrho_3 + \varrho_2\varrho_3 + \varrho_1\varrho_2 - 2\varrho_1\varrho_2\varrho_3$	(not known)
XX.*	$(z_2 - \mathrm{i}z_1)^2 z_1^2 + z_2^2 + z_3^2 = \varrho_1\varrho_2\varrho_3$ $\quad - 2z_3(z_2 - \mathrm{i}z_1) = \varrho_1\varrho_2 + \varrho_1\varrho_3 + \varrho_2\varrho_3 - \varrho_1\varrho_2\varrho_3$ $z_1^2 + z_2^2 + z_3^2$ $\quad = \varrho_1\varrho_2\varrho_3 - \varrho_1\varrho_2 - \varrho_1\varrho_3 - \varrho_2\varrho_3 + \varrho_1 + \varrho_2 + \varrho_3$ $z_4^2 = -(\varrho_1 - 1)(\varrho_2 - 1)(\varrho_3 - 1)$	(not known)
XXI.*	$(z_1 + \mathrm{i}z_2)^2 = 2\varrho_1\varrho_2\varrho_3$ $(z_1 + \mathrm{i}z_2)((z_3 + \mathrm{i}z_4) = -(\varrho_1\varrho_2 + \varrho_2\varrho_3 + \varrho_1\varrho_3)$ $-(z_1 + \mathrm{i}z_2)(z_3 - \mathrm{i}z_4) + \frac{1}{2}(z_3 + \mathrm{i}z_4)^2 = \varrho_1 + \varrho_2 + \varrho_3$	(not known)

* In view of the absence of commonly agreed names of the various hyperboloids/spheres appearing here, we present only the equations of the respective families of surfaces in the cases XVIII.-XXI on $\Lambda^{(3)}$. ·

whether one considers a compact or non-compact variable range. In the compact case, the abstract complex space allows the interpretation of the real three-dimensional sphere with its discrete spectrum. In the non-compact case allows the interpretation of the three-dimensional $\Lambda^{(3)}$ and $O(2,2)$-hyperboloid, respectively, with a continuous spectrum. Therefore the eigenvalues of the complex sphere

$$\frac{\hbar^2}{2m}\sigma(\sigma+2) \longleftarrow \begin{cases} \dfrac{\hbar^2}{2m}l(l+2) & l=0,1,2,\ldots & \text{sphere} \\[2mm] \dfrac{\hbar^2}{2m}(p^2+1) & p>0 & \text{hyperboloid.} \end{cases} \tag{8.35}$$

This includes the replacement of the summation of the discrete principal quantum number l, say, by the principal continuous quantum number p, i.e., $\sum_l \to \int_0^\infty dp$. Furthermore, the wave-functions have to analytically continued, say the discrete wave-functions for the spherical coordinate system on the real sphere:

$$\Psi_{J,m_1,m_2}(\chi,\vartheta,\varphi)=N^{-1/2}e^{im_1\varphi}(\sin\chi)^{m_1}C_{J-m_1}^{m_1+2}(\cos\chi)(\sin\vartheta)^{m_2}C_{m_1-m_2}^{m_2+3/2}(\cos\vartheta)\ , \tag{8.36}$$

must be replaced by the continuous wave-functions for the spherical system on the hyperboloid $\Lambda^{(3)}$ ($Y_l^m(\vartheta,\varphi)$ are the usual spherical harmonics on the two-dimensional sphere, c.f. Section 2.3):

$$\Psi_{p,l,m}(\tau,\vartheta,\varphi)=Y_l^m(\vartheta,\varphi)(\sinh\tau)^{-1/2}\sqrt{\frac{p\sinh\pi p}{\pi}}\Gamma(ip+l+1)\mathcal{P}_{ip-1/2}^{-\frac{1}{2}-l}(\cosh\tau)\ . \tag{8.37}$$

Also, the invariant distance (under rotations) on the real sphere must be replaced by the invariant distance on the hyperboloid, respectively.

Chapter 9

Path Integrals on Hermitian Hyperbolic Space

9.1 Hermitian Hyperbolic Space HH(2)

We have the following path integral representations on the four-dimensional Hermitian hyperbolic space [232]

I. Spherical, $\omega > 0$, $\beta \in [0, \pi/2)$, $\varphi_1, \varphi_2 \in [0, 2\pi)$: ($\Phi_n^{(k_1,k_2)}(\beta)$ are the Pöschl–Teller functions, $\Psi_p^{(\mu,\nu)}(\omega)$ are the modified Pöschl–Teller functions)

$$
\int_{\omega(t')=\omega'}^{\omega(t'')=\omega''} \mathcal{D}\omega(t) \int_{\beta(t')=\beta'}^{\beta(t'')=\beta''} \mathcal{D}\beta(t) \int_{\varphi_1(t')=\varphi_1'}^{\varphi_1(t'')=\varphi_1''} \mathcal{D}\varphi_1(t) \int_{\varphi_2(t')=\varphi_2'}^{\varphi_2(t'')=\varphi_2''} \mathcal{D}\varphi_2(t) \sinh^3 \omega \cosh \omega \sin \beta \cos \beta
$$

$$
\times \exp\left(\frac{\mathrm{i}}{\hbar} \int_0^T \left\{ \frac{m}{2} \left[\dot{\omega}^2 + \sinh^2 \omega \dot{\beta}^2 + (\dot{\varphi}_1, \dot{\varphi}_2)(\hat{g}_{ab}) \begin{pmatrix} \dot{\varphi}_1 \\ \dot{\varphi}_2 \end{pmatrix} \right] \right. \right.
$$

$$
\left. \left. + \frac{\hbar^2}{8m} \left[\left(\frac{1}{\sinh^2 \omega} - \frac{1}{\cosh^2 \omega} - 16 \right) + \frac{1}{\sinh^2 \omega} \left(\frac{1}{\sin^2 \beta} + \frac{1}{\cos^2 \beta} \right) \right] \right\} dt \right)
$$

$$
= \sum_{k_1 \in \mathbb{Z}} \sum_{k_2 \in \mathbb{Z}} \sum_{n \in \mathbb{N}} \int_0^\infty dp \, \Psi_{p,n,k_1 k_2}(\omega'', \beta'', \varphi_1'', \varphi_2'') \Psi_{p,n,k_1 k_2}^*(\omega', \beta', \varphi_1', \varphi_2') \, \mathrm{e}^{-\mathrm{i}\hbar T(p^2+4)/2m} \, ,
$$

$$(9.1)$$

$$
\Psi_{p,n,k_1 k_2}(\omega, \beta, \varphi_1, \varphi_2)
$$

$$
= (\tfrac{1}{4} \sinh 2\omega \sin 2\beta)^{-1/2} \frac{\mathrm{e}^{\mathrm{i}(k_1\varphi_1+k_2\varphi_2)}}{2\pi} \Phi_n^{(k_1,k_2)}(\beta) \Psi_p^{(k_1+k_2+2n-1,k_1+k_2)}(\omega) \, . \tag{9.2}
$$

II. Equidistant I, $\tau_{1,2} > 0$, $\varphi_1, \varphi_2 \in [0, 2\pi)$:

$$
\int_{\tau_1(t')=\tau_1'}^{\tau_1(t'')=\tau_1''} \mathcal{D}\tau_1(t) \int_{\tau_2(t')=\tau_2'}^{\tau_2(t'')=\tau_2''} \mathcal{D}\tau_2(t) \int_{\varphi_1(t')=\varphi_1'}^{\varphi_1(t'')=\varphi_1''} \mathcal{D}\varphi_1(t) \int_{\varphi_2(t')=\varphi_2'}^{\varphi_2(t'')=\varphi_2''} \mathcal{D}\varphi_2(t) \sinh \tau_1 \cosh^3 \tau_1 \tfrac{1}{2} \sinh 2\tau_2
$$

147

$$\times \exp\left(\frac{i}{\hbar}\int_0^T \left\{\frac{m}{2}\left[\dot{\tau}_1^2 + \cosh^2\tau_1\dot{\tau}_2^2 + (\dot{\varphi}_1,\dot{\varphi}_2)(\widehat{g}_{ab})\begin{pmatrix}\dot{\varphi}_1\\\dot{\varphi}_2\end{pmatrix}\right]\right.\right.$$
$$\left.\left.+\frac{\hbar^2}{8m}\left[\frac{1}{\sinh^2\tau_1}-\frac{1}{\cosh^2\tau_1}-16+\frac{1}{\cosh^2\tau_1}\left(\frac{1}{\sinh^2\tau_2}+\frac{1}{\cosh^2\tau_2}\right)\right]\right\}dt\right)$$

$$=\int_0^\infty dp\, e^{-i\hbar T(p^2+4)/2m}$$
$$\times\left\{\sum_{k_1\in\mathbb{Z}}\sum_{k_2\in\mathbb{Z}}\int_0^\infty dk_{\tau_2}\Psi_{p,k_{\tau_2},k_1,k_2}(\tau_1'',\tau_2'',\varphi_1'',\varphi_2'')\Psi^*_{p,k_{\tau_2},k_1,k_2}(\tau_1',\tau_2',\varphi_1',\varphi_2')\right.$$
$$\left.+\sum_{k_1\in\mathbb{Z}}\sum_{k_2\in\mathbb{Z}}\sum_{n_{\tau_2}=0}^{M_{\max}}\Psi_{p,n_{\tau_2},k_1,k_2}(\tau_1'',\tau_2'',\varphi_1'',\varphi_2'')\Psi^*_{p,n_{\tau_2},k_1,k_2}(\tau_1',\tau_2',\varphi_1',\varphi_2')]\right\},\quad (9.3)$$

$$\Psi_{p,k_{\tau_2},k_1,k_2}(\tau_1,\tau_2,\varphi_1,\varphi_2)=\frac{e^{i(k_1\varphi_1+k_2\varphi_2)}}{2\pi}\frac{\Psi_{k_{\tau_2}}^{(k_1,k_1+k_2)}(\tau_2)\Psi_p^{(k_1,ik_{\tau_2})}(\tau_1)}{(\frac{1}{4}\sinh 2\tau_1\sinh 2\tau_2)^{1/2}},\quad (9.4)$$

$$\Psi_{p,n_{\tau_2},k_1,k_2}(\tau_1,\tau_2,\varphi_1,\varphi_2)=\frac{e^{i(k_1\varphi_1+k_2\varphi_2)}}{2\pi}\frac{\Psi_{n_{\tau_2}}^{(k_1,k_1+k_2)}(\tau_2)\Psi_p^{(k_1,n_{\tau_2})}(\tau_1)}{(\frac{1}{4}\sinh 2\tau_1\sinh 2\tau_2)^{1/2}}.\quad (9.5)$$

III. Equidistant II, $\tau_1>0$, $\tau_2,u\in\mathbb{R}$, $\varphi\in[0,2\pi)$:

$$\int_{\tau_1(t')=\tau_1'}^{\tau_1(t'')=\tau_1''}\mathcal{D}\tau_1(t)\int_{\tau_2(t')=\tau_2'}^{\tau_2(t'')=\tau_2''}\mathcal{D}\tau_2(t)\int_{u(t')=u'}^{u(t'')=u''}\mathcal{D}u(t)\int_{\varphi(t')=\varphi'}^{\varphi(t'')=\varphi''}\mathcal{D}\varphi(t)\sinh\tau_1\cosh^3\tau_1\cosh 2\tau_2$$
$$\times\exp\left[\frac{i}{\hbar}\int_0^T\left(\frac{m}{2}\left\{\dot{\tau}_1^2+\cosh^2\tau_1\left[\dot{\tau}_2^2+(\dot{u},\dot{\varphi})(\widehat{g}_{ab})\begin{pmatrix}\dot{u}\\\dot{\varphi}\end{pmatrix}\right]\right\}\right.\right.$$
$$\left.\left.+\frac{\hbar^2}{8m}\left(\frac{1}{\sinh^2\tau_1}-\frac{1}{\cosh^2\tau_1}-16\right)\right)dt\right]$$

$$=\int_0^\infty dp\, e^{-i\hbar T(p^2+4)/2m}$$
$$\times\left\{\sum_{k_\varphi\in\mathbb{Z}}\int dk_u\int_0^\infty dk_{\tau_2}\Psi_{p,k_{\tau_2},k_u,k_\varphi}(\tau_1'',\tau_2'',u'',\varphi'')\Psi^*_{p,k_{\tau_2},k_u,k_\varphi}(\tau_1',\tau_2',u',\varphi')\right.$$
$$\left.+\sum_{k_\varphi\in\mathbb{Z}}\int dk_u\sum_{n_{\tau_2}=0}^{M_{\max}}\Psi_{p,n_{\tau_2},k_u,k_\varphi}(\tau_1'',\tau_2'',u'',\varphi'')\Psi^*_{p,n_{\tau_2},k_u,k_\varphi}(\tau_1',\tau_2',u',\varphi')\right\},\quad (9.6)$$

$$\Psi_{p,k_{\tau_2},k_u,k_\varphi}(\tau_1,\tau_2,u,\varphi)=\frac{e^{i(k_u u+k_\varphi\varphi)}}{2\pi}\frac{\Psi_{k_{\tau_2}}^{(\mathrm{HBP})}(\tau_2)\Psi_p^{(k_u,ik_{\tau_2})}(\tau_1)}{(\frac{1}{2}\sinh 2\tau_1\cosh 2\tau_2)^{-1/2}},\quad (9.7)$$

$$\Psi_{p,n_{\tau_2},k_u,k_\varphi}(\tau_1,\tau_2,u,\varphi)=\frac{e^{i(k_u u+k_\varphi\varphi)}}{2\pi}\frac{\Psi_{n_{\tau_2}}^{(\mathrm{HBP})}(\tau_2)\Psi_p^{(k_u,n_{\tau_2})}(\tau_1)}{(\frac{1}{2}\sinh 2\tau_1\cosh 2\tau_2)^{-1/2}},\quad (9.8)$$

$$\Psi_n^{(\mathrm{HBP})}(\tau_2)=\left[\frac{(2\lambda_R-2n-1)n!\,\Gamma(\lambda-n)}{2\Gamma(2\lambda_R-n)\Gamma(n+1-\lambda^*)}\right]^{1/2}$$

$$\times\left(\frac{1+\mathrm{i}\sinh x}{2}\right)^{\frac{1}{2}(\frac{1}{2}-\lambda)}\left(\frac{1-\mathrm{i}\sinh x}{2}\right)^{\frac{1}{2}(\frac{1}{2}-\lambda^*)}P_n^{(-\lambda^*,-\lambda)}(\mathrm{i}\sinh 2\tau_2) \quad (9.9)$$

$$\left(n_{\tau_2}=n+\tfrac{1}{2}-\sqrt{\tfrac{1}{2}\left[\sqrt{(\tfrac{1}{4}+V_2)^2+V_1^2}+\tfrac{1}{4}+V_2\right]},V_2=-(k_u^2-k_\varphi^2),V_1=-2k_uk_\varphi\right) ,$$

$$\Psi_{k_{\tau_2}}^{(\mathrm{HBP})}(\tau_2)=\frac{\Gamma(\frac{1}{2}-\lambda_R-\mathrm{i}k_{\tau_2})|}{\pi\Gamma(1-\lambda^*)}\sqrt{k_{\tau_2}\sinh(2\pi k_{\tau_2})\Gamma\left(\tfrac{1}{2}+\mathrm{i}(k_{\tau_2}-\lambda_I)\right)\Gamma\left(\tfrac{1}{2}+\mathrm{i}(k_{\tau_2}+\lambda_I)\right)}$$

$$\times {}_2F_1\left(\tfrac{1}{2}+\mathrm{i}(\lambda_I-k_{\tau_2}),\tfrac{1}{2}-\lambda_R-\mathrm{i}k_{\tau_2};1-\lambda^*;\frac{\mathrm{i}\sinh 2\tau_2-1}{\mathrm{i}\sinh 2\tau_2+1}\right). \quad (9.10)$$

IV. Equidistant III, $\tau_{1,2}>0$, $u\in\mathbb{R}$, $\varphi\in[0,2\pi)$:

$$\int_{\tau_1(t')=\tau_1'}^{\tau_1(t'')=\tau_1''}\mathcal{D}\tau_1(t)\int_{\tau_2(t')=\tau_2'}^{\tau_2(t'')=\tau_2''}\mathcal{D}\tau_2(t)\int_{u(t')=u'}^{u(t'')=u''}\mathcal{D}u(t)\int_{\varphi(t')=\varphi'}^{\varphi(t'')=\varphi''}\mathcal{D}\varphi(t)\sinh\tau_1\cosh^3\tau_1$$

$$\times\exp\left[\!\left[\frac{\mathrm{i}}{\hbar}\int_0^T\left(\frac{m}{2}\left\{\dot\tau_1^2+\cosh^2\tau_1\left[\dot\tau_2^2+\mathrm{e}^{-4\tau_2}\cosh^2\tau_1\dot u^2+\sinh^2\dot\varphi^2-2\mathrm{e}^{-4\tau_2}\sinh^2\tau_1\dot u\dot\varphi\right]\right\}\right.\right.$$

$$\left.\left.+\frac{\hbar^2}{8m}\left(\frac{1}{\sinh^2\tau_1}-\frac{1}{\cosh^2\tau_1}-16\right)\right)dt\right]\!\right]$$

$$=(\sinh\tau_1'\cosh\tau_1'\sinh\tau_1''\cosh\tau_1'')^{-1/2}$$

$$\times\mathrm{e}^{(\tau_2'+\tau_2'')/2}\sum_{k_\varphi\in\mathbb{Z}}\int\mathrm{d}k_u\frac{\mathrm{e}^{\mathrm{i}[k_\varphi(\varphi''-\varphi')+k_u(u''-u')]}}{(2\pi)^2}\int_0^\infty\mathrm{d}p\,\mathrm{e}^{-\mathrm{i}\hbar T(p^2+4)/2m}$$

$$\times\left\{\sum_{n_{\tau_2}}\int_0^\infty\mathrm{d}p\Psi_{n_{\tau_2}}(\tau_2'')^{(\mathrm{MP})}\Psi_{n_{\tau_2}}^{(\mathrm{MP})}(\tau_2')\Psi_p^{(n_{\tau_2},k_\varphi)\,*}(\tau_1'')\Psi_p^{(n_{\tau_2},k_\varphi)}(\tau_1')\right.$$

$$\left.+\int_0^\infty\mathrm{d}k_{\tau_2}\int_0^\infty\mathrm{d}p\Psi_{k_{\tau_2}}^{(\mathrm{MP})*}(\tau_2'')\Psi_{k_{\tau_2}}^{(\mathrm{MP})}(\tau_2')\Psi_p^{(\mathrm{i}k_{\tau_2},k_\varphi)\,*}(\tau_1'')\Psi_p^{(\mathrm{i}k_{\tau_2},k_\varphi)}(\tau_1')\right\}, \quad (9.11)$$

$$\Psi_{n_{\tau_2}}^{(\mathrm{MP})}(\tau_2)=\sqrt{\frac{2n!(k_{\varphi,k_u}-2n_{\tau_2}-1)}{\Gamma(k_{\varphi,k_u}-n)}}(|k_u|\mathrm{e}^{2\tau_2})^{k_{\varphi,k_u}-n}\mathrm{e}^{-\frac{1}{2}|k_u|\mathrm{e}^{2\tau_2}}L_n^{(k_{\varphi,k_u}-2n-1)}(|k_u|\mathrm{e}^{2\tau_2}) ,$$

$$\Psi_{k_{\tau_2}}^{(\mathrm{MP})}(\tau_2)=\sqrt{\frac{k_{\tau_2}\sinh\pi k_{\tau_2}}{2\pi^2|k_u|}}\Gamma[\tfrac{1}{2}(1+\mathrm{i}k_{\tau_2}+k_{\varphi,k_u})]W_{k_{\varphi,k_u}/2,k_{\tau_2}/2}(|k_u|\mathrm{e}^{2\tau_2}) . \quad (9.12)$$

V. Horicyclic I, $r>0$, $u,q\in\mathbb{R}$, $\varphi\in[0,2\pi)$:

$$\mathrm{e}^{-2\mathrm{i}\hbar T/m}\int_{q(t')=q'}^{q(t'')=q''}\mathcal{D}q(t)\int_{r(t')=r'}^{r(t'')=r''}\mathcal{D}r(t)\int_{u(t')=u'}^{u(t'')=u''}\mathcal{D}u(t)\int_{\varphi(t')=\varphi'}^{\varphi(t'')=\varphi''}\mathcal{D}\varphi(t)r\,\mathrm{e}^{-4q}$$

$$\times\exp\left(\frac{\mathrm{i}}{\hbar}\int_0^T\left\{\frac{m}{2}\left[\dot q^2+\frac{\dot r^2}{\mathrm{e}^{2q}}-\mathrm{e}^{-4q}\left(\dot u^2+(\mathrm{e}^{2q}+r^2)\dot\varphi^2-2r^2\dot u\dot\varphi\right)\right]+\mathrm{e}^{2q}\frac{\hbar^2}{2mr^2}\right\}dt\right)$$

$$=\int_0^\infty\mathrm{d}k_u\sum_{k_\varphi\in\mathbb{Z}}\sum_{n=0}^\infty\int_0^\infty\mathrm{d}p\,\mathrm{e}^{-\mathrm{i}\hbar T(p^2+4)/2m}\Psi_{p,n,k_u,k_\varphi}(q'',r'',u'',\varphi'')\Psi_{p,n,k_u,k_\varphi}^*(q',r',u',\varphi'),$$

$$(9.13)$$

$$\Psi_{p,n,k_u,k_\varphi}(y,r,u,\varphi) = \frac{e^{i(k_u u + k_\varphi \varphi)}}{2\pi} \sqrt{\frac{2n!}{\Gamma(n+|k_u|+1)}} (|k_u|r)^{|k_u|} e^{-\frac{1}{2}|k_u|r^2} L_n^{(|k_u|)}(|k_u|r)$$

$$\times \sqrt{\frac{p\sinh\pi p}{2\pi^2}} \Gamma(\tfrac{1}{2}(1+ip+E_n))W_{-E_n/2,ip/2}(|k_u|e^{2q}) \ . \tag{9.14}$$

VI. Horicyclic II, $q,u,x,z \in \mathbb{R}$:

$$e^{-2i\hbar T/m} \int_{q(t')=q'}^{q(t'')=q''} \mathcal{D}q(t) \int_{x(t')=x'}^{x(t'')=x''} \mathcal{D}x(t) \int_{u(t')=u'}^{u(t'')=u''} \mathcal{D}u(t) \int_{z(t')=z'}^{z(t'')=z''} \mathcal{D}z(t)\, e^{-4q}$$

$$\times \exp\left\{ \frac{i}{\hbar}\int_0^T \left[\frac{m}{2}\left(\dot{q}^2 + \frac{\dot{x}^2}{e^{2q}}\right) + e^{-4q}\left((4x^2+e^{2q})\dot{z}^2 + \dot{u}^2 + 4x\dot{u}\dot{z}\right) \right]dt \right\}$$

$$= \int_0^\infty dk_u \int_0^\infty dk_u \sum_{n=0}^\infty \int_0^\infty dp\, e^{-iE_p T/\hbar} \Psi_{p,n,k_u,k_\varphi}(q'',x'',u'',z'')\Psi^*_{p,n,k_u,k_\varphi}(q',x',u',z') \ , \tag{9.15}$$

$$\Psi_{q,n,k_u,k_z}(y,x,u,z) = \frac{e^{i(k_u u + k_z z)}}{2\pi} \frac{\sqrt[4]{2|k_u|/\pi}}{\sqrt{2^n n!}}\, e^{-|k_u|x^2} H_n\left[2|k_u|\left(x - \frac{k_z}{2|k_u|}\right)\right]$$

$$\times \sqrt{\frac{p\sinh\pi p}{4\pi^2|k_u|}} \Gamma[\tfrac{1}{2}(1+ip+n+\tfrac{1}{2})]W_{-(n+\frac{1}{2})/2,ip/2}(2|k_u|e^{2q}) \ . \tag{9.16}$$

9.2   Path Integral Evaluations on HH(2)

Hermitian Hyperbolic Space.

The Hermitian hyperbolic space HH(n) is defined by $SU(n,1)/S[U(1)\times U(n)]$ (see e.g. Helgason [273] or Venkov [506]). $SU(n,1)$ is the isometry group of HH(n) that leaves the Hermitian form invariant, and $S[U(1)\times U(n)] = SU(n,1)\,[U(1)\times U(n)]$ is an isotropy subgroup of the isometry group. For HH(2), Boyer et al. [84] found twelve coordinate systems which allow separation of variables in the Helmholtz, respectively the Schrödinger equation, and the path integral.

The present system is of interest due to the structure of the metric which has the form $(-,+,\ldots,+)$, i.e. it is of the Minkowski-type, and the Hamiltonian system under consideration is integrable and relativistic with non-trivial interaction after integrating out the ignorable variables [84]. This feature of constructing interaction, respectively potential forces, is also known from examples of quantum motion on other group spaces [66, 139, 168, 353].

I do not want to go into the details of the construction of the Hermitian hyperbolic space HH(n) or HH(2), see [84] for details. The Hamiltonian for HH(2) has the form

$$\mathcal{H} = \frac{4}{2m}(1-|z_1|^2-|z_2|^2)\Big[(|z_1|^2-1)|p_{z_1}|^2 + (|z_2|^2-1)|p_{z_2}|^2 + z_1\bar{z}_2 p_{z_1}\bar{p}_{z_2} + \bar{z}_1 z_2 \bar{p}_{z_1} p_{z_2}\Big] \tag{9.17}$$

and this information will be sufficient for our purposes.

The Spherical Coordinate System.

The spherical coordinate system want to present in some detail. The spherical coordinate system on HH(2) is given by

$$
\left.\begin{array}{l}
z_1 = \tanh \omega \cos \beta \, e^{i\varphi_1} \\
z_2 = \tanh \omega \sin \beta \, e^{i\varphi_2}
\end{array}\right\} \qquad (\omega > 0, \beta \in (0, \tfrac{\pi}{2}), \varphi_1, \varphi_2 \in [0, 2\pi)) \ . \qquad (9.18)
$$

Plugging $z_{1,2}$ into $\mathcal{H}$ gives for the Hamiltonian

$$
\mathcal{H}(\omega, p_\omega, \beta, p_\beta, p_{\varphi_1}, p_{\varphi_2}) = \frac{1}{2m} \left[p_\omega^2 + \frac{p_\beta^2}{\sinh^2 \omega} + A(\omega, \beta) p_{\varphi_1}^2 + B(\omega, \beta) p_{\varphi_2}^2 + \frac{2 p_{\varphi_1} p_{\varphi_2}}{\cosh^2 \omega} \right] ,
$$
$$(9.19)$$

with the quantities $A(\omega, \beta)$ and $B(\omega, \beta)$ given by

$$
A(\omega, \beta) = \frac{1}{\sinh^2 \omega \cos^2 \beta} - \frac{1}{\cosh^2 \omega} \ , \qquad B(\omega, \beta) = \frac{1}{\sinh^2 \omega \sin^2 \beta} - \frac{1}{\cosh^2 \omega} \ . \qquad (9.20)
$$

Vice versa, this yields for the corresponding Lagrangian

$$
\mathcal{L} = \frac{m}{2} \left[\dot\omega^2 + \sinh^2 \omega \dot\beta^2 + (\dot\varphi_1, \dot\varphi_2)(\hat{g}_{ab}) \begin{pmatrix} \dot\varphi_1 \\ \dot\varphi_2 \end{pmatrix} \right] \qquad (9.21)
$$

and we have abbreviated

$$
(\hat{g}_{ab}) = \begin{pmatrix} \sinh^2 \omega \cos^2 \beta F(\omega, \beta) & \sinh^4 \omega \sin^2 \beta \cos^2 \beta \\ \sinh^4 \omega \sin^2 \beta \cos^2 \beta & \sinh^2 \omega \sin^2 \beta F(\omega, \beta) \end{pmatrix} , \qquad (9.22)
$$

and $F(\omega, \beta) = (\cosh^2 \omega - \sinh^2 \omega \cos^2 \beta)$. It follows the Lagrangian path integral representation (9.1). The path integration for the $\varphi_{1,2}$ variables can be separated off via a Fourier expansion and the general Gaussian integral in D dimensions. The emerging path integral is first in the variable β a path integral for the Pöschl–Teller potential with a discrete spectrum and quantum number n, and second in the variable ω a path integral for the modified Pöschl–Teller potential with a continuous spectrum and the quantum number p, and the representation (9.1) [232] follows.

The Equidistant Coordinate Systems.

The three path integrals corresponding to the equidistant coordinate systems are evaluated in a very similar way [232]:

1. For the equidistant-I coordinates we have

$$
(\hat{g}_{ab}) = \sinh^2 \tau_1 \cosh^2 \tau_1 \sinh^2 \tau_2 \begin{pmatrix} \coth^2 \tau_1 \cosh^2 \tau_2 - 1 & 1 \\ 1 & \dfrac{1}{\sinh^2 \tau_2} \end{pmatrix} \qquad (9.23)
$$

and the $\tau_{1,2}$ path integration is again of the modified Pöschl–Teller potential type.

2. For the equidistant-II coordinates we have similarly as for the equidistant-I coordinates

$$
(\widehat{g}_{ab}) = \begin{pmatrix} \cosh^4 \tau_1 \cosh^2 2\tau_2 - \sinh^2 \tau_1 \cosh^2 \tau_1 & \sinh^2 \tau_1 \cosh^2 \tau_1 \sinh 2\tau_2 \\ \sinh^2 \tau_1 \cosh^2 \tau_1 \sinh 2\tau_2 & \sinh^2 \tau_1 \cosh^2 \tau_1 \end{pmatrix}.
$$
(9.24)

Again, the path integration in the variables u and φ can be separated off, and the remaining path integrations in the variables $\tau_{1,2}$ is of the modified Pöschl–Teller potential type(τ_1) and for the hyperbolic barrier potential (τ_2) [219, 232], respectively.

3. For the equidistant-III coordinates we find

$$
(\widehat{g}_{ab}) = \begin{pmatrix} e^{-4\tau_2} \cosh^4 \tau_1 & -e^{-2\tau_2} \sinh^2 \tau_1 \cosh^2 \tau_1 \\ -e^{-2\tau_2} \sinh^2 \tau_1 \cosh^2 \tau_1 & \sinh^2 \tau_1 \cosh^2 \tau_1 \end{pmatrix}
$$
(9.25)

and after Gaussian path integrations for the variables u and φ the remaining path integrals in the variables $\tau_{1,2}$ are of the modified Pöschl–Teller potential and the Morse-potential, respectively.

The Horicyclic Coordinate Systems.

The two path integrals corresponding to the horicyclic coordinate systems are also evaluated in a very similar way [232]:

1. For the horicyclic I system we have

$$
(\widehat{g}_{ab}) = e^{-4q} \begin{pmatrix} 1 & -r^2 \\ -r^2 & e^{2q} + r^2 \end{pmatrix}.
$$
(9.26)

The path integrals in the variables u and φ are of Gaussian type, the remaining path integrals in q and r correspond to a radial harmonic oscillator (variable r) and for a "oscillator-like" potential on the hyperbolic plane [204, 232].

2. For the horicyclic II system we have

$$
(\widehat{g}_{ab}) = e^{-4q} \begin{pmatrix} 4x^2 + e^{2q} & 2x \\ 2x & 1 \end{pmatrix},
$$
(9.27)

the path integrals in the variables u and z are Gaussian, and the path integrals for x and q are for a shifted harmonic oscillator and for Liouville quantum mechanics [232, 247], respectively.

The Remaining Coordinate Systems.

Let us shortly discuss the six remaining coordinate systems.

1. The elliptic and semi-hyperbolic coordinate systems are parametric coordinate system, and very little is known, let alone that a path integral evaluation is possible.

2. in the elliptic- and hyperbolic-parabolic and semicircular-parabolic coordinate systems it is possible to evaluate the corresponding Gaussian integrals of the "ignorable" coordinates, however, in the remaining path integrals there appear potential terms which are of higher order, i.e., one obtains quartic and sextic terms. The potential terms are such that the belong to a type of so-called "quasi-exactly-solvable" potentials [330]. These path integrals cannot be further evaluated.

Table 9.1: Coordinates on the Four-Dimensional Hermitian Hyberbolic Space

Coordinate System	Coordinates	Path Integral Solution
I. Spherical	$z_1 = \tanh \omega \cos \beta \, e^{i\varphi_1}$ $z_2 = \tanh \omega \sin \beta \, e^{i\varphi_2}$	(9.1, 9.2)
II. Equidistant-I	$z_1 = \tanh \tau_1 \, e^{i\varphi_1}$ $z_2 = \dfrac{\tanh \tau_1}{\cosh \tau_2} e^{i\varphi_2}$	(9.3–9.4)
III. Equidistand-II	$z_1 = \dfrac{i \sinh \tau_2 \cosh u - \cosh \tau_2 \sinh u}{i \cosh \tau_2 \cosh u + \sinh \tau_2 \sinh u}$ $z_2 = \dfrac{i \tanh \tau_1}{i \cosh \tau_2 \cosh u + \sinh \tau_2 \sinh u} e^{i\varphi}$	(9.6–9.10)
IV. Equidistand-III	$z_1 = \dfrac{\sinh \tau_2 + iu \, e^{-\tau_2}}{\cosh \tau_2 + iu \, e^{-\tau_2}}$ $z_2 = \dfrac{\tanh \tau_1}{\cosh \tau_2 + iu \, e^{-\tau_2}} e^{i\varphi}$	(9.11–9.12)
V. Horicyclic-I	$z_1 = \dfrac{-1 + e^{2q} + r^2 + 2iu}{1 + e^{2q} + r^2 + 2iu}$ $z_2 = \dfrac{r}{1 + e^{2q} + r^2 + 2iu} e^{i\varphi}$	(9.13–9.14)

Table 9.1 (cont.) Coordinate System	Coordinates	Path Integral Solution
VI. Horicyclic-II	$z_1 = \dfrac{2(u + xz) - i(e^{2q} + x^2 + z^2 - 1)}{2(u + xz) - i(e^{2q} + x^2 + z^2 + 1)}$ $z_2 = \dfrac{-2(z + ix))}{2(u + xz) - i(e^{2q} + x^2 + z^2 + 1)}$	(9.15–9.16)
VII. Elliptic-I $(1 \le \varrho \le a \le \nu < \infty)$	$z_1^2 = \dfrac{a(\nu - 1)(\varrho - 1)}{(a - 1)\nu\varrho}\, e^{2i\varphi_1}$ $z_2^2 = \dfrac{(\nu - a)(a - \varrho)}{(a - 1)\nu\varrho}\, e^{2i\varphi_2}$	(not known)
VIII. Elliptic-II $(1 < a \le \nu,\ \varrho \le 0)$	$z_1^2 = -\dfrac{(a - 1)\nu\varrho}{a(\nu - 1)(1 - \varrho)}\, e^{2i\varphi_1}$ $z_2^2 = \dfrac{(\nu - a)(a - \varrho)}{a(\nu - 1)(1 - \varrho)}\, e^{2i\varphi_2}$	(not known)
IX. Semi-hyperbolic $(u \in \mathbb{R}, \varphi \in [0, 2\pi))$ $(\nu < 0 < \varrho,\ a = \alpha + i\beta)$	$z_1 = \dfrac{is_1 \cosh u - s_0 \sinh u}{is_0 \cosh u + s_1 \sinh u}$ $z_2 = \dfrac{is_2 e^{i\varphi}}{is_0 \cosh u + s_1 \sinh u}$ $\tfrac{1}{2}(s_0 + is_1)^2 = \dfrac{(\nu - a)(\varrho - a)}{a(a - a^*)},\ s_2^2 = -\dfrac{\nu\varrho}{|a|^2}$	(not known)
X. Elliptic Parabolic $(0 < \nu < 1 < \varrho, u \in \mathbb{R})$	$z_1 = \dfrac{\nu + \varrho - 2\nu\varrho + 2i\nu\varrho u}{\nu + \varrho + 2i\nu\varrho u}$ $z_2 = \dfrac{2e^{i\varphi}\sqrt{\nu\varrho(1 - \nu)(\varrho - 1)}}{\nu + \varrho + 2i\nu\varrho u}$	(not known)
XI. Hyperbolic-Parabolic $(0 < \nu < 1 < \varrho, u \in \mathbb{R})$	$z_1 = \dfrac{\nu + \varrho + 2i\nu\varrho u}{\nu + \varrho - 2\nu\varrho - 2i\nu\varrho u}$ $z_2 = \dfrac{2ie^{i\varphi}\sqrt{\nu\varrho(1 - \nu)(\varrho - 1)}}{\nu + \varrho - 2\nu\varrho - 2i\nu\varrho u}$	(not known)
XII. Semi-Hyperbolic $(\nu < 0 < \varrho, u_1, u_2 \in \mathbb{R})$	$z_1 = \dfrac{2\varrho^2\nu^2 u_1 - 2\varrho\nu(\varrho + \nu)u_2 - i[(\varrho - \nu)^2 + \nu^2\varrho^2(u_2^2 - 1)]}{2\varrho^2\nu^2 u_1 - 2\varrho\nu(\varrho + \nu)u_2 - i[(\varrho - \nu)^2 + \nu^2\varrho^2(u_2^2 + 1)]}$ $z_2 = \dfrac{2\varrho\nu u_2 - 2i(\varrho + \nu)}{2\varrho^2\nu^2 u_1 - 2\varrho\nu(\varrho + \nu)u_2 - i[(\varrho - \nu)^2 + \nu^2\varrho^2(u_2^2 + 1)]}$ (Path integral solution not known)	

Chapter 10

Path Integrals on Darboux Spaces

10.1 Two-Dimensional Darboux Spaces

We have the following path integral representations on the two-dimensional Darboux-spaces [233, 244] (Systems III and XI are omitted because they are not solvable).

I. Darboux Space D_{I}: (u,v)-Coordinates, $u > a > 0$, $v \in [0, 2\pi)$:

$\left(x_< \text{ and } x_> \text{ denote the smaller and larger of } x' \text{ and } x'', \tilde{I}_\nu(z) = I_\nu\left(\frac{4\sqrt{-mE}}{3\hbar} z^{3/2}\right)\right)$

$$
\int_{u(t')=u'}^{u(t'')=u''} \mathcal{D}u(t) \int_{v(t')=v'}^{v(t'')=v''} \mathcal{D}v(t)\, 2u \exp\left[\frac{im}{\hbar}\int_0^T u(\dot{u}^2 + \dot{v}^2)dt\right]
$$

$$
= \int_{-\infty}^{\infty} \frac{dE}{2\pi\hbar}\, e^{-iET/\hbar} \sum_{l=-\infty}^{\infty} \frac{e^{il(v''-v')}}{2\pi} \frac{4m}{3\hbar}\left[\left(u' - \frac{l^2\hbar^2}{4mE}\right)\left(u'' - \frac{l^2\hbar^2}{4mE}\right)\right]^{1/2}
$$

$$
\times \left[\tilde{I}_{1/3}\left(u_< - \frac{l^2\hbar^2}{4mE}\right)\tilde{K}_{1/3}\left(u_> - \frac{l^2\hbar^2}{4mE}\right)\right.
$$

$$
\left. - \frac{\tilde{I}_{1/3}\left(a - \frac{l^2\hbar^2}{4mE}\right)}{\tilde{K}_{1/3}\left(a - \frac{l^2\hbar^2}{4mE}\right)}\tilde{K}_{1/3}\left(u' - \frac{l^2\hbar^2}{4mE}\right)\tilde{K}_{1/3}\left(u'' - \frac{l^2\hbar^2}{4mE}\right)\right]. \quad (10.1)
$$

II. Darboux Space D_{I}: Rotated (r,q)-Coordinates, $r, q \in \mathbb{R}$:

$$
\int_{r(t')=r'}^{r(t'')=r''} \mathcal{D}r(t) \int_{q(t')=q'}^{q(t'')=q''} \mathcal{D}q(t)\, 2(r\cos\vartheta + q\sin\vartheta) \exp\left[\frac{im}{\hbar}\int_0^T (r\cos\vartheta + q\sin\vartheta)(\dot{r}^2 + \dot{q}^2)dt\right]
$$

$$
= \int_{-\infty}^{\infty} \frac{dE}{2\pi\hbar}\, e^{-iET/\hbar} \int d\mathcal{E}\left(\frac{m}{\hbar^2}\sqrt{\frac{2}{E\cos\vartheta}}\right)^{2/3}
$$

$$\times \mathrm{Ai}\left[-\left(r'' - \frac{\mathcal{E}}{2E\cos\vartheta}\right)\left(\frac{4mE\cos\vartheta}{\hbar^2}\right)^{1/3}\right]\mathrm{Ai}\left[-\left(r' - \frac{\mathcal{E}}{2E\cos\vartheta}\right)\left(\frac{4mE\cos\vartheta}{\hbar^2}\right)^{1/3}\right]$$

$$\times\left(\frac{m}{\hbar^2}\sqrt{\frac{2}{E\sin\vartheta}}\right)^{2/3}$$

$$\times\mathrm{Ai}\left[-\left(q'' + \frac{\mathcal{E}}{2E\sin\vartheta}\right)\left(\frac{4mE\sin\vartheta}{\hbar^2}\right)^{1/3}\right]\mathrm{Ai}\left[-\left(q' + \frac{\mathcal{E}}{2E\sin\vartheta}\right)\left(\frac{4mE\sin\vartheta}{\hbar^2}\right)^{1/3}\right].$$

$$(10.2)$$

IV. Darboux Space D_{II}: (u,v)-Coordinates, $u > 0, v \in \mathbb{R}$ $\left(\lambda = \sqrt{1/4 - 2m|a|E/\hbar^2}\right)$:

$$\int_{u(t')=u'}^{u(t'')=u''}\mathcal{D}u(t)\int_{v(t')=v'}^{v(t'')=v''}\mathcal{D}v(t)\frac{bu^2 - a}{u^2}\exp\left[\frac{\mathrm{i}m}{2\hbar}\int_0^T\frac{bu^2 - a}{u^2}(\dot{u}^2 + \dot{v}^2)dt\right]$$

$$= \int_{-\infty}^{\infty}\frac{\mathrm{d}E}{2\pi\hbar}\,\mathrm{e}^{-\mathrm{i}ET/\hbar}$$

$$\times\frac{2m\sqrt{u'u''}}{\mathrm{i}\hbar}\int_{-\infty}^{\infty}\mathrm{d}k\,\mathrm{e}^{\mathrm{i}k(v''-v')}I_\lambda\left(\sqrt{k^2 - \frac{2mbE}{\hbar^2}}\,u_<\right)K_\lambda\left(\sqrt{k^2 - \frac{2mbE}{\hbar^2}}\,u_>\right) \quad (10.3)$$

$$= \int_{-\infty}^{\infty}\frac{\mathrm{d}E}{2\pi\hbar}\,\mathrm{e}^{-\mathrm{i}ET/\hbar}\frac{\hbar}{\pi^3}\int_{-\infty}^{\infty}\mathrm{d}k\,\mathrm{e}^{\mathrm{i}k(v''-v')}$$

$$\times\int_0^{\infty}\frac{p\sinh\pi p\,dp}{\frac{\hbar^2}{2m|a|}(p^2 + \frac{1}{4}) - E}K_{\mathrm{i}p}\left(\sqrt{k^2 - \frac{2mbE}{\hbar^2}}\,u'\right)K_{\mathrm{i}p}\left(\sqrt{k^2 - \frac{2mbE}{\hbar^2}}\,u''\right) \quad . \quad (10.4)$$

V-a. Darboux Space D_{II}: Polar Coordinates, $\varrho > 0, \vartheta \in [0, 2\pi)$:

$$\int_{\varrho(t')=\varrho'}^{\varrho(t'')=\varrho''}\mathcal{D}\varrho(t)\int_{\vartheta(t')=\vartheta'}^{\vartheta(t'')=\vartheta''}\mathcal{D}\vartheta(t)\varrho\left(b - \frac{a}{\varrho^2\cos^2\vartheta}\right)$$

$$\times\exp\left\{\frac{\mathrm{i}}{\hbar}\int_0^T\left[\frac{m}{2}\left(b - \frac{a}{\varrho^2\cos^2\vartheta}\right)(\dot{\varrho}^2 + \varrho^2\dot{\vartheta}^2) + \left(b - \frac{a}{\varrho^2\cos^2\vartheta}\right)^{-1}\frac{\hbar^2}{8m\varrho^2}\right]dt\right\}$$

V-b. Darboux Space D_{II}: Polar Coordinates, $\varrho = \mathrm{e}^{\tau_2}, \cos\vartheta = 1/\cosh\tau_1$ $(\tau_{1,2} \in \mathbb{R})$:

$$= \int_{\tau_1(t')=\tau_1'}^{\tau_1(t'')=\tau_1''}\mathcal{D}\tau_1(t)\int_{\tau_2(t')=\tau_2'}^{\tau_2(t'')=\tau_2''}\mathcal{D}\tau_2(t)\cosh\tau_1\left(\frac{b\,\mathrm{e}^{2\tau_2}}{\cosh^2\tau_1} - a\right)$$

$$\times\exp\left\{\frac{\mathrm{i}}{\hbar}\int_0^T\left[\frac{m}{2}\left(\frac{b\,\mathrm{e}^{2\tau_2}}{\cosh^2\tau_1} - a\right)(\dot{\tau}_1^2 + \cosh^2\tau_1\dot{\tau}_2^2)\right.\right.$$

$$\left.\left. -\left(\frac{b\,\mathrm{e}^{2\tau_2}}{\cosh^2\tau_1} - a\right)^{-1}\frac{\hbar^2}{8m}\left(1 + \frac{1}{\cosh^2\tau_1}\right)\right]dt\right\}$$

$$= \int_{-\infty}^{\infty}\frac{\mathrm{d}E}{2\pi\hbar}\,\mathrm{e}^{-\mathrm{i}ET/\hbar}\sum_{\pm}\int_0^{\infty}\mathrm{d}p\int_0^{\infty}\mathrm{d}k\frac{\Psi_{p,k,\pm}(\tau_1'',\tau_2'')\Psi_{p,k,\pm}^*(\tau_1',\tau_2')}{\frac{\hbar^2}{2m|a|}(p^2 + \frac{1}{4}) - E} \quad , \quad (10.5)$$

$$\Psi_{p,k,\pm}(\tau_1,\tau_2) = \frac{\sqrt{2\cosh\tau_1}}{\pi}\sqrt{k\sinh\pi k}$$

$$\times K_{ik}\left(i\sqrt{\frac{b}{|a|}\left(p^2+\frac{1}{4}\right)}\,e^{\tau_2}\right)\sqrt{\frac{p\sinh\pi p\,dp}{\cosh^2 k+\sinh^2\pi p}}\,P_{ik-1/2}^{ip}(\pm\tanh\tau_1)\;, \qquad (10.6)$$

$$\Psi_{p,k,\pm}(\varrho,\vartheta) = \frac{\sqrt{2\sin\vartheta}}{\pi}\sqrt{k\sinh\pi k}$$

$$\times K_{ik}\left(i\sqrt{\frac{b}{|a|}\left(p^2+\frac{1}{4}\right)}\,\varrho\right)\sqrt{\frac{p\sinh\pi p\,dp}{\cosh^2 k+\sinh^2\pi p}}\,P_{ik-1/2}^{ip}(\pm\sin\vartheta)\;. \qquad (10.7)$$

VI. Darboux Space D_{II}: Parabolic Coordinates, $\xi,\eta>0$ $(\sqrt{b(p^2+\frac{1}{4})/|a|}\equiv\tilde{p})$:

$$\int_{\xi(t')=\xi'}^{\xi(t'')=\xi''}\mathcal{D}\xi(t)\int_{\eta(t')=\eta'}^{\eta(t'')=\eta''}\mathcal{D}\eta(t)\frac{b\xi^2\eta^2-a}{\xi^2\eta^2}(\xi^2+\eta^2)$$

$$\times\exp\left[\frac{im}{2\hbar}\int_0^T\frac{b\xi^2\eta^2-a}{\xi^2\eta^2}(\xi^2+\eta^2)(\dot\xi^2+\dot\eta^2)dt\right]$$

$$=\int_{-\infty}^{\infty}\frac{dE}{2\pi\hbar}\,e^{-iET/\hbar}\frac{i\hbar}{4\pi^2}(\xi'\xi''\eta'\eta'')^{-1/2}\int d\mathcal{E}$$

$$\times\int_0^{\infty}\frac{dp\,p\sinh\pi p}{\frac{\hbar^2}{2m|a|}(p^2+\frac{1}{4})-E}\frac{|\Gamma[\frac{1}{2}(1+ip-\mathcal{E})]|^2}{\tilde{p}^2}W_{\mathcal{E}/2,ip/2}\left(i\tilde{p}\xi''^2\right)W_{\mathcal{E}/2,ip/2}^*\left(i\tilde{p}\xi'^2\right)$$

$$\times|\Gamma[\tfrac{1}{2}(1+ip-\mathcal{E})]|^2 W_{\mathcal{E}/2,ip/2}(i\tilde{p}\eta_>^2)\frac{W_{\mathcal{E}/2,ip/2}(i\tilde{p}\eta_<^2)}{\Gamma[\frac{1}{2}(1-ip-\mathcal{E})]\Gamma(1+ip)} \qquad (10.8)$$

$$=\int_{-\infty}^{\infty}\frac{dE}{2\pi\hbar}\,e^{-iET/\hbar}\hbar(\xi'\xi''\eta'\eta'')^{-1/2}\int d\mathcal{E}\int_0^{\infty}\frac{dp\,p\sinh\pi p}{\frac{\hbar^2}{2m|a|}(p^2+\frac{1}{4})-E}\frac{|\Gamma[\frac{1}{2}(1+ip-\mathcal{E})]|^4}{2\pi\tilde{p}^2}$$

$$\times W_{\mathcal{E}/2,ip/2}\left(i\tilde{p}\xi''^2\right)W_{\mathcal{E}/2,ip/2}^*\left(i\tilde{p}\xi'^2\right)W_{\mathcal{E}/2,ip/2}\left(i\tilde{p}\xi''^2\right)W_{\mathcal{E}/2,ip/2}^*\left(i\tilde{p}\eta'^2\right)\;. \qquad (10.9)$$

VII. Darboux Space D_{II}: Elliptic Coordinates, $\omega\in\mathbb{R},\varphi\in(-\pi/2,\pi/2)$:

$$\int_{\omega(t')=\omega'}^{\omega(t'')=\omega''}\mathcal{D}\omega(t)\int_{\varphi(t')=\varphi'}^{\varphi(t'')=\varphi''}\mathcal{D}\varphi(t)\frac{bd^2\cosh^2\omega\cos^2\varphi-a}{\cosh^2\omega\cos^2\varphi}(\cosh^2\omega-\cos^2\varphi)$$

$$\times\exp\left[\frac{im}{2\hbar}\int_0^T\frac{bd^2\cosh^2\omega\cos^2\varphi-a}{\cosh^2\omega\cos^2\varphi}(\cosh^2\omega-\cos^2\varphi)(\dot\omega^2+\dot\varphi^2)dt\right]$$

$$=\int_{-\infty}^{\infty}\frac{dE}{2\pi\hbar}\,e^{-iET/\hbar}\sqrt{\cos\varphi'\cos\varphi''}\int_0^{\infty}\frac{dp\,\sinh\pi p}{\frac{\hbar^2}{2m|a|}(p^2+\frac{1}{4})-E}\int_0^{\infty}\frac{dk\,k\sinh\pi k}{(\cosh^2\pi k+\sinh^2\pi p)^2}$$

$$\times\sum_{\epsilon,\epsilon'=\pm1}S_{ip-1/2}^{ik\,(1)}(\epsilon\tanh\omega'';-\tilde{p})S_{ip-1/2}^{ik\,(1)\,*}(\epsilon\tanh\omega';-\tilde{p})$$

$$\times\mathrm{ps}_{ik-1/2}^{ip}(\epsilon'\sin\varphi'';\tilde{p}^2)\mathrm{ps}_{ik-1/2}^{ip\,*}(\epsilon'\sin\varphi';\tilde{p}^2)\;. \qquad (10.10)$$

VIII. Darboux Space D_{III}: (u,v)-Coordinates $u > 0, v \in [0, 2\pi)$:

$$\int\limits_{u(t')=u'}^{u(t'')=u''} \mathcal{D}u(t) \int\limits_{v(t')=v'}^{v(t'')=v''} \mathcal{D}v(t)(a\,e^{-u} + b\,e^{-2u}) \exp\left[\frac{im}{2\hbar}\int_0^T (a\,e^{-u} + b\,e^{-2u})(\dot{u}^2 + \dot{v}^2)dt\right]$$

$$= \int_{-\infty}^{\infty} \frac{dE}{2\pi\hbar}\,e^{-iET/\hbar} \sum_{l=-\infty}^{\infty} \frac{e^{il(v''-v')}}{2\pi} \frac{m\Gamma(\frac{1}{2} + l + a\sqrt{-2mbE}/2b\hbar)}{\hbar\sqrt{-2mbE}\,\Gamma(1 + 2l)}\,e^{(u'+u'')/2}$$

$$\times W_{-a\sqrt{-2mbE}/2b\hbar,l}\left(2\frac{\sqrt{-2mbE}}{\hbar}\,e^{-u_<}\right) M_{-a\sqrt{-2mbE}/2b\hbar,l}\left(2\frac{\sqrt{-2mbE}}{\hbar}\,e^{-u_>}\right) \quad (10.11)$$

$$= \int_{-\infty}^{\infty} \frac{dE}{2\pi\hbar}\,e^{-iET/\hbar} \sum_{l=-\infty}^{\infty} \frac{e^{il(v''-v')}}{2\pi}e^{(u'+u'')/2}$$

$$\times \int_0^{\infty} \frac{e^{\pi p/2}dp}{\frac{\hbar^2 p^2}{2m} - E} \frac{|\Gamma(\frac{1}{2} + l + ip)|^2}{2\pi\Gamma^2(1 + 2l)} M_{ip/2,l}\left(-2ip\,e^{-u'}\right) M_{-ip/2,l}\left(2ip\,e^{-u''}\right)$$

$$+ \sum_{l=-\infty}^{\infty} \frac{e^{il(v''-v')}}{2\pi} \sum_{n=0}^{\infty} \frac{1}{E_{nl} - E}\Psi_{nl}(u'')\Psi_{nl}(u') \quad, \quad (10.12)$$

$$\Psi_{nl}(u) = N_{nl}\frac{(2\mathcal{E} - 2n - 1)n!}{\Gamma(2\mathcal{E} - n)}\left(\frac{\sqrt{-8mbE_{nl}}}{\hbar}\right)^{\mathcal{E}-n-1/2}$$

$$\times e^{(\mathcal{E}-n-1/2)u - \sqrt{-8mbE_{nl}}\,e^u/\hbar}L_n^{(2\mathcal{E}-2n-1)}\left(\frac{\sqrt{-8mbE_{nl}}}{\hbar}\,e^u\right) \quad, \quad (10.13)$$

$$E_{nl} = -\frac{\hbar^2}{2m}\frac{b}{a^2}(2n + 2l + 1)^2 \qquad \left(\mathcal{E} = \tfrac{1}{2} + l + a\sqrt{-2mbE_{nl}}/2b\hbar\right) \quad . \quad (10.14)$$

IX. Darboux Space D_{III}: Polar Coordinates $\varrho > 0, \varphi \in [0, 2\pi)$:

$$\int\limits_{\varrho(t')=\varrho'}^{\varrho(t'')=\varrho''} \mathcal{D}\varrho(t) \int\limits_{\varphi(t')=\varphi'}^{\varphi(t'')=\varphi''} \mathcal{D}\varphi(t)(a + \tfrac{b}{4}\varrho^2)\varrho$$

$$\times \exp\left\{\frac{i}{\hbar}\int_0^T \left[\frac{m}{2}(a + \tfrac{b}{4}\varrho^2)(\dot{\varrho}^2 + \varrho^2\dot{\varphi}^2) + (a + \tfrac{b}{4}\varrho^2)^{-1}\frac{\hbar^2}{8m\varrho^2}\right]dt\right\}$$

$$= \int_{-\infty}^{\infty} \frac{dE}{2\pi\hbar}\,e^{-iET/\hbar} \sum_{l=-\infty}^{\infty} \frac{e^{il(\varphi''-\varphi')}}{2\pi} \cdot \sqrt{-\frac{m}{2E}}\frac{\Gamma[\frac{1}{2}(1 + l - a\sqrt{-2mE/b}/\hbar)]}{\Gamma(1 + l)}$$

$$\times W_{a\sqrt{-2mE/b}/2\hbar,\frac{l}{2}}\left(b\sqrt{-\frac{2mE}{\hbar^2}}\,\varrho_>\right) M_{a\sqrt{-2mE/b}/2\hbar,\frac{l}{2}}\left(b\sqrt{-\frac{2mE}{\hbar^2}}\,\varrho_<\right) \quad (10.15)$$

$$= \int_{-\infty}^{\infty} \frac{dE}{2\pi\hbar}\,e^{-iET/\hbar} \sum_{l=-\infty}^{\infty} \frac{e^{il(\varphi''-\varphi')}}{2\pi} \frac{1}{2\pi b\sqrt{\varrho'\varrho''}}$$

$$\times \int_0^{\infty} \frac{dp\,e^{\pi p}}{\frac{\hbar^2 p^2}{2m} - E} \frac{|\Gamma[\frac{1}{2}(1 + l + iap)]|^2}{\Gamma^2(1 + l)} M_{iap/2,l/2}(-ibp\varrho') M_{-iap/2,l/2}(ibp\varrho'')$$

$$+ \sum_{l=-\infty}^{\infty} \frac{e^{il(\varphi''-\varphi')}}{2\pi b \sqrt{\varrho' \varrho''}} \sum_{n=0}^{\infty} \frac{1}{E_{nl} - E} \Psi_{nl}^{(D\,\mathrm{III})}(\varrho'') \Psi_{nl}^{(D\,\mathrm{III})}(\varrho') \ . \tag{10.16}$$

X. Darboux Space D_{III}: Parabolic Coordinates $\xi, \eta > 0$:

$$\int_{\xi(t')=\xi'}^{\xi(t'')=\xi''} \mathcal{D}\xi(t) \int_{\eta(t')=\eta'}^{\eta(t'')=\eta''} \mathcal{D}\eta(t)(a + \tfrac{b}{4}(\xi^2 + \eta^2)) \exp\left[\frac{im}{2\hbar} \int_0^T (a + \tfrac{b}{4}(\xi^2 + \eta^2))(\dot\xi^2 + \dot\eta^2)dt\right]$$

$$= \int_{-\infty}^{\infty} \frac{dE}{2\pi\hbar}\, e^{-iET/\hbar} \int d\mathcal{E}\, \frac{m}{\pi\hbar^2 b} \sqrt{-\frac{m}{2E}}\, \Gamma\left(\frac{1}{2} + \frac{aE - \mathcal{E}}{b\hbar}\sqrt{-\frac{m}{2E}}\right) \Gamma\left(\frac{1}{2} + \frac{\mathcal{E}}{b\hbar}\sqrt{-\frac{m}{2E}}\right)$$

$$\times D_{-\frac{1}{2}+\frac{aE-\mathcal{E}}{b\hbar}\sqrt{-\frac{m}{2E}}} \left(\sqrt[4]{-\frac{8mEb^2}{\hbar^2}}\, \xi_>\right) D_{-\frac{1}{2}+\frac{aE-\mathcal{E}}{b\hbar}\sqrt{-\frac{m}{2E}}} \left(-\sqrt[4]{-\frac{8mEb^2}{\hbar^2}}\, \xi_<\right)$$

$$\times D_{-\frac{1}{2}+\frac{\mathcal{E}}{b\hbar}\sqrt{-\frac{m}{2E}}} \left(\sqrt[4]{-\frac{8mEb^2}{\hbar^2}}\, \eta_>\right) D_{-\frac{1}{2}+\frac{\mathcal{E}}{b\hbar}\sqrt{-\frac{m}{2E}}} \left(-\sqrt[4]{-\frac{8mEb^2}{\hbar^2}}\, \eta_<\right) \tag{10.17}$$

$$= \sum_{e,o} \int_{\mathrm{IR}} d\zeta \int_{\mathrm{IR}} dp\, e^{-i\hbar p^2 T/2m}\Psi_{p,\zeta}^{(e,o)\,*}(\xi',\eta')\Psi_{p,\zeta}^{(e,o)}(\xi'',\eta'') \ , \tag{10.18}$$

$$\Psi_{p,\zeta}^{(e,o)}(\xi,\eta) = \frac{e^{\pi/2ap}}{\sqrt{24\pi^2}}$$

$$\times \begin{pmatrix} |\Gamma(\tfrac{1}{4} - \tfrac{i}{2p}(1/a_B + \zeta))|^2 E_{-\frac{1}{2}+\frac{i}{p}(1/a_B+\zeta)}^{(0)}(e^{-i\pi/4}\sqrt{2p}\,\xi) E_{-\frac{1}{2}-\frac{i}{p}(1/a_B+\zeta)}^{(0)}(e^{-i\pi/4}\sqrt{2p}\,\eta) \\ |\Gamma(\tfrac{3}{4} - \tfrac{i}{2p}(1/a_B + \zeta))|^2 E_{-\frac{1}{2}+\frac{i}{p}(1/a_B+\zeta)}^{(1)}(e^{-i\pi/4}\sqrt{2p}\,\xi) E_{-\frac{1}{2}-\frac{i}{p}(1/a_B+\zeta)}^{(1)}(e^{-i\pi/4}\sqrt{2p}\,\eta) \end{pmatrix}$$

$$+ \sum_{n_\xi=0}^{\infty} \sum_{n_\eta=0}^{\infty} \frac{N_{n_\xi n_\eta}^2}{E_{n_\xi n_\eta} - E} \Psi_{n_\xi}^{(HO)}(\xi'') \Psi_{n_\xi}^{(HO)}(\xi'') \Psi_{n_\eta}^{(HO)}(\eta'') \Psi_{n_\eta}^{(HO)}(\eta'') \ . \tag{10.19}$$

XII. Darboux Space D_{III}: Hyperbolic Coordinates $\mu, \nu > 0$:

$$\int_{\mu(t')=\mu'}^{\mu(t'')=\mu''} \mathcal{D}\mu(t) \int_{\nu(t')=\nu'}^{\nu(t'')=\nu''} \mathcal{D}\nu(t) \frac{(a + \tfrac{b}{2}(\mu - \nu))(\mu + \nu)}{\mu\nu}$$

$$\times \exp\left[\frac{im}{2\hbar} \int_0^T (a + \tfrac{b}{2}(\mu - \nu))(\mu + \nu)\left(\frac{\dot\mu^2}{\mu^2} - \frac{\dot\nu^2}{\nu^2}\right)dt\right]$$

$$= \int_{\mathrm{IR}} dp\, e^{-i\hbar p^2 T/2m} \int \frac{\lambda d\lambda}{\mu'\mu''\nu'\nu''} \frac{|\Gamma(\tfrac{1}{2} + \lambda + ip)|^4}{4\pi p^2 \Gamma^4(1 + 2\lambda)} e^{2\pi p}$$

$$\times M_{ip,\lambda}\left(-2ip\mu'^2\right) M_{-ip,\lambda}\left(2ip\mu''^2\right) M_{ip,\lambda}\left(-2ip\nu'^2\right) M_{-ip,\lambda}\left(2ip\nu''^2\right)$$

$$+ \sum_{n_\mu=0}^{\infty} \sum_{n_\nu=0}^{\infty} \frac{N_{n_\mu n_\nu u}^2}{E_{n_\mu n_\nu} - E} \Psi_{n_\mu}^{(MP)}(\mu'') \Psi_{n_\mu}^{(MP)}(\mu'') \Psi_{n_\nu}^{(MP)}(\nu'') \Psi_{n_\nu}^{(MP)}(\nu'') \ . \tag{10.20}$$

XIII. Darboux Space $D_{\rm IV}$: (u,v)-Coordinates $u \in [0, \pi/2], v \in \mathbb{R}$:

$$\int_{u(t')=u'}^{u(t'')=u''} \mathcal{D}u(t) \int_{v(t')=v'}^{v(t'')=v''} \mathcal{D}v(t) \left(\frac{a_+}{\sin^2 u} + \frac{a_-}{\cos^2 u} \right)$$

$$\times \exp\left[\frac{im}{2\hbar} \int_0^T \left(\frac{a_+}{\sin^2 u} + \frac{a_-}{\cos^2 u} \right) (\dot{u}^2 + \dot{v}^2) dt \right]$$

$$= \int_{-\infty}^{\infty} dk_v \int_0^{\infty} dp \, \exp\left[\frac{i\hbar T}{2ma_+} \left(p^2 + \frac{1}{4} \right) \right] \Psi_{p,k_v}(u'', v'') \Psi_{p,k_v}^*(u', v') \ , \tag{10.21}$$

$$\Psi_{p,k_v}(u,v) = \frac{e^{ik_v v}}{\sqrt{2\pi a_+ \cosh \tau}} \Psi_p^{(\eta, ik)}(\cos u) \ . \tag{10.22}$$

XIV. Darboux Space $D_{\rm IV}$: Equidistant Coordinates $\alpha, \beta \in \mathbb{R}$ $(p_\pm \equiv (p^2 + \frac{1}{4})/a_\pm)$:

$$\int_{\alpha(t')=\alpha'}^{\alpha(t'')=\alpha''} \mathcal{D}\alpha(t) \int_{\beta(t')=\beta'}^{\beta(t'')=\beta''} \mathcal{D}\beta(t) \frac{a - 2b \tanh \alpha}{4} \cosh \alpha$$

$$\times \exp\left\{ \frac{i}{\hbar} \int_0^T \left[\frac{m}{2} \frac{a - 2b \tanh \alpha}{4} (\dot{\alpha}^2 + \cosh^2 \alpha \dot{\beta}^2) \right. \right.$$

$$\left. \left. - \left(\frac{a - 2b \tanh \alpha}{4} \right)^{-1} \frac{\hbar^2}{8m} \left(1 + \frac{1}{\cosh^2 \alpha} \right) \right] dt \right\}$$

$$= \int_{-\infty}^{\infty} dk_\beta \int_0^{\infty} dp \, \exp\left[\frac{i\hbar T}{2ma_+} \left(p^2 + \frac{1}{4} \right) \right] \Psi_{p,k_\beta}^{(\pm)*}(\alpha', \beta') \Psi_{p,k_\beta}^{(\pm)}(\alpha'', \beta'') \ , \tag{10.23}$$

$$\Psi_{p,k_\beta}^{(\pm)}(\alpha, \beta) = \frac{e^{ik_\beta \beta}}{\sqrt{2\pi}} \sqrt{\frac{m}{2\pi \hbar^2}} \frac{|\sinh[\frac{\pi}{2}(p_+ \pm p_-)]|}{\cosh[\frac{\pi}{2}(p_+ + p_-)]}$$

$$\times \left(\frac{1 + \tanh \alpha}{2} \right)^{i(p_+ + p_-)/4} \left(\frac{1 - \tanh \alpha}{2} \right)^{i(p_+ - p_-)/4}$$

$$\times {}_2F_1 \left(\frac{ip_+}{2} + ik_\beta + \frac{1}{2}, \frac{ip_+}{2} - ik_\beta + \frac{1}{2}; 1 + ip_+ \pm ip_-; 1 \pm \tanh \alpha \right) \ . \tag{10.24}$$

XV. Darboux Space $D_{\rm IV}$: Horospherical Coordinates $\mu, \nu > 0$:

$$\int_{\mu(t')=\mu'}^{\mu(t'')=\mu''} \mathcal{D}\mu(t) \int_{\nu(t')=\nu'}^{\nu(t'')=\nu''} \mathcal{D}\nu(t) \left(\frac{a_+}{\nu^2} + \frac{a_-}{\mu^2} \right) \exp\left[\frac{im}{2\hbar} \int_0^T \left(\frac{a_+}{\nu^2} + \frac{a_-}{\mu^2} \right) (\dot{\mu}^2 + \dot{\nu}^2) dt \right]$$

$$= \int_{-\infty}^{\infty} \frac{dE}{2\pi\hbar} \, e^{-iET/\hbar} \frac{4m^2}{\hbar^3} \sqrt{\mu'\mu''\nu'\nu''} \int \frac{d\mathcal{E}}{2\pi i}$$

$$\times I_\lambda\left(\sqrt{2m\mathcal{E}} \frac{\nu_<}{\hbar} \right) K_\lambda\left(\sqrt{2m\mathcal{E}} \frac{\nu_>}{\hbar} \right) I_{\tilde{\lambda}}\left(\sqrt{-2m\mathcal{E}} \frac{\mu_<}{\hbar} \right) K_{\tilde{\lambda}}\left(\sqrt{-2m\mathcal{E}} \frac{\mu_>}{\hbar} \right) \tag{10.25}$$

$$= \int_{-\infty}^{\infty} \frac{dE}{2\pi\hbar} \, e^{-iET/\hbar} \frac{1}{8\pi^2} \sqrt{\mu'\mu''\nu'\nu''} \int d\kappa \int_0^{\infty} \frac{dp \, p \sinh \pi p \sinh \pi \tilde{p}}{\frac{\hbar^2}{2ma_+}(p^2 + \frac{1}{4}) - E}$$

$$\times \left[K_{ip}(\sqrt{\kappa}\,\nu') K_{ip}(\sqrt{\kappa}\,\nu') H_{-i\tilde{p}}^{(1)}(\sqrt{\kappa}\,\mu') H_{-i\tilde{p}}^{(1)}(\sqrt{\kappa}\,\mu'') + (\mu \leftrightarrow \nu) \right]$$

$$\left(\lambda^2 = \frac{1}{4} - \frac{2mE}{\hbar^2}a_+, \quad \tilde\lambda^2 = \frac{1}{4} - \frac{2mE}{\hbar^2}a_- , \quad \tilde{p}^2 = \frac{a_-}{a_+}\left(p^2 + \frac{1}{4}\right) - \frac{1}{4}\right) . \tag{10.26}$$

XVI. Darboux Space D_{IV}: Elliptic Coordinates $\omega > 0, \varphi \in [0, \pi/2)$:

$$\int_{\omega(t')=\omega'}^{\omega(t'')=\omega''} \mathcal{D}\omega(t) \int_{\varphi(t')=\varphi'}^{\varphi(t'')=\varphi''} \mathcal{D}\varphi(t) \left(\frac{a_+}{\sin^2\varphi} + \frac{a_-}{\cos^2\varphi} + \frac{a_+}{\sinh^2\omega} - \frac{a_-}{\cosh^2\omega}\right)$$

$$\times \exp\left[\frac{im}{2\hbar}\int_0^T \left(\frac{a_+}{\sin^2\varphi} + \frac{a_-}{\cos^2\varphi} + \frac{a_+}{\sinh^2\omega} - \frac{a_-}{\cosh^2\omega}\right)(\dot\omega^2 + \dot\varphi^2)dt\right]$$

$$= \int_{-\infty}^\infty dk_v \int_0^\infty dp \exp\left[-\frac{i\hbar T}{2ma_+}\left(p^2 + \frac{1}{4}\right)\right]\Psi_{p,k}^*(\omega',\varphi')\Psi_{p,k}(\omega'',\varphi'') , \tag{10.27}$$

$$\Psi_{p,k}(\omega,\varphi) = (1 - \cos^2\varphi)^{1/4}\,\Psi_k^{(\eta,\nu)}(\omega)\Psi_p^{(\eta,ik)}(\cos\varphi)$$

$$\left(\eta^2 = \tfrac{1}{4} + 2mE_p a_+/\hbar^2, \quad \nu^2 = \tfrac{1}{4} + 2mE_p a_-/\hbar^2, \quad E_p = \frac{\hbar^2}{2ma_+}\left(p^2 + \tfrac{1}{4}\right)\right) . \tag{10.28}$$

10.2 Path Integral Evaluations

10.2.1 Darboux Space D_{I}

The (u, v)-Coordinate System.

For the evaluation of the path integral one has to perform a time-transformation yielding (with corresponding energy-dependent Green-functions $G_u(E;\mathcal{E}_u)$ and $G_v(E;\mathcal{E}_v)$)

$$K(u'', u', v'', v'; T)$$

$$= \int_{-\infty}^\infty \frac{dE}{2\pi\hbar} e^{-iET/\hbar} \int_0^\infty ds'' K_v(v'', v'; s'') \cdot K_u(u'', u'; s'')$$

$$= \int_{-\infty}^\infty \frac{dE}{2\pi\hbar} e^{-iET/\hbar} \sum_{l=-\infty}^\infty \frac{e^{il(v''-v')}}{2\pi} G_u\left(E; u'', u'; -\frac{l^2\hbar^2}{2m}\right) . \tag{10.29}$$

Furthermore we must exploit the solution of the linear potential

$$G^{(k)}(x'', x'; \mathcal{E}) = \frac{4m}{3\hbar}\left[\left(x' - \frac{\mathcal{E}}{k}\right)\left(x'' - \frac{\mathcal{E}}{k}\right)\right]^{1/2}$$

$$\times I_{1/3}\left[\frac{\sqrt{8mk}}{3\hbar}\left(x_< - \frac{\mathcal{E}}{k}\right)^{3/2}\right] K_{1/3}\left[\frac{\sqrt{8mk}}{3\hbar}\left(x_> - \frac{\mathcal{E}}{k}\right)^{3/2}\right] . \tag{10.30}$$

In addition, we have to recall that the motion in u takes place only in the half-space $u > a$. In order to construct the Green's function in the half-space $x > a$ we have to put Dirichlet boundary-conditions at $x = a$ [216, 217]:

$$G_{(x=a)}(u'', u'; \mathcal{E}) = G^{(k)}(u'', u'; \mathcal{E}) - \frac{G^{(k)}(u'', a; \mathcal{E})G^{(k)}(a, u'; \mathcal{E})}{G^{(k)}(a, a; \mathcal{E})} . \tag{10.31}$$

Inserting the linear-potential Green's function yields (10.1).

The Rotated (r, q)-Coordinate System.

The path integral evaluation for the rotated (r, q)-coordinate system is very similar as for the (u, v)-coordinate system. The only difference consists of the re-writing of the Bessel-function in terms of the Airy-functions ($\xi = \frac{2}{3} z^{3/2}$):

$$\mathrm{Ai}(z) = \frac{1}{3}\sqrt{z}\left[I_{-1/3}(\xi) - I_{1/3}(\xi)\right] = \frac{1}{\pi}\sqrt{\frac{z}{3}}\,K_{1/3}(\xi) \ , \tag{10.32}$$

$$\mathrm{Ai}(-z) = \frac{1}{3}\sqrt{z}\left[J_{-1/3}(\xi) - J_{1/3}(\xi)\right] \ , \tag{10.33}$$

and we are done (10.2).

Let us finally note that the path integral in displaced parabolic coordinates gives quartic terms in the variables ξ and η, which cannot be treated.

10.2.2　Darboux Space D_{II}

The (u, v)-Coordinate System.

Let us first note that the Darboux Space D_{II} has special limiting cases which is summarized in the following table:

Table 10.1: Some Limiting Cases of Coordinate Systems on D_{II}

Metric:	D_{II}	$\Lambda^{(2)}$ $(a{=}{-}1, b{=}0)$	E_2 $(a{=}0, b{=}1)$
$\dfrac{bu^2 - a}{u^2}(\mathrm{d}u^2 + \mathrm{d}v^2)$	(u, v)-System	Horicyclic	Cartesian
$\dfrac{b\varrho^2 \cos^2\vartheta - a}{\varrho^2 \cos^2\vartheta}(\mathrm{d}\varrho^2 + \mathrm{d}\vartheta^2)$	Polar	Equidistant	Polar
$\dfrac{b\xi^2\eta^2 - a}{\xi^2\eta^2}(\xi^2 + \eta^2)(\mathrm{d}\xi^2 + \mathrm{d}\eta^2)$	Parabolic	Semi-circular parabolic	Parabolic
$\dfrac{bd^2 \cosh^2\omega \cos^2\varphi - a}{\cosh^2\omega \cos^2\varphi}$ $\times(\cosh^2\omega - \cos^2\varphi)(\mathrm{d}\omega^2 + \mathrm{d}^2\varphi^2)$	Elliptic	Elliptic-parabolic	Elliptic

We find for the Gaussian curvature in the (u, v)-system

$$K = \frac{a(a - 3bu^2)}{(a - 2bu^2)^3} \ . \tag{10.34}$$

For $b = 0$ we find $K = 1/a$ which is indeed a space of constant curvature, and the quantity a measures the curvature. In particular, for the unit-two-dimensional hyperboloid we have $K = 1/a$, with $a = -1$ as the special case of $\Lambda^{(2)}$. In the following we will assume that $a < 0$ in order to assure the positive definiteness of the metric (1.2). Note also the energy-spectrum:

$$E = \frac{\hbar^2}{2m|a|}\left(p^2 + \frac{1}{4}\right) . \tag{10.35}$$

From this observation it is not surprising that the path integral evaluation in the various coordinate system in D_{II} is similar to the corresponding path integral evaluations on the two-dimensional hyperboloid. In fact, the path integral calculation in the (u, v)-coordinate system follows the same line of reasoning as in [247], finally yielding (10.3, 10.4) [233].

The Polar-Coordinate System.

For the path integral evaluation in polar coordinates it is useful to switch to new coordinates according to $\varrho = \mathrm{e}^{\tau_2}$ and $\cos\vartheta = 1/\cosh\tau_1$. This then gives a combination of a path integral similar to the equidistant system on the two-dimension hyperboloid and for Liouville quantum mechanics. Of course, the energy-spectrum is same as in the previous case (10.5–10.7). Also, the limiting cases as in the above table can be recovered [233].

The Parabolic-Coordinate System.

In the path integral solution in parabolic coordinates one uses the path integral solution for the radial harmonic oscillator in both coordinates ξ and η, respectively. In comparison to the path integral evaluation in semi-hyperbolic coordinates on the two-dimensional hyperboloid one has an additional structure yielding as wave-function Whittaker-function instead of Bessel-functions (10.8, 10.9). Again, the limiting cases as indicated in the above table can be recovered by an appropriate choice of the parameters [233].

The Elliptic-Coordinate System.

In the path integral for the elliptic-coordinate system one has to consider a path integral representation for the prolate-spheroidal coordinate system (c.f. Chapter 5). By respecting a proper insertion of the parameters one obtains (10.10).

10.2.3 Darboux Space D_{III}

The (u, v)-Coordinate and Polar Coordinate System.

Let us first note that the Darboux Space D_{III} has special limiting cases which is summarized in the table 10.2. For the Gaussian curvature we find

$$K = -\frac{ab\,\mathrm{e}^{-3u}}{(b\,\mathrm{e}^{-2u} + a\,\mathrm{e}^{-u})^4} . \tag{10.36}$$

For e.g. $a = 1, b = 0$ we recover the two-dimensional flat space with the corresponding coordinate systems. To assure the positive definiteness of the metric (1.3), we can require $a, b > 0$. Note that a constant curvature limiting case with $K \neq 0$ is not possible because of the different powers in the nominator and denominator (except either $a = 0$ or $b = 0$, respectively, yielding $K = 0$).

For the path integral solution in the (u, v)-coordinate system in D_{III} one has again perform a time transformation. This gives a path integral for a free particle and another one for the Morse-potential. Surprisingly, the calculation is similar as for the usual Coulomb potential in flat space. In order to extract the wave-functions, one has to use the identity

$$
\frac{1}{\sin \alpha} \exp \left[- (x + y) \cot \alpha \right] I_{2\mu}\left(\frac{2\sqrt{xy}}{\sin \alpha} \right) \tag{10.37}
$$
$$
= \frac{1}{2\pi\sqrt{xy}} \int_{-\infty}^{\infty} \frac{\Gamma(\frac{1}{2} + \mu + \mathrm{i}p)\Gamma(\frac{1}{2} + \mu - \mathrm{i}p)}{\Gamma^2(1 + 2\mu)} e^{-2\alpha p + \pi p} M_{+\mathrm{i}p,\mu}(-2\mathrm{i}x) M_{-\mathrm{i}p,\mu}(+2\mathrm{i}y) dp
$$

which can be derived by using an integral representation as given by Buchholz [88, p.158].

In addition, we have a discrete spectrum. This is found by analyzing the poles of the Green function

$$
n + l + \frac{1}{2} - \frac{a}{2\hbar}\sqrt{-\frac{2m}{bE_{nl}}} = 0 \; , \tag{10.38}
$$

with the solution

$$
E_{nl} = -\frac{\hbar^2}{2m}\frac{b}{a^2}(2n + 2l + 1)^2 \; . \tag{10.39}
$$

yielding for $b > 0$ an infinite number of bound states. The constant N_{nl} is determined by taking the Green function at the residuum E_{nl}. The wave-functions vanish for $u \to \infty$ due to $e^{-\sqrt{-8mbE_{nl}}\, e^u/\hbar} = e^{-2b\hbar(2n+2l+1)e^u/a} \to 0$ for $u \to \infty$, provided $b/a > 0$ for all $n \in \mathbb{N}$, which shows that the discrete spectrum is indeed infinite.

The calculation for the path integral (10.15, 10.16). in polar coordinates is very similar to the (u, v)-system, the principal difference being another counting in l and the replacement $\varrho = e^{-u}$. The same holds for the discrete wave-functions with energy spectrum (10.39).

The Parabolic Coordinate System.

In the path integral evaluation for the parabolic coordinates in D_{III} one uses the path integral of the harmonic oscillator, in particular in terms of the Green's function for the harmonic oscillator

$$
G(x'', x'; E) = \sqrt{\frac{m}{\pi\omega\hbar^2}}\Gamma\left(\frac{1}{2} - \frac{E}{\hbar\omega}\right) D_{-\frac{1}{2}+E/\hbar\omega}\left(\sqrt{\frac{2m\omega}{\hbar}}x_>\right) D_{-\frac{1}{2}+E/\hbar\omega}\left(-\sqrt{\frac{2m\omega}{\hbar}}x_<\right). \tag{10.40}
$$

The $D_{\nu}(z)$ are parabolic cylinder-functions [187, p.1064]. Actually, one observes that it has the same form as the path integral for the Coulomb potential in two dimensions

Table 10.2: Some Limiting Cases of Coordinate Systems on D_{III}

Metric:	D_{III}	$\mathbb{R}^2$ $(a=0)$	$\mathbb{R}^2$ $(b=0)$
$(a + \frac{b}{4}\varrho^2)(\mathrm{d}\varrho^2 + \varrho^2 \mathrm{d}\varphi^2)$	Polar	–	Polar
$(a + \frac{b}{4}(\xi^2 + \eta^2))(\mathrm{d}\xi^2 + \mathrm{d}\eta^2)$	Parabolic	Parabolic	Cartesian
$(a + \frac{b}{4}d^2(\sinh^2\omega + \cos^2\varphi))$ $\times d^2(\sinh^2\omega + \sin^2\varphi)(\mathrm{d}\omega^2 + \mathrm{d}\varphi^2)$	Elliptic	–	Elliptic
$(a + \frac{b}{2}(\mu - \nu))(\mu + \nu)\left(\frac{\mathrm{d}\mu^2}{\mu^2} - \frac{\mathrm{d}\nu^2}{\nu^2}\right)$	Hyperbolic	Hyperbolic 3 in E(1,1)	–

in parabolic coordinates which was solved in [142, 223, 241, 251]. Introducing the "Bohr"-radius $a_B = \hbar^2/m\alpha = 2\hbar^2/maE$ gives the wave-functions

$$\Psi_{p,\zeta}^{(e,o)}(\xi,\eta) = \frac{e^{\pi/2ap}}{\sqrt{24\pi^2}}$$

$$\times \left(\begin{array}{c} |\Gamma(\frac{1}{4} - \frac{i}{2p}(1/a_B+\zeta))|^2 E_{-\frac{1}{2}+\frac{i}{p}(1/a_B+\zeta)}^{(0)}\left(e^{-i\pi/4}\sqrt{2p}\,\xi\right) E_{-\frac{1}{2}-\frac{i}{p}(1/a_B+\zeta)}^{(0)}\left(e^{-i\pi/4}\sqrt{2p}\,\eta\right) \\ |\Gamma(\frac{3}{4} - \frac{i}{2p}(1/a_B+\zeta))|^2 E_{-\frac{1}{2}+\frac{i}{p}(1/a_B+\zeta)}^{(1)}\left(e^{-i\pi/4}\sqrt{2p}\,\xi\right) E_{-\frac{1}{2}-\frac{i}{p}(1/a_B+\zeta)}^{(1)}\left(e^{-i\pi/4}\sqrt{2p}\,\eta\right) \end{array} \right),$$

(10.41)

which are δ-normalized according to [443]

$$\int_0^\infty d\nu \int_{\mathbb{R}} d\xi (\xi^2 + \eta^2) \Psi_{p',\zeta'}^{(e,o)\,*}(\xi,\eta) \Psi_{p,\zeta}^{(e,o)}(\xi,\eta) = \delta(p' - p)\delta(\zeta' - \zeta) ,$$

(10.42)

and ζ is the parabolic separation constant. The functions $E_\nu^{(0)}(z)$ and $E_\nu^{(1)}(z)$ are even and odd parabolic cylinder functions in the variable z. Putting everything together gives finally (10.17–10.19).

The discrete wave-functions are the wave-functions of the harmonic oscillator (HO), where $E_{n_\xi n_\eta}$ is determined by the equation

$$(n_\xi + n_\eta + 1) - \frac{1}{\hbar}aE\sqrt{-\frac{2m}{bE}} = 0$$

(10.43)

which is (up to a different counting in the quantum numbers) identical with (10.39). This concludes the discussion.

The Hyperbolic Coordinate System.

For the path integral solution for the hyperbolic system one has to perform a space-time transformation and obtains in both coordinates a path integral for the Morse-potential. This is solved in the usual way giving finally (10.20). The discrete spectrum is given by wave-functions for the Morse potential (MP) [252] with the same energy-spectrum as before (10.39).

10.2.4   Darboux Space D_{IV}

The (u, v)-Coordinate and Equidistant Coordinate System.

We start with some general remarks about the space D_{IV}. We observe, c.f. in the list of path integrals, that the diagonal term in the metric corresponds to a Pöschl–Teller potential, a Rosen–Morse potential, an inverse-square radial potential, and a Pöschl–Teller and modified Pöschl–Teller, respectively. In particular, the (u, v) and the equidistant systems are the same, they just differ in the parameterization. The limiting cases $a = 2b$ and $b = 0$ give particular cases for the metric on the two-dimensional hyperboloid.

Table 10.3: Some Limiting Cases of Coordinate Systems on D_{IV}

Metric:	D_{IV}	$\Lambda^{(2)}$ $(a = 2b)$	$\Lambda^{(2)}$ $(b = 0)$
$\dfrac{2b\cos u + a}{4\sin^2 u}(\mathrm{d}u^2 + \mathrm{d}v^2)$	(u, v)-Coordinates	Equidistant	Equidistant
$\dfrac{a - 2b\tanh\alpha}{4}(\mathrm{d}\alpha^2 + \cosh^2\alpha\,\mathrm{d}\beta^2)$	Equidistant	Equidistant	Equidistant
$\left(\dfrac{a_-}{\mu^2} + \dfrac{a_+}{\nu^2}\right)(\mathrm{d}\mu^2 + \mathrm{d}\nu^2)$	Horospherical	Horicyclic	Semi-circular parabolic
$\left(\dfrac{a_-}{\cosh^2\omega\cos^2\varphi} + \dfrac{a_+}{\sinh^2\omega\sin^2\varphi}\right)$ $\times(\cosh^2\omega - \cos^2\varphi)(\mathrm{d}\omega^2 + \mathrm{d}\varphi^2)$	Elliptic	Elliptic-parabolic	Hyperbolic-parabolic

For the Gaussian curvature we obtain e.g. in the (u, v)-system

$$K = -\frac{\dfrac{a_+^2}{\sin^6 u} + \dfrac{a_-^2}{\cos^6 u} + \dfrac{a_- a_+}{\sin^4 u \cos^4 u}}{\left(\dfrac{a_+}{\sin^2 u} + \dfrac{a_-}{\cos^2 u}\right)^3} \ . \tag{10.44}$$

The case $a = 2b$ yields $a_- = 0$, and

$$K = -\frac{1}{b} \ , \tag{10.45}$$

and therefore again a space of constant curvature, the hyperboloid $\Lambda^{(2)}$ is given for $b > 0$. We have set the sign in the metric (1.4) in such a way that from $a = 2b > 0$ the hyperboloid $\Lambda^{(2)}$ emerges. We could also choose the metric (1.4) with the opposite sign, then $a = 2b < 0$ would give the same result. In the following it is understood that we make this restriction of positive definiteness of the metric and we do not dwell into the problem of continuation into non-positive definiteness. Because the (u, v)-coordinates and the equidistant system are the same, we do not evaluate the path integral in the equidistant system. In the following we assume $a_+ > 0$ and $a_+ > a_-$.

In order to solve the path integral, the formulation in (u, v)-coordinates is unconvenient. Following [200] we perform the coordinate transformation $\cos u = \tanh \tau$. Further, we separate off the v-path integration, and additionally we make a time-transformation with the time-transformation function $f = a_+/\sin^2 u + a_-/\cos^2 u$. The special case $a_- = 0$ gives the wave-functions on the two-dimensional hyperboloid in equidistant coordinates, respectively on the hyperbolic strip. Inserting the solution for the modified Pöschl–Teller potential and evaluating the Green's function on the cut yields the path integral solution on $D_{\rm IV}$ (10.21–10.24).

The Horospherical Coordinate System.

The path integral evaluation in the horospherical coordinate system can be solved by means of the solution for the inverse-square radial potential for the semi-circular-parabolic system on the two-dimensional hyperboloid. However, one must also take into account the integral representation [187, p.725]

$$\int_0^\infty e^{-x/2-(z^2+w^2)/2x} K_\nu\left(\frac{zw}{x}\right) \frac{dx}{x} = 2K_\nu(z)K_\nu(w) \qquad (10.46)$$

and the rewriting of the I- and K-Bessel-functions in term of J-Bessel and $H^{(1)}$-Hankel functions, in addition with

$$\left[J_{-{\rm i}p}(k\mu'')H^{(1)}_{-{\rm i}p}(k\mu') - J_{{\rm i}p}(k\mu'')H^{(1)}_{{\rm i}p}(k\mu')\right] = \sinh \pi p H^{(1)}_{{\rm i}p}(k\mu'')H^{(1)}_{-{\rm i}p}(k\mu'). \qquad (10.47)$$

Note that $a_- = 0$ gives the horicyclic path integral on the two-dimensional hyperboloid. Putting the various results together gives (10.25, 10.26).

The Elliptic Coordinate System.

In the path integral for the elliptic coordinate system one uses the path integral solutions for the modified Pöschl–Teller potential. Note that on $D_{\rm IV}$ the energy spectrum is given by

$$E_p = \frac{\hbar^2}{2ma_+}\left(p^2 + \frac{1}{4}\right) . \qquad (10.48)$$

This concludes the discussion for the two-dimensional Darboux-spaces.

Table 10.4: Coordinates on Two-Dimensional Darboux-Spaces

Coordinate System	Coordinates	Path Integral Solution
Darboux Space D_{I}: I. (u, v)-Coordinates	$x = u + \mathrm{i}v$ $y = u - \mathrm{i}v$	(10.1)
II. Rotated (r, q)-Coordinates	$u = r\cos\vartheta + q\sin\vartheta$ $v = -r\sin\vartheta + q\cos\vartheta$	(10.2)
III. Displaced parabolic $(\xi \in \mathbb{R}, \eta > 0, a > 0)$	$u = \frac{1}{2}(\xi^2 - \eta^2) + a$ $v = \xi\eta$	(not known)
Darboux Space D_{II}: IV. (u, v)-Coordinates	$x = \frac{1}{2}(v + \mathrm{i}u)$ $y = \frac{1}{2}(v - \mathrm{i}u)$	(10.3, 10.4)
V. Polar	$u = \varrho\cos\vartheta$ $v = \varrho\sin\vartheta$	(10.5–10.7)
VI. Parabolic	$u = \xi\eta$ $v = \frac{1}{2}(\xi^2 - \eta^2)$	(10.8, 10.9)
VII. Elliptic	$u = d\cosh\omega\cos\varphi$ $v = d\sinh\omega\sin\varphi$	(10.10)
Darboux Space D_{III}: VIII. (u, v)-Coordinates	$x = u + \mathrm{i}v$ $y = u - \mathrm{i}v$	(10.11–10.14)
IX. Polar	$\xi = \varrho\cos\varphi$ $\eta = \varrho\sin\varphi$	(10.15, 10.16)
X. Parabolic	$\xi = 2\,\mathrm{e}^{-u/2}\cos\frac{v}{2}$ $\eta = 2\,\mathrm{e}^{-u/2}\sin\frac{v}{2}$ $u = \ln\dfrac{4}{\xi^2 + \eta^2}$ $v = \arcsin\dfrac{2\xi\eta}{\xi^2 + \eta^2}$	(10.17–10.19)
XI. Elliptic	$\xi = d\cosh\omega\cos\varphi$ $\eta = d\sinh\omega\sin\varphi$	(not known)
XII. Hyperbolic	$\xi = \dfrac{\mu - \nu}{2\sqrt{\mu\nu}} + \sqrt{\mu\nu}$ $\eta = \mathrm{i}\dfrac{\mu - \nu}{2\sqrt{\mu\nu}} - \sqrt{\mu\nu}$	(10.20)

Table 10.4 (cont.) Coordinate System	Coordinates	Path Integral Solution
Darboux Space D_{IV}: XIII. (u, v)-Coordinates	$x = v + iu$ $y = v - iu$	(10.21, 10.22)
XIV. Equidistant	$u = \arctan(e^{\alpha})$ $v = \dfrac{\beta}{2}$	(10.23, 10.24)
XV. Horospherical	$x = \log \dfrac{\mu - i\nu}{2}$ $y = \log \dfrac{\mu + i\nu}{2}$	(10.25, 10.26)
XVI. Elliptic	$\mu = d \cosh \omega \cos \varphi$ $\nu = d \sinh \omega \sin \varphi$	(10.27, 10.28)

10.3 Three-Dimensional Darboux Spaces

10.3.1 The Three-Dimensional Darboux Space $D_{3d-\mathrm{I}}$

We have the following path integral representations on the three-dimensional Darboux Space $D_{3d-\mathrm{I}}$ [234]) (System III is omitted because it is not solvable).

I. (u, v)-Coordinates, $u > a > 0, v, w \in [0, 2\pi)$:

$\left(\tilde{I}_\nu(z) \text{ and } \tilde{I}_\nu(z) \text{ as in the previous section, } \mathbf{L}^2 = l_v^2 + l_w^2 \right)$

$$
\int_{u(t')=u'}^{u(t'')=u''} \mathcal{D}u(t) \int_{v(t')=v'}^{v(t'')=v''} \mathcal{D}v(t) \int_{w(t')=w'}^{w(t'')=w''} \mathcal{D}w(t)(2u)^{3/2} \exp\left[\frac{im}{\hbar} \int_0^T u(\dot{u}^2 + \dot{v}^2 + \dot{w}^2)dt \right]
$$

$$
= \int_{-\infty}^{\infty} \frac{dE}{2\pi\hbar} e^{-iET/\hbar} \sum_{l_v, l_w = -\infty}^{\infty} \frac{e^{il_v(v''-v')+il_w(w''-w')}}{(2\pi)^2}
$$

$$
\times \frac{4m}{3\hbar}(4u'u'')^{-1/2}\left[\left(u' - \frac{\mathbf{L}^2\hbar^2}{4mE} \right)\left(u'' - \frac{\mathbf{L}^2\hbar^2}{4mE} \right) \right]^{1/2}
$$

$$
\times \left[\tilde{I}_{1/3}\left(u_< - \frac{\mathbf{L}^2\hbar^2}{4mE} \right) \tilde{K}_{1/3}\left(u_> - \frac{\mathbf{L}^2\hbar^2}{4mE} \right) \right.
$$

$$
\left. - \frac{\tilde{I}_{1/3}\left(a - \dfrac{\mathbf{L}^2\hbar^2}{4mE} \right)}{\tilde{K}_{1/3}\left(a - \dfrac{\mathbf{L}^2\hbar^2}{4mE} \right)} \tilde{K}_{1/3}\left(u' - \frac{\mathbf{L}^2\hbar^2}{4mE} \right) \tilde{K}_{1/3}\left(u'' - \frac{\mathbf{L}^2\hbar^2}{4mE} \right) \right] . \quad (10.49)
$$

II. Circular Rotated (r,q)-Coordinates, $r,q,w \in \mathbb{R}$:

$$\int_{r(t')=r'}^{r(t'')=r''} \mathcal{D}r(t) \int_{q(t')=q'}^{q(t'')=q''} \mathcal{D}q(t) \int_{w(t')=w'}^{w(t'')=w''} \mathcal{D}w(t)[2(r\cos\vartheta+q\sin\vartheta)]^{3/2}$$

$$\times \exp\left[\frac{im}{\hbar}\int_0^T (r\cos\vartheta+q\sin\vartheta)(\dot r^2+\dot q^2+\dot w^2)dt\right]$$

$$= \int_{-\infty}^{\infty} \frac{dE}{2\pi\hbar}\, e^{-iET/\hbar}\sum_{l_w=-\infty}^{\infty}\frac{e^{il_w(w''-w')}}{2\pi}\int d\mathcal{E}$$

$$\times\left\{\left(\frac{m}{\hbar^2}\sqrt{\frac{2}{E\cos\vartheta}}\right)^{2/3}\mathrm{Ai}\left[-\left(r''-\frac{\mathcal{E}-\hbar^2 l_w^2/2m}{2E\cos\vartheta}\right)\left(\frac{4mE\cos\vartheta}{\hbar^2}\right)^{1/3}\right]\right.$$

$$\times\left(\frac{m}{\hbar^2}\sqrt{\frac{2}{E\sin\vartheta}}\right)^{2/3}\mathrm{Ai}\left[-\left(r'-\frac{\mathcal{E}-\hbar^2 l_w^2/2m}{2E\cos\vartheta}\right)\left(\frac{4mE\cos\vartheta}{\hbar^2}\right)^{1/3}\right]$$

$$\times\mathrm{Ai}\left[-\left(q''+\frac{\mathcal{E}}{2E\sin\vartheta}\right)\left(\frac{4mE\sin\vartheta}{\hbar^2}\right)^{1/3}\right]\mathrm{Ai}\left[-\left(q'+\frac{\mathcal{E}}{2E\sin\vartheta}\right)\left(\frac{4mE\sin\vartheta}{\hbar^2}\right)^{1/3}\right]$$

$$\left.+(r\leftrightarrow q)\right\}\;. \tag{10.50}$$

IV.–VI. Circular Coordinates from $\mathbb{R}^2$

$$\int_{u(t')=u'}^{u(t'')=u''} \mathcal{D}u(t) \int_{v(t')=v'}^{v(t'')=v''} \mathcal{D}v(t) \int_{w(t')=w'}^{w(t'')=w''} \mathcal{D}w(t)(2u)^{3/2}\exp\left[\frac{im}{\hbar}\int_0^T u(\dot u^2+\dot v^2+\dot w^2)dt\right]$$

IV. Circular Polar

$$= \int_{-\infty}^{\infty}\frac{dE}{2\pi\hbar}\,e^{-iET/\hbar}\sum_{\nu\in\mathbb{Z}}e^{i\nu(\varphi''-\varphi')}\int_0^{\infty}dk\,kJ_\nu(k\varrho')J_\nu(k\varrho'')G_k(u'',u';E)\;, \tag{10.51}$$

V. Circular Elliptic

$$= \int_{-\infty}^{\infty}\frac{dE}{2\pi\hbar}\,e^{-iET/\hbar}\frac{1}{2\pi}\sum_{n\in\mathbb{Z}}\int_0^{\infty}kdk$$

$$\times me_n^*(\nu';\tfrac{d^2k^2}{4})me_n(\nu'';\tfrac{d^2k^2}{4})M_n^{(1)*}(\mu';\tfrac{dk}{2})M_n^{(1)}(\mu'';\tfrac{dk}{2})G_k(u'',u';E)\;, \tag{10.52}$$

VI. Circular Parabolic

$$= \int_{-\infty}^{\infty}\frac{dE}{2\pi\hbar}\,e^{-iET/\hbar}\int_{\mathbb{R}}d\zeta\int_{\mathbb{R}}\frac{dk}{32\pi^4}$$

$$\times\begin{pmatrix}|\Gamma(\tfrac{1}{4}+\tfrac{i\zeta}{2k})|^2 E_{-1/2+i\zeta/k}^{(0)}(e^{-i\pi/4}\sqrt{2k}\,\xi'')E_{-1/2-i\zeta/k}^{(0)}(e^{-i\pi/4}\sqrt{2k}\,\eta'')\\ |\Gamma(\tfrac{3}{4}+\tfrac{i\zeta}{2p})|^2 E_{-1/2+i\zeta/k}^{(1)}(e^{i\pi/4}\sqrt{2k}\,\xi'')E_{-1/2-i\zeta/k}^{(1)}(e^{i\pi/4}\sqrt{2k}\,\eta'')\end{pmatrix}G_k(u'',u';E),$$

$$\tag{10.53}$$

$$G_k(u'',u';E) = \frac{4m}{3\hbar}(4u'u'')^{-1/2},\left[\left(u'-\frac{k^2\hbar^2}{4mE}\right)\left(u''-\frac{k^2\hbar^2}{4mE}\right)\right]^{1/2}$$

$$\times \left[\tilde{I}_{1/3}\left(u_< - \frac{k^2\hbar^2}{4mE}\right)\tilde{K}_{1/3}\left(u_> - \frac{k^2\hbar^2}{4mE}\right) \right.$$

$$\left. - \frac{\tilde{I}_{1/3}\left(a - \dfrac{k^2\hbar^2}{4mE}\right)}{\tilde{K}_{1/3}\left(a - \dfrac{k^2\hbar^2}{4mE}\right)}\tilde{K}_{1/3}\left(u' - \frac{k^2\hbar^2}{4mE}\right)\tilde{K}_{1/3}\left(u'' - \frac{k^2\hbar^2}{4mE}\right) \right] . \quad (10.54)$$

$\tilde{I}_\nu(z)$ denotes $\tilde{I}_\nu(z) = I_\nu\left(\frac{4\sqrt{-mE}}{3\hbar}z^{3/2}\right)$, with $\tilde{K}_\nu(z)$ similarly. Due to the relation to the Airy-function [1] $K_{\pm 1/3}(\zeta) = \pi\sqrt{3/z}\,\mathrm{Ai}(z)$, $z = (3\zeta/2)^{2/3}$, and the observation that for $E < 0$ the argument of $\mathrm{Ai}(z)$ is always greater than zero, there are no bound states.

Table 10.5: Coordinates on Three-Dimensional Darboux-Space $D_{3d-\mathrm{I}}$

Coordinate System	Metric-Term Coordinates	Path Integral Solution
I. (u,v)-Coordinates	$ds^2 = 2u(du^2 + dv^2 + dw^2)$ $u = u',\ v = v',\ w = w'$	(10.49)
II. Circular-Rotated-(r,q)-Coordinates	$ds^2 = 2(r\cos\vartheta + q\sin\vartheta)(dr^2 + dq^2 + dw^2)$ $u = r\cos\vartheta + q\sin\vartheta$ $v = -r\sin\vartheta + q\cos\vartheta,\ w'_w$	(10.50)
III. Displaced Parabolic	$ds^2 = (\xi^2 - \eta^2 + 2a)((\xi^2 + \eta^2)(d\xi^2 + d\eta^2) + dw^2)$ $u = \frac{1}{2}(\xi^2 - \eta^2) + a,\ v = \xi\eta,\ w = w'$	(not known)
IV. Circular-Polar	$ds^2 = 2u(du^2 + d\varrho^2 + \varrho^2 d\varphi^2)$ $u = u',\ v = \varrho\cos\varphi,\ w = \varrho\sin\varphi$	(10.51, 10.54)
V. Circular-Elliptic	$ds^2 = 2u(du^2 + d^2(\cosh^2\xi + \sin^2\eta)(d\xi^2 + r^2 d\eta^2))$ $u = u',\ v = d\cosh\xi\cos\eta,\ w = d\sinh\eta\sin\eta$	(10.52, 10.54)
VI. Circular-Parabolic	$ds^2 = 2u(du^2 + (\xi^2 + \eta^2)(d\eta^2 + d\xi^2))$ $u = u',\ v = \frac{1}{2}(\eta^2 - \xi^2),\ w = \xi\eta$	(10.53, 10.54)
VII. Parabolic	$ds^2 = (\eta^2 - \xi^2)((\xi^2 + \eta^2)(d\eta^2 + d\xi^2) + \xi^2\eta^2 d\varphi^2)$ $u = \frac{1}{2}(\eta^2 - \xi^2)$ $v = \xi\eta\cos\varphi$ $w = \xi\eta\sin\varphi$	(not known)

10.3.2 The Three-Dimensional Darboux Space $D_{3d-\mathrm{II}}$

We have the following path integral representations on the three-dimensional Darboux Space $D_{3d-\mathrm{II}}$ [234]

I. (u,v)-Coordinates, $u > a > 0, v, w \in \mathbb{R}$:

$(f = h/u$ and $h = \sqrt{bu^2 - a})$

$$\int_{u(t')=u'}^{u(t'')=u''} \mathcal{D}u(t) \int_{v(t')=v'}^{v(t'')=v''} \mathcal{D}v(t) \int_{w(0)=w'}^{w(s'')=w''} \mathcal{D}w(s)$$

$$\times \left(\frac{bu^2-a}{u^2}\right)^{3/2} \exp\left\{\frac{i}{\hbar}\int_0^T \left[\frac{m}{2}\frac{bu^2-a}{u^2}(\dot{u}^2+\dot{v}^2+\dot{w}^2) - \frac{3\hbar^2}{8mf^2u^2}\right]dt\right\}$$

$$= \int_{-\infty}^{\infty} \frac{dE}{2\pi\hbar} e^{-iET/\hbar} \frac{1}{[f(u')f(u')]^{1/4}} \frac{\hbar}{\pi^2} \int_{\mathbb{R}^2} d\mathbf{K} \frac{e^{ik_v(v''-v')+ik_w(w''-w')}}{(2\pi)^2}$$

$$\times \int_0^\infty \frac{2p\sinh\pi p\,dp}{\frac{\hbar^2}{2m|a|}(p^2+1)-E} K_{ip}\left(\sqrt{\mathbf{K}^2-\frac{2mbE}{\hbar^2}}\,u'\right) K_{ip}\left(\sqrt{\mathbf{K}^2-\frac{2mbE}{\hbar^2}}\,u''\right) \quad . \quad (10.55)$$

II.-IV. Cylindrical Coordinates from $D_{3d-\mathrm{II}}$

II. Cylindrical Polar

$$\int_{\varrho(t')=\varrho'}^{\varrho(t'')=\varrho''} \mathcal{D}\varrho(t) \int_{\varphi(t')=\varphi'}^{\varphi(t'')=\varphi''} \mathcal{D}\varphi(t) \int_{w(0)=w'}^{w(s'')=w''} \mathcal{D}w(s)\left(b-\frac{a}{\varrho^2\cos^2\vartheta}\right)^{3/2}\varrho$$

$$\times \exp\left\{\frac{i}{\hbar}\int_0^T\left[\frac{m}{2}\left(b-\frac{a}{\varrho^2\cos^2\vartheta}\right)(\dot{\varrho}^2+\varrho^2\dot{\vartheta}^2+\dot{w}^2)+\frac{\hbar^2}{8mf^2\varrho^2}-\frac{3\hbar^2}{8mf^2\varrho^2\cos^2\vartheta}\right]dt\right\}$$

$$= \int_{-\infty}^{\infty} \frac{dE}{2\pi\hbar} e^{-iET/\hbar}$$

$$\times \sum_{\pm} \int_{\mathbb{R}} dk_w \int_0^\infty dp \int_0^\infty dk \frac{\Psi_{p,k,\pm,k_w}(\varrho'',\vartheta'',w'')\Psi^*_{p,k,\pm,k_w}(\varrho',\vartheta',w')}{\frac{\hbar^2}{2m|a|}(p^2+1)-E} \quad , \quad (10.56)$$

$$\Psi_{p,k,\pm,k_w}(\varrho,\vartheta,w) = \frac{\sqrt{\sin\vartheta}}{\pi^{3/2}f^{1/4}(u)}e^{ik_w w}$$

$$\times \sqrt{k\sinh\pi k}\, K_{ik}\left(i\sqrt{\frac{b}{|a|}(p^2+1+k_w^2)}\,\varrho\right)\sqrt{\frac{p\sinh\pi p\,dp}{\cosh^2 k+\sinh^2\pi p}}P^{ip}_{ik-1/2}(\pm\sin\vartheta). \quad (10.57)$$

III. Cylindrical Parabolic $\left(\sqrt{b(p^2+1+k_w^2)/|a|}\equiv\tilde{p}\right)$

$$\int_{\xi(t')=\xi'}^{\xi(t'')=\xi''} \mathcal{D}\xi(t) \int_{\eta(t')=\eta'}^{\eta(t'')=\eta''} \mathcal{D}\eta(t) \int_{w(0)=w'}^{w(s'')=w''} \mathcal{D}w(s)\left(\frac{b\xi^2\eta^2-a}{\xi^2\eta^2}\right)^{3/2}(\xi^2+\eta^2)$$

$$\times \exp\left\{\frac{i}{\hbar}\int_0^T\left[\frac{m}{2}\frac{b\xi^2\eta^2-a}{\xi^2\eta^2}((\xi^2+\eta^2)(\dot{\xi}^2+\dot{\eta}^2)+\dot{w}^2) - \frac{3\hbar^2}{8mf^2\xi^2\eta^2}\right]dt\right\}$$

$$= \int_{-\infty}^{\infty} \frac{dE}{2\pi\hbar} \, e^{-iET/\hbar} \int_{\mathbb{R}} dk_w \frac{e^{ik_w(w''-w')}}{2\pi}$$
$$\times \frac{\hbar(\xi'\xi''\eta'\eta'')^{-1/2}}{[f(u')f(u')]^{1/4}} \int d\mathcal{E} \int_0^{\infty} \frac{dp\, p \sinh \pi p}{\frac{\hbar^2}{2m|a|}(p^2 + \frac{1}{4}) - E} \frac{|\Gamma[\frac{1}{2}(1 + ip - \mathcal{E})]|^4}{2\pi \tilde{p}^2}$$
$$\times W_{\mathcal{E}/2,ip/2}\left(i\tilde{p}\xi''^2\right) W_{\mathcal{E}/2,ip/2}^*\left(i\tilde{p}\xi'^2\right) W_{\mathcal{E}/2,ip/2}\left(i\tilde{p}\xi''^2\right) W_{\mathcal{E}/2,ip/2}^*\left(i\tilde{p}\eta'^2\right) \ . \tag{10.58}$$

IV. Cylindrical Elliptic

$$\left(d\sqrt{b(p^2+1+k_w^2)/|a|} \equiv \tilde{p}, \ f^2 = (bd^2 \cosh^2 \omega \cos^2 \varphi - a)/d^2 \cosh^2 \omega \cos^2 \varphi\right)$$

$$\int_{\omega(t')=\omega'}^{\omega(t'')=\omega''} \mathcal{D}\omega(t) \int_{\varphi(t')=\varphi'}^{\varphi(t'')=\varphi''} \mathcal{D}\varphi(t) \int_{w(t')=w'}^{w(t'')=w''} \mathcal{D}w(t) \left(\frac{bd^2 \cosh^2 \omega \cos^2 \varphi - a}{d^2 \cosh^2 \omega \cos^2 \varphi}\right)^{3/2} d^2(\cosh^2 \omega - \cos^2 \varphi)$$

$$\times \exp\left\{\frac{i}{\hbar}\int_0^T \left[\frac{m}{2}f^2(d^2(\cosh^2 \omega - \cos^2 \varphi)(\dot{\omega}^2 + \dot{\varphi}^2) + \dot{w}^2) - \frac{3\hbar^2}{8mf^2\xi^2\eta^2}\right]dt\right\}$$

$$= \int_{-\infty}^{\infty} \frac{dE}{2\pi\hbar} \, e^{-iET/\hbar} \int_{\mathbb{R}} dk_w \frac{e^{ik_w(w''-w')}}{2\pi}$$
$$\times \frac{\sqrt{\cos \varphi' \cos \varphi''}}{[f(u')f(u')]^{1/4}} \int_0^{\infty} \frac{dp \sinh \pi p}{\frac{\hbar^2}{2m|a|}(p^2 + \frac{1}{4}) - E} \int_0^{\infty} \frac{dk\, k \sinh \pi k}{(\cosh^2 \pi k + \sinh^2 \pi p)^2}$$
$$\times \sum_{\epsilon,\epsilon'=\pm 1} S_{ip-1/2}^{ik\,(1)}(\epsilon \tanh \omega''; -\tilde{p}) S_{ip-1/2}^{ik\,(1)\,*}(\epsilon \tanh \omega'; -\tilde{p})$$
$$\times ps_{ik-1/2}^{ip}(\epsilon' \sin \varphi''; \tilde{p}^2) ps_{ik-1/2}^{ip\,*}(\epsilon' \sin \varphi'; \tilde{p}^2) \ . \tag{10.59}$$

V.-VII. Circular Coordinates from $\mathbb{R}^2$

$$\int_{u(t')=u'}^{u(t'')=u''} \mathcal{D}u(t) \int_{v(t')=v'}^{v(t'')=v''} \mathcal{D}v(t) \int_{w(0)=w'}^{w(s'')=w''} \mathcal{D}w(s)$$

$$\times \left(\frac{bu^2 - a}{u^2}\right)^{3/2} \exp\left\{\frac{i}{\hbar}\int_0^T \left[\frac{m}{2}\frac{bu^2 - a}{u^2}(\dot{u}^2 + \dot{v}^2 + \dot{w}^2) - \frac{3\hbar^2}{8mf^2u^2}\right]dt\right\}$$

V. Circular Polar

$$= \int_{-\infty}^{\infty} \frac{dE}{2\pi\hbar} \, e^{-iET/\hbar} \sum_{\nu \in \mathbb{Z}} e^{i\nu(\varphi''-\varphi')} \int_0^{\infty} dk\, k J_\nu(k\varrho') J_\nu(k\varrho'') G_k(u'', u'; E) \ , \tag{10.60}$$

VI. Circular Elliptic

$$= \int_{-\infty}^{\infty} \frac{dE}{2\pi\hbar} \, e^{-iET/\hbar} \frac{1}{2\pi} \sum_{n \in \mathbb{Z}} \int_0^{\infty} kdk$$
$$\times me_n^*(\nu'; \tfrac{d^2k^2}{4}) me_n(\nu''; \tfrac{d^2k^2}{4}) M_n^{(1)\,*}(\mu'; \tfrac{dk}{2}) M_n^{(1)}(\mu''; \tfrac{dk}{2}) G_k(u'', u'; E) \ , \tag{10.61}$$

VII. Circular Parabolic

$$= \int_{-\infty}^{\infty} \frac{\mathrm{d}E}{2\pi\hbar}\, \mathrm{e}^{-\mathrm{i}ET/\hbar} \int_{\mathbb{R}} d\zeta \int_{\mathbb{R}} \frac{dk}{32\pi^4}$$

$$\times \begin{pmatrix} |\Gamma(\tfrac{1}{4} + \tfrac{\mathrm{i}\zeta}{2k})|^2 \, E^{(0)}_{-1/2+\mathrm{i}\zeta/k}(\mathrm{e}^{-\mathrm{i}\pi/4}\sqrt{2k}\,\xi'')\, E^{(0)}_{-1/2-\mathrm{i}\zeta/k}(\mathrm{e}^{-\mathrm{i}\pi/4}\sqrt{2k}\,\eta'') \\ |\Gamma(\tfrac{3}{4} + \tfrac{\mathrm{i}\zeta}{2p})|^2 \, E^{(1)}_{-1/2+\mathrm{i}\zeta/k}(\mathrm{e}^{\mathrm{i}\pi/4}\sqrt{2k}\,\xi'')\, E^{(1)}_{-1/2-\mathrm{i}\zeta/k}(\mathrm{e}^{\mathrm{i}\pi/4}\sqrt{2k}\,\eta'') \end{pmatrix}$$

$$\times G_k(u'', u'; E) \;, \tag{10.62}$$

$$G_k(u'', u'; E) = \frac{1}{[f(u')f(u')]^{1/4}} \frac{\hbar}{\pi^2}$$

$$\times \int_0^\infty \frac{2p \sinh \pi p \, dp}{\frac{\hbar^2}{2m|a|}(p^2+1) - E} K_{\mathrm{i}p}\left(\sqrt{k^2 - \frac{2mbE}{\hbar^2}}\, u'\right) K_{\mathrm{i}p}\left(\sqrt{k^2 - \frac{2mbE}{\hbar^2}}\, u''\right) \;. \tag{10.63}$$

VIII. Spherical, $r > 0$, $\vartheta \in [0,\pi)$, $\varphi \in [0,2\pi)$ $(f(\tau_1,\tau_2) = b\,\mathrm{e}^{2\tau_2}/\cosh^2\tau_1 - a)$:

$$\int_{\tau_1(t')=\tau_1'}^{\tau_1(t'')=\tau_1''} \mathcal{D}\tau_1(t) \int_{\tau_2(t')=\tau_2'}^{\tau_2(t'')=\tau_2''} \mathcal{D}\tau_2(t) \int_{\varphi(t')=\varphi'}^{\varphi(t'')=\varphi''} \mathcal{D}\varphi(t) \sinh\tau_1 \cosh\tau_1 \left(\frac{b\,\mathrm{e}^{2\tau_2}}{\cosh^2\tau_1} - a\right)^{3/2}$$

$$\times \exp\left\{\frac{\mathrm{i}}{\hbar}\int_0^T \left[\frac{m}{2} f(\tau_1,\tau_2)(\dot\tau_1^2 + \cosh^2\tau_1 \dot\tau_2^2 + \sinh^2\tau_1 \dot\varphi^2)\right.\right.$$

$$\left.\left. - \frac{1}{f(\tau_1,\tau_2)} \frac{\hbar^2}{2m}\left(4 + \frac{1}{\cosh^2\tau_1} - \frac{1}{\sinh^2\tau_1}\right)\right]dt\right\}$$

$$= \int_{-\infty}^{\infty} \frac{\mathrm{d}E}{2\pi\hbar}\, \mathrm{e}^{-\mathrm{i}ET/\hbar} \int_0^\infty \mathrm{d}p \int_0^\infty \mathrm{d}k_{\tau_2} \sum_{k_\varphi \in \mathbb{Z}} \frac{\Psi_{p,k_{\tau_2},k_\varphi}(\tau_1'', \tau_2'', \varphi'')\Psi^*_{p,k_{\tau_2},k_\varphi}(\tau_1'', \tau_2'', \varphi'')}{\frac{\hbar^2}{2m|a|}(p^2+1) - E} \;,$$

$$\tag{10.64}$$

$$\Psi_{p,k_{\tau_2},k_\varphi}(\tau_1,\tau_2,\varphi) = \frac{\sqrt{2\sinh\tau_1 \cosh\tau_1}}{f(\tau_1,\tau_2)^{1/4}}$$

$$\times \frac{\mathrm{e}^{\mathrm{i}k_\varphi(\varphi''-\varphi')}}{\sqrt{2\pi}} \frac{\sqrt{k_{\tau_2}\sinh\pi k_{\tau_2}}}{\pi} K_{\mathrm{i}k_{\tau_2}}\left(\mathrm{i}\sqrt{\frac{b}{|a|}}(p^2+1)\,\mathrm{e}^{\tau_2}\right)\Psi^{(k_\varphi,\mathrm{i}k_{\tau_2})}(\tau_3) \;. \tag{10.65}$$

10.3.3 Path Integral Evaluations on Three-Dimensional Darboux Space

Actually the path integral evaluations on the three-dimensional Darboux-Spaces are not very different in comparison to the corresponding two-dimensional spaces. In each case the same techniques apply. This is the case in particular for the space $D_{3d-\mathrm{I}}$. However, one must take into account another important feature.

We start with the three-dimensional Darboux space $D_{3d-\mathrm{I}}$ and consider the metric:

$$\mathrm{d}s^2 = 2u(\mathrm{d}u^2 + \mathrm{d}v^2 + \mathrm{d}w^2) \;. \tag{10.66}$$

The proper definition of the range of the variables (u,v,w) depends on the proper definition of the space we in fact consider. We assume in the following that $u > a$, where $a > 0$ and that there is no restriction on the variables v, w. They can be cyclic or range within the entire real line. According to the general theory we have $g = \det(g_{ab}) = (2u)^3$, therefore $\Gamma_u = 3/2u, \Gamma_v = \Gamma_w = 0$. The Laplace-Beltrami operator has the form

$$\Delta_{LB} = \frac{1}{2u}\left(\frac{\partial^2}{\partial u^2} - \frac{1}{2u}\frac{\partial}{\partial u} + \frac{\partial^2}{\partial v^2} + \frac{\partial^2}{\partial w^2}\right), \qquad p_u = \frac{\hbar}{i}\left(\frac{\partial}{\partial u} + \frac{3}{4u}\right), \qquad (10.67)$$

and p_u are the corresponding momentum operator for the coordinate u. According to our theory we can calculate the corresponding quantum potential by means of [252]

$$\Delta V = \frac{\hbar^2}{8m}\frac{D-2}{f^4}\left[(D-4)f'^2 + 2ff''\right], \qquad (10.68)$$

provided the metric is proportional to the unit tensor $(g_{ab}) = f^2\mathbb{1}_3$. Indeed $f = \sqrt{2u}$ and $D = 3$, which yields

$$\Delta V = -\frac{3\hbar^2}{64mu^3} . \qquad (10.69)$$

This gives an effective Lagrangian in the corresponding path integral in the product form definition

$$\mathcal{L}_{\text{eff}}(u, \dot{u}, v, \dot{v}, w, \dot{w}) = \frac{m}{2}(2u)(\dot{u}^2 + \dot{v}^2 + \dot{w}^2) + \frac{3\hbar^2}{64mu^3} . \qquad (10.70)$$

The quantum potential (10.69) has actually the form of a Schwarzian derivative. Performing a space-time transformation in the path integral where $u^2 = 4r$ cancels ΔV, and produces in turn in the transformed Lagrangian $\mathcal{L}_E = \mathcal{L}_{\text{eff}} + E$ a potential $\propto 2E/\sqrt{r}$ (coupling constant metamorphosis). Potentials like this are called "conditionally solvable" [330]. However, in order that they are in fact conditionally solvable, requires that an additional potential of the form of ΔV is present. This is not the case here after the transformation into the new variable r and the corresponding time-transformation; we are left with an intractable path integral.

In order to obtain a proper quantum theory on D_{3d-I}, we therefore *define* our quantum theory for the free motion on D_{3d-I} as follows:

$$\begin{aligned}
H_{D_{3d-I}} &= -\frac{\hbar^2}{2m}\frac{1}{2u}\left(\frac{\partial^2}{\partial u^2} - \frac{1}{2u}\frac{\partial}{\partial u} + \frac{\partial^2}{\partial v^2} + \frac{\partial^2}{\partial w^2}\right) + \frac{3\hbar^2}{64mu^3} \\
&= \frac{1}{2m}\frac{1}{\sqrt{2u}}(p_u^2 + p_v^2 + p_w^2)\frac{1}{\sqrt{2u}} .
\end{aligned} \qquad (10.71)$$

This gives in turn a proper definition of the Lagrangian on D_{3d-I}

$$\mathcal{L}_{\text{eff}}^{(D_{3d-I})}(u, \dot{u}, v, \dot{v}, w, \dot{w}) := \frac{m}{2}(2u)(\dot{u}^2 + \dot{v}^2 + \dot{w}^2) , \qquad (10.72)$$

and we have a well-defined path integral.

In a similar way, we can define a proper quantum theory on the three-dimensional Darboux-space $D_{3d-\mathrm{II}}$. We consider the metric

$$\mathrm{d}s^2 = \frac{bu^2 - a}{u^2}(\mathrm{d}u^2 + \mathrm{d}v^2 + \mathrm{d}w^2), \qquad p_u = \frac{\hbar}{\mathrm{i}}\left(\frac{\partial}{\partial u} + \frac{\Gamma_u}{2}\right) , \qquad (10.73)$$

and w is the new variable. We can write the metric tensor according to $(g_{ab}) = f^2 \mathbb{1}_3$ with $f = h/u$, and $h = \sqrt{bu^2 - a}$. The general theory yields $g = (h/u)^6$, $\Gamma_u = 3h'/h - 3/u$, and

$$\begin{aligned}
\Delta V &= \Delta V_1 + \Delta V_2 \\
\Delta V_1 &= \frac{\hbar^2}{8mh^6}\left(2ab(u^2 - 1) - 3b^2 u^4\right), \qquad \Delta V_2 = \frac{3\hbar^2}{8mf^2 u^2} \ .
\end{aligned} \qquad (10.74)$$

The quantum potential ΔV_2 is necessary in order to obtain the correct energy spectrum, the quantum potential ΔV_1 is interpreted as a curvature term which we add in the metric for our proper quantum theory on $D_{3d-\mathrm{II}}$. Therefore similar as for $D_{3d-\mathrm{I}}$:

$$\mathcal{L}_{\mathrm{eff}}^{(D_{3d-\mathrm{II}})}(u, \dot{u}, v, \dot{v}, w, \dot{w}) := \frac{m}{2}\frac{bu^2 - a}{u^2}(\dot{u}^2 + \dot{v}^2 + \dot{w}^2) + \Delta V_1 \ . \qquad (10.75)$$

Endowed with a proper quantum theory and a path integral we can evaluate the corresponding path integrals. In the systems I.–IV. the additional variable w in the three-dimensional systems can be separated off by circular, respectively plane waves, modifying the parameter $\tilde{p}$.

In the systems V.–VII. the two-dimensional path integral solutions for the free motion in $\mathbb{R}^2$ are separeted off, and it only remains to calculate the Green's function $G_k(u'', u'; E)$ (10.63), which is of the same form as for the two-dimensional Darboux space D_{II} in the (u, v)-coordinate system.

The case of the spherical system in $D_{3d-\mathrm{II}}$ consists of successive path integrations of circular waves, Liouville quantum mechanics, and a modified Pöschl-Teller potential with only continuous states (10.64, 10.65). In all cases the continuous spectrum is given by

$$E = \frac{\hbar^2}{2m|a|}(p^2 + 1) \ . \qquad (10.76)$$

Again, the well-known feature of a zero-point energy $E_0 = \hbar^2/2m|a|$ appears. This concludes the discussion.

Table 10.6: Coordinates on Three-Dimensional Darboux-Space $D_{3d-\mathrm{II}}$

Coordinate System	Metric-Term Coordinates	Path Integral Solution
I. (u,v)- Coordinates	$ds^2 = \dfrac{bu^2 - a}{u^2}(du^2 + dv^2 + dw^2)$ u v w	(10.55)
II. Cylindrical Polar	$ds^2 = \dfrac{b\varrho^2 \cos^2\vartheta - a}{\varrho^2 \cos^2\vartheta}(d\varrho^2 + \varrho^2 d\vartheta^2 + dw^2)$ $u = \varrho\cos\vartheta$ $v = \varrho\sin\vartheta$ w	(10.56, 10.57)
III. Cylindrical Parabolic	$ds^2 = \dfrac{b\xi^2\eta^2 - a}{\xi^2\eta^2}((\xi^2 + \eta^2)(d\xi^2 + d\eta^2) + dw^2)$ $u = \xi\eta$ $v = \frac{1}{2}(\xi^2 - \eta^2)$ w	(10.58)
IV. Cylindrical Elliptic	$ds^2 = \dfrac{bd^2 \cosh^2\omega \cos^2\varphi - a}{\cosh^2\omega \cos^2\varphi}$ $\times((\cosh^2\omega - \cos^2\varphi)(d\omega^2 + d\varphi^2) + dw^2)$ $u = d\cosh\omega\cos\varphi$ $v = d\sinh\omega\sin\varphi$ w	(10.59)
V. Circular- Polar	$ds^2 = \dfrac{bu^2 - a}{u^2}(du^2 + d\varrho^2 + \varrho^2 d\varphi^2)$ u $v = \varrho\cos\varphi$ $w = \varrho\sin\varphi$	(10.60, 10.63)
VI. Circular- Elliptic	$ds^2 = \dfrac{bu^2 - a}{u^2}(du^2$ $\qquad + d^2(\cosh^2\xi + \sin^2\eta)(d\xi^2 + r^2 d\eta^2))$ u $v = d\cosh\xi\cos\eta$ $w = d\sinh\eta\sin\eta$	(10.61, 10.63)
VII. Circular- Parabolic	$ds^2 = \dfrac{bu^2 - a}{u^2}(du^2 + (\xi^2 + \eta^2)(d\eta^2 + d\xi^2))$ u $v = \frac{1}{2}(\eta^2 - \xi^2)$ $w = \xi\eta$	(10.62, 10.63)

Table 10.6 (cont.) Coordinate System	Metric-Term Coordinates	Path Integral Solution
VIII. Spherical	$ds^2 = \left(b - \dfrac{a}{r^2 \cos^2 \vartheta}\right)(dr^2 + r^2 d\vartheta^2 + r^2 \sin^2 \vartheta d\varphi^2)$ $u = r \cos \vartheta$ $v = r \sin \vartheta \cos \varphi$ $w = r \sin \vartheta \sin \varphi$	(10.64, 10.65)
IX. Sphero- Conical	$ds^2 = \left(b - \dfrac{a}{r^2 \mathrm{dn}^2 \alpha \mathrm{sn}^2 \beta}\right)$ $\quad \times (dr^2 + r^2(k^2 \mathrm{cn}^2 \alpha + k'^2 \mathrm{cn}^2 \beta)(d\alpha^2 + d\beta^2)$ $u = r \mathrm{dn}(\alpha, k)\mathrm{sn}(\beta, k')$ $v = r \mathrm{sn}(\alpha, k)\mathrm{dn}(\beta, k')$ $w = r \mathrm{cn}(\alpha, k)\mathrm{cn}(\beta, k')$	(not known)
X. Oblate- Spheroidal	$ds^2 = \left(b - \dfrac{a}{d^2 \sinh^2 \mu \cos^2 \nu}\right)$ $\quad \times d^2\left((\cosh^2 \bar{\mu} - \sin^2 \bar{\nu})(\dot{\mu}^2 + \dot{\nu}^2) + \cosh^2 \bar{\mu} \sin^2 \bar{\nu} \dot{\phi}^2\right)$ $u = d \sinh \mu \cos \nu$ $v = d \cosh \mu \sin \nu \cos \varphi$ $w = d \cosh \mu \sin \nu \sin \varphi$	(not known)
XI. Prolate- Spheroidal	$ds^2 = \left(b - \dfrac{a}{d^2 \cosh^2 \mu \cos^2 \nu}\right)$ $\quad \times d^2\left((\sinh^2 \mu + \sin^2 \nu)(\dot{\mu}^2 + \dot{\nu}^2) + \sinh^2 \mu \sin^2 \nu \dot{\phi}^2\right)$ $u = d \cosh \mu \cos \nu$ $v = d \cosh \mu \sin \nu \cos \varphi$ $w = d \cosh \mu \sin \nu \sin \varphi$	(not known)
XII. Ellipsoidal	$ds^2 = \left(b - \dfrac{a/(k')^2}{(a^2 - c^2)\mathrm{dn}^2 \alpha \mathrm{dn}^2 \beta \mathrm{dn}^2 \gamma}\right) g_{\rho_i \rho_i} d\rho_i^2$ $u = (\mathrm{i}/k')\sqrt{a^2 - c^2}\, \mathrm{dn}\alpha \mathrm{dn}\beta \mathrm{dn}\gamma$ $v = k^2 \sqrt{a^2 - c^2}\, \mathrm{sn}\alpha \mathrm{sn}\beta \mathrm{sn}\gamma$ $w = -(k^2/k')\sqrt{a^2 - c^2}\, \mathrm{cn}\alpha \mathrm{cn}\beta \mathrm{cn}\gamma$	(not known)

Chapter 11

Path Integrals
on Single-Sheeted Hyperboloids

11.1 The Two-Dimensional
Single-Sheeted Hyperboloid

Path integral solutions in imaginary Lobachevsky space, respectively on the single-sheeted hyperboloid have been presented in [220]. Together with the results of the path integral representations on $O(2,2)$ we are able to present two solutions which are in accordance with [176, 177, 299, 318, 510].

Equidistant, $\tau > 0$, $\varphi \in [0, 2\pi)$:

$$\int_{\tau(t')=\tau'}^{\tau(t'')=\tau''} \mathcal{D}\tau(t) \cosh\tau \int_{\varphi(t')=\varphi'}^{\varphi(t'')=\varphi''} \mathcal{D}\varphi(t)$$

$$\times \exp\left\{ \frac{i}{\hbar} \int_{t'}^{t''} \left[\frac{m}{2}(\dot{\tau}^2 - \cosh^2\tau\,\dot{\varphi}^2) - \frac{\hbar^2}{8m}\left(1 + \frac{1}{\cosh^2\tau}\right)\right] dt \right\}$$

$$= (\cosh\tau' \cosh\tau'')^{-1/2} \sum_{l=-\infty}^{\infty} \frac{e^{il(\varphi''-\varphi')}}{2\pi} \left[\sum_{n\in\mathbb{N}_0} \left(n - |l| - \frac{1}{2}\right)\frac{\Gamma(2|l|-n)}{n!}\right.$$

$$\times P_{|l|-1/2}^{n-|l|+\frac{1}{2}}(\tanh\tau') P_{|l|-1/2}^{n-|l|+\frac{1}{2}}(\tanh\tau'') \exp\left\{ -\frac{i\hbar T}{2m}\left[\left(n - |l| + \tfrac{1}{2}\right)^2 - \frac{1}{4}\right]\right\}$$

$$+ \int_{\mathbb{R}} dp\, p \sinh\pi p\, e^{-i\hbar T(p^2+\frac{1}{4})/2m} \left. \frac{P_{|l|-1/2}^{ip}(\tanh\tau'') P_{|l|-1/2}^{-ip}(\tanh\tau')}{\cos^2\pi l + \sinh^2\pi p}\right]. \quad (11.1)$$

Horicyclic, $\varrho, x \in \mathbb{R}$:

$$= \int_{\varrho(t')=\varrho'}^{\varrho(t'')=\varrho''} \mathcal{D}\varrho(t)e^{-\varrho} \int_{x(t')=x'}^{x(t'')=x''} \mathcal{D}x(t) \exp\left\{ \frac{i}{\hbar} \int_{t'}^{t''}\left[\frac{m}{2}(\dot{\varrho}^2 - e^{-2\varrho}\dot{x}^2) + \frac{\hbar^2}{8m}e^{2\varrho}\right]dt\right\}$$

$$= e^{\frac{1}{2}(\varrho'+\varrho'')} \int_{\mathbb{R}} dk \frac{e^{ik(x''-x')}}{2\pi} \left\{ \sum_{n=0}^{\infty} e^{2i\hbar T n^2/m} 2(2n+\alpha) J_{2n+\alpha}(|k|e^{\varrho'}) J_{2n+\alpha}(|k|e^{\varrho''}) \right.$$

$$\left. + \int_0^{\infty} \frac{p\, dp\, e^{-i\hbar T(p^2+\frac{1}{4})/2m}}{2\sinh \pi p} \left[J_{ip}(|k|e^{\varrho'}) + J_{-ip}(|k|e^{\varrho'}) \right] \left[J_{ip}(|k|e^{\varrho''}) + J_{-ip}(|k|e^{\varrho''}) \right] \right\} . \quad (11.2)$$

The invariant distance on $\Lambda_S^{(2)}$, e.g., in the spherical system is given by

$$\cosh d_{\Lambda_S^{(2)}}(\mathrm{u}'', \mathrm{u}') = \sinh \tau'' \sinh \tau' - \cosh \tau'' \cosh \tau' \cos(\varphi'' - \varphi') . \quad (11.3)$$

Equidistant Coordinates.

The two-dimensional single-sheeted hyperboloid is defined by the equation

$$u_0^2 - u_1^2 - u_2^2 = -1 , \qquad R = 1 . \quad (11.4)$$

In comparison to the double-sheeted hyperboloid we have now a minus sign in the quadratic form. The ambiguity in the case of the double-sheeted hyperboloid that we have to take one of the two sheets is not present here. The path integral in equidistant coordinates has been calculated in [220] (see also Kalnins and Miller [318], and Niederle et al. [166, 485] for the Schrödinger approach in general hyperbolic spaces defined by their corresponding quadratic forms). In [220] this coordinate system has been called spherical, however due to the nomenclature for the double-sheeted pseudosphere the notion "equidistant" seems to be the more appropriate one.

Actually, we have the same number of coordinate systems on the single-sheeted hyperboloid as on the double-sheeted hyperboloid. However, there is only little information about the corresponding wavefunctions of the Hamiltonian in the corresponding coordinate systems, let alone the path integral. This is due to the fact that the Hamiltonian usually requires a self-adjoint extension, giving rise to a continuous and a *discrete* spectrum. The handling of the discrete spectrum is not difficult in the case of the equidistant coordinate systems. After the separation of the φ-path integration we have an attractive $1/\cosh^2$-potential for which the propagator is known. The spectrum of the Hamiltonian on the single-sheeted pseudosphere, and its D-dimensional generalization have the property that there is in addition to the continuous spectrum a discrete spectrum with an infinite number of negative energy levels. See [220] for some more details, and for the cases where a "Kepler-like" problem and magnetic fields were incorporated.

Horicyclic Coordinates.

Whereas in the case of the equidistant coordinates the discrete spectrum is a simple consequence of the $1/\cosh^2$-potential well, the discrete spectrum emerges in the case of the horicyclic coordinates from the self-adjoint extension of the inverted Liouville potential $V(\varrho) = -(k^2 + \frac{1}{4})e^{2\varrho}$. The self-adjoint extension has been given in [299, 405]. It is a consequence of the self-adjoint extension of the horicyclic coordinate system on

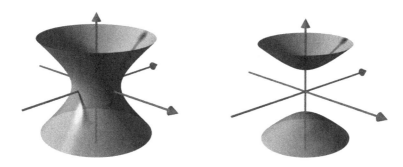

Figure 11.1: (a,b) A one- (left) and double-sheeted hyperboloid (right).

the $O(2,2)$ hyperboloid or the Hamiltonian on $SU(1,1)$, respectively. In the present case we can separate the x-path integration quite easily and we are left with a ϱ-path integration which has the form of the path integral of the inverted Liouville potential. Using the path integral identity (2.85) we obtain (11.2).

Coordinates on the One-Sheeted Two-Dimensional Hyperboloid.

Usually, one is more familiar with the double-sheeted hyperboloid. In Figure 11.1 I have displayed for visualization a one- and double-sheeted hyperboloid, respectively.

Other Coordinate Systems.

Self-adjoint extensions and the corresponding path integral representations are not to our disposal for the remaining other seven coordinate systems. In the spherical

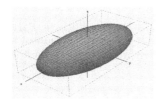  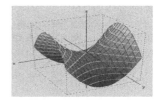

Figure 11.2: (a–c) An ellipsoid (left), a paraboloid (middle), and a hyperbolic paraboloid (right).

system, for instance, we have

$$K(\mathbf{u}'', \mathbf{u}'; T) = \int\limits_{\tau(t')=\tau'}^{\tau(t'')=\tau''} \mathcal{D}\tau(t)\, \sinh\tau \int\limits_{\varphi(t')=\varphi'}^{\varphi(t'')=\varphi''} \mathcal{D}\varphi(t)$$

$$\times \exp\left\{\frac{i}{\hbar} \int_{t'}^{t''} \left[\frac{m}{2}(\dot\tau^2 - \sinh^2\tau\,\dot\varphi^2) - \frac{\hbar^2}{8m}\left(1 - \frac{1}{\sinh^2\tau}\right)\right] dt\right\} \ . \qquad (11.5)$$

For the emerging attractive singular $-1/\sinh^2\tau$ term I have not found any self-adjoint extension in the literature. However, this does not alter the fact that the spectrum is the same as displayed in the two cases of equidistant and horicyclic coordinates. The same is true for the other six coordinate systems which we do not discuss here any further.

The double-sheeted hyperboloid, denoted by $\Lambda^{(2)}$, is defined by the set of points related by the equation

$$u_0^2 - u_1^2 - u_2^2 = R^2 \ , \qquad u_0 > 0 \ , \qquad (11.6)$$

whereas the restriction $u_0 > 0$ gives the upper-sheet, say, of the hyperboloid. The single-sheeted hyperboloid, denoted by $\Lambda_{-1}^{(2)}$, in turn is defined by the set of points related by the equation

$$u_0^2 - u_1^2 - u_2^2 = -R^2 \ . \qquad (11.7)$$

In the following we restrict ourself to unit curvature $R = 1$. A complete classi-fication of the coordinate systems which separate the Schrödinger equation on the two-dimensional hyperboloid is due to [425]. In [243] we have given a comprehensive approach to superintegrable systems on $\Lambda^{(2)}$. We have displayed the algebraic defini-tion of the coordinate systems, which is more convenient for the characterization of the ρ_i-const.-lines, $(\rho_i,\ i = 1, 2,$ any coordinate system on $\Lambda^{(2)})$, and in each case a convenient coordinate system for setting up the path integral formulation.

For completeness, I have listed in table 11.1 the coordinate on the two-dimensional hyperboloid according to [425]. This includes a description of the lines $\varrho_i = $ const. $(\varrho_i,\ i = 1, 2$ any coordinate, and we have taken the normalization $k = -1$, i.e. the unit-hyperboloid). In the more complicated systems I have displayed a convenient coordinate representation as well as the algebraic form which allows for a better description of the $\varrho_i = $ const.-lines, for details see e.g. [243].

For the characterization of the coordinate systems on the two-dimensional hyper-boloid we usualy also need the characteristic operator L. It is a combination of the three operators (boots and angular momentum, respectively, omitting factors $\hbar$ and i)

$$N_2 = u_0\frac{\partial}{\partial u_2} - u_2\frac{\partial}{\partial u_0} \ , \qquad N_3 = u_0\frac{\partial}{\partial u_1} - u_1\frac{\partial}{\partial u_0} \ , \qquad M_a = u_2\frac{\partial}{\partial u_1} - u_1\frac{\partial}{\partial u_2} \ , \quad (11.8)$$

which satisfy the commutation relations

$$[N_3, N_2] = -M_1 \ , \qquad [N_3, M_1] = -N_2 \ , \qquad [N_2, M_1] = N_3 \ . \qquad (11.9)$$

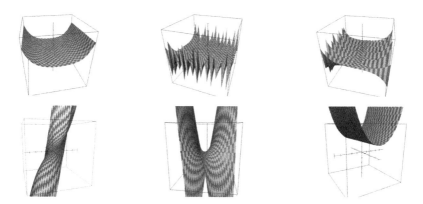

Figure 11.3: Spherical coordinates (u_0, u_1, u_2) on the double-sheeted hyperboloid (top) and for the single-sheeted hyperboloid (bottom).

It is quite useful to visualize the various surfaces. In Figure 11.2 I have displayed the contour of an ellipsoid, of a paraboloid, and a hyperbolic paraboloid. The paraboloid can be a rotational paraboloid and an elliptic paraboloid, respectively.

The Spherical System.

In order to analyze the various path integral representations on the single-sheeted hyperboloid, we start with the spherical system. It is given by

$$u_0 = \sinh \tau_1 \cosh \tau_2 \; , \qquad u_1 = \sinh \tau_1 \sinh \tau_2 \; , \qquad u_2 = \cosh \tau_1 \; . \qquad (11.10)$$

The characteristic operator of the spherical system reads: $L = M_1^2$. The spherical system is characcrerized by $\tau_1 = $ const. which are concentric circles with centre at 0, and $\tau_2 = $ const. wich are a family of straight lines with vertex at 0.

In Figure 11.3 I have displayed for the hyperboloid the coordinates (u_0, u_1, u_2) in spherical form.

By the usual procedure we set up the classical Lagrangian, the classical Hamiltonian, the momentum operators, and finally the quantum Hamiltonian. This will not repeated all over again. Important for us is the effective Lagrangian to be used in the path integral, where an additional quantum potential $\propto \hbar^2$ appears. Hence

$$ds^2 = d\tau_1^2 - \sinh^2 \tau_1 d\tau_2^2 \; , \qquad\qquad\qquad (11.11)$$

$$\mathcal{L}_{\text{eff}} = \frac{m}{2}(\dot{\tau}_1^2 - \sinh^2 \tau_1 \dot{\tau}_2^2) - \frac{\hbar^2}{8m}\left(1 + \frac{1}{\sinh^2 \tau}\right) \; . \qquad (11.12)$$

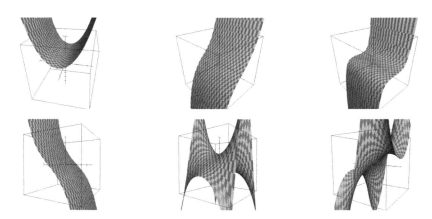

Figure 11.4: Equidistant coordinates (u_0, u_1, u_2) on the double-sheeted hyperboloid (top) and for the single-sheeted hyperboloid (bottom).

The Equidistant System.

Next, the equidistant system is defined by

$$u_0 = \sinh \tau \ , \qquad u_1 = \cosh \tau \cos \varphi \ , \qquad u_2 = \cosh \tau \sin \varphi \ . \qquad (11.13)$$

We easily verify $u_0^2 - u_1^2 - u_2^2 = -1$, as it should be. The characteristic operator of the equidistant system reads: $L = N_3^2$. The equidistant system is characterized by $\tau_1 =$ const. which are equidistant from the x-axis, and $\tau_2 =$ const. which are perpendicular to the x-axis. The line-element has the form:

$$ds^2 = d\tau^2 - \cosh^2 \tau d\varphi^2 \ . \qquad (11.14)$$

Note the minus sign. The effective Lagrangian has the form

$$\mathcal{L}_{\text{eff}} = \frac{m}{2}(d\tau^2 - \cosh^2 \tau d\varphi^2) + \frac{\hbar^2}{8m}\left(1 + \frac{1}{\cosh^2 \tau}\right) \ . \qquad (11.15)$$

In Figure 11.4 I have displayed for the hyperboloid the coordinates (u_0, u_1, u_2) in equidistant form.

The Horicyclic System.

Next we consider the horicyclic system. It is characterized by the operator $L = (N_2 - M_1)^2$ and is given by:

$$u_0 = \tfrac{1}{2}\big[e^u(1 + x^2) - e^{-u}\big] \ , \qquad u_1 = xe^u \ , \qquad u_2 = \tfrac{1}{2}\big[e^u(1 - x^2) + e^{-u}\big] \ . \qquad (11.16)$$

The horicyclic system is characterized by $\varrho = \ln y = $ const. which are coaxial lines, and $x = $ const. which is a family of straight lines. The line-element has the form:

$$ds^2 = du^2 - e^{2u}dx^2 , \tag{11.17}$$

and for the effective Lagrangian we get:

$$\mathcal{L}_{\text{eff}} = \frac{m}{2}(\dot{u}^2 - e^{2u}\dot{x}^2) + \frac{\hbar^2}{8m}e^{-2u} . \tag{11.18}$$

In Figure 11.5 I have displayed for the hyperboloid the coordinates (u_0, u_1, u_2) in horicyclic form.

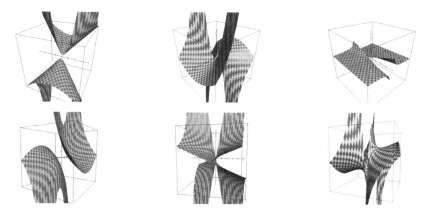

Figure 11.5: Horicyclic coordinates (u_0, u_1, u_2) on the double-sheeted hyperboloid (top) and for the single-sheeted hyperboloid (bottom).

The Elliptic System.

The elliptic system in algebraic form is given by

$$\frac{\pm u_0^2}{\varrho_i - b} + \frac{\pm u_1^2}{\varrho_i - a} - \frac{\pm u_2^2}{\varrho_i - c} = 0 , \tag{11.19}$$

for $i = 1, 2$, and $(c < b < \varrho_2 < a < \varrho_1)$ on $\Lambda^{(2)}$ with the $+$-sign and $(\varrho_2 < c < b < \varrho_1 < a)$ on $\Lambda^{(2)}_{-1}$ with the $-$-sign. The characteristic operator has the form: $L = N_3^2 + aN_2^2$. It is characterized by $\varrho_1 = $ const. which is a family of confocal ellipses, and $\varrho_2 = $ const. which is a family of confocal hyperbolas.

In Figure 11.6 I have displayed for the hyperboloid the coordinates (u_0, u_1, u_2) in elliptic form.

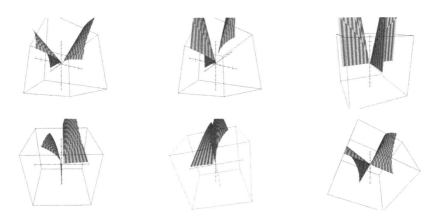

Figure 11.6: Elliptic coordinates (u_0, u_1, u_2) on the double-sheeted hyperboloid (top) and for the single-sheeted hyperboloid (bottom).

The Hyperbolic System.

In algebraic form the elliptic-parabolic system has the form

$$\frac{u_0^2}{\varrho_i - c} + \frac{u_1^2}{\varrho_i - a} - \frac{u_2^2}{\varrho_i - b} = 0 \ , \tag{11.20}$$

for $i = 1, 2$, and $(\varrho_2 < c < b < a < \varrho_1)$ on $\Lambda^{(2)}$ with the +-sign and $(\varrho_1, \varrho_2 > a > b > c)$, respectively $(c < b < \varrho_1, \varrho_2 < a)$ on $\Lambda_{-1}^{(2)}$ with the −-sign. The characteristic operator has the form: $L = N_3^2 - aM_1^2$. It is characterized by $\varrho_i = $ const. which are two mutually orthogonal familes of confocal hyperbolas $(i = 1, 2)$.

In Figure 11.7 I have displayed for the hyperboloid the coordinates (u_0, u_1, u_2) in hyperbolic form.

The Semi-Hyperbolic System.

In algebraic form the elliptic-parabolic system has the form

$$\frac{\pm u_0^2}{\varrho_i - a} - \frac{2\delta \pm u_1 u_2 + (\varrho_i - \gamma)(\pm u_1^2 \mp u_2^2)}{(\varrho_i - \gamma)^2 + \delta^2} = 0 \ , \qquad (b = \gamma + \mathrm{i}\delta, c = \gamma - \mathrm{i}\delta) \ \ (11.21)$$

for $i = 1, 2$. The characteristic operator reads: $L = \alpha(M_1^2 - n_2^2) - \beta(\{M_1, N_2\}$. It is characterized by $\varrho_i = $ const. which are two mutually orthogonal familes of orthogonal confocal semi-hyperbolas $(i = 1, 2$, a semi-hyperbola is the set of all points u_0, u_1, u_2 which are at equal distance from the focus and a given equidistant.).

In Figure 11.8 I have displayed for the hyperboloid the coordinates (u_0, u_1, u_2) in semi-hyperbolic form.

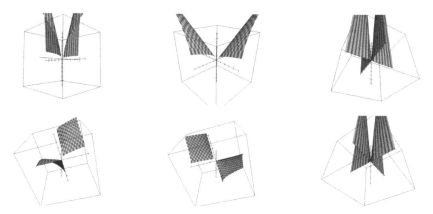

Figure 11.7: Hyperbolic coordinates (u_0, u_1, u_2) on the double-sheeted hyperboloid (top) and for the single-sheeted coordinates (bottom).

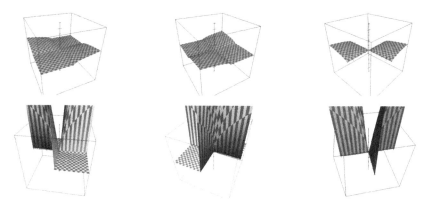

Figure 11.8: Semi-hyperbolic coordinates (u_0, u_1, u_2) on the double-sheeted hyperboloid (top) and for the single-sheeted hyperboloid (bottom).

The Elliptic-Parabolic System.

The elliptic-parabolic system has the characteristic operator $L = N_3^2 + (N_2 - M_1)^2$ and is defined in a convenient way by $(a \in \mathbb{R}, b > 0)$

$$u_0 = \frac{\cosh^2 a - \sinh^2 b}{2 \cosh a \sinh b} \quad , \qquad u_1 = \frac{\sinh^2 a - \cosh^2 b}{2 \cosh a \sinh b} \quad , \qquad u_2 = \tanh a \coth b \ .$$

$$(11.22)$$

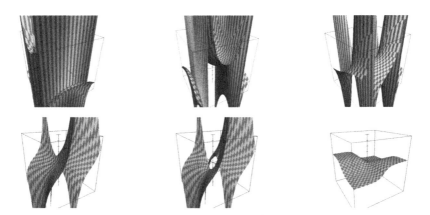

Figure 11.9: Elliptic-parabolic coordinates (u_0, u_1, u_2) on the double-sheeted hyperboloid (top) and on the single-sheeted hyperboloid (bottom).

In algebraic form the elliptic-parabolic system has the form

$$\frac{u_0^2}{\varrho_i - a} - \frac{u_1^2 - u_2^2}{\varrho_i - b} + \frac{(u_1 - u_2)^2}{(\varrho_i - c)^2} = 0 \ , \qquad (c = b < \varrho_2 < a < \varrho_1) \tag{11.23}$$

and is characterized by $\varrho_i = $ const. which are two mutually orthogonal familes of confocal elliptic parabolas ($i = 1, 2$, an elliptic-parabola is the set of all points u_0, u_1, u_2 which are at equal distance from the focus and a given line). We obtain in the usual way

$$\mathrm{d}s^2 \ = \ \left(\frac{\cosh^2 a + \sinh^2 b}{\cosh^2 a \sinh^2 b} \right) (\mathrm{d}b^2 + \mathrm{d}a^2) \ , \tag{11.24}$$

$$\mathcal{L}_{\mathrm{eff}} \ = \ \frac{m}{2} \left(\frac{\cosh^2 a + \sinh^2 b}{\cosh^2 a \sinh^2 b} \right) (\mathrm{d}b^2 + \mathrm{d}a^2) \ . \tag{11.25}$$

There is no additional quantum potential.

In Figure 11.9 I have displayed for the hyperboloid the coordinates (u_0, u_1, u_2) in elliptic-parabolic form.

The Hyperbolic-Parabolic System.

The hyperbolic-parabolic system has the characteristic operator $L = N_3^2 - (N_2 - M_3)^2$ and is defined in a convenient way by ($\alpha, \beta \in (0, \pi)$)

$$u_0 = \frac{\cos^2 \alpha + \cos^2 \beta}{2 \sin \alpha \sin \beta} \ , \qquad u_1 = \frac{\sin^2 \alpha + \sin^2 \beta}{2 \sin \alpha \sin \beta} \ , \qquad u_2 = \cot \alpha \cot \beta \ . \tag{11.26}$$

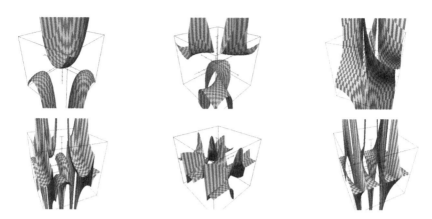

Figure 11.10: Hyperbolic-parabolic coordinates (u_0, u_1, u_2) on the double-sheeted hyperboloid (top) and on the single-sheeted hyperboloid (bottom).

In algebraic form the hyperbolic-parabolic system has the form

$$\frac{u_0^2}{\varrho_i - a} - \frac{u_1^2 - u_2^2}{\varrho_i - b} \pm \frac{(u_1 - u_2)^2}{(\varrho_i - c)^2} = 0 \ , \qquad (\varrho_2 < c = b < a < \varrho_1) \tag{11.27}$$

and is characterized by $\varrho_i = $ const. are two mutually orthogonal familes of confocal hyperbolic parabolas ($i = 1, 2$, a hyperbolic-parabola is the set of al points u_0, u_1, u_2 which are at equal distance from a given focal-line and a given line). We obtain in the usual way

$$\mathrm{d}s^2 = \left(\frac{\sin^2 \alpha - \sin^2 \beta}{\sin^2 \alpha \sin^2 \beta}\right) (\mathrm{d}\alpha^2 + \mathrm{d}\beta^2) \ , \tag{11.28}$$

$$\mathcal{L}_{\mathrm{eff}} = \frac{m}{2}\left(\frac{\sin^2 \alpha - \sin^2 \beta}{\sin^2 \alpha \sin^2 \beta}\right)(\dot\alpha^2 + \dot\beta^2) \ . \tag{11.29}$$

There is no additional quantum potential.

In Figure 11.10 I have displayed for the hyperboloid the coordinates (u_0, u_1, u_2) in hyperbolic-parabolic form.

The Semi-Circular System.

The semi-circular system has the characteristic operator $L = \{N_3, N_2 + M_1\}$ and is defined in a convenient way by ($\xi, \eta > 0$)

$$u_0 = \frac{4 + (\xi^2 - \eta^2)^2}{8\xi\eta} \ , \qquad u_1 = \frac{4 - (\xi^2 - \eta^2)^2}{8\xi\eta} \ , \qquad u_2 = \frac{\xi^2 + \eta^2}{2\xi\eta} \ . \tag{11.30}$$

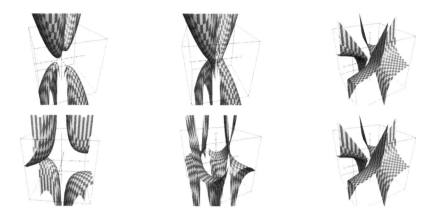

Figure 11.11: Semi-circular-hyperbolic coordinates (u_0, u_1, u_2) on the double-sheeted hyperboloid (top) and on the single-sheeted hyperboloid (bottom).

In algebraic form the semi-circular system has the form

$$\left(u_0 + \frac{u_1 - u_2}{\varrho_i - a}\right)^2 = u_1^2 - u_2^2 \ , \qquad (\varrho_2 < c = b = a < \varrho_1) \tag{11.31}$$

and is characterized by $\varrho_i = $ const. which are two mutually orthogonal familes of confocal semi-circular parabolas ($i = 1, 2$, a semi-circular-parabola is the set of all points u_0, u_1, u_2 which are at equal distance from a given line and one of its axes). We obtain in the usual way

$$\mathrm{d}s^2 = \left(\frac{\xi^2 - \eta^2}{\xi^2 \eta^2}\right)(\mathrm{d}\xi^2 + \mathrm{d}\eta^2) \ , \qquad \mathcal{L}_{\mathrm{eff}} = \frac{m}{2}\left(\frac{\xi^2 - \eta^2}{\xi^2 \eta^2}\right)(\dot{\xi}^2 + \dot{\eta}^2) \ . \tag{11.32}$$

There is no additional quantum potential.

In Figure 11.11 I have displayed for the hyperboloid the coordinates (u_0, u_1, u_2) in semi-circular-hyperbolic form.

Table 11.1: Coordinates on the Two-Dimensional Single-Sheeted Hyperboloid

Coordinate System	Coordinates	Path Integral Solution
I. Spherical	$u_0 = \sinh \tau_1 \cosh \tau_2$ $u_1 = \sinh \tau_1 \sinh \tau_2$ $u_2 = \cosh \tau_1$	(not known)
II. Equidistant	$u_0 = \sinh \tau_1$ $u_1 = \cosh \tau \cos \varphi$ $u_2 = \cosh \tau \sin \varphi$	(11.1)
III. Horicyclic	$u_0 = (x^2 - y^2 + 1)/2y$ $u_1 = x/y$ $u_2 = (1 - x^2 + y^2)/2y$	(11.2)
IV. Elliptic	$u_0 = k\,\mathrm{sn}(\nu, k)\mathrm{sn}(\eta, k)$ $u_1 = (k/k')\mathrm{cn}(\nu, k)\mathrm{cn}(\eta, k)$ $u_2 = (i/k')\mathrm{dn}(\nu, k)\mathrm{dn}(\eta, k)$	(not known)
V. Hyperbolic	$u_0 = (ik/k')\mathrm{cn}(\alpha, k)\mathrm{cn}(\beta, k)$ $u_1 = (i/k')\mathrm{dn}(\alpha, k)\mathrm{dn}(\beta, k)$ $u_2 = ik\,\mathrm{sn}(\alpha, k)\mathrm{sn}(\beta, k)$	(not known)
VI. Semi-Hyperbolic	$u_0 = \frac{1}{\sqrt{2}}(\sqrt{(1 + \mu_1^2)(1 + \mu_2^2)} + \mu_1\mu_2 - 1)^{1/2}$ $u_1 = \frac{1}{\sqrt{2}}(\sqrt{(1 + \mu_1^2)(1 + \mu_2^2)} - \mu_1\mu_2 + 1)^{1/2}$ $u_2 = \sqrt{\mu_1\mu_2}$	(not known)
VII. Elliptic-Parabolic	$u_0 = \dfrac{\cosh^2 a - \sinh^2 b}{2\cosh a \sinh b}$ $u_1 = \dfrac{\sinh^2 a - \cosh^2 b}{2\cosh a \sinh b}$ $u_2 = \tanh a \tanh b$	(not known)
VIII. Hyperbolic- Parabolic	$u_0 = \dfrac{\cos^2 \alpha + \cos^2 \beta}{2\sin \alpha \sin \beta}$ $u_1 = \dfrac{\sinh^2 a - \sin^2 \vartheta}{2\sin \alpha \sin \beta}$ $u_2 = \tan \alpha \tan \beta$	(not known)
IX. Semi-Circular- Parabolic	$u_0 = \dfrac{(4 + \xi^2\eta^2)^2}{8\xi\eta}$ $u_1 = \dfrac{4 - (\xi^2 - \eta^2)^2}{8\xi\eta}$ $u_2 = \dfrac{\eta^2 + \xi^2}{2\xi\eta}$	(not known)

Chapter 12

Miscellaneous Results on Path Integration

12.1   The D-Dimensional Pseudosphere

For the path integral for the quantum motion on the $(D-1)$-dimensional pseudosphere, I want to cite just three coordinate systems which generalize the coordinate systems on the two- and three-dimensional pseudosphere in the most obvious way. These coordinate systems may be called horicyclic, spherical and equidistant, respectively. In each of them the corresponding subsystems are contained in a simple way because the subgroup structure is very simple. We have for the horicyclic system $SO(D-1,1) \supset E(D-2)$, for the spherical system $SO(D-1,1) \supset SO(D-2)$, and for the equidistant system $SO(D-1,1) \supset SO(D-2,1)$. This has the consequence that in order to count all coordinate systems which separate the Hamiltonian and the path integral, all coordinate systems in the corresponding subsystems must be counted.

The three path integral solutions corresponding to the three subgroup coordinate systems are now obtained in the following way. For the horicyclic system we separate the path integral solution of the $(D-2)$-Euclidean space which gives us the path integral of the Liouville potential as already encountered in the case of the two- and three-dimensional pseudosphere [202, 223, 247]. In the spherical system we separate the path integration of the motion on the $(D-2)$-dimensional sphere, and we are left with the path integral of the repulsive $1/\sinh^2 \tau$-potential. This is again the same as in the case of the two- and three-dimensional pseudospheres, with the only change in the strength of the potential. In the equidistant system we must solve interrelated $1/\cosh^2 \tau$ (repulsive) potentials which are, of course, straightforward. For completeness we have added the general form of the Green's function [223, 249]. We have the path integral representations [66, 210, 223, 249] ($\mathbf{k} = (k_1, \ldots, k_{D-2})$)

Horicyclic, $\mathbf{x} \in \mathbb{R}^{D-2}, y > 0$:

$$
\int_{\mathbf{x}(t')=\mathbf{x}'}^{\mathbf{x}(t'')=\mathbf{x}''} \mathcal{D}\mathbf{x}(t) \int_{y(t')=y'}^{y(t'')=y''} \frac{\mathcal{D}y(t)}{y^{D-1}} \exp\left[\frac{\mathrm{i}}{\hbar} \int_{t'}^{t''} \left(\frac{m}{2} \frac{\dot{\mathbf{x}}^2 + \dot{y}^2}{y^2} - \frac{\hbar^2}{8m}(D-1)(D-3) \right) dt \right]
$$

$$
= (y'y'')^{\frac{D-2}{2}} \int \frac{d\mathbf{k}}{2\pi} \mathrm{e}^{\mathrm{i}\mathbf{k}\cdot(\mathbf{x}''-\mathbf{x}')} \frac{2}{\pi^2} \int_0^\infty dp\, p \sinh \pi p\, \mathrm{e}^{-\mathrm{i}E_p T/\hbar} K_{\mathrm{i}p}(|\mathbf{k}|y') K_{\mathrm{i}p}(|\mathbf{k}|y'') . \quad (12.1)
$$

Spherical, $\tau > 0, \boldsymbol{\Omega} \in S^{(D-2)}$:

$$
\int_{\tau(t')=\tau'}^{\tau(t'')=\tau''} \mathcal{D}\tau(t)(\sinh \tau)^{D-2} \int_{\boldsymbol{\Omega}(t')=\boldsymbol{\Omega}'}^{\boldsymbol{\Omega}(t'')=\boldsymbol{\Omega}''} \mathcal{D}\boldsymbol{\Omega}(t)
$$

$$
\times \exp\left\{ \frac{\mathrm{i}}{\hbar} \int_{t'}^{t''} \left[\frac{m}{2}\left(\dot{\tau}^2 + \sinh^2 \tau \dot{\boldsymbol{\Omega}}^2 \right) - \hbar^2 \frac{(D-2)^2}{8m} \right] dt \right\}
$$

$$
= \int_0^\infty dp \sum_{l,\mu} H_{p,l,\mu}^{(D)*}(\boldsymbol{\Omega}',\tau') H_{p,l,\mu}^{(D)}(\boldsymbol{\Omega}'',\tau'')\, \mathrm{e}^{-\mathrm{i}E_p T/\hbar} . \quad (12.2)
$$

Equidistant, $\tau_1,\ldots,\tau_{D-1} \in \mathbb{R}^{D-1}$:

$$
\int_{\tau_1(t')=\tau_1'}^{\tau_1(t'')=\tau_1''} \mathcal{D}\tau_1(t)(\cosh \tau_1)^{D-2} \ldots \int_{\tau_{D-1}(t')=\tau_{D-1}'}^{\tau_{D-1}(t'')=\tau_{D-1}''} \mathcal{D}\tau_{D-1}(t)
$$

$$
\times \exp\left\{ \frac{\mathrm{i}}{\hbar} \int_{t'}^{t''} \left[\frac{m}{2}\left(\dot{\tau}_1^2 + \cosh^2 \tau_1 \dot{\tau}_2^2 + \ldots + (\cosh^2 \tau_2 \ldots \cosh^2 \tau_{D-2})\dot{\tau}_{D-1}^2 \right) \right.\right.
$$

$$
\left.\left. - \frac{\hbar^2}{8m}\left((D-2)^2 + \frac{1}{\cosh^2 \tau_1} + \ldots + \frac{1}{\cosh^2 \tau_1 \ldots \cosh^2 \tau_{D-2}} \right) \right] dt \right\}
$$

$$
= \left(\cosh^{D-2} \tau_1' \cosh^{D-2} \tau_1'' \times \ldots \times \cosh \tau_{D-2}' \cosh \tau_{D-2}'' \right)^{-1/2}
$$

$$
\times \int_{\mathbb{R}} \frac{dp_0}{2\pi} \mathrm{e}^{\mathrm{i}p_0(\tau_{D-1}'' - \tau_{D-1}')} \prod_{j=1}^{D-2} \int_{\mathbb{R}} \frac{dp_j\, p_j \sinh \pi p_j}{(\cosh^2 \pi p_{j-1} + \sinh^2 \pi p_j)} \mathrm{e}^{-\mathrm{i}E_{p_{D-2}} T/\hbar}
$$

$$
\times P_{\mathrm{i}p_{j-1}-1/2}^{\mathrm{i}p_j}(\tanh \tau_{D-1-j}) P_{\mathrm{i}p_{j-1}-1/2}^{-\mathrm{i}p_j}(\tanh \tau_{D-1-j}) . \quad (12.3)
$$

General expression for the Green's function:

$$
= \frac{m}{\pi \hbar^2} \left(\frac{\mathrm{e}^{-\mathrm{i}\pi}}{2\pi \sinh d} \right)^{(D-3)/2} \mathcal{Q}_{-1/2-\mathrm{i}\sqrt{2m(E-E_0)}/\hbar}^{(D-3)/2}\left(\cosh d(\mathbf{q}'',\mathbf{q}') \right) . \quad (12.4)
$$

$(\mathbf{x} = \{x_i\} \equiv (x_1,\ldots,x_{D-2}), r^2 = \sum_{i=1}^{D-2} x_i^2)$ with $E_p = E_0 + \hbar^2 p^2/2m$, $E_0 = \hbar^2(D-2)^2/8m$, and

$$
\cosh d(\mathbf{q}'',\mathbf{q}') = \frac{|\mathbf{x}'' - \mathbf{x}'|^2 + y'^2 + y''^2}{2y'y''} . \quad (12.5)
$$

The wavefunctions $H_{p,l,\mu}^{(D)}$ are given by

$$
H_{p,l,\mu}^{(D)}(\boldsymbol{\Omega},\tau) = S_{l,\mu}^{(D-1)}(\boldsymbol{\Omega}) \frac{\Gamma(\mathrm{i}p + l + \frac{D-2}{2})}{\Gamma(\mathrm{i}p)} (\sinh \tau)^{\frac{3-D}{2}} \mathcal{P}_{\mathrm{i}p-\frac{1}{2}}^{\frac{3-D}{2}-l}(\cosh \tau) . \quad (12.6)
$$

Let us make some remarks concerning the three displayed path integral representations. They represent important sample path integral solutions:

1. In the horicyclic system, we can insert in the $\mathbf{x}$-path integration any accessible path integration in $\mathbb{R}^{D-2}$, $D = 3, 4, \ldots$. However, a systematic enumeration of the number of orthogonal coordinate systems in $\mathbb{R}^D$ seems not to be known.

2. In the spherical system, we can insert in the $\mathbf{\Omega}$-path integration any accessible path integration on $S^{(D-2)}$, $D = 3, 4, \ldots$. Again, a systematic enumeration of the number of orthogonal coordinate systems on $S^{(D-2)}$ seems not to be known.

3. In the equidistant system, we can insert any path integration on $\Lambda^{(D-2)}$, i.e., we can alternatively state the path integral representations

Equidistant, $\tau \in \mathbb{R}, \mathbf{v} \in \Lambda^{(D-2)}$:

$$
\int_{\tau(t')=\tau'}^{\tau(t'')=\tau''} \mathcal{D}\tau(t)(\cosh \tau)^{D-2} \int_{\mathbf{v}(t')=\mathbf{v}'}^{\mathbf{v}(t'')=\mathbf{v}''} \frac{\mathcal{D}\mathbf{v}(t)}{v_0}
$$

$$
\times \exp\left\{ \frac{i}{\hbar} \int_{t'}^{t''} \left[\frac{m}{2}(\dot{\tau}^2 + \cosh^2 \tau \dot{\mathbf{v}}^2) - \frac{\hbar^2}{8m}(D-2)^2 - \frac{\Delta V(\mathbf{v})}{\cosh^2 \tau} \right] dt \right\}
$$

$$
= (\cosh \tau' \cosh \tau_1)^{-(D-2)/2} \int d\mathbf{k}\, \Psi_{\mathbf{k}}^{\Lambda^{(D-2)}}(\mathbf{v}'')\Psi_{\mathbf{k}}^{\Lambda^{(D-2)}\,*}(\mathbf{v}')
$$

$$
\times \int_{\mathbb{R}} e^{-iE_p T/\hbar} \frac{dp\, p \sinh \pi p}{\cosh^2 \pi k + \sinh^2 \pi p} P_{ik-1/2}^{ip}(\tanh \tau) P_{ik-1/2}^{-ip}(\tanh \tau) \ . \quad (12.7)
$$

Here the $\Psi_{\mathbf{k}}^{\Lambda^{(D-2)}}(\mathbf{v})$ are the wavefunctions on the hyperboloid $\Lambda^{(D-2)}$ with the set of the quantum numbers $\mathbf{k}$, and the principal quantum number k, i.e., $E_k^{\Lambda^{(D-2)}} = \frac{\hbar^2}{2m}[k^2 + (D-3)^2/4]$, $D = 3, 4, \ldots$. Again, a systematic enumeration of the number of orthogonal coordinate systems on $\Lambda^{(D-2)}$ seems not to be known.

12.2 Hyperbolic Rank-One Spaces

Beside the single- and double-sheeted pseudospheres there are some more hyperbolic spaces with a sufficient simple structure. A sufficient simple structure means that it is possible to measure distances in these hyperbolic spaces by just one quantity, the hyperbolic distance. These hyperbolic spaces are being called of rank one. Coordinate systems on hermitian hyperbolic spaces have been studied by Boyer et al. [84]. A discussion of trace formulæ in these spaces is due to Venkov [506].

It is possible to extend the path integral analysis on all hyperbolic spaces of rank one, and we cite some path integral results achieved in [208]. We consider a hyperbolic space X as a quotient space of a Gelfand pair (G, K), $X = G/K$. This property allows an Iwasawa decomposition according to the direct sum of the algebra on G, $\mathfrak{g}$, such that $\mathfrak{g} = \mathfrak{k} + \mathfrak{a} + \mathfrak{n}$ (e.g., Hashizume et al. [265], Helgason [273, 274], Sekiguchi [463], Terras [495], and Venkov [506]; for more details on notation and relevant references c.f. [210, 265] and section 14.1). The root system of the pair $(\mathfrak{g}, \mathfrak{a})$ is denoted by P. The Laplace-Beltrami operator (a Casimir operator) on X then is the invariant operator on X with respect to the group actions. The root system in these spaces can be taken with all roots positive (restricted set P^+) and decomposed into two systems, α and β, such that if $\mu \in P_\beta$ then $\mu/2$ is not a root. The subspaces $\mathfrak{g}(\alpha)$ and $\mathfrak{g}(\beta)$ have the dimensions m_α and m_β, respectively. Furthermore, the algebra $\mathfrak{g}$ can be written as a Cartan decomposition $\mathfrak{g} = \mathfrak{k} + \mathfrak{p}$, $\mathfrak{p}$ being the orthogonal complement of $\mathfrak{k}$. In the cases in question the subspace corresponding to the algebra P^+ can be represented as a $(m_\alpha + m_\beta + 1)$-dimensional sphere, denoted as $S^{(m_\alpha + m_\beta)}$.

The crucial observation is that a separation in polar and angular variables is always possible. For our purposes it is sufficient to consider the relevant Laplacian which can be cast into the form (e.g., Camporesi [94], Hashizume et al. [265], Helgason [272])

$$\Delta_{LB}^{G/K} = \frac{\partial^2}{\partial \tau^2} + (m_\alpha \coth \tau + 2 m_\beta \coth 2\tau) \frac{\partial}{\partial \tau}$$
$$- \left[\frac{\mathcal{L}^{(\mu^+)}}{\sinh^2 \tau} + \left(\frac{1}{\sinh^2 2\tau} - \frac{1}{\sinh^2 \tau} \right) \mathcal{L}^{(2\mu)} \right] . \qquad (12.8)$$

The operators $\mathcal{L}^{(\mu^+)}$ and $\mathcal{L}^{(2\mu)}$ act on the space of rootsystems $\mathfrak{g}(\alpha^+)$ (all positive roots) and $\mathfrak{g}(\beta)$, respectively. It is found that the operators $\mathcal{L}^{(\mu^+)}$ and $\mathcal{L}^{(2\mu)}$ have eigenvalues $4l(l + m_\beta - 1) + l m_\alpha$ and $4l(l + m_\beta - 1)$, respectively, with common quantum number $l \in \mathbb{N}_0$. Therefore we have by means of the path integral solution for the modified Pöschl-Teller potential the path integral solution in the hyperbolic polar coordinates $\tau > 0$

$$\frac{\mathrm{i}}{\hbar} \int_0^\infty dT \, \mathrm{e}^{\mathrm{i}ET/\hbar} \int_{\tau(t')=\tau'}^{\tau(t'')=\tau''} \mathcal{D}\tau(t)$$

$$\times \exp\left\{ \frac{\mathrm{i}}{\hbar} \int_{t'}^{t''} \left[\frac{m}{2} \dot{\tau}^2 - \frac{\hbar^2}{8m} \left(\frac{(2l + m_\alpha + m_\beta - 1)^2 - 1}{\sinh^2 \tau} + \frac{(2l + m_\beta - 1)^2 - 1}{\cosh^2 \tau} \right) \right] dt \right\}$$

$$= \frac{m}{\hbar^2} \frac{\Gamma(m_1 - L_\nu)\Gamma(L_\nu + m_1 + 1)}{\Gamma(m_1 + m_2 + 1)\Gamma(m_1 - m_2 + 1)}$$
$$\times (\cosh r' \cosh r'')^{-(m_1 - m_2)} (\tanh r' \tanh r'')^{m_1 + m_2 + 1/2}$$
$$\times {}_2F_1\left(-L_\nu + m_1, L_\nu + m_1 + 1; m_1 - m_2 + 1; \frac{1}{\cosh^2 r_<} \right)$$
$$\times {}_2F_1\left(-L_\nu + m_1, L_\nu + m_1 + 1; m_1 + m_2 + 1; \tanh^2 r_> \right) \qquad (12.9)$$

$$= \int_0^\infty dp \frac{\Psi_p^{G/K *}(\tau')\Psi_p^{G/K}(\tau'')}{E_p^{G/K} - E} , \qquad (12.10)$$

$(m_{1,2} = \frac{1}{2}(\eta \pm \sqrt{-2mE}/\hbar),\ L_\nu = \frac{1}{2}(1 - \nu),\ \eta = 2l + m_\alpha + m_\beta - 1,\ \nu = 2l + m_\beta - 1),$

with the energy-spectrum

$$E_p^{G/K} = \frac{\hbar^2}{2m}p^2 + E_0^{G/K}, \qquad E_0^{G/K} = \frac{\hbar^2}{8m}(m_\alpha + 2m_\beta)^2, \qquad (12.11)$$

and the radial wavefunctions are given by

$$\Psi_p^{G/K}(\tau) = N_p^{G/K}(\tanh\tau)^{l+\frac{1}{2}(m_\alpha+m_\beta)}(\cosh\tau)^{\mathrm{i}p}$$

$$\times {}_2F_1\left[l + \frac{1}{2}\left(\frac{m_\alpha}{2} + m_\beta - \mathrm{i}p\right), \frac{1}{2}\left(\frac{m_\alpha}{2} + 1 - \mathrm{i}p\right); \frac{1}{2}(m_\alpha + m_\beta + 1); \tanh^2\tau\right], \quad (12.12)$$

$$N_p^{G/K} = \sqrt{\frac{p\sinh\pi p}{2\pi^2}} \frac{\Gamma[l + \frac{1}{2}(\frac{m_\alpha}{2} + m_\beta + \mathrm{i}p)]\Gamma[\frac{1}{2}(\frac{m_\alpha}{2} + 1 + \mathrm{i}p)]}{\Gamma[l + \frac{1}{2}(m_\alpha + m_\beta + 1)]} . \qquad (12.13)$$

Note that in the case of the hyperboloids $\Lambda^{(D-1)}$ we have $m_\alpha = D - 1, m_\beta = 0$.

The Hermitian Hyperbolic Space.

The space $S_2 \simeq SU(n,1)/S[U(1) \times U(n)]$ is an example for a hermitian hyperbolic space [84]. It is possible to perform the path integration explicitly in at least two coordinate systems [210]. We have for the metric

$$ds^2 = \frac{dy^2}{y^2} + \frac{1}{y^2}\sum_{k=2}^{n} dz_k dz_k^* + \frac{1}{y^4}\left(dx_1 + \Im\sum_{k=2}^{n} z_k^* dz_k\right)^2, \qquad (12.14)$$

$(z_k = x_k + \mathrm{i}y_k \in \mathbb{C}\ (k = 2,\ldots,n), x_1 \in \mathbb{R}, y > 0)$, with the hyperbolic distance given by $(\mathbf{x} = \{x_k\}_{k=2}^n, \mathbf{y} = \{y_k\}_{k=2}^n)$

$$\cosh d(\mathbf{q}'', \mathbf{q}') = \frac{1}{4(y'y'')^4}\left[\left((\mathbf{x}'' - \mathbf{x}')^2 + y'^2 + y''^2\right)^2 + 4\left(x_1'' - x_1' + (\mathbf{x}''\cdot\mathbf{y}' - \mathbf{y}''\cdot\mathbf{x}')\right)^2\right].$$
$$(12.15)$$

The symmetry properties of the space give rise to two important coordinates systems, where the problem is separable, namely $(n-1)$-fold two-dimensional polar coordinates according to $x_k = r_k\cos\varphi_k, y_k = r_k\sin\varphi_k, (r_k > 0, 0 \le \varphi_k \le 2\pi, k = 2,\ldots,n)$, respectively, $(2n-1)$-dimensional $SU(n-1)/SU(n-2)$ polar coordinates

$$\begin{aligned}
z_n &= r\mathrm{e}^{\mathrm{i}\varphi_n}\cos\vartheta_{n-1} \\
z_{n-1} &= r\mathrm{e}^{\mathrm{i}\varphi_{n-1}}\sin\vartheta_{n-1}\cos\vartheta_{n-2} \\
&\vdots \\
z_3 &= r\mathrm{e}^{\mathrm{i}\varphi_3}\sin\vartheta_{n-1}\ldots\sin\vartheta_3\cos\vartheta_2 \\
z_2 &= r\mathrm{e}^{\mathrm{i}\varphi_2}\sin\vartheta_{n-1}\ldots\sin\vartheta_3\sin\vartheta_2 .
\end{aligned} \qquad \left(\begin{array}{ll} 0 \le \varphi_i \le 2\pi & i = 2,\ldots,n, \\ 0 \le \vartheta_j \le \pi & j = 2,\ldots,n-1, \\ r > 0 & \end{array}\right)$$
$$(12.16)$$

Here, of course, any coordinate system of the $S^{(n-2)}$-sphere can be attached.

In $(n-1)$-fold two-dimensional polar coordinates we find the path integral representation [210] ($\mathbf{r} = \{r_k\}_{k=2}^n$, $\boldsymbol{\varphi} = \{\varphi_k\}_{k=2}^n$)

$$K^{S_2}(\mathbf{x}'',\mathbf{y}'',\mathbf{x}',\mathbf{y}',x_1'',x_1',y'',y';T) \equiv K^{S_2}(\mathbf{r}'',\mathbf{r}',\boldsymbol{\varphi}'',\boldsymbol{\varphi}',x_1'',x_1',y'',y';T)$$

$$= \exp\left[-\frac{i\hbar T}{8m}(4n^2-1)\right] \int_{y(t')=y'}^{y(t'')=y''} \frac{\mathcal{D}y(t)}{y^{2n+1}} \int_{x_1(t')=x_1'}^{x_1(t'')=x_1''} \mathcal{D}x_1(t) \int_{\mathbf{x}(t')=\mathbf{x}'}^{\mathbf{x}(t'')=\mathbf{x}''} \mathcal{D}\mathbf{x}(t) \int_{y(t')=y'}^{y(t'')=y''} \mathcal{D}\mathbf{y}(t)$$

$$\times \exp\left\{\frac{i}{\hbar}\int_{t'}^{t''}\left[\frac{m}{2}\frac{\dot{y}^2+\dot{\mathbf{x}}^2+\dot{\mathbf{y}}^2}{y^2} + \frac{\dot{x}_1^2+2\dot{x}_1(\mathbf{x}\dot{\mathbf{y}}-\mathbf{y}\dot{\mathbf{x}})+(\mathbf{x}\dot{\mathbf{y}}-\mathbf{y}\dot{\mathbf{x}})^2}{y^4}\right]dt\right\}$$

$$= \exp\left[-\frac{i\hbar T}{8m}(4n^2-1)\right] \int_{y(t')=y'}^{y(t'')=y''} \frac{\mathcal{D}y(t)}{y^{2n+1}} \int_{x_1(t')=x_1'}^{x_1(t'')=x_1''} \mathcal{D}x_1(t)$$

$$\times \prod_{k=2}^n \int_{r_k(t')=r_k'}^{r_k(t'')=r_k''} \mathcal{D}r_k(t)r_k \cdot \int_{\boldsymbol{\varphi}(t')=\boldsymbol{\varphi}'}^{\boldsymbol{\varphi}(t'')=\boldsymbol{\varphi}''} \mathcal{D}\boldsymbol{\varphi}(t) \exp\left[\frac{i}{\hbar}\int_{t'}^{t''}\left(\frac{m}{2}\left\{\frac{1}{y^2}\left[\dot{y}^2+\sum_{k=2}^n(\dot{r}_k^2+r_k^2\dot{\varphi}_k^2)\right]\right.\right.$$

$$\left.\left.+\frac{1}{y^4}\left[\dot{x}_1^2+2\dot{x}_1\sum_{k=2}^n r_k^2\dot{\varphi}_k + \left(\sum_{k=2}^n r_k^2\dot{\varphi}_k\right)^2\right]\right\} + \frac{\hbar^2 y^2}{8m}\sum_{k=2}^n\frac{1}{r_k^2}\right)dt\right]$$

$$= \int_{\mathbb{R}} dk_1 \prod_{k=2}^n \sum_{l_k\in\mathbb{Z}} \sum_{n_k=0}^\infty \int_0^\infty dp\, e^{-iE_p^{S_2}T/\hbar} \Psi_{k_1,\mathbf{L},\mathbf{N},p}^*(\mathbf{r}',\boldsymbol{\varphi}',x_1',y')\Psi_{k_1,\mathbf{L},\mathbf{N},p}(\mathbf{r}'',\boldsymbol{\varphi}'',x_1'',y'') \ ,$$

$$(12.17)$$

with the energy-spectrum ($m_\alpha = 2(n-1), m_\beta = 1$) $E_p^{S_2} = \hbar^2(p^2+n^2)/2m$, and the radial wavefunctions

$$\Psi_{k_1,\mathbf{L},\mathbf{N},p}(\mathbf{r},\boldsymbol{\varphi},x_1,y) = \frac{e^{i(l_k\varphi_k+k_1x_1)}}{(2\pi)^{n/2}}\frac{R_{n_k}^{l_k}(r_k)}{\sqrt{r_k}}\varphi_p(y) \qquad (12.18)$$

with

$$R_n^l(r) = \sqrt{\frac{2|k_1|n!}{\Gamma(n+|l|+1)}}\,(|k_1|r)^{|l|}\exp(-|k_1|r)L_n^{(|l|)}(|k_1|r^2) \ , \qquad (12.19)$$

$$\varphi_p(y) = \sqrt{\frac{p\sinh\pi p}{2\pi^2|k_1|}}\,\Gamma\left[\frac{1}{2}\left(1+ip+\frac{mE_\lambda}{|k_1|}\right)\right]y^{n-1}W_{-mE_\lambda/2|k_1|,ip/2}(|k_1|y^2) \ , \quad (12.20)$$

$$E_\lambda = \frac{\hbar^2}{m}\sum_{k=2}^n[|k_1|(2n_k+1)+|k_1l_k|-k_1l_k] \ . \qquad (12.21)$$

$\mathbf{L},\mathbf{N}$ denote the set of quantum numbers ($l_k\in\mathbb{Z}, n_k\in\mathbb{N}_0, k=2,3,\ldots$). Each step in the respective path integration is obvious: A φ_k-path integration gives circular exponentials, a r_k-path integration gives radial harmonic oscillator path integrals, and the final y-path integration then is equivalent to a pure scattering Morse potential.

In SU(n)/ SU$(n-1)$-spherical polar coordinates we obtain [210] ($\boldsymbol{\vartheta} = \{\vartheta_k\}_{k=2}^{n-1}$)

$$K^{S_2}(\mathbf{x}'', \mathbf{y}'', \mathbf{x}', \mathbf{y}', x_1'', x_1', y'', y'; T) \equiv K^{S_2}(r'', r', \boldsymbol{\vartheta}'', \boldsymbol{\vartheta}', \boldsymbol{\varphi}'', \boldsymbol{\varphi}', x_1'', x_1', y'', y'; T)$$

$$= \frac{y'y''}{2\pi} \int_{\mathbb{R}} dk_1 \, e^{ik(x_1''-x_1')} \exp\left[-\frac{i\hbar T}{8m}(4n^2-1)\right] \int_{y(t')=y'}^{y(t'')=y''} \frac{\mathcal{D}y(t)}{y^{2n-1}} \int_{r(t')=r'}^{r(t'')=r''} \mathcal{D}r(t) r^{2n-3}$$

$$\times \int_{\boldsymbol{\vartheta}(t')=\boldsymbol{\vartheta}'}^{\boldsymbol{\vartheta}(t'')=\boldsymbol{\vartheta}''} \mathcal{D}\boldsymbol{\vartheta}(t) \cdot \prod_{k=2}^{n-1} \cos\vartheta_k (\sin\vartheta_k)^{2k-3} \cdot \int_{\boldsymbol{\varphi}(t')=\boldsymbol{\varphi}'}^{\boldsymbol{\varphi}(t'')=\boldsymbol{\varphi}''} \mathcal{D}\boldsymbol{\varphi}(t)$$

$$\times \exp\Bigg[\frac{i}{\hbar} \int_{t'}^{t''} \Bigg(\frac{m}{2y^2}\Big\{\dot{y}^2 + \dot{r}^2 + r^2\Big[\dot{\vartheta}_{n-1}^2 + \cos^2\vartheta_{n-1}\dot{\varphi}_n^2 + \dots$$

$$\dots + \sin^2\vartheta_3\left(\dot{\vartheta}_3^2 + \cos^2\vartheta_3\dot{\varphi}_3^2 + \sin^2\vartheta_2\dot{\varphi}_2^2\right)\dots\Big]\Big\}$$

$$- \frac{\hbar^2 k_1^2}{2m}y^4 + \hbar k_1 r^2\Big\{\dot{\varphi}_n \cos^2\vartheta_{n-1} + \sin^2\vartheta_{n-1}\Big[\dot{\varphi}_{n-1}\sin^2\vartheta_{n-2} + \dots$$

$$\dots + \sin^2\vartheta_3(\dot{\varphi}_3\cos^2\vartheta_2 + \dot{\varphi}_2\sin^2\vartheta_2)\dots\Big]\Big\}$$

$$+ \frac{\hbar^2 y^2}{8mr^2}\Bigg[1 + \frac{1}{\cos^2\vartheta_{n-1}} + \dots + \frac{1}{\sin^2\vartheta_3}\left(1 + \frac{1}{\cos^2\vartheta_2} + \frac{1}{\sin^2\vartheta_2}\right)\dots\Bigg]\Bigg) dt\Bigg]$$

$$= \int_{\mathbb{R}} dk_1 \sum_{\mathbf{L}} \sum_{N=0}^{\infty} \int_0^{\infty} dp \, e^{-iE_p T/\hbar} \Psi_{k_1,\mathbf{L},N,p}^*(x_1', \boldsymbol{\vartheta}', \boldsymbol{\varphi}', r', y') \Psi_{k_1,\mathbf{L},N,p}(x_1'', \boldsymbol{\vartheta}'', \boldsymbol{\varphi}'', r'', y'') \,,$$

$$(12.22)$$

with the same energy spectrum as before and the wavefunctions

$$\Psi_{k_1,\mathbf{L},N,p}(x_1, \{\vartheta, \varphi\}, r, y) = \frac{e^{ik_1 x_1}}{\sqrt{2\pi}} \Psi_{\mathbf{L}}^{S^{(n-2)}}(\mathbf{s}_{S^{(n-2)}}) R_N^{L+n-2}(r)\varphi_p(y) \,, \qquad (12.23)$$

in the notation of the previous section and as before for $R_n^l(r)$ and $\varphi_p(y)$, respectively. The path integration for r- and y-variables is the same as in the previous case. I have written $\Psi_{\mathbf{L}}^{S^{(n-2)}}(\mathbf{s}_{S^{(n-2)}})$ for the angular variable wavefunctions to underline the possibility of taking any coordinate system on $S^{(n-2)}$ into account.

The Space S_3.

In the space S_3 we have $m_\alpha = 4(n-1)$ and $m_\beta = 3$. The metric in the space $S_3 \cong \mathrm{Sp}(n,1)/[\mathrm{Sp}(1) \times \mathrm{Sp}(n)]$ is given by

$$ds^2 = \frac{dy^2}{y^2} + \frac{1}{y^2} \sum_{k=2}^{n}(dz_k dz_k^* + dz_{n+k} dz_{k+n}^*) + \frac{1}{y^4}\left(dx_1 + \Im \sum_{k=2}^{n}(z_k^* dz_k + z_{n+k}^* dz_{k+n})\right)^2$$

$$+ \frac{1}{y^4}\left|dz_{n+1} + \sum_{k=2}^{n}(z_{n+k} dz_k - z_k dz_{k+n})\right|^2 \,. \qquad (12.24)$$

Here $y > 0$, $x_1 \in \mathbb{R}$, and $z_k = x_k + iy_k \in \mathbb{C}$ $(k = 2, \ldots, 2n)$. Hence $E_p^{S_3} = \frac{\hbar^2}{2m}[p^2 + (2n+1)^2]$. The radial wavefunctions have the form

$$\Psi_{p,l}^{S_3}(\tau) = N_p^{S_3} (\tanh \tau)^{2n+l-\frac{1}{2}}(\cosh \tau)^{ip}$$
$$\times {}_2F_1\left(l + n + \frac{1 - ip}{2}, n - \frac{1 + ip}{2}; 2n + l; \tanh^2 \tau\right) , \tag{12.25}$$

$$N_p^{S_3} = \frac{1}{\Gamma(2n + l)} \sqrt{\frac{p \sinh \pi p}{2\pi^2}} \Gamma\left(l + n + \frac{1 + ip}{2}\right)\Gamma\left(n - \frac{1 + ip}{2}\right) . \tag{12.26}$$

The Exceptional Case.

The exceptional space is defined as $S_4 \cong F_{4(-20)}/Spin(9)$. Here $m_\alpha = 8$ and $m_\beta = 7$. Consequently, the radial wavefunctions are given by

$$\Psi_{p,l}^{S_4}(\tau) = N_p^{S_4} (\tanh \tau)^{l+8-\frac{1}{2}}(\cosh \tau)^{ip}{}_2F_1\left(l + \frac{11 - ip}{2}, \frac{5 - ip}{2}; l + 8; \tanh^2 \tau\right) , \tag{12.27}$$

$$N_p^{S_4} = \frac{1}{\Gamma(l + 8)} \sqrt{\frac{p \sinh \pi p}{2\pi^2}} \Gamma\left(l + 5 + \frac{1 + ip}{2}\right)\Gamma\left(2 + \frac{1 + ip}{2}\right) . \tag{12.28}$$

$E_p^{S_4} = \hbar^2(p^2 + 121)/2m$ and the zero-point energy is $E_0^{S_4} = 121\hbar^2/2m$.

12.3   Path Integral on $SU(n)$ and $SU(n-1,1)$

It is straightforward to find a path integral representation for the spaces $SU(n)$ and $SU(n-1,1)$ in bi-spherical polar coordinates, respectively [208].

12.3.1   Path Integral on $SU(n)$

For the homogeneous space $SU(n)$ a bi-spherical coordinate system is given by [51, 65, 66, 208] (12.15). For r as a variable we have $SU(n)$ polar coordinates and for $r = 1$ we describe motion on the pure $SU(n)$ manifold, respectively. Introducing (note: $\sqrt{g} = \sqrt{\det(g_{ab})} = r^{2n-1} \prod_{i=1}^{n-1} \cos \vartheta_i (\sin \vartheta_i)^{2i-1}$):

$$\mathcal{L}_{Cl}^{(n)}(\{\vartheta, \dot{\vartheta}\}, \{\varphi, \dot{\varphi}\}) = \frac{m}{2} \left\{ \dot{\vartheta}_{n-1}^2 + \cos^2 \vartheta_{n-1}\dot{\varphi}_n^2 + \sin^2 \vartheta_{n-1} \left[\dot{\vartheta}_{n-2}^2 \right. \right.$$
$$+ \cos^2 \vartheta_{n-2}\dot{\varphi}_{n-1}^2 + \ldots + \sin^2 \vartheta_2 \left(\dot{\vartheta}_1^2 + \cos^2 \vartheta_1\dot{\varphi}_2^2 + \sin^2 \vartheta_1\dot{\varphi}_1^2 \right) \ldots \right] \Big\} , \tag{12.29}$$

$$\Delta V^{(n)}(\{\vartheta, \varphi\}) = -\frac{\hbar^2}{8m} \left\{ (2n-2)^2 + \frac{1}{\cos^2 \vartheta_{n-1}} \right.$$
$$+ \frac{1}{\sin^2 \vartheta_{n-1}} \left[1 + \frac{1}{\cos^2 \vartheta_{n-2}} + \ldots + \frac{1}{\sin^2 \vartheta_2} \left(1 + \frac{1}{\cos^2 \vartheta_1} + \frac{1}{\sin^2 \vartheta_1} \right) \ldots \right] \Big\} . \tag{12.30}$$

We have the path integral representation [208]:

$$\int \sqrt{g^{(n)}} \mathcal{D}\vartheta^{(n)}(t) \int \mathcal{D}\varphi^{(n)}(t) \exp\left\{\frac{i}{\hbar}\int_{t'}^{t''}\left[\mathcal{L}_{Cl}^{(n)}(\{\vartheta,\dot\vartheta\},\{\varphi,\dot\varphi\}) - \Delta V^{(n)}(\{\vartheta\})\right]dt\right\}$$

$$= \sum_{\{L\}} \Psi_{\{L\}}^{(n)*}(\{\vartheta'\},\{\varphi'\})\Psi_{\{L\}}^{(n)}(\{\vartheta''\},\{\varphi''\})e^{iTE_L/\hbar} , \tag{12.31}$$

$$\Psi_{\{L\}}^{(n)}(\{\vartheta\},\{\varphi\}) = \left[\prod_{j=1}^{n-1} \cos\vartheta_j(\sin\vartheta_j)^{2j-1}\right]^{-\frac{1}{2}}$$

$$\times (2\pi)^{-\frac{n}{2}} \exp\left(i\sum_{j=1}^{n} k_i\varphi_i\right)\varphi_{n_1}^{(k_1,k_2)}(\vartheta_1) \times \ldots \times \varphi_{n_{n-1}}^{(L_{n-2}+n-2,k_n)}(\vartheta_{n-1}) , \tag{12.32}$$

$$E_L = \frac{\hbar^2}{2m}L(L+2n-2) , \qquad L \in \mathbb{N}_0 . \tag{12.33}$$

Special cases are:

1. Path integral integral on a circle ($n=1$)

$$K^{S^{(1)}}(\varphi'',\varphi';T) = \frac{1}{2\pi}\sum_{k=-\infty}^{\infty} e^{ik(\varphi''-\varphi')-i\hbar Tk^2/2m} = \frac{1}{2\pi}\vartheta\left(\frac{\varphi''-\varphi'}{2}\middle| -\frac{\hbar T}{2\pi m}\right) . \tag{12.34}$$

2. $SU(2)$ Path integral integral

$$K^{SU(2)}(\vartheta'',\vartheta',\{\varphi'',\varphi'\};T) = \int \cos\vartheta\sin\vartheta\mathcal{D}\vartheta(t)\int\mathcal{D}\varphi_1(t)\int\mathcal{D}\varphi_2(t)$$

$$\times \exp\left\{\frac{i}{\hbar}\int_{t'}^{t''}\left[\frac{m}{2}\left(\dot\vartheta^2+\cos^2\vartheta\dot\varphi_2^2+\sin^2\vartheta\dot\varphi_1^2\right)+\frac{\hbar^2}{8m}\left(4+\frac{1}{\cos^2\vartheta}+\frac{1}{\sin^2\vartheta}\right)\right]dt\right\}$$

$$= \frac{1}{(2\pi)^2}[\sin\vartheta'\sin\vartheta''\cos\vartheta'\cos\vartheta'']^{-\frac{1}{2}}$$

$$\times \sum_{k_1,k_2=-\infty}^{\infty} e^{ik_1(\varphi_1''-\varphi_1')+ik_2(\varphi_2''-\varphi_2')}\sum_{N=0}^{\infty}\varphi_N^{(k_1,k_2)*}(\vartheta')\varphi_N^{(k_1,k_2)}(\vartheta'')e^{-iE_LT/\hbar} , \tag{12.35}$$

$$= \sum_{J=0,\frac{1}{2}}^{\infty} \frac{2J+1}{2\pi^2}C_{2J}^1\left(\cos\frac{\Omega}{2}\right)\exp\left[-\frac{2i\hbar T}{m}J(J+1)\right] , \tag{12.36}$$

$$\cos\frac{\Omega}{2} = \sin\vartheta'\sin\vartheta''\cos(\varphi_1''-\varphi_1') + \cos\vartheta'\cos\vartheta''\cos(\varphi_2''-\varphi_2') \tag{12.37}$$

$$E_L = \frac{\hbar^2}{2m}L(L+2), \qquad L = 2N+|k_1|+|k_2| \in \mathbb{N}_0 . \tag{12.38}$$

3. $SU(3)$ Path integral integral

$$K^{SU(3)}(\{\vartheta'', \vartheta'\}, \{\varphi'', \varphi'\}; T)$$

$$= \sum_{L=0}^{\infty} \sum_{k_3=-L}^{L} \sum_{L_1=0}^{L} \sum_{k_1,k_2=-L_1}^{L_1} \frac{(L+2)(L_1+1)}{2\pi^3} e^{ik_1(\varphi_1''-\varphi_1')+ik_2(\varphi_2''-\varphi_2')+ik_3(\varphi_3''-\varphi_3')}$$

$$\times D^{\frac{L+1}{2}*}_{\frac{k_3+L_1+1}{2}, \frac{k_3-L_1-1}{2}}(\cos 2\vartheta_2') D^{\frac{L+1}{2}}_{\frac{k_3+L_1+1}{2}, \frac{k_3-L_1-1}{2}}(\cos 2\vartheta_2'')$$

$$\times D^{L_1/2*}_{\frac{k_1+k_2}{2}, \frac{k_1-k_2}{2}}(\cos 2\vartheta_1') D^{L_1/2}_{\frac{k_1+k_2}{2}, \frac{k_1-k_2}{2}}(\cos 2\vartheta_1'')$$

$$\times \exp\left[-\frac{i\hbar T}{2m} L(L+4)\right], \tag{12.39}$$

where the D^l_{mn} are the Wigner-polynomials, which are e.g. given by

$$C^1_{2J}\left(\cos\frac{\Omega}{2}\right) = \frac{\sin(J+\frac{1}{2})\frac{\Omega}{2}}{\sin\frac{\Omega}{2}} = \sum_{\mu,\nu=-J}^{J} e^{-i\mu(\alpha''-\alpha')-i\nu(\beta''-\beta')} D^{J*}_{\mu\nu}(\cos\gamma') D^J_{\mu\nu}(\cos\gamma''), \tag{12.40}$$

and the C^1_n is a Gegenbauer-polynomial.

12.3.2   Path Integral on $SU(n-1,1)$

For the homogeneous space $SU(n-1,1)$ a bi-spherical coordinate system is given by

$$
\begin{aligned}
z_n &= r e^{i\varphi_n} \cosh\tau \\
z_{n-1} &= r e^{i\varphi_{n-1}} \sinh\tau \cos\vartheta_{n-2} \\
z_{n-2} &= r e^{i\varphi_{n-2}} \sinh\tau \sin\vartheta_{n-2} \cos\vartheta_{n-3} \\
\cdots & \quad \cdots \\
z_2 &= r e^{i\varphi_2} \sinh\tau \ldots \sin\vartheta_2 \cos\vartheta_1 \\
z_1 &= r e^{i\varphi_1} \sinh\tau \ldots \sin\vartheta_2 \sin\vartheta_1,
\end{aligned}
\qquad
\left(
\begin{aligned}
&0 \le \varphi_i \le 2\pi, \quad i = 1,\ldots,n, \\
&0 \le \vartheta_j \le \tfrac{\pi}{2}, \quad j = 1,\ldots,n-2, \\
&\tau > 0, \qquad\qquad r \ge 0.
\end{aligned}
\right)
$$

$$\tag{12.41}$$

In the following we set $r = 1$. Motion in terms of these coordinates is constrained by the condition $|z_1|^2 + |z_2|^2 + \ldots + |z_{n-1}|^2 - |z_n|^2 = -1$. Thus the signature of the metric reads $(+1,\ldots,+1,-1)$. We find

$$K^{SU(n-1,1)}(\tau'', \tau', \{\vartheta'', \vartheta'\}, \{\varphi'', \varphi'\}; T)$$

$$= \int \cosh\tau (\sinh\tau)^{2n-3} \mathcal{D}\tau(t) \int \prod_{k=1}^{n-2} \cos\vartheta_k (\sin\vartheta_k)^{2k-1} \mathcal{D}\vartheta_k(t) \int \prod_{j=1}^{n} \mathcal{D}\varphi_j(t)$$

$$\times \exp\left[\frac{i}{\hbar} \int_{t'}^{t''} \left(\frac{m}{2}\left\{\dot{\tau}^2 - \cosh^2\tau \dot{\varphi}_n^2\right.\right.\right.$$

$$\left.\left.\left. + \sinh^2\tau \left[\dot{\vartheta}_{n-2}^2 + \cos^2\vartheta_{n-2}\dot{\varphi}_{n-1}^2 + \ldots + \sin^2\vartheta_2\left(\dot{\vartheta}_1^2 + \cos^2\vartheta_1\dot{\varphi}_2^2 + \sin^2\vartheta_1\dot{\varphi}_1^2\right)\ldots\right]\right\}\right.\right.$$

$$-\frac{\hbar^2}{8m}\left\{(2n-2)^2 + \frac{1}{\cosh^2\tau} - \frac{1}{\sinh^2\tau}\left[1 + \frac{1}{\cos^2\vartheta_{n-2}} + \dots\right.\right.$$

$$\left.\left.\dots + \frac{1}{\sin^2\vartheta_2}\left(1 + \frac{1}{\cos^2\vartheta_1} + \frac{1}{\sin^2\vartheta_1}\right)\dots\right]\right\}\right)dt\Bigg]\qquad(12.42)$$

$$= \left[\cosh\tau'\cosh\tau''(\sinh\tau'\sinh\tau'')^{2n-3}\prod_{j=1}^{n-2}\cos\vartheta'_j\cos\vartheta''_j(\sin\vartheta'_j\sin\vartheta''_j)^{2j-1}\right]^{-\frac{1}{2}}$$

$$\times\left(\prod_{i=1}^{n}\frac{1}{2\pi}\sum_{k_i=-\infty}^{\infty}e^{ik_i(\varphi''_i-\varphi'_i)}\right)$$

$$\times\sum_{n_1,\dots,n_{n-2}}\left[\varphi_{n_1}^{(k_1,k_2)}(\vartheta'_1)\varphi_{n_1}^{(k_1,k_2)}(\vartheta''_1)\varphi_{n_{n-2}}^{(L_{n-3}+n-3,k_{n-1})}(\vartheta'_{n-2})\varphi_{n_{n-2}}^{(L_{n-3}+n-3,k_{n-1})}(\vartheta''_{n-2})\right]$$

$$\times\left\{\sum_{N=0}^{N_M}\varphi_n^{(L_{n-2}+n-2,k_n)\,*}(\tau')\varphi_n^{(L_{n-2}+n-2,k_n)}(\tau'')e^{-iTE_N/\hbar}\right.$$

$$\left.+\int_0^\infty dp\,\Psi_p^{(L_{n-2}+n-2,k_n)\,*}(\tau')\Psi_p^{(L_{n-2}+n-2,k_n)}(\tau'')\,e^{-iTE_p/\hbar}\right\}\,,\qquad(12.43)$$

$$E_N = -\frac{\hbar^2}{2m}\left(|k_n| - L_{n-2} - n - 2N + 1\right)^2\,,\qquad(12.44)$$

$$E_p = \frac{\hbar^2}{2m}[p^2 + (n-1)^2]\,,\qquad p > 0\,,\qquad(12.45)$$

where $N = 0, 1, \dots, N_M < \frac{1}{2}(|k_n| - L_{n-2} - n + 1)$, and with the largest lower bound $E_0 = \inf_p E_p = \frac{\hbar^2}{2m}(n-1)^2$.

Chapter 13

Billiard Systems
and Periodic Orbit Theory

13.1 Some Elements of Periodic Orbit Theory

Let us outline the derivation of the periodic orbit formula of Gutzwiller. We start with the semiclassical propagator which has the following form [124, 414, 438, 504]

$$K(\mathbf{q}'', \mathbf{q}'; T) \simeq \sum_{\text{classical orbits}} \frac{\sqrt{M}}{2\pi i\hbar} \exp\left[\frac{i}{\hbar} \int_0^T \mathcal{L}(\mathbf{q}, \dot{\mathbf{q}}; t)dt - \frac{i\pi}{2}\nu\right] . \tag{13.1}$$

We consider for simplicity only two-dimensional systems. Two-dimensional systems which are classically chaotic have only the energy as the observable. The phase index ν is equal to the number of conjugate points in the system plus twice the number of reflections [372]. Roughly speaking we have to take into account a phase factor $e^{-i\pi/2}$ for every reduction in the rank of M by one at the focal points. M is the van Vleck-Morette-Pauli determinant defined by

$$M = \left| -\frac{\partial S^2(\mathbf{q}'', \mathbf{q}')}{\partial \mathbf{q}' \partial \mathbf{q}''} \right| , \tag{13.2}$$

and $S(\mathbf{q}'', \mathbf{q}')$ is the classical action. The semiclassical propagator satisfies the Schrödinger equation up to terms of order $O(\hbar^2)$. Let us restrict ourselves to billiard systems. The corresponding semiclassical Green's function has the form

$$G(\mathbf{q}'', \mathbf{q}'; E) \simeq \sum_{\text{classical orbits}} \frac{i}{\hbar}\sqrt{\frac{D}{2\pi i\hbar}} \exp\left[\frac{i}{\hbar}\tilde{S}(\mathbf{q}'', \mathbf{q}'; E) - \frac{i\pi}{2}\mu\right] . \tag{13.3}$$

Here μ is the Maslov index, $\tilde{S}(\mathbf{q}'', \mathbf{q}''; E) = S(\mathbf{q}'', \mathbf{q}''; T)|_{T=mL/p} + E$ with L the length of the classical orbit and p its momentum, and D denotes the determinant

$$
\mathrm{D} = - \begin{vmatrix} \dfrac{\partial^2 \tilde{S}}{\partial \mathbf{q}' \partial \mathbf{q}''} & \dfrac{\partial^2 \tilde{S}}{\partial E \partial \mathbf{q}''} \\[2ex] \dfrac{\partial^2 \tilde{S}}{\partial \mathbf{q}' \partial E} & \dfrac{\partial^2 \tilde{S}}{\partial E^2} \end{vmatrix} . \tag{13.4}
$$

Let us introduce the monodromy matrix $\mathbf{M}$. It is defined as

$$
\mathbf{M} = \left(\frac{\partial^2 \tilde{S}}{\partial \mathbf{q}' \partial \mathbf{q}''} \right)^{-1} \begin{pmatrix} -\dfrac{\partial^2 \tilde{S}}{\partial \mathbf{q}'^2} & -1 \\[2ex] \left[\left(\dfrac{\partial^2 \tilde{S}}{\partial \mathbf{q}' \partial \mathbf{q}''} \right)^2 - \dfrac{\partial^2 \tilde{S}}{\partial \mathbf{q}'^2} \right] & -\dfrac{\partial^2 \tilde{S}}{\partial \mathbf{q}''^2} \end{pmatrix} = \begin{pmatrix} M_{11} & M_{12} \\ M_{21} & M_{22} \end{pmatrix} , \tag{13.5}
$$

and $\mathbf{M}$ describes the effect in phase space of an infinitesimal variation of the initial state $(\mathbf{q}', \mathbf{p}')$ into $(\mathbf{q}'', \mathbf{p}'')$, i.e.,

$$
\begin{pmatrix} d\mathbf{q}'' \\ d\mathbf{p}'' \end{pmatrix} = \mathbf{M} \begin{pmatrix} d\mathbf{q}' \\ d\mathbf{p}' \end{pmatrix} . \tag{13.6}
$$

One often also uses the modified monodromy matrix $\tilde{\mathbf{M}}$

$$
\tilde{\mathbf{M}} = \begin{pmatrix} M_{11} & p M_{12} \\ M_{21}/p & M_{22} \end{pmatrix} . \tag{13.7}
$$

For a two-dimensional flat billiard system the semiclassical Green's function becomes

$$
G(\mathbf{q}'', \mathbf{q}'; E) \simeq \frac{\mathrm{i}}{\hbar} \frac{m}{2\hbar} \sum_{\text{cl. orbits}} \sqrt{\frac{L}{|\tilde{M}_{12}|}} \, e^{-\mathrm{i}\pi\nu/2} H_0^{(1)} \left(\frac{S}{\hbar} \right) . \tag{13.8}
$$

We now want to consider the trace $g(E)$ of the semiclassical Green's function. Using the method of stationary phase one has $\mathrm{D} = m^2/p\tilde{M}_{12}$ and must only take into account the subset of the *periodic orbits* (po) of all classical orbits (Albeverio et al. [2], Gutzwiller [258, 260])

$$
g_{po}(E) \simeq \frac{\mathrm{i}}{\hbar} \sum_{\text{po}} \frac{L_0}{v} \frac{1}{\sqrt{2 - \mathrm{tr}(\mathbf{M})}} \exp \left(\frac{\mathrm{i}}{\hbar} S - \mathrm{i} \frac{\pi}{2} \sigma \right) . \tag{13.9}
$$

Here L_0 is a primitive periodic orbit, v its velocity, and σ denotes the maximal number of conjugate points.

However, this periodic orbit formula is in general not well-defined. This is due to the fact that for a general system it is not clear whether the trace of the resolvent exists. Furthermore, one wants to learn something about the spectrum of the system, and the resolvent is known to have poles, respectively a cut if $E \in \sigma(E)$, the spectral set. In order to circumvent this problem Sieber and Steiner [469] have developed a regularized periodic orbit formula (which is actually an analogue of the Selberg trace formula) for billiard systems. It has the following form:

Let us consider a classical system with periodic orbits, where the notion periodic orbits means that these orbits are periodic in phase space. The periodic orbits are labeled by $\gamma \in \Gamma$ with some classification Γ, and the l_γ denote their lengths. The corresponding quantum system has a Hamiltonian H with countable energy eigenvalues E_n and eigenfunctions Ψ_n, $n \in \mathbb{N}$. Let us consider further a testfunction $h(\sqrt{H})$. Then the general form of a periodic orbit formula for a billiard system is given by (Gutzwiller [260], Sieber [467], and Sieber and Steiner [469])

$$\sum_n h(p_n) \simeq \int_0^\infty dp\, h(p)\bar{d}(E) + C_0 h(0) + \frac{1}{\hbar} \sum_{\gamma \in \Gamma} \sum_{k=1}^\infty l_\gamma \frac{g(kl_\gamma/\hbar)}{2\sinh(ku_\gamma/2)} \quad . \tag{13.10}$$

Here $g(u) = (1/\pi) \int_0^\infty h(p)\cos(up - \pi k\nu_\gamma/2)dp$, C_0 is the constant appearing in Weyl's law, $\bar{d}(E) = (p/m) < d(E) >$ denotes the average energy level-density in the Thomas-Fermi approximation with $< d(E) > = dN(E)/dE$, $E = p^2/2m$, $N(E)$ the spectral staircase, and k is the number of the multiple traversals of the orbits l_γ ("winding number"). Here one has used that in the simplest (i.e. strictly hyperbolic) case [467] $\sqrt{|2 - \mathrm{tr}(\mathbf{M})|} = 2\sinh(ku_\gamma/2)$, where u_γ is the stability exponent. One assumes the asymptotic behaviour $\mathrm{e}^{-u_\gamma/2} \propto O(\mathrm{e}^{-\bar{u}l_\gamma/2})$ for $l_\gamma \to \infty$ for some $\bar{u} \geq 0$. Then furthermore one has defined $\sigma = \tau - \bar{u}/2$, where τ is the topological entropy. This means that the asymptotic proliferation of the orbits l_γ is according to $N(l) \propto \mathrm{e}^{\tau l}/\tau l$, as $l \to \infty$. The testfunction must satisfy the conditions

- $h(p)$ is an even function in p,

- $h(p)$ is analytic in the strip $|\Im(p)| \leq \sigma + \epsilon$, $\epsilon > 0$,

- $h(p) \leq a|p|^{-2-\delta}$ for $|p| \to \infty$, $a, \delta > 0$.

Periodic orbit theory has been very successful in describing several physical systems which could not be understood by other semiclassical quantization methods. Let us note triangular billiards in hyperbolic geometry which do not correspond to a Fuchsian group by Balazs and Voros [37], and Aurich and Steiner et al. [23], a Coulombic muffin-tin potential by Brandis [85] (here the original periodic orbit theory has to be extended to include scattering orbits), the hydrogen atom in a uniform magnetic field by Friedrich and Wintgen [170], the cardioid billiard system by Robnik et al. (c.f. [453] and references therein), and Bäcker et al. [35], the stadium billiard (Heller et al. [275], and, e.g., Sieber at al. [468] and references therein), the wedge billiard by Szeredi and Gooding [486], the anisotropic Kepler problem by Tanner and Wintgen [493], the hyperbola billiard by Hesse [19], and Sieber and Steiner [467, 469], and the semiclassical Helium atom by Wintgen et al. [522].

13.2   A Billiard System in a Hyperbolic Rectangle

In this section I want to present a particular billiard system in hyperbolic geometry. The study of this system developed for several reasons. First, the study of billiard systems in quantum chaos was initiated by the search for an answer to the question what does quantum chaos actually mean. In classical mechanics it is well-known that only in integrable systems the time-evolution is not sensitive with respect to the initial conditions. In comparison, in classical chaotic systems this is the case, which has the consequence that nearby orbits may exponentially diverge in their time evolution, i.e., they have a positive Lyapunov exponent. In an idealized experiment, where this exponentially divergence does not take place, the initial conditions are required to consist of just a point in phase-space. In quantum mechanics the finiteness of $\hbar$ provides a lower limit of the volume in phase space on the initial conditions of a physical system. Therefore the usual argument of classical mechanics is not valid. Furthermore, the time-dependent Schrödinger equation is a first order linear differential equation which does not allow for "chaotic time evolution" of the wavefunctions. Therefore one has to look for features in the quantum system which are remnants of the chaotic behaviour of the corresponding classically chaotic system.

Second, motion in hyperbolic space in (compact or non-compact) domains serve as fairly good understood models for classically chaotic systems. Here Aurich and Steiner et al. [15]-[32] have undertaken a comprehensive investigation of the classical and quantum systems. One could hope that in the study of a simple integrable approximation of these chaotic models at least some information about the latter might be extracted, say, for the low lying energy levels.

Usually the classical motion in a finite domain in hyperbolic geometry is chaotic. Consequently, the quantum motion reflects some of the chaotic properties of the chaotic classical motion, e.g., there is in general level repulsion. However, are there simple models which are nevertheless separable and resemble some properties of hyperbolic space? Let us for instance consider the motion in a bounded domain in a hyperbolic geometry where the boundaries are geodesics (with zero curvature). The everywhere negative curvature of hyperbolic space causes that near lying geodesics diverge exponentially in time evolution, i.e., they have a positive Lyapunov exponent which is equal to one. This means that the property of the space defocuses the classical trajectories. If it is possible to choose boundaries which have a focusing property it may be possible that the defocusing property of the hyperbolic space and the focusing property of the boundaries interact in such a way that the system remains separable and not chaotic. The billiard system presented in this section exactly has this property.

In [211] this system was numerically investigated and the statistical properties of the eigenvalues were studied. It was found that the system behaves perfectly as it is expected for an integrable system: The spectral staircase for the energy levels is in agreement with Weyl's law, the nearest neighbour statistics is Poisson-like (however not perfect), and the number variance and the spectral rigidity obey in a good approximation the predictions of the periodic orbit analysis of Berry [56, 58].

However, the comparison with the corresponding chaotic system showed no similarities in the energy spectrum and its statistical behaviour. Nevertheless the idea remains to study billiards in unusual settings, e.g., the one presented here which serves as an example of a separable quantum billiard in hyperbolic geometry. Another model was studied by, e.g., Graham et al. [112] for an integrable approximation of a cosmological billiard (Artin's billiard).

Because the rectangular billiard has been investigated already in [211] with respect to some aspects in periodic orbit theory it is sufficient to report only the main results. It is also a model which allows to test in an integrable system some predictions concerning the different behaviour of the energy level statistics of classically chaotic systems and classically integrable systems, respectively. However, no periodic orbit theory analysis has been undertaken yet, and will also not be presented here. Such a thorough analysis would give on the one hand a crosscheck of the numerically calculated energy levels, and on the other would explain similarly as in [32] the behaviour of the number variance and rigidity, respectively.

Before going into some details of the analysis let us shortly describe the model.

The rectangle in the hyperbolic plane is constructed as follows. Let us start with the Poincaré upper half-plane which is defined as

$$\mathcal{H} = \{\zeta = x + iy | x \in \mathbb{R}, \, \Re(y) > 0\} \,, \tag{13.11}$$

endowed with the hyperbolic metric $ds^2 = (dx^2 + d^2y)/y^2$. By means of the transformation (Cayley-transformation)

$$\zeta = \frac{-iz + i}{z + 1} \,, \qquad z = x_1 + ix_2 = \frac{-\zeta + i}{\zeta + i} \,, \tag{13.12}$$

the Poincaré upper half-plane is mapped onto the Poincaré disc with metric $g_{ab} = [2/(1 - r^2)]\mathrm{diag}(1, r^2)$ $(r^2 = x_1^2 + x_2^2)$, and by means of

$$\eta = X + iY = -\ln(-i\zeta) = 2\arctan(z) \tag{13.13}$$

onto the hyperbolic strip with metric $g_{ab} = \delta_{ab}/\cos^2 Y$.

In figures 13.1–13.3 I have displayed how the rectangle in the three realizations of the hyperbolic plane looks alike. In the hyperbolic strip it consists of two vertical straight lines which are geodesics (solid lines), and two horizontal straight lines which are not geodesics (dotted lines). In figures 13.2, 13.3, respectively, it is also displayed how this rectangle looks alike in the Poincaré upper half-plane and the Poincaré disc (dotted lines), respectively, and how the hyperbolic strip, the Poincaré upper half-plane and disc are tesselated by the symmetric octagon and hyperbolic rectangles (see below), respectively.

I also have displayed the regular octagon which corresponds to a Riemann surface of genus two, the simplest Riemann surface tesselating the hyperbolic plane. The regular octagon can be conveniently constructed in the Poincaré disc with eight

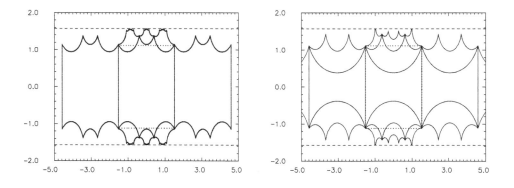

Figure 13.1: Tesselations in the hyperbolic strip.

quarter-circles described by [37]

$$\begin{pmatrix} x_1 \\ x_2 \end{pmatrix} = \begin{pmatrix} \tilde{\varrho} - \hat{\varrho}\cos\alpha \\ -\hat{\varrho}\cos\alpha \end{pmatrix} \ , \quad -\frac{\pi}{4} \le \alpha \le \frac{\pi}{4} \ , \quad \tilde{\varrho} = \sqrt{\frac{\sqrt{2}+1}{2}} \ , \quad \hat{\varrho} = \sqrt{\frac{\sqrt{2}-1}{2}} \ ,$$

(13.14)

which gives rise to the eight generators of the symmetric octagon in the Poincaré disc [24, 26, 37]

$$\gamma_k = \begin{pmatrix} \cosh(L_0/2) & \sinh(L_0/2)\,e^{ik\pi/4} \\ \sinh(L_0/2)\,e^{-ik\pi/4} & \cosh(L_0/2) \end{pmatrix} \ , \qquad (k=0,1,2,3) \ , \qquad (13.15)$$

including the inverse γ_k^{-1}, with $\cosh\frac{L_0}{2} = \cot\frac{\pi}{8} = 1 + \sqrt{2} = 2.414, 213, 562\ldots$. L_0 is the length of the shortest closed periodic geodesic in the regular octagon [24]. From the geometry it is clear that the area of a rectangle in the hyperbolic strip is given by

$$A = \int_{X_a}^{X_b} \int_{Y_a}^{Y_b} \frac{dX\,dY}{\cos^2 Y} = (X_b - X_a)(\tan Y_b - \tan Y_a) \qquad (13.16)$$

with some numbers (X_a, X_b, Y_a, Y_b).

Since the vertical lines coincide with the corresponding lines of the regular octagon we choose as $-X_a = X_b = L_0/2$ with L_0 given by

$$L_0 = 2\ln\left(\frac{1+\sqrt{\sqrt{2}-1}}{1-\sqrt{\sqrt{2}-1}}\right) = 3.057, 141, 839\ldots \qquad (13.17)$$

From the definition of a rectangle $[-L_0/2, L_0/2]\times[-Y_0, Y_0]$ we have several possibilities in choosing a particular one. One can either choose for Y_0 the point Y_A given by the

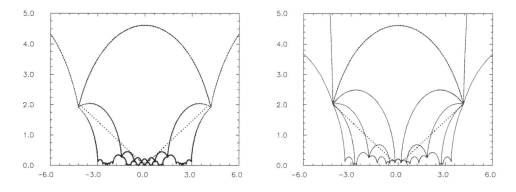

Figure 13.2: Tesselations in the Poincaré upper half-plane.

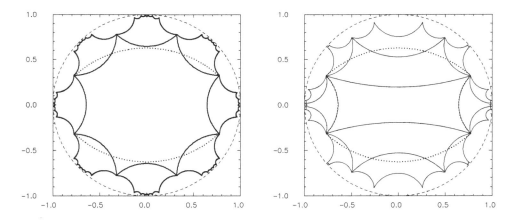

Figure 13.3: Tesselations in the Poincaré disc.

right upper corner of the octagon, the point Y_B given by the maximum value of the variable Y the octagon takes on in the strip, the point Y_C given by the minimum value of the variable Y the octagon takes on in the strip, or by some intermediate value Y_M given if the area of the rectangle equals, say $A = 4\pi$. Whereas the separability of the rectangle in the hyperbolic strip is obvious, the separability in the Poincaré upper half-plane and the Poincaré disc, respectively, is not so obvious. However, some particular polar coordinate system (ϱ, φ) does exist.

The two horizontal lines in figure 13.1 of the rectangle are not geodesics. It is nevertheless possible to construct a hyperbolic square bounded by geodesics. We consider the rectangle in the Poincaré upper half-plane and look for the arcs of a circle connecting the points $\zeta_A = e^{L_0/2}(\sin Y_0 + i \cos Y_0)$ and $\zeta_A' = e^{-L_0/2}(\sin Y_0 + i \cos Y_0)$. The circles described by $(x \mp x_\varrho)^2 + y^2 = \varrho^2$ with

$$x_\varrho = \pm \frac{\cosh \frac{L_0}{2}}{\sin Y_0} , \qquad \varrho = \sqrt{\frac{\cosh^2 \frac{L_0}{2}}{\sin^2 Y_0} - 1} \qquad (13.18)$$

do the job. The emerging domain in $\mathcal{H}$ can be seen as generated by the matrices

$$\gamma_1 = \begin{pmatrix} e^{L_0/2} & 0 \\ 0 & e^{-L_0/2} \end{pmatrix} , \qquad \gamma_2 = \frac{1}{\sqrt{\cosh^2 \frac{L_0}{2} - \sin^2 Y_0}} \begin{pmatrix} \cosh \frac{L_0}{2} & \sin Y_0 \\ \sin Y_0 & \cosh \frac{L_0}{2} \end{pmatrix} .$$
$$(13.19)$$

On the left hand side in figures 13.1–13.3 I have displayed the action of applying the hyperbolic boosts to the fundamental domain of the regular octagon. Also shown is the "rectangle approximation" (dotted lines). On the right the two-fold action of the generators is shown. Due to the general *formula* for a polygon in the hyperbolic plane $A = [(V - 2)\pi - \sum \alpha]$, where V denotes the number of vertices and α the corresponding angles [37], we get for the area of the hyperbolic square generated by the matrices (13.19)

$$\alpha = Y_0 + \arctan\left(\tan Y_0 - \frac{1 + e^{-L_0}}{\sin 2Y_0} \right), \qquad A = 4\arctan\left(\frac{e^{L_0} - 1}{\sqrt{2e^{L_0}(x_\varrho^2 + r^2) - e^{2L_0} - 1}} \right),$$
$$(13.20)$$

which gives $\alpha = 0.490, 923, 447, \ldots$, $A = 4.319, 491, 516, \ldots$ for $Y_0 = Y_M$ with $x_\varrho = \pm 2.684, 818, 117 \ldots$, $\varrho = 2.491, 635, 672 \ldots$, respectively.

For reasons of practicability and simplicity we now make the choice

$$Y_0 = Y_M = \arctan\left(\frac{2\pi}{L_0} \right) = 1.117, 959, 030 \ldots , \qquad (13.21)$$

and denote by the notion "rectangle in the hyperbolic plane" the rectangle in the strip bounded by the four straight lines as described above. This particular choice makes a reasonable compromise between the total number of levels which can be calculated within the region of stability of the numerical investigation by a simple FORTRAN-program. Also the width and length along the X- and Y-axis of this rectangle are almost equal.

In order to set up our quantization conditions and Weyl's law we must distinguish between four parity classes in the hyperbolic rectangle and the lengths of its boundaries. The rectangle is symmetric with respect to the X- and Y-axis, hence we get the four classes $P_1 = (+, +)$, $P_2 = (+, -)$, $P_3 = (-, +)$ and $P_4 = (-, -)$.

Let us set up the quantization condition. The free wavefunctions in the entire strip are given by [196]

$$\Psi_{p,k}(X, Y) = \sqrt{\frac{p \sinh \pi p \cos Y}{4\pi^2(\cosh^2 \pi k + \sinh^2 \pi p)}} e^{ikX} P_{ik-\frac{1}{2}}^{ip}(\pm \sin Y) , \qquad (13.22)$$

which are normalized solutions of the Schrödinger equation

$$-\cos^2 Y (\partial_X^2 + \partial_Y^2) \Psi(X,Y) = E \Psi(X,Y) \tag{13.23}$$

and the energy spectrum is $E = p^2 + \frac{1}{4}$ and I have used dimensionless units ($\hbar = 2m = 1$). Even and odd parity with respect to the X-coordinate yield the quantization condition with respect to the X-dependence

$$\text{even:} \qquad \cos \frac{L_0 k_l}{2} = 0 \;, \quad \rightarrow \quad k_l = \frac{2\pi(l + \frac{1}{2})}{L_0} \;, \quad l = 0, 1, 2, \dots \;, \tag{13.24}$$

$$\text{odd:} \qquad \sin \frac{L_0 k_l}{2} = 0 \;, \quad \rightarrow \quad k_l = \frac{2\pi l}{L_0} \;, \qquad l = 1, 2, 3, \dots \;. \tag{13.25}$$

Even and odd parity with respect to the Y-coordinate then give the quantization conditions

$$\text{even:} \qquad P_{ik_l - \frac{1}{2}}^{ip_n}(\sin Y_0) + P_{ik_l - \frac{1}{2}}^{ip_n}(-\sin Y_0) = 0 \;, \tag{13.26}$$

$$\text{odd:} \qquad P_{ik_l - \frac{1}{2}}^{ip_n}(\sin Y_0) - P_{ik_l - \frac{1}{2}}^{ip_n}(-\sin Y_0) = 0 \;. \tag{13.27}$$

The last two equations are transcendental equations for p_n, $n = 1, 2, \dots$ and must be solved numerically. Actually one uses the representation

$$P_{ik - \frac{1}{2}}^{ip}(\sin Y) = \frac{1}{\Gamma(1 - ip)} \left(\frac{1 + \sin Y}{1 - \sin Y} \right)^{\frac{ip}{2}} {}_2F_1\left(\frac{1}{2} - ik, \frac{1}{2} + ik; 1 - ip; \frac{1 - \sin Y}{2} \right) \;, \tag{13.28}$$

and omits the $1/\Gamma(1 - ip)$-factor. The energy of the nth-level finally has the form

$$E_n = p_n^2 + \frac{1}{4} \;. \tag{13.29}$$

Actually, the quantization rule follows from the path integral representation incorporating boundary conditions at $X = \pm L_0, Y = \pm Y_0$ according to [217]

$$G_{(|Y|<Y_0,|X|<L_0)}(X'', X', Y'', Y'; E)$$

$$= \frac{i}{\hbar} \int_0^\infty dT\, e^{iET/\hbar} \int_{X(t')=X'}^{X(t'')=X''} \mathcal{D}_{|X|<L_0} X(t) \int_{Y(t')=Y'}^{Y(t'')=Y''} \mathcal{D}_{|Y|<Y_0} Y(t) \exp\left(\frac{im}{2\hbar} \int_{t'}^{t''} \frac{\dot{X}^2 + \dot{Y}^2}{\cos^2 Y} dt \right)$$

$$= \frac{2\pi}{L_0} \sum_{l=0}^\infty \begin{pmatrix} \sin(k_l X') \sin(k_l X'') \\ \cos(k_l X') \cos(k_l X'') \end{pmatrix}$$

$$\times \frac{\begin{vmatrix} G^{(V)}(Y'', Y'; E) & G^{(V)}(Y'', -Y_0; E) & G^{(V)}(Y'', Y_0; E) \\ G^{(V)}(-Y_0, Y'; E) & G^{(V)}(-Y_0, -Y_0; E) & G^{(V)}(-Y_0, Y_0; E) \\ G^{(V)}(Y_0, Y'; E) & G^{(V)}(Y_0, -Y_0; E) & G^{(V)}(Y_0, Y_0; E) \end{vmatrix}}{\begin{vmatrix} G^{(V)}(-Y_0, -Y_0; E) & G^{(V)}(-Y_0, Y_0; E) \\ G^{(V)}(Y_0, -Y_0; E) & G^{(V)}(Y_0, Y_0; E) \end{vmatrix}} \;, \tag{13.30}$$

with k_l chosen accordingly for odd/even states in X, and $G^{(V)}(E)$ for the unrestricted motion in Y is given by [353]

$$G^{(V)}(Y'', Y'; E) = \frac{m}{\hbar^2} \Gamma\left(\frac{1}{\hbar}\sqrt{-2mE} - ik_l + \frac{1}{2}\right) \Gamma\left(\frac{1}{\hbar}\sqrt{-2mE} + ik_l + \frac{1}{2}\right)$$
$$\times P_{ik_l-1/2}^{-\sqrt{-2mE}/\hbar}(\sin Y_>) P_{ik_l-1/2}^{-\sqrt{-2mE}/\hbar}(-\sin Y_<) \ . \tag{13.31}$$

Let us note that the factor $\left(\frac{1+\sin Y}{1-\sin Y}\right)^{ip/2}$ alone in the quantization conditions would give a semi-classical quantization of the system. Indeed, the asymptotic behaviour of the energy levels show this feature in general for all k. However, the semiclassical quantization takes into account only the quantum number p and not k, and thus gives actually rather bad results.

Let us first concentrate on the $(-,-)$-case, the others are similar. We have $A = \pi$ for $Y_0 = Y_M$ in all parity classes. Weyl's law [518] describes the mean number of energy levels up to a certain energy, i.e., $\bar{N}(E) = \{\#N(\text{levels with energy } E_N)|E < E_N\}$ such that

$$N(E) = \bar{N}(E) + N_{osc.}(E) \ , \tag{13.32}$$

and N_{osc} describes the oscillating of the step function about $\bar{N}(E)$. According to Baltes and Hilf [38] and Stewartson and Waechter [483] one obtains for a two dimensional quantum system with Dirichlet boundary conditions on all boundaries, e.g., following [36, 112]

$$\bar{N}(E) = \frac{AE}{4\pi} - \frac{\partial A}{4\pi}\sqrt{E} + \frac{1}{24}\sum_{corners}\left(\frac{\pi}{\alpha_r} - \frac{\alpha_r}{\pi}\right)$$
$$+ \frac{1}{12\pi}\iint_A K(\sigma)d\sigma - \frac{1}{24\pi}\oint_{\partial A}\kappa(s)ds + O\left(\frac{1}{\sqrt{E}}\right) \ . \tag{13.33}$$

Here α_r denotes the angle of the rth-corner, A the area of the system and ∂A the length of its boundary. K is the Gaussian curvature (here $K = -1$), $d^2\sigma$ the surface integral, and the boundary mean curvature κ is given by

$$\kappa(s) = -2t^a(s)g_{ab}(s)\frac{D}{Ds}n^b(s) \ . \tag{13.34}$$

$n(s)$ is the normal, $t(s)$ is the tangential vector along the boundary, and the covariant derivative D/Ds is given by

$$\frac{D}{Ds}n^a(s) = \frac{d}{ds}\frac{dn^a}{ds} + \Gamma_{bc}^a(s)\frac{dt^b}{ds}\frac{dn^c}{ds} \ , \tag{13.35}$$

(Γ_{bc}^a: Christoffel symbol). In the case that one considers, i.e., the shifted Laplacian $-\Delta - 1/4$, the E-independent terms in $\bar{N}(E)$ change into [36]

$$C_0 = \frac{1}{24}\sum_{corners}\left(\frac{\pi}{\alpha_r} - \frac{\alpha_r}{\pi}\right) + \frac{1}{48\pi}\iint_A K(\sigma)d\sigma - \frac{1}{96\pi}\oint_{\partial A}\kappa(s)ds \ . \tag{13.36}$$

This suffices for our purposes. Using the general formula for the length of a curve in a curved space

$$s = \int_{t_i}^{t_f} \sqrt{g_{11}\dot{x}^2 + g_{22}\dot{y}^2 + g_{12}\dot{x}\dot{y}} \, dt \tag{13.37}$$

we obtain for the length of the boundary of a quarter rectangle

$$\partial A = 2\ln\left(\frac{1 + \sin Y_0}{\cos Y_0}\right) + \frac{L_0}{2}\left(1 + \frac{1}{\cos Y_0}\right) . \tag{13.38}$$

Note that the vertical lines are geodesics and therefore their length can also be evaluated by the two-point *formula* in the strip

$$\cosh d(q_1, q_2) = \frac{\cosh(X'' - X')}{\cos Y' \cos Y''} - \tan Y' \tan Y'' . \tag{13.39}$$

We have $\alpha_r = \pi/2$ $(r = 1, 2, 3, 4)$ and therefore for the (X, Y) (odd-odd) states

$$\bar{N}(E)^{(-,-)} \simeq \frac{L_0 \tan Y_0}{8\pi} E - \frac{\sqrt{E}}{4\pi}\left[2\ln\left(\frac{1 + \sin Y_0}{\cos Y_0}\right) + \frac{L_0}{2}\left(1 + \frac{1}{\cos Y_0}\right)\right]$$
$$+ \frac{1}{4} - \frac{1}{48} + \frac{L_0 \tan Y_0}{192 \cos^2 Y_0} , \tag{13.40}$$

where we have allowed some arbitrary Y_0; for (X, Y) (even, even) states we have

$$\bar{N}(E)^{(+,+)} \simeq \frac{L_0 \tan Y_0}{8\pi} E - \frac{L_0}{8\pi}\left(\frac{1}{\cos Y_0} - 1\right)\sqrt{E} - \frac{1}{48} + \frac{L_0 \tan Y_0}{192 \cos^2 Y_0} . \tag{13.41}$$

Similar results hold for the (even, odd) and (odd, even) states. For the entire rectangle we have

$$\bar{N}(E) \simeq \frac{L_0 \tan Y_0}{2\pi} E - \frac{\sqrt{E}}{4\pi}\left[4\ln\left(\frac{1 + \sin Y_0}{\cos Y_0}\right) + \frac{2L_0}{\cos Y_0}\right] + \frac{1}{4} - \frac{1}{12} + \frac{L_0 \tan Y_0}{48 \cos^2 Y_0} . \tag{13.42}$$

Hence the odd parities in X and Y, respectively, give Dirichlet boundary conditions on the lines $Y = 0$ and $X = 0$, respectively, and even parities X and Y, respectively, give Neumann boundary conditions on the lines $Y = 0$ and $X = 0$. The Schrödinger equation restricted to our particular rectangle is not invariant with respect to translations in the Y-direction (it is in the X-direction). The Schrödinger operator is only invariant with respect to elements of a Fuchsian group, i.e., the generators of hyperbolic polygons tesselating the hyperbolic plane.

The check with Weyl's law confirms the data down to the lowest eigenvalues. The spectral staircase (or step function) $N(E)$ (solid line) and Weyl's law (dotted line) are hardly distinguishable from each other.

Let us introduce the fluctuations $(N_0(E) = \frac{1}{2}\lim_{\epsilon \to 0}[N(E + \epsilon) + N(E - \epsilon)])$

$$\delta_n = N_0(E_n) - \bar{N}(E_n) - 0.5 . \tag{13.43}$$

A first analysis gives the level-spacing distribution $P(S)$ of spacing between neighboring levels. Classically integrable systems belong to the universality class of uncorrelated level sequences. $P(S)$ is calculated for the scaled energy spectrum, which has a mean level spacing of one ($= \hbar$). One applies Weyl's law onto the calculated energy levels and obtains the normalized levels E'_n by

$$E'_n = N(E_n) \qquad (13.44)$$

and quantities for the scaled spectrum are denoted by a prime in the following.

In integrable systems one typically has level clustering, which is expressed by $P(S) \to 1$, as $S \to 0$, whereas chaotic systems show level repulsion, i.e. $P(S) \to 0$, as $S \to 0$. The functional form of the nearest neighbour level spacing $P(S)$ for classically integrable systems is assumed (but not proven) to behave like

$$P(S) = e^{-S} , \qquad (13.45)$$

which is a Poisson distribution; the result from random matrix theory, a theory developed in the phenomenology of energy levels for nuclei, for the level spacing distribution of a GOE-ensemble is approximated by a Wigner distribution

$$P(S) = \frac{\pi}{2} S e^{-\pi S^2 / 4} , \qquad (13.46)$$

and the corresponding level spacing distribution of a GUE-ensemble is given by

$$P(S) = \frac{32}{\pi^2} S^2 e^{-4S^2 / \pi} . \qquad (13.47)$$

Figure 13.4 shows the analysis of our system and the consistency with a Poisson distribution (dotted line) is evident. The corresponding level spacing distributions for GUE (Gaussian Unitarian Ensemble) is denoted by the dashed line, and for GOE (Gaussian Orthogonal Ensemble) by the dashed-dotted line. Clearly GUE and GOE distributions are excluded.

A similar feature was first observed by Casati, Chirikov and Guarneri [100] for the flat rectangular billiard. With the chosen ΔS, the actual level distribution shows nevertheless fluctuations about the Poisson distribution. Making ΔS smaller would increase these fluctuations. The calculated χ^2-test gives for $P(S)$ $\chi^2 = 77.9$, $\chi^2 = 118$ and $\chi^2 = 290$ for $\Delta S = 0.25, 0.20$ and 0.10, respectively, with confidence levels of $\alpha = O(10^{-7})$ and smaller, with respect to a Poisson distributed sequence which is negligible. This feature shows that the sequence of energy-levels is not completely random.

The level spacing distribution $P(S)$ is a short range statistics. Another important tool in the analysis of the spectrum is the number variance $\Sigma^2(L)$ and the spectral rigidity $\Delta_3(L)$ [143], respectively. $\Sigma^2(L)$ is defined as the local variance of the number $n(E', L)$ of scaled energy levels in the interval from $E' - L/2$ to $E' + L/2$. It has the form

$$\Sigma^2(L) = \left\langle n(E', L) - L^2 \right\rangle . \qquad (13.48)$$

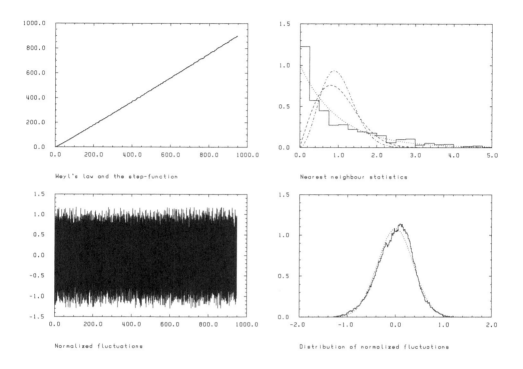

Figure 13.4: Analysis of the entire square.

The Δ_3 statistics of Metha and Dyson [143] is defined as the local average of the mean square deviation of the staircase $N'(E')$ from the best fitting straight line over an energy range corresponding to L mean level spacings, namely

$$\Delta_3(L) = \left\langle \min_{(a,b)} \frac{1}{L} \int_{-L/2}^{L/2} d\epsilon [N'(E' + \epsilon) - a - b\epsilon]^2 \right\rangle \ . \qquad (13.49)$$

Whenever $L \ll 1$, the very fact that $N(E)$ is a staircase leads in this limit to [56]

$$\Sigma^2(S) = L, \qquad \Delta_3(L) = \frac{L}{15} \ , \qquad (13.50)$$

and both statistics are linear and show the so-called Poisson behaviour, i.e., in the case of a genuine Poisson distributed level sequence, these results are exact.

The spectral rigidity gives no information about the very finest scales corresponding to the spacings between neighboring levels, whether they are Poisson distributed or not. Its usefulness lies in the way it describes level sequences larger than the inner energy scale ($L = 1$) of a system.

Berry [56, 58] has developed a semiclassical analysis of the spectral rigidity and has shown that one must discriminate between at least three universality classes of rigidity, depending on whether one deals with classically integrable systems or classically chaotic systems. The first universality class occurs for classically integrable systems. Here the Poisson $L/15$-form for the spectral rigidity extends from $L = 0$ to L_{Max}. L_{Max} corresponds to an outer energy scale $\propto 1/T_{Min}$ (the inner energy scale corresponds to $L \simeq 1$), where T_{Min} is the period of the shortest classical orbit and $L_{Max} \propto \hbar^{N-1}$, and $\propto 1/\hbar$ for $N = 2$ (i.e., a two-dimensional system). The properties of the rigidity are determined by the contributions of the very short orbits. These orbits have a non-universal behaviour, which differ from system to system. A consequence of this fact is that there is a shortest orbit, thus the spectral rigidity saturates and approaches a non-universal constant Δ_∞ as $L \to \infty$ (and the same line of reasoning is true for the number variance Σ^2) [26, 56, 467].

We can explain our results in terms of the periodic orbit analysis of Berry [56]. We do not go here into the details of the analysis for the four symmetry classes (odd–odd, odd–even), (even–odd) and (even–even), respectively, of our model. This has been done for the nearest neighbour statistics, Weyl's law, number variance, and the spectral rigidity in [213]. I give only a short summary for the entire rectangle, including some new aspects. Of course, L_0 is the length on a closed periodic orbit which corresponds to motion along the X-axis. L_1 with $L_1 = 2\ln(\frac{1+\sin Y_0}{\cos Y_0}) = 2.936, 147, 388 \ldots$ is the length of an orbit along the Y axis and is slightly shorter than the former one. However, these two shortest closed orbits are almost equal such that we can apply Berry's analysis of the spectral rigidity in its simplest way.

For a classically integrable system the spectral rigidity approaches for $L > L_{Max}$ a value Δ_∞ which is given by $\Delta_\infty = const.\sqrt{\mathcal{E}}$, where $\mathcal{E}$ denotes the scaled energy range over which the spectral rigidity has been evaluated. In the case of a flat square with sides $a = b = 1$, Δ_∞ can be evaluated as

$$\Delta_\infty \simeq \pi^{-\frac{5}{2}}\left[\zeta\left(\frac{3}{2}\right)\beta\left(\frac{3}{2}\right) - \frac{1}{2}\zeta(3)\right]\sqrt{\mathcal{E}} \simeq 0.0947\sqrt{\mathcal{E}} \ . \tag{13.51}$$

Similarly, $L_{Max} \simeq \sqrt{\pi\mathcal{E}}$. According to, e.g., [26, 467] this gives a prediction for the number variance $\Sigma^2(L)$ for $L > L_{Max}$, i.e., $\Sigma_\infty = 2\Delta_\infty + O(1/L^3)$ (fluctuations neglected, this behaviour explains why Σ^2 saturates much faster than Δ_3). Assuming now that our square can be further approximated by a flat square with sides $a = L_0 \simeq b = L_1$, we obtain

$$L_{Max} \simeq 55 \ , \qquad \Sigma_\infty^{theory} \simeq 5.84 \ , \qquad \Delta_\infty^{theory} \simeq 2.92 \ . \tag{13.52}$$

The numerical results are (Σ_∞^{num} is the mean value for $L > 20$)

$$\Sigma_\infty^{num} \simeq 6.66 \ , \qquad \Delta_\infty^{num} \simeq 3.118 \ , \tag{13.53}$$

and the fit parameters for Δ_∞^{num} are $a = 4.650$ and $b = 170.24$, respectively. Here $\Sigma_\infty^{num} - 2\Delta_\infty^{num} \simeq 0.43$.

In table 13.1 I have listed the comparison with the numerical results for the spectral rigidity and the theoretical predictions à la Berry for the entire rectangle. A maximum of 900 levels have been taken into account with $E \leq 950$. In the first column I have indicated the number of energy levels $N \simeq \mathcal{E}$ taken into account for the calculation of the spectral rigidity, in the second column L_{Max} according to $L_{Max} \simeq \sqrt{\pi \mathcal{E}}$, in the third Δ_∞ according to (13.51), in the fourth the numerical result for the calculated maximum $\Delta_{Max} = \Delta_3(950)$ of the spectral rigidity, in the fifth the resulting quotient of Δ_{Max} and $\sqrt{\mathcal{E}}$, and in the sixth and seventh the analogue of the former two columns for the corrected numerical Δ_∞. Our numerical results are in agreement with Berry's theory.

Table 13.1: The Spectral Rigidity and Berry's Theory.

N	L_{Max}	Δ_∞^{theor}	Δ_{Max}	$\Delta_{Max}/\sqrt{\mathcal{E}}$	Δ_∞	$\Delta_\infty/\sqrt{\mathcal{E}}$
100	17.1	0.947	0.825	0.083	0.825	0.083
200	25.3	1.350	1.50	0.105	1.49	0.105
300	30.6	1.634	1.82	0.105	1.85	0.107
400	35.3	1.885	2.13	0.106	2.15	0.108
500	40.0	2.118	2.29	0.102	2.32	0.104
600	43.4	2.318	2.28	0.093	2.35	0.096
700	46.9	2.503	2.25	0.085	2.35	0.089
800	50.1	2.676	2.72	0.096	2.80	0.090
900	53.2	2.840	2.85	0.095	3.10	0.101

The mean value for $\Delta_{Max}/\sqrt{\mathcal{E}}$ is given by 0.097 ± 0.003, whereas for $\Delta_\infty/\sqrt{\mathcal{E}}$ by 0.099 ± 0.003, in excellent agreement with the theoretical value 0.0947. Note that Δ_∞^{num} gives somewhat larger values, which can be explained by the fact that our rectangle in the hyperbolic geometry is a distorted geometrical object with only almost equal sides ($L_0 \simeq L_1$, see above), whereas the semiclassical analysis for the spectral rigidity of Berry deals with a flat square.

The present billiard system does not only give reasonable results concerning Berry's semiclassical analysis, but it also serves to check a universal behaviour which appears in the study of quantum mechanical billiard systems, may they be integrable or chaotic. It concerns the statistical properties of the energy fluctuations. In 1984 it was conjectured by Bohigas et al. [63] that these fluctuations of chaotic systems are described by the universal laws of random matrix theory [401]. However, it is known today that the predictions of random matrix theory agree only for short- and medium range correlations of the quantal spectra, but fail completely for long-range correlations.

This was first analyzed by Berry [56] using the semiclassical trace formula. Moreover, it was found recently that there exists a very special class of chaotic systems, showing *arithmetical chaos*, e.g., [69, 480] and references therein, which violate universality in energy level statistics even in the short-range regime. This was investigated in several classically chaotic systems in hyperbolic geometry, i.e., in the hyperbolic octagon [15]–[18, 25]–[32] and in Artin's billiard [17, 22, 71, 395].

Returning to our billiard in the rectangle in the hyperbolic plane we can, of course, not check the conjectures as far as their statements concerning chaotic billiard systems are concerned. What can be checked is the statistical behaviour of the normalized fluctuations $\alpha_n = \delta_n / E_n^{1/4}$.

In order to obtain due to the limited number of eigenvalues a useful statistic, I have compared the fluctuations of the step-functions with respect to Weyl's law not at the location of the eigenvalues themselves, but rather at randomly chosen values. This has the advantage that arbitrarily many fluctuations can be determined. In figure 13.4 I have displayed the fluctuations obtained in this way for 50,000 randomly chosen values. It is obvious that in the energy range the fluctuations are bounded with $|\alpha_n| . 1$.

Using these fluctuations I have determined the corresponding limit distribution which is shown in figure 13.5. The distribution has a variance of $\sigma = 0.35\ldots$ and a skewness of $\eta = -0.2\ldots$, and is therefore definitely skew and non-Gaussian. The most striking feature is that the slope on the right hand side of the distribution is definitely steeper than on the left hand side and has a long tail on the left hand side. Although it may be argued that this may be an artefact of the limited number of eigenvalues, it must be noted that this feature is sensitive neither with respect to the number of eigenvalues nor with the number of randomly chosen values if the entire energy range taken into account. In addition, a Kolmogorov-Smirnov test gives a negligible confidence level of $< 10^{-6}$ (10, 000 supporting points) in comparison to a Gaussian distribution [34] (usually a value of 0.05 is interpreted as an acceptance). Therefore we may say that the skewness of our distribution is indeed a genuine effect of our billiard system.

Let us finally discuss another aspect of the periodic orbit analysis which may be called "Inverse Quantum Chaology". Instead of having as input the lengths $\{l_n\}$ of periodic orbits in (13.10) and as output semiclassical energy levels $\{E_n\}$, one makes a numerical investigation the other way round and has as input the energy levels and as output the semiclassical lengths spectrum. Although this theory has been developed for classically chaotic systems, I try to apply it nevertheless to this integrable system. To do this by means of the periodic orbit formula one considers as the testfunction the cosine modulated heat-kernel function $h(p, L) = \cos(pL)e^{-p^2 t}$ with a suitable chosen small t, e.g., [30]

$$\sum_{n=1}^{\infty} \cos(p_n L)e^{-p_n^2 t} \simeq \int_0^{\infty} dp\, p \cos(pL)e^{-p^2 t}\bar{d}(E) + C_0$$

$$+ \frac{1}{8\sqrt{\pi t}} \sum_{\gamma \in \Gamma} \sum_{k=1}^{\infty} \frac{l_\gamma}{\sinh(ku_\gamma/2)} \left(e^{(L-kl_\gamma)^2/4t} - e^{(L+kl_\gamma)^2/4t} \right) , \qquad (13.54)$$

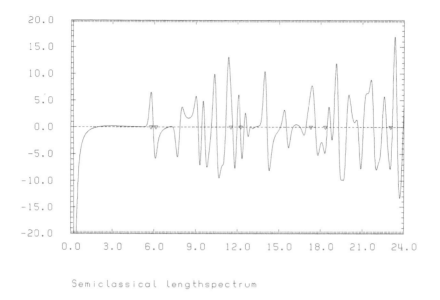

Semiclassical lengthspectrum

Figure 13.5: Semiclassical length spectrum.

which is absolutely convergent for any $t > 0$. The periodic orbit term on the right hand side shows that at fixed $t > 0$ as a function of L Gaussian peaks of width $\Delta L \sim 2\sqrt{2t}$ appear at the length l_n of the classical periodic orbits. We apply (13.54) to our integrable billiard system. For this billiard system no systematic search for periodic orbits has been done yet. Taking $t = 10/E_{max} \simeq 0.01$ gives a resolution of $\Delta L \simeq 0.3$. In figure 13.5 I have on the right hand side of (13.54) taken 900 energy levels into account. Multiples of the lengths L_1 and L_0 are marked by triangles. We see that at the locations of the triangles Gaussian peaks are clearly visible; however, the resolution cannot be considered as satisfactory which is due to the relatively small number of eigenvalues.

13.3 Other Integrable Billiards in Two and Three Dimensions

Following [230] I include some more integrable billiard systems which can be studied without much difficulty and serve as simple models to investigate the most important features of integrable billiards.

13.3.1   Flat Billiards

The Square.

The simplest systems are *rectangular billiards* in flat space (Euclidean rectangles) [100], where the energy levels are simply given by

$$E_{n_1,n_2} = \pi^2 \left(\frac{n_1^2}{a^2} + \frac{n_2^2}{b^2} \right) \ , \tag{13.55}$$

with n_1, n_2 natural numbers and a, b the sides of the rectangle, and I have taken natural units $\hbar = 2m = 1$. Weyl's law has the form [38]

$$\bar{N}(E) \simeq \frac{ab}{4\pi} E - \frac{a+b}{2\pi} \sqrt{E} + \frac{1}{4} \ . \tag{13.56}$$

Usually one uses $a = 1$, with b varying. It can be proven that the limit distribution $P(\alpha)$ is non-Gaussian [267, 357]. The system has been investigated by Casati et al. [100] and by Berry [56].

The Sphere.

The second system in flat space is the *circular billiard* with radius R and area $A = \pi R^2$. The energy levels are given by

$$E_{l,n} = j_{l,n}^2 / R^2 \ , \qquad l = 0, 1, \ldots, \qquad n = 1, 2, \ldots, \tag{13.57}$$

where $x_n = j_{l,m}$ is the n-th zero of the Bessel function $J_l(x)$. Weyl's law has the form [38]

$$\bar{N}(E) \simeq \frac{R^2}{4} E - \frac{R}{2} \sqrt{E} + \frac{1}{6} \ . \tag{13.58}$$

The circle billiard has been studied by, e.g., Kim [346] it has been shown that the distribution of the energy fluctuations is non-Gaussian and skew.

The Parallelepiped.

The simplest three-dimensional system is a *rectangular parallelepiped* with sides a, b, c in flat space, where the energy levels are simply given by

$$E_{n_1,n_2} = \pi^2 \left(\frac{n_1^2}{a^2} + \frac{n_2^2}{b^2} + \frac{n_3^2}{c^2} \right) \ , \tag{13.59}$$

with n_1, n_2, n_3 natural numbers. Weyl's law has the form [38]

$$\bar{N}(E) \simeq \frac{abc}{6\pi^2} E^{3/2} - \frac{ab+bc+ac}{8\pi} E + \frac{a+b+c}{2\pi} \sqrt{E} - \frac{1}{8} \ . \tag{13.60}$$

13.3.2   Hyperbolic Billiards

In this subsection I present some more integrable billiard systems in hyperbolic geometry. The hyperbolic rectangle have been discussed in some length before, but furthermore it is possible to investigate an hyperbolic circle and an hyperbolic triangle [230].

The Hyperbolic Circle.

The *hyperbolic circle* is defined as a circle in the Poincaré disc $D = \{(r, \varphi)|r < 1, \varphi \in [0, 2\pi)\}$ endowed with the corresponding hyperbolic geometry $ds^2 = (dr^2 + r^2 d\varphi^2)/(1 - r^2)^2$ [202, 243]. Choosing $r = 1/\sqrt{2}$ yields $A = 4\pi$, and the quantization condition is given by the transcendental equation (assuming Dirichlet boundary conditions at $r = 1/\sqrt{2}$, i.e., $\cosh R = 3$, and $R = \text{arcosh}(3) = 1.762, 747, 174, \ldots$)

$$\mathcal{P}_{-1/2 + ip_n}^{-l}(3) = 0 , \qquad l = 0, 1, \ldots , \qquad n = 1, 2 \ldots , \qquad (13.61)$$

where $\mathcal{P}_\nu^\mu(z)$ are Legendre functions. The quantization condition follows from considering the path integral representation of the two-dimensional hyperboloid in the domain $\tau < R$ which in (pseudo-) spherical coordinates has the form (note the boundary conditions in comparison to the free motion!)

$$K(\tau'', \tau', \varphi'', \varphi'; T) = \int_{\tau(t')=\tau'}^{\tau(t'')=\tau''} \mathcal{D}_{\tau<R}\tau(t) \sinh \tau \int_{\varphi(t')=\varphi'}^{\varphi(t'')=\varphi''} \mathcal{D}\varphi(t)$$

$$\times \exp\left\{\frac{i}{\hbar} \int_{t'}^{t''} \left[\frac{m}{2}(\dot{\tau}^2 + \sinh^2 \tau \dot{\varphi}^2) - \frac{\hbar^2}{8m}\left(1 - \frac{1}{\sinh^2 \tau}\right)\right] dt\right\}$$

$$= (\sinh \tau' \sinh \tau'')^{-1/2} \sum_{\nu \in \mathbb{Z}} \frac{e^{i\nu(\varphi'' - \varphi')}}{2\pi}$$

$$\times \int_{\tau(t')=\tau'}^{\tau(t'')=\tau''} \mathcal{D}_{\tau<R}\tau(t) \exp\left\{\frac{i}{\hbar} \int_{t'}^{t''} \left[\frac{m}{2}\dot{\tau}^2 - \frac{\hbar^2}{2m}\frac{\nu^2 - 1/4}{\sinh^2 \tau}\right] dt\right\} . \qquad (13.62)$$

The remaining τ-path integration gives for the free (unrestricted) motion for the corresponding Green function ($E = \frac{\hbar^2}{2m}(p_E^2 + 1/4)$)

$$G_\nu(\tau'', \tau'; E) = \frac{2m}{\hbar^2}\sqrt{\sinh \tau' \sinh \tau''}\, e^{-i\pi\nu}\mathcal{P}_{-1/2+ip_E}^{-\nu}(\cosh \tau_<)\mathcal{Q}_{-1/2+ip_E}^\nu(\cosh \tau_>) ,$$

$$(13.63)$$

where $\tau_{<,>}$ denotes the smaller, respectively larger of τ', τ''. Denoting by $A(\tau) = \mathcal{Q}_{-1/2+ip_E}(\cosh \tau)$, $B(\tau) = \mathcal{P}_{-1/2+ip_E}(\cosh \tau)$, we obtain according to [216, 217] the Green function $G_{(\nu, \tau<R)}(E)$ for the restricted system with $\tau < R$

$$G_{(\nu,\tau<R)}(\tau'',\tau';E)$$
$$= \frac{2m}{\hbar^2}e^{-i\pi\nu}\sqrt{\sinh\tau'\sinh\tau''}\left\{A_E(\tau_>)B_E(\tau_<) - \frac{A_E(R)}{B_E(R)}B_E(\tau')B_E(\tau'')\right\} . \quad (13.64)$$

For $\nu \geq 1$ the energy levels $E_n = p_n^2 + 1/4$ are two-fold degenerate, and Weyl's law has the form

$$\bar{N}(E) \simeq E - \sqrt{2E} + \frac{\sqrt{2}-1}{3} . \quad (13.65)$$

For the numerical investigation of the zeros of the transcendental equation (13.61) one uses the representation [187, p.1010]

$$\mathcal{P}_\nu^\mu(z) = \frac{\Gamma(-\nu-1/2)}{2^{\nu+1}\sqrt{\pi}\,\Gamma(-\nu-\mu)}(z^2-1)^{-(\nu+1)/2}{}_2F_1\left(\frac{\nu-\mu+1}{2},\frac{\nu+\mu+1}{2};\nu+\frac{3}{2};\frac{1}{1-z^2}\right)$$
$$+ \frac{2^\nu\Gamma(\nu+1/2)}{\sqrt{\pi}\,\Gamma(\nu-\mu+1)}(z^2-1)^{\nu/2}{}_2F_1\left(\frac{\mu-\nu}{2},\frac{\nu-\mu}{2};\frac{1}{2}-\nu;\frac{1}{1-z^2}\right), \quad (13.66)$$

and in addition to avoid numerical overflow effects the multiplication formula of the Γ-function [187, p.937]

$$\Gamma(nz) = (2\pi)^{(1-n)/2}n^{nz-1/2}\prod_{k=0}^{n-1}\Gamma\left(z+\frac{k}{n}\right) . \quad (13.67)$$

The Hyperbolic Triangle.

The last two-dimensional system in hyperbolic geometry is a *hyperbolic triangle* [113] in the Poincaré upper half-plane $\mathcal{H} = \{(x,y)|y > 0, x \in \mathbb{R}\}$, endowed with the corresponding geometry $ds^2 = (dx^2 + dy^2)/y^2$ [202, 243], defined by $y > 1, |x| < 1/2$ [113]. The hyperbolic triangle is a non-compact domain with area $A = 1/2$. The energy levels $E_n = p_n^2 + 1/4$ are implicitly defined by

$$K_{ip_n}(l\pi) , \qquad l = 2,4,\dots, \qquad n = 1,2,\dots, \quad (13.68)$$

where $K_\nu(x)$ is a modified Bessel function. Weyl's law has the form (assuming Dirichlet boundary conditions at $y = 1$ and $x = \pm 1/2$) [113]

$$\bar{N}(E) \simeq \frac{E}{8\pi} - \frac{\ln E\sqrt{E}}{4\pi} + \frac{3/2 - 2\ln 2}{4\pi}\sqrt{E} + \frac{\pi}{2} . \quad (13.69)$$

The $\ln E\sqrt{E}$-term is typically for non-compact billiards. The quantization condition can be derived form the path integral representation [113]

$$K(y'',y',x'',x';T) = \int_{y(t')=y'}^{y(t'')=y''}\frac{\mathcal{D}_{(y>1)}y(t)}{y^2}\int_{x(t')=x'}^{x(t'')=x''}\mathcal{D}x(t)\exp\left(\frac{im}{2\hbar}\int_{t'}^{t''}\frac{\dot{y}^2+\dot{x}^2}{y^2}dt\right)$$

$$= \sqrt{y'y''} \sum_{l=1,3,\ldots} \sin\left[l\pi(x - \tfrac{1}{2})\right]$$

$$\times \int_{y(t')=y'}^{y(t'')=y''} \frac{\mathcal{D}_{(y>1)}y(t)}{y} \exp\left[\frac{i}{\hbar} \int_{t'}^{t''} \left(\frac{\dot{y}^2}{y^2} - \frac{\hbar^2 l^2 \pi^2}{2my^2}\right) dt - \frac{i\hbar T}{4m}\right] . \tag{13.70}$$

The remaining y-path integration gives for the free unrestricted motion the Green function

$$G(y'', y'; E) = \frac{2m}{\hbar^2} I_{\sqrt{-2mE/\hbar^2+1/4}}(l\pi y_<) K_{\sqrt{-2mE/\hbar^2+1/4}}(l\pi y_>) , \tag{13.71}$$

therefore with $A_E(y) = I_{\sqrt{-2mE/\hbar^2+1/4}}(l\pi y), B_E(y) = K_{\sqrt{-2mE/\hbar^2+1/4}}(l\pi y)$ for the Green's function in $y > 1$

$$G_{(y>1)}(y'', y'; E) = \frac{2m}{\hbar^2}\left\{A_E(y_>)B_E(y_<) - \frac{A_E(1)}{B_E(1)}B_E(y')B_E(y'')\right\} . \tag{13.72}$$

This hyperbolic triangle billiard has been investigated by Graham et al. [113] in some detail, i.e., energy level statistics, spectral rigidity and the transition from a classical integrable system, i.e., the hyperbolic triangle billiard, to a classically chaotic system, i.e., motion in the modular domain. However, the determination of $P(\alpha)$ has not been done.

The Hyperbolic Sphere.

The *hyperbolic sphere* is defined as a sphere in the Poincaré sphere $B = \{(r, \vartheta, \varphi)|r < 1, \vartheta \in [0, \pi], \varphi \in [0, 2\pi)\}$ endowed with the corresponding hyperbolic geometry $ds^2 = dr^2 + \sinh^2 r(d\vartheta^2 + \sin^2 \vartheta d\varphi^2)$ [202, 243]. Choosing $r = 1/\sqrt{2}$ yields $V = 4\pi$, and the quantization condition is given by the transcendental equation (assuming Dirichlet boundary conditions at $r = 1/\sqrt{2}$)

$$\mathcal{P}^{-l-1/2}_{-1/2+ip_n}(3) = 0 , \qquad l = 0, 1, \ldots, \qquad n = 1, 2\ldots . \tag{13.73}$$

This quantization condition follows from the path integral representation in (pseudo-) spherical coordinates

$$\int_{\tau(t')=\tau'}^{\tau(t'')=\tau''} \mathcal{D}_{\tau<R}\tau(t) \sinh^2 \tau \int_{\vartheta(t')=\vartheta'}^{\vartheta(t'')=\vartheta''} \mathcal{D}\vartheta(t) \sin\vartheta \int_{\varphi(t')=\varphi'}^{\varphi(t'')=\varphi''} \mathcal{D}\varphi(t)$$

$$\times \exp\left\{\frac{i}{\hbar} \int_{t'}^{t''} \left[\frac{m}{2}\left(\dot\tau^2 + \sinh^2\tau(\dot\vartheta^2 + \sin^2\vartheta\dot\varphi^2)\right)\right.\right.$$

$$\left.\left. -\frac{\hbar^2}{8m}\left(4 - \frac{1}{\sinh^2\tau}\left(1 + \frac{1}{\sin^2\vartheta}\right)\right)\right] dt\right\}$$

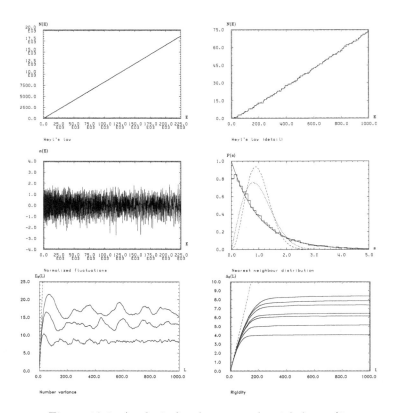

Figure 13.6: Analysis for the rectangle with $b = \pi/3$.

$$= (\sinh \tau' \sinh \tau'')^{-1} \sum_{l=0}^{\infty} \sum_{m=-l}^{l} Y_l^{m*}(\vartheta', \varphi') Y_l^m(\vartheta'', \varphi'')$$

$$\times \int_{\tau(t')=\tau'}^{\tau(t'')=\tau''} \mathcal{D}_{\tau<R}\tau(t) \exp\left\{ \frac{i}{\hbar} \int_{t'}^{t''} \left[\frac{m}{2}\dot{\tau}^2 - \frac{\hbar^2}{2m} \frac{l(l+1)}{\sinh^2 \tau} \right] dt - \frac{i\hbar T}{2m} \right\} . \quad (13.74)$$

Here the $Y_l^m(\vartheta, \varphi)$ are the spherical harmonics on the sphere. The quantization condition then follows from the corresponding Green function $G_{(\tau<R)}(E)$ by replacing in (13.64) $\nu \to l + 1/2$. For $l > 0$ the energy levels $E_n = p_n^2 + 1$ are $(2l + 1)$-fold degenerate, and Weyl's law has the form

$$\bar{N}(E) \simeq \frac{6\sqrt{2} - \mathrm{arcosh}(3)}{3\pi} E^{3/2} - 2E + \frac{4\sqrt{2} - 1}{3\pi} E^{1/2} - 1 . \quad (13.75)$$

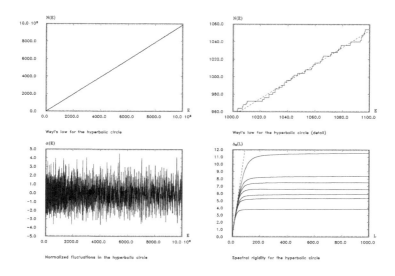

Figure 13.7: Analysis for the hyperbolic circle.

13.4 Numerical Investigation of Integrable Billiard Systems

13.4.1 Two-Dimensional Systems

In figure 13.6 I have displayed for completeness the result for the rectangle billiard with $b = \pi/3$. We see that every feature is nicely confirmed, i.e., Weyl's law, the distribution of the normalized fluctuations, the number variance and the spectral rigidity, where $\Sigma_2(L)$ (average value) and $\Delta_3(L)$ saturate and $\Sigma_2(L) \simeq 2\Delta_3(L)$, as $L \to \infty$. For more details, c.f. [56, 100]. In table 13.2 I have also listed for comparison the effect of different values of b, i.e., $A = b$. For the investigation of the circle-billiard, which scales, see [346].

In figure 13.7 I have displayed for the hyperbolic circle billiard the nice fitting of the spectral staircase (solid line) with Weyl's law (dashed line, c.f., the enlargement), the normalized fluctuations α_n, and the spectral rigidity $\Delta_3(L)$ for the number of energy levels of $\# = 1000, 2000, 2500, 3000, 4000, 5000$ and 10056, respectively, from which the number c_{Δ_∞} can be determined. Note that I have also incorporated the spectral rigidity for a Poissonian distribution with twofold degeneration (dashed line) $\Delta_3(L) = 2L/15$. All typical features of a classical integrable system are found. I omit the level-spacing statistics due to the degeneracy of the energy levels. Needless to say, $\Sigma_2(L) \simeq 2\Delta_3(L)$, as $L \to \infty$ (also omitted).

Table 13.2: Comparison of the Two-Dimensional Billiard Systems.

Billiard System	Area	# levels	E_{Max}	L_{Max}	c_{Δ_∞}	σ	κ
Euclidean Rectangle	4π	1 500	1 584	69	0.54	0.99	-0.91
Euclidean Rectangle	2π	2 500	5 178	89	0.200	1.00	-0.60
Euclidean Rectangle	π	6 100	24 810	138	0.094	1.01	-0.77
Euclidean Rectangle	$\pi/2$	13 000	105 030	202	0.065	1.01	-0.63
Euclidean Rectangle	$\pi/3$	19 000	229 890	244	0.0610	1.03	-0.62
Euclidean Square	1	19 000	240 710	244	0.113	1.01	-0.58
Euclidean Rectangle	$1/20$	1 000	271 770	56	1.14	0.97	-0.83
Euclidean Circle	π	12 488	50 996	198	0.13	1.00	$+0.20$
Hyperbolic Rectangle	4π	943	1 011	56	0.102	1.02	-0.75
Hyperbolic Rectangle	2π	466	1 011	39	0.093	1.02	-1.03
Hyperbolic Rectangle	π	230	1 000	27	0.079	0.98	-0.30
Hyperbolic Circle	4π	10 056	10 201	178	0.120	1.00	$+0.40$
Hyperbolic Circle	2π	750	1 566	49	0.14	0.99	$+0.40$
Hyperbolic Triangle	$1/2$	1 100	31 300	59	0.09	1.01	-0.72

In figure 13.8a (top) I display the distribution of the normalized fluctuations for the cases of the Euclidean rectangle with area $A = 1, \pi, 4\pi$ and the Euclidean circle with $R = 1$ (solid, dashed, dotted, dashed-dotted). The distributions have mean zero, standard deviation one (within the error margins), and are all skew, c.f. the value of κ, table 13.2; in particular the Euclidean square with $a = 4\pi$ stands out with an extremely fast decay for $\alpha \to \infty$, as is expected from the $e^{-c_2\alpha^4}$ behaviour, as $\alpha \to +\infty$.

In figure 13.8b (bottom) I display the distribution of the normalized fluctuations for the cases of the hyperbolic rectangles with area $A = 4\pi, \pi$, the hyperbolic triangle with $A = 1/2$, and the hyperbolic circle with $A = 4\pi$ (dashed, solid and dotted, dashed-dotted line), respectively. Note that the distribution $P(\alpha_n)$ in hyperbolic geometry seems to be more regular, at least in the investigated systems.

The numerical results for the two-dimensional systems are summarized in table 13.2. # denotes the number of energy levels taken into account with E_{Max} the maximal energy. We observe nice confirmation of the semiclassical theory within an error margin of 3% in a wide energy range for billiard systems of different shape and geometry. The skewness κ can be considered as significant by means of a Kolmogorov-Smirnov test in comparison with a Gaussian distribution. Note that in all cases we observe a typical slow increase on the left side and a fast decay on the right side in the distributions in comparison to a Gaussian (which is omitted for graphical clarity), and the circle billiards have $\kappa > 0$.

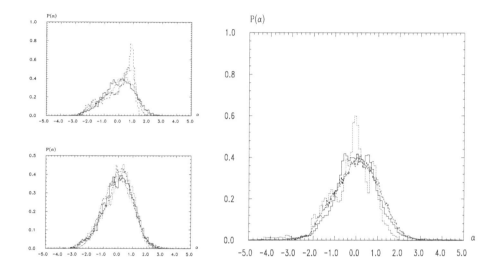

Figure 13.8: Left. Distribution of deviations in the Euclidean plane (a, top) and in the hyperbolic plane (b, bottom). Right: Distribution of deviations in three dimensions.

13.4.2 Three-Dimensional Systems

In figure 13.9 I have displayed the results for the $\pi/3$-parallelepiped with $a = \pi/3, b = a^2, c = a^3$. Due to the irrational relation of the lengths of the sides we have no energy level degeneracy, and the nearest neighbour statistics turns out to be Poissonian indeed in comparison to GOE (dotted) and GUE (dashed). Also, Weyl's law (and detail down to ground state) shows nice confirmation of the semiclassical theory. The fluctuations grow according to $\sqrt{E}$, which is confirmed by the shape of the normalized fluctuations. The number variance $\Sigma_2(L)$, which is displayed for $\# = 2500, \# =$

Table 13.3: Comparison of the Three-Dimensional Billiard Systems

Billiard System	Volume	# levels	E_{Max}	L_{Max}	c_{Δ_∞}	σ	κ
Euclidean Cube	1	13 400	9 021	205	0.014	1.08	-0.5
Euclidean Epiped	$(\pi/3)^6$	11 500	7 443	190	0.0022	1.01	-0.6
Euclidean Epiped	$(\pi/2)^6$	3 000	1 400	97	0.0040	0.93	-0.8
Euclidean Epiped	π^6	500	122	40	0.045	1.02	-1.2
Hyperbolic sphere	4π	27 780	1 156	285	0.17	0.97	$+0.2$

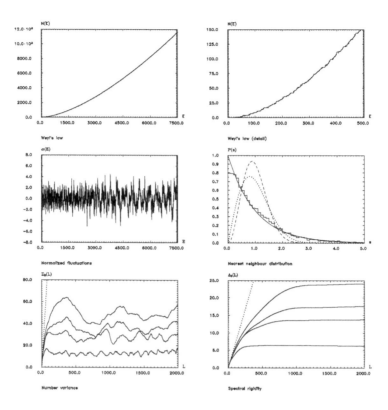

Figure 13.9: Analysis for the parallelepiped with $a = \pi/3$.

$5000 \# = 10000$ and $\# = 11500$, and the spectral rigidity $\Delta_3(L)$, which is displayed for a maximum number of levels $\# = 2500$, $\# = 5000$, $\# = 10000$ and $\# = 11500$, respectively, from which c_{Δ_∞} can be determined, show saturation depending on L, where $\Sigma_2(L) \simeq 2\Delta_3(L)$, as $L \to \infty$, and with the correct L- and $L/15$-behaviour for $L \to 0$.

In figure 13.10 I have displayed the corresponding results for the hyperbolic sphere-billiard, i.e., the comparison with Weyl's law, the normalized energy fluctuations, the number variance $\Sigma_2(L)$ (solid line), and the spectral rigidity (dotted line) in the entire range of the energy levels. Note the large jumps in the number of levels in comparison to Weyl's law due to the degeneracy of levels. I have displayed $\Sigma_2(L)$ for the maximum number of levels (dashed line), and $\Delta_3(L)$ for $\# = 5000$, $\# = 10000$, $\# = 20000$ and $\# = 27780$ (solid Lines), respectively. Again, saturation for $\Sigma_2(L)$ and $\Delta_3(L)$ is observed, and $\Sigma_2(L) \simeq 2\Delta_3(L)$, as $L \to \infty$. The numerical results for the three-dimensional systems are summarized in table 13.3, including other parallelepipeds

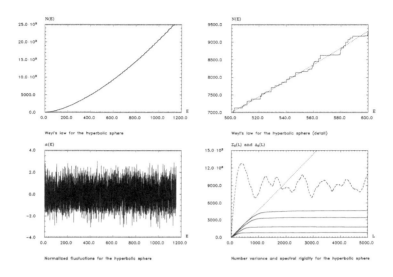

Figure 13.10: Analysis for the hyperbolic sphere.

with $a = 1, \pi/2$ and $a = \pi$. Within an error margin of 8% all features of the semiclassical theory or confirmed. The large value for Δ_3 for the hyperbolic sphere is easily explained by the high degeneracy of the levels which grows in average as $(2l + 1)/2$, as can be easily checked (the dotted line denotes $\Delta_3(L) = 72L/15$ for a Poissonian distribution with a 72-fold degeneration because $l_{Max} = 72$ is the maximal angular momentum number within the considered energy range). The average value for c_{Δ_∞} turns out to be $c_{\Delta_\infty} = 0.17 \pm .01$.

Figure 13.8 (right) finally shows the display of the deviations of the fluctuations of the various systems in three dimensions for the parallelepiped with $a = \pi/3$ (solid), $a = \pi$ (dotted), and the hyperbolic sphere (dashed). The distributions display a more regular (i.e. almost Gaussian behaviour), except that the distribution for the $a = \pi$ parallelepiped turns out to be more irregular, i.e. non-Gaussian as it should be. All distributions show the typical slow increase on the left side and the fast decay on the right side, as in two dimensions.

Chapter 14

The Selberg Trace Formula

Until now we have only dealt with integrable systems. However, particularly in the case of hyperbolic spaces, the very essence of motion in fundamental domains of discrete subgroups of $SL(2, \mathbb{R})$ is that the classical motion is completely chaotic. In this Chapter I focus on the Selberg trace formula and in the next on its super generalization. Let us note that trace formulæ can be formulated in a very general context for pseudo-differential operators [103, 137]. I present the most important results which will be stated as theorems. No derivations and no proofs will be given.

14.1 The Selberg Trace Formula in Mathematical Physics

The Selberg trace formula looks exactly like the semiclassical periodic orbit formula: But it is an *exact* formula. However, one must keep in mind that the equivalence of orbits and conjugacy classes (including winding numbers) is only valid for strictly hyperbolic Fuchsian groups. This one-to-one correspondence is no longer true if elliptic elements are present. The norms are the properly defined quantities in all cases. The summation over norms of the conjugacy classes is then the summation over group elements corresponding to these norms and then the Selberg trace formula can be interpreted in terms of the "mirror-principle" for quantum mechanical path integrals, i.e., propagators. We have for the propagator on a manifold $\mathbb{M}$ ($\mathbf{z} \in \mathbb{M}$)

$$K_\Gamma(\mathbf{z}'', \mathbf{z}'; T) = \sum_{\gamma \in \Gamma} K_0(\mathbf{z}'', \gamma \mathbf{z}'; T) \tag{14.1}$$

and $K_0(T)$ is the free propagator on $\mathbb{M}$, i.e., in the present case on the hyperbolic plane $\mathcal{H}$. Taking the trace is in general not well-defined because the test-function corresponding to the time-evolution operator does not belong to the classes of functions which are allowed in the Selberg trace formula. In order that the emerging trace formulæ are mathematically well-defined, one must use a proper regularization function with corresponding kernel $K(T)$. For instance, this regularization function can be the heat kernel function corresponding to the Laplacian on the manifold [2]. Actually (14.1) is a very general formula for the quantum motion in a coset space $\mathfrak{M} = \mathfrak{N}\backslash\Gamma$ with the requirement that the Hamiltonian in $\mathfrak{N}$ is invariant under transformations belonging to the group Γ. This approach is valid for any compact Lie group, e.g., [393]. The identification in the case of Riemann surfaces is obvious.

Generally, the operators $\gamma \in \Gamma$ denote some transformation of the coordinates $\mathbf{z}$, and Γ is some group. Equation (14.1) therefore describes how we can evaluate the path integral on the quotient manifold $\Gamma\backslash\mathbb{M}$ if we know the propagator on $\mathbb{M}$. The derivation of the Selberg trace formula starts from the very fact that we have an (at least implicitly given) representation of the propagator or the Green's function, respectively. For the Poincaré upper half-plane [202, 247, 361] as well as for the super Poincaré upper half-plane the Green's function can be evaluated by means of path integrals [203, 397, 427, 499]. By means of an appropriate reshuffling of domains in the Poincaré upper half-plane under the action of the (Fuchsian) group Γ it is then possible to perform the summation in (14.1) in such a way that it can be reformulated in terms of the norms of the group elements and their winding numbers. For regular surfaces the norms of the (hyperbolic) group elements can be set into a one-to-one correspondence with the lengths of closed periodic orbits on the surface, and the Selberg trace formula turns out to be an *exact* periodic orbit formula. This is true for both the usual Selberg trace formula on Riemann surfaces and the Selberg super trace formula on super Riemann surfaces. It is quite remarkable that this kind of identification can be made which is not possible in the periodic orbit formula (13.10).

Furthermore, the introduction of the Selberg zeta-function can also be appropriately interpreted in the language of (13.10). The emerging zeta-functions (including the Selberg one) are called "dynamical zeta-functions". All these zeta-functions have, according to the system they correspond to, a characteristic analytical structure, i.e., "trivial" poles and zeros and "non-trivial" poles and zeros. In the Selberg theory, this analytical structure can be derived from the trace formula. But it is not possible to derive from (13.10) similar conclusions, except "trivial" zeros and "nontrivial" zeros on a critical line, where the latter correspond to the eigen-values of the Hamiltonian of the problem. But structures like this seem to be important in the study of chaotic systems and poles are numerically observed, e.g., Wintgen et al. [493, 522]. This means that in the Gutzwiller formula as it is presently known the contributions to the semi-classical periodic orbit formula from a continuous spectrum are completely neglected. However, if a continuous spectrum exists, additional contributions to the semi-classical trace formula must exist, and these additional contributions should generate a pole structure in the corresponding dynamical zeta-functions. This underlines the relevance of developing trace formulæ and the study of the analytic properties of zeta-functions, be they Selberg-like or not.

14.2 Applications and Generalizations

Let us discuss some examples where Selberg trace formulæ do serve as tools in various branches in Mathematical Physics:

Quantum Chaos.

I have already outlined in the previous Chapter some features of the problem of the classical and quantum behaviour of a generic physical system. In order that a classical system is integrable, the phase space must separate into invariant tori. Then also the systems can be quantized canonically. Therefore the question arises, how can a classically chaotic system be properly quantized? The answer lies in the periodic orbit formula (13.9) of Gutzwiller, and in the periodic orbit formula for billiard systems (13.10) of Sieber and Steiner. The periodic orbit formula quantizes a classical system (integrable or non-integrable) semiclassically by means of its periodic orbits. Energy levels and periodic orbits are intimately connected. It should therefore be possible to obtain the energy levels (at least semiclassically) of any system by the sole knowledge of its classical dynamics. The striking similarity of the Selberg trace formula with the Gutzwiller-Sieber-Steiner (GSS) formula suggests that the same statement is true for the quantum motion on Riemann surfaces: The energy levels of the Laplacian on a Riemann surface are completely determined and implicitly known by the knowledge of the periodic orbits on the surface, and hence by the norms of the hyperbolic conjugacy classes of the Fuchsian group which corresponds to it. Therefore, in quantum chaos the Selberg trace formula is used to quantize the motion on Riemann surfaces.

Steiner et al. [18, 480] have formulated two conjectures and have claimed that it can be justified that quantum chaos has been found. They argue that there are unique fluctuation properties in quantum mechanics which are universal and maximally random if the corresponding classical system is strongly chaotic. The two conjectures are the following [480] (note that the limiting behaviour of the spectral fluctuations of $a(E)$, $E \to \infty$, is *proportional* to the limiting behaviour of $\sqrt{\Delta_\infty}$, $E \to \infty$).

Conjecture 1   *1. Let $\bar{\mathcal{N}}(p)$ be the perturbative contribution to the total number $\mathcal{N}(p)$ of energy levels $E_n = p_n^2$, $p_n < p$, for a typical quantum system, including all terms of the Laurent expansion in $\hbar$ up to $\mathcal{O}(\hbar^0)$ (Weyl's asymptotic formula [518] in the case of two-dimensional billiards). Then the arithmetical function $\delta_n := n - 1/2 - \bar{\mathcal{N}}(p_n) =: \mathcal{N}_{osc}(p_n)$ fluctuates about zero with increasing average amplitude $a_n := a(p_n^2)$, in the sense that*

$$\left\langle \mathcal{N}_{osc}(p) \right\rangle = \frac{1}{\mathcal{N}(p)} \sum_{p_n \le p} \delta_n = \mathcal{O}(p^{-1}) \tag{14.2}$$

$$\left\langle \mathcal{N}_{osc}^2(p) \right\rangle = \frac{1}{\mathcal{N}(p)} \sum_{p_n \le p} \delta_n^2 = \mathcal{O}(a^2(p^2)) \tag{14.3}$$

as $E = p^2 \to \infty$, where

$$a(E) = \begin{cases} E^{1/4} & \text{for integrable systems,} \\ (\log E)^{1/2} & \text{for generic chaotic systems,} \\ E^{1/4}(\log E)^{-1/2} & \text{for chaotic systems with arithmetical chaos.} \end{cases}$$

$$(14.4)$$

2. *The normalized fluctuations*

$$\alpha_n := \delta_n / a_n \ , \tag{14.5}$$

considered as random numbers have, as n tends to infinity, a limit distribution $\mu(d\alpha)$ which is a probability distribution in $\mathbb{R}$ and is absolutely continuous with respect to a Lebesgue measure with density $f(\alpha)$, such that for every piecewise continuous bounded function $\Phi(\alpha)$ on $\mathbb{R}$, the following mean value converges

$$\lim_{p \to \infty} \frac{1}{\mathcal{N}(p)} \sum_{p_n \le p} \Phi(p_n) = \int_{\mathbb{R}} \Phi(\alpha) f(\alpha) d\alpha \tag{14.6}$$

and is given by the above integral, where $f(\alpha)$ does not depend on $\Phi(\alpha)$. Moreover, the density $f(\alpha)$ satisfies

$$\int_{\mathbb{R}} f(\alpha) d\alpha = 1 \ , \qquad \int_{\mathbb{R}} \alpha f(\alpha) d\alpha = 0 \ , \qquad \int_{\mathbb{R}} \alpha^2 f(\alpha) d\alpha = \sigma^2 \ , \tag{14.7}$$

where the variance σ^2 is strictly positive.

3. *For strongly chaotic systems, the central limit theorem is satisfied, that is the function $f(\alpha)$ is universal and is given by the Gaussian (normal distribution)*

$$f(\alpha) = \frac{1}{\sqrt{2\pi}\,\sigma} e^{-\alpha^2/2\sigma^2} \tag{14.8}$$

with mean zero and standard deviation $\sigma = 1/\sqrt{2}\,\pi$ or $\sigma = 1/2\pi$ in the non-arithmetic case corresponding to systems with time-reversal or without time-reversal invariance, respectively, and $\sigma = \sqrt{A/2\pi^2}$ in the arithmetic case of hyperbolic billiards with area A. In particular, all higher moments of the sequence $\{\alpha_n\}$ exist, where the odd moments vanish, and the even moments satisfy $(k \in \mathbb{N})$

$$\lim_{p \to \infty} \frac{1}{\mathcal{N}(p)} \sum_{p_n \le p} \alpha_n^{2k} = \frac{(2k)!}{2^k k!} \sigma^{2k} \ . \tag{14.9}$$

4. *In contrast to the above universal situation for chaotic systems, for integrable systems there is in general no central limit theorem for the fluctuations, and the profile of the density $f(\alpha)$ can be very different for different systems. The higher moments of $\{\alpha_n\}$ may not converge to the moments of the limit distribution and the odd moments may not be zero such that $f(\alpha)$ is usually skew and can be both unimodal and multimodal.*

Conjecture 2   *1. Let $\psi_n(\mathbf{q})$, $n \in \mathbb{N}$, be the normalized eigenfunction of a strongly chaotic quantum system. Then $\Psi_n(\mathbf{q})$ has, as n tends to infinity, a limit distribution with density $P(\psi)$, such that for every piecewise continuous bounded function $\Phi(\psi)$ on $\mathbb{R}$, the following limit converges*

$$\lim_{n \to \infty} \frac{1}{A} \int_\Omega \Phi(\psi_n(\mathbf{q}))d\mathbf{q} = \int_\mathbb{R} \Phi(\psi)P(\psi)d\psi \tag{14.10}$$

and is given by the above integral, where $P(\psi)$ does not depend on $\Phi(\psi)$.

2. For strongly chaotic systems, the central limit theorem is satisfied, that is the function $P(\psi)$ is universal and is given by a Gaussian

$$P(\psi) = \frac{1}{\sqrt{2\pi}\,\sigma}e^{-\psi^2/2\sigma^2} \tag{14.11}$$

with mean zero and standard deviation $\sigma^2 = 1/A$, where A denotes the area of Ω.

Whereas these two conjectures have been formulated for planar billiards, hyperbolic billiard systems can be easily incorporated by taking into account the hyperbolic geometry with obvious modifications.

The fact that the classical motion on a Riemann surface is chaotic has been known since the end of the last century. The study of this kind of system is closely connected with the names of Hadamard [261] and Poincaré [445]. In these dynamical systems where a particle is allowed to move freely on a surface of negative constant curvature, the geodesics are the classical orbits, and in ergodic theory Hadamard's dynamical system is called the geodesic flow on the surface. Hadamard's great achievement has been that he could prove that all trajectories in his system are unstable and that neighboring trajectories diverge in time at a rate $e^{\omega t}$, where $\omega = \sqrt{2E/m}$ is the Lyapunov exponent. Thus, the motion on a Riemann surface is very sensitive to the initial conditions and therefore unpredictable, even though the system is governed by the deterministic laws of motion of Newton, hence the notion *deterministic chaos*.

The motion on a Riemann surface seems a purely mathematical feature with no application in physics. However, the mathematicians studying these systems had in mind the question of the stability of planet orbits evolving around the sun, and the stability of the solar system is still an open question. Hence, the Hadamard model, respectively the motion on Riemann surfaces, has a direct application in physics. Sinai [471] translated the problem of the Boltzmann-Gibbs gas into the study of the nowadays known Sinai-billiard which in turn can be related precisely to Hadamard's model. Actually, the motion in a planar square, opposite sides identified, with a cut-out hole, is equivalent to the motion on a torus with a disc removed, i.e., a bordered torus. A bordered torus can now be identified with a closed Riemann surface of genus two, i.e., a double-torus (c.f. section 14.4 about the Selberg trace formula for bordered Riemann surfaces).

Another model of motion on a Riemann surface is the motion on the modular domain. On the Poincaré upper half-plane $\mathcal{H}$ the modular group $\Gamma = \text{PSL}(2, \mathbb{Z})$ operates via fractional linear transformations, i.e., by $\gamma = \begin{pmatrix} a & b \\ c & d \end{pmatrix}$, $z \in \mathcal{H}$, $\gamma z = (az + b)/(cz + d)$. The modular group consists of the two generating matrices $E = \begin{pmatrix} 0 & -1 \\ 1 & 0 \end{pmatrix}$, $R = \begin{pmatrix} 0 & -1 \\ 1 & 1 \end{pmatrix}$, say [270, p.507], subject to the conditions $E^2 = R^3 = \mathbb{1}$. The fundamental region $\mathcal{F}$ of Γ looks familiar and can be chosen to be $\mathcal{F} = \{|z| \geq 1; -1/2 \leq x \leq 1/2\}$, and has a non-Euclidean area of $\mathcal{A} = \pi/3$. Gutzwiller used this model to study the scattering of particles coming from an infinite horn [260]. Artin [13] desymmetrized this system and obtained motion in the half-domain $\tilde{\mathcal{F}} = \{|z| \geq 1; 0 \leq x \leq 1/2\}$, and since the line $x = 0$ is totally reflecting, all other sides must be also totally reflecting and we end up with a billiard system, the so-called Artin's billiard.

Some intensive numerical studies have been devoted to these two systems, the Hadamard model of a Riemann surface of genus two (the hyperbolic octagon) and the modular domain, respectively Artin's billiard. Aurich and Steiner started a numerical investigation of the hyperbolic octagon in order to use the Selberg trace formula to determine the energy levels of the Laplacian in the octagon. Because the generators of the hyperbolic octagon are explicitly known, the norms of the hyperbolic conjugacy classes must only be inserted into the trace formula [25].

It provided a vast amount of numerical data to study the chaotic motion on a Riemann surface, and how the classical chaos influences the quantum mechanical system. Later on, they were able to obtain the quantal levels directly by solving the Schrödinger equation numerically. This provided a detailed analysis of the system concerning the level statistics, and by particularly chosen sum-rules in the Selberg trace formula, i.e., by choosing various test-functions, a numerical comparison of the left and right-hand side of the Selberg trace formula, e.g., [18, 26]–[32], could be made. They also performed a numerical study of highly excited eigenfunctions in a sample of hyperbolic octagons [31]. It is expected that the eigenfunctions of a classical ergodic system should exhibit Gaussian random behaviour as the wavenumber n tends to infinity, c.f. Conjecture 2. The numerical evidence strongly supports the hypothesis of a random Gaussian behaviour. A similar analysis was performed by Hejhal and Rackner [271] who studied waveforms in the modular domain. Their findings were similar. A peculiarity of the modular domain and the regular octagon [25] comes into play: They represent arithmetic Fuchsian groups. This has the consequence that the eigenvalue spectrum shows at reasonable high energies Poissonian fluctuation, i.e., fluctuations similar to those of a sequence of equidistributed random numbers, whereas the time-reversal invariant chaotic systems exhibit GOE statistics. This feature has been, e.g., discussed by Bolte [69].

The original motivation of Aurich and Steiner [25] to obtain the quantal energies from the knowledge of the periodic orbits and the use of a properly chosen testfunction in the Selberg trace formula has not proved very successful because the accuracy of the obtained energy levels was not very good. To overcome this difficulty they have considered the periodic orbit formula in the following form

$$g(E) \simeq \bar{g}(E) + \frac{1}{2p}\frac{\mathrm{d}}{\mathrm{d}p}\log Z(\mathrm{i}p) \ , \tag{14.12}$$

where $g(E)$ is the trace of the Green's function, $\bar{g}(E)$ is a smooth function which contains information about the mean number of the quantal energies, and $Z(s)$ is the dynamical zeta-function of the (classically chaotic) system. In the case of the Selberg trace formula, the relation is exact (regularization arguments let aside) and not only valid semiclassically. It follows that the fluctuations $N_{fl}(E)$ of the spectral staircase $N(E)$ about the mean number of energy levels $\bar{N}(E)$ (Weyl's law, which is encoded in $\bar{g}(E)$) can be described by [18] (in natural units $E = p^2$)

$$N_{fl}(E) = \frac{1}{\pi}\arg Z(\mathrm{i}p) \ , \tag{14.13}$$

and $\bar{N}(E)$ is such that $Z(\mathrm{i}p)\mathrm{e}^{\mathrm{i}\pi\bar{N}(E)}$ is real valued for $p \in \mathbb{R}$. The dynamical zeta-function $Z(s)$ is thus taken at the critical line $s = 0$. In the case of the Selberg zeta function, the critical line is located at $s = \frac{1}{2}$ and the equations have to be modified accordingly. Using this, one derives a quantization condition for classically chaotic systems which has the form

$$\cos\left(\pi N(E)\right) = 0 \ . \tag{14.14}$$

In $N(E)$ now both contributions from the mean number of energies $\bar{N}(E)$ and the fluctuations $N_{fl}(E)$ are taken into account. The advantage of this equation lies in the very fact that it varies only in the interval $[-1, 1]$ and all classical quantities can be evaluated (the periodic orbits, respectively the norms of the hyperbolic conjugacy classes). However, the difficulty is that through the fluctuation part the dynamical zeta-function is taken in a domain, where it is not defined in the usual infinite product definition. Usually the zeta function is defined only in a half-region away from the critical line. In the case of the Selberg zeta-function the critical line is $s = \frac{1}{2}$, and the Selberg zeta-function is defined only in the half-space $\Re(s) > 1$. This difficulty can be overcome by a proper reformulation of the problem which is based on the argument of conditionally convergence of Dirichlet series (Aurich and Bolte [16], Aurich et al. [17]–[22], Aurich and Steiner [29], Sieber [467], and Sieber and Steiner [469]).

Summarizing, we can say that the quantization rule (14.14) has been successfully used in various systems. This is true for the semiclassically periodic orbit formula (13.9) as said in the previous section, and for the quantum motion on Riemann surfaces [17, 22, 29].

The Selberg Zeta-Function and the Riemann ζ-Function.

To fully appreciate the value and importance of the Selberg trace formula one needs to know something about the classical Riemann zeta function $\zeta(s)$. We cite a short summary from Hejhal [268]. As one knows

$$\zeta(s) = \sum_{n=1}^{\infty}\frac{1}{n^s} = \prod_p\frac{1}{1 - p^{-s}} \ , \qquad \Re(s) > 1 \ , \tag{14.15}$$

and the product is taken over all prime numbers p. We have the relation

$$\frac{\zeta'(s)}{\zeta(s)} = -\sum_{n=1}^{\infty} \frac{\Lambda(n)}{n^s} , \qquad \Re(s) > 1 , \tag{14.16}$$

where $\Lambda(n)$ is the Mangoldt function defined by

$$\Lambda(n) = \left\{ \begin{array}{ll} \ln p , & \text{if } n = p^k \text{ for some } k \geq 1 \\ 0 , & \text{otherwise} \end{array} \right\} . \tag{14.17}$$

As an analytic function, $\zeta(s)$ can be continued to the left of $\Re(s) = 1$. More precisely, $\zeta(s)$ will be analytic on all of $\mathbb{C}$ except for a simple pole at $s = 1$. The functional relation for the Riemann zeta-function has the form

$$\zeta(s) = \zeta(1-s)\pi^{s-1/2}\frac{\Gamma(\frac{1-s}{2})}{\Gamma(\frac{s}{2})} . \tag{14.18}$$

Using this functional equation, we see that $\zeta(-2n) = 0$ for $n \geq 1$. These zeros are called *trivial* zeros. All the *nontrivial* zeros are located inside $0 \leq \Re(s) \leq 1$; they are denoted by ϱ. The Riemann Hypothesis [452] states that $\Re(\varrho) = \frac{1}{2}$ for *all* ϱ. This is after 126 years still an unsolved problem! There is an intimate connection between the nontrivial zeros of the Riemann zeta-function and the prime numbers which can be summarized in the Weil explicit formula [517]. Set $\varrho = \frac{1}{2} + i\gamma$, $\gamma \in \mathbb{C}$, and the set of all γ is denoted by Γ. Let $h(p)$ be a testfunction with Fourier-transform $g(u)$ (for the proper requirements on the function $h(p)$, c.f. theorem 14.2). We then have the following explicit formula

$$\sum_{\gamma \in \Gamma} h(\gamma) = \frac{1}{2\pi} \int_{\mathbb{R}} h(p)\Psi(\tfrac{1}{4} + \tfrac{ip}{2}) + h(\tfrac{i}{2}) + h(-\tfrac{i}{2})$$

$$- g(0)\ln \pi - 2\sum_{n=1}^{\infty} \frac{\Lambda(n)}{\sqrt{n}}g(\ln n) . \tag{14.19}$$

It is obvious that the similarity to the Selberg trace formula is striking. However, does the Weil explicit formula have anything to do with a trace formula? Selberg noticed this similarity in 1950–1951 and was led to a deeper study of trace formulæ. Among other things he formulated a zeta-function $Z(s)$, nowadays called Selberg zeta function, which mirrors some of the analytic properties of the Riemann zeta-function. In particular he could show that the Riemann Hypothesis is *true* for $Z(s)$. This hypothesis is also true for all generalizations in the subsequent studies in other Selberg-like trace formulæ, i.e., on all symmetric space forms of rank one, and the corresponding super generalizations.

Although, some of the similarities of the Weil formula and the Selberg trace formula seem striking, there are also some difficulties. Let us, e.g., mention the following:

1. Assume that we are allowed to interpret $\ln p$ as "length of orbits" respectively p as "norms of hyperbolic transformations": The sign of the sums over all

"conjugacy classes" is the wrong one. In the language of periodic orbit theory the overall minus sign in front of this sum would be interpreted as an overall phase factor of π. No such system is known.

2. The sum over all zeros of the Riemann zeta-function would correspond in the case of the Selberg trace formula to a sum over all eigenvalues of the Laplacian on a Riemann surface. Therefore one is supposed to look for a Hamiltonian H with the property that its eigenvalues are just the zeros of the Riemann zeta-function. Although some numerical studies exist in this direction [57], no hard facts seem to be known. One only knows from the level spacing distribution of the zeros of the Riemann ζ function, which is GUE, that if such an H exists, it is not invariant with respect to time-reversal.

In conclusion, it is completely unknown if the Weil formula is a trace formula in the sense of periodic orbit or the Selberg theory at all. For instance, Hejhal performed some studies concerning the so-called congruence groups, and in particular the modular group $SL(2, \mathbb{Z})$, as promising candidates [270, 271]. However, "if the Riemann Hypothesis is false, this whole program is probably a waste of time" [268].

String Theory.

In string theory [191]–[194, 446, 447, 462] the theory of the Selberg trace formula enables one to express the determinants of Laplacians on Riemann surfaces in terms of the Selberg zeta-function. Basically, the Polyakov approach [125]–[128, 446, 447] to (bosonic-, fermionic- and super-) string theory is a quantum field theory on Riemann surfaces. In the perturbative expansion of the Polyakov path integral one is left with a summation over all topologies of world sheets a string can sweep out, and an integral over the moduli space of Riemann surfaces. This picture is true for bosonic strings as well as for fermionic strings. The partition function for open as well as closed bosonic strings corresponding to a topology without conformal Killing vectors (D'Hoker and Phong [125, 128]) turns out to be

$$Z_0^{(BS)} = \sum_g \int d\mu_{WP} [\det(P_1^\dagger P_1)]^{1/2} (\det{}'\Delta_0)^{-D/2} , \qquad (14.20)$$

where P_1 and Δ_0 are the symmetrized traceless covariant derivative and scalar Laplacian with Dirichlet boundary conditions, respectively, and $d\mu_{WP}$ denotes the Weil-Petersson measure. D denotes the critical dimension which equals 26 for the bosonic string. For the fermionic-, respectively the super-string all quantities have to be replaced by their appropriate super versions, and the critical dimension is $D = 10$. In the theory of the fermionic string the partition function for open as well as closed fermionic strings can be expressed as [127, 128]

$$Z_0^{(FS)} = \sum_g \int d\mu_{sWP} [\text{sdet}(P_1^\dagger P_1)]^{1/2} (\text{sdet}'\square_0^2)^{-5/2} , \qquad (14.21)$$

where P_1 and $\Box_0$ are the super-analogues of the symmetrized traceless covariant derivative and scalar Dirac-Laplace operator with Dirichlet boundary conditions, respectively, and $d\mu_{sWP}$ denotes the super Weil-Peterson measure. In order to deal with the vector Dirac-Laplace operator $P_1^\dagger P_1$ the incorporation of m-weighted super automorphic forms into the formalism is required.

The calculation of determinants of Laplacians on Riemann surfaces is due to several authors. Let us mention the evaluation of these determinants in terms of Selberg zeta-functions by, e.g., Bolte and Steiner [72], D'Hoker and Phong [126, 128], Efrat [146], Gilbert [181], Namazie and Rajeev [418], Sarnak [458], Steiner [479], and Voros [515], and in terms of the period matrix and theta-functions by, e.g., Alvarez-Gaumé et al. [6] and Manin [388]. Let us note that the Selberg zeta-function approach has enabled Gross and Periwal [253] to show that the bosonic string perturbation theory is not (Borel-) summable and hence not finite.

In the perturbative expansion of the bosonic string the classical Selberg trace formula could be applied in a straightforward way, whereas the perturbation theory for the fermionic string required the introduction of the Selberg super-trace formula. Here Baranov, Manin et al. [42] originally started this activity, and it has been further developed by Aoki [8] and in [203, 213, 221].

Of course, while dealing with open strings one has to distinguish Dirichlet and Neumann boundary conditions, respectively. In particular, relations between the determinants $\det\Delta_\Sigma^{(D)}$ and $\det'\Delta_\Sigma^{(N)}$ corresponding to Dirichlet and Neumann boundary conditions on the bordered surface Σ, and the determinant of the scalar Laplacian $\det'\Delta_{\hat\Sigma}$ for the doubled (closed) surface $\hat\Sigma$ could be derived, i.e.

$$\det\Delta_\Sigma^{(D)} \cdot \det'\Delta_\Sigma^{(N)} = \det'\Delta_{\hat\Sigma} \ . \tag{14.22}$$

In the sequel I am not going into details of string theory any further.

The Selberg Trace Formula on Symmetric Space Forms of Rank One.

As already noted in section 12.2 there are also higher-dimensional hyperbolic spaces which admit the formulation of a Selberg trace formula and a Selberg zeta-function. This is on the one hand the D-dimensional hyperboloid as the universal covering space (Bérard-Bergery [52], Elstrodt et al. [152], Subia [487], Takahashi [489], Tomaschitz [497], and Venkov [505]), and on the other the symmetric rank one spaces, as for instance the hermitian hyperbolic spaces (Hashizume et al. [265], Takahashi [489], Takas [492], and Venkov [505, 506]), and the $\mathrm{Sp}(1,D)/\mathrm{Sp}(D-1)$ hyperbolic space [490, 491, 506]. Generally these hyperbolic spaces have a common structure. One considers the hermitian $p + q$ form, e.g., [273, 274, 358, 529] $(x, y \in \mathbb{F}^{(p+q)}, e = \pm 1)$

$$\mathcal{Q}_e^{(p,q)} = y_1^* x_1 + \ldots y_p^* x_p - y_{p+1}^* x_{p+1} - \ldots - y_{p+q}^* x_{p+q} = e \ , \tag{14.23}$$

and asks for the Lie group of linear operators acting on $\mathbb{F}^{p+q}$ which leaves it invariant. Here $\mathbb{F}$ can be $\mathbb{F} = \mathbb{R}$, $\mathbb{F} = \mathbb{C}$ or $\mathbb{F} = \mathbb{H}$, respectively, where $\mathbb{H}$ denotes the field of quaternions. For the hyperboloids leading to the study of hyperbolic spaces, we have $p = n$, $q = 1$ and $e = 1$, say. Depending on the choice of $\mathbb{F}$ one deals with the four multi-dimensional hyperbolic spaces of rank one, namely $S_1 \equiv \mathcal{H}^D \cong$ $SO(D,1)/SO(D)$, $S_2 \cong SU(D,1)/S[U(1) \times U(D)]$ and $S_3 \cong Sp(D,1)/[Sp(1) \times Sp(D)]$, and the so-called exceptional space $S_4 \cong F_{4(-20)}/Spin(9)$ [358, 491]. The group pairs $(SO(D,1), SO(D))$, $(SU(1,D), S[U(1) \times U(D)])$, $(Sp(n,1), [Sp(1) \times Sp(D)])$ and $(F_{4(-20)}, Spin(9))$ form so-called Gelfand pairs.

Let us cite some basic results of this theory which is due to, e.g., Chavel [102], Gangoli [173], Gangoli and Warner [174], and Venkov [505, 506]. In the theory of the usual Selberg zeta function, we consider a Riemann surface, $M = \Gamma \backslash \mathcal{H}$, of genus $g \geq 2$, where $\mathcal{H}$ is the Poincaré upper half-plane and Γ is a discrete subgroup of $PSL(2, \mathbb{R})$ acting freely on $\mathcal{H}$ via fractional linear transformations. Furthermore, usually a finite dimensional unitary representation T of Γ is also introduced with character χ.

In order to set up a general Selberg trace formula I will shortly sketch and introduce the relevant notions. This will go a little bit further as in section 12.2 and I will rely on [173]. Let G be a connected semisimple Lie group with finite center, K a maximal compact subgroup, and $\mathbb{H}$ the symmetric space G/K and we assume that $\text{rank}(G/K) = 1$; we endow $\mathbb{H}$ with a G-invariant metric. Let Γ be a discrete torsion-free subgroup of G such that $\Gamma \backslash G$ is compact, say. The manifold $M = \Gamma \backslash \mathbb{H}(\Gamma \backslash G/K)$ is a compact Riemannian manifold, whose simply connected covering manifold is $\mathbb{H}$. M is a compact space form of $\mathbb{H}$. Corresponding to G and K let $\mathfrak{g}$ and $\mathfrak{k}$ be their respective Lie algebras, and let $\mathfrak{g} = \mathfrak{k} + \mathfrak{p}$ be a Cartan decomposition of $\mathfrak{g}$. Let $\mathfrak{a}_\mathfrak{p}$ be a maximal Abelian subspace of $\mathfrak{p}$ and we will assume that $\dim(\mathfrak{a}_\mathfrak{p}) = 1$. Extend $\mathfrak{a}_\mathfrak{p}$ to a maximal Abelian subalgebra $\mathfrak{a}$ of $\mathfrak{g}$, so that $\mathfrak{a} = \mathfrak{a}_\mathfrak{k} + \mathfrak{a}_\mathfrak{p}$, with $\mathfrak{k} = \mathfrak{a} \cap \mathfrak{k}, \mathfrak{p} = \mathfrak{g} \cap \mathfrak{p}$. Then $\mathfrak{a}$ is a Cartan subalgebra of $\mathfrak{g}$. Denote by $\mathfrak{a}^\mathbb{C}, \mathfrak{g}^\mathbb{C}$ the complexification of $\mathfrak{a}, \mathfrak{g}$, and let $\Phi(\mathfrak{g}^\mathbb{C}, \mathfrak{a}^\mathbb{C})$ denote the set of roots of $(\mathfrak{g}^\mathbb{C}, \mathfrak{a}^\mathbb{C})$. Order the dual spaces of $\mathfrak{a}$ and $\mathfrak{a}_\mathfrak{p} + i\mathfrak{a}_\mathfrak{k}$ compatibly, and let Φ^+ be the set of positive roots under this order. Let

$$P_+ = \{\alpha \in \Phi^+; \alpha \neq 0 \text{ on } \mathfrak{a}\} \,, \qquad P_- = \{\alpha \in \Phi^+; \alpha \equiv 0 \text{ on } \mathfrak{a}\} \,. \qquad (14.24)$$

Put $\varrho = \frac{1}{2} \sum_{\alpha \in P^+} \alpha$. For $\alpha \in \Phi_+$, let X_α be a root vector belonging to α and put $\mathfrak{n}^\mathbb{C} = \sum_{\alpha \in P^+} \mathbb{C} X_\alpha$. Then if $\mathfrak{n} = \mathfrak{n}^\mathbb{C} \cap \mathfrak{g}$ we have the Iwasawa decomposition $\mathfrak{g} = \mathfrak{k} + \mathfrak{a}_\mathfrak{p} + \mathfrak{n}$, $G = K A_\mathfrak{p} N$, where $A_\mathfrak{a} = e^{\mathfrak{a}_\mathfrak{p}}$, $N = e^\mathfrak{n}$.

We denote by Λ the real dual of $\mathfrak{a}$, by $\Lambda^\mathbb{C}$ its complexification $\Lambda + i\Lambda$. Denote by $C^\infty(K \backslash G/K)$ the space of differentiable spherical functions, i.e, those that satisfy $f(k_1 x k_2) = f(x), x \in G, k_1, k_2 \in K$. For any $\nu \in \Lambda^\mathbb{C}$, we denote by ϕ_ν the elementary spherical function corresponding to ν.

Let Σ be the set of restrictions to $\mathfrak{a}_\mathfrak{p}$ of elements of R^+. Since $\dim(\mathfrak{a}_\mathfrak{p}) = 1$, we can find $\beta \in \Sigma$ such that 2β is the only other possible element in Σ. Let m_α be the number of roots in P^+ whose restriction to $\mathfrak{a}_\mathfrak{p}$ is β, and let m_β be the number of the remaining elements in P^+. We fix the element $H_0 \in \mathfrak{a}_\mathfrak{p}$ by the property $\beta(H_0) = 1$. One then has $\varrho_0 = \varrho(H_0) = \frac{1}{2}(m_\alpha + 2m_\beta)$.

With these preliminaries we can set up the Selberg trace formula on M. Let Γ be a discrete subgroup of G such that $\Gamma \backslash G$ is compact. We may assume that Γ has no elements of finite order. Then every element $\gamma \in \Gamma$ is conjugate in G to an element of the Cartan subgroup $A = $ centralizer of $\mathfrak{a}$ in G: $A = A_{\mathfrak{k}}A_{\mathfrak{p}}$; choose an element $h(\gamma)$ of A to which γ is conjugate, and let $h(\gamma) = h_{\mathfrak{k}}(\gamma)h_{\mathfrak{p}}(\gamma)$. We define $u_\gamma = \beta(\log h_{\mathfrak{p}}(\gamma))$. Thus $u_\gamma = u(h_{\mathfrak{p}}(\gamma))$. Though u_γ will depend on the choice of $h(\gamma)$, its absolute value depends only on γ; $|u_\gamma|$ is essentially the length of the shortest geodesic in the free homotopy class associated to γ on the manifold $\Gamma\backslash G/K$, respectively $N_\gamma = e^{|u_\gamma|}$ is the norm of the smallest element in the free homotopy class associated to γ on the manifold $\Gamma\backslash G/K$.

An element $\gamma \in \Gamma$, $\gamma \neq \mathbb{1}$ is called primitive if it cannot be expressed as δ^n, for some $n > 1, \delta \in \Gamma$. Every $\gamma \neq \mathbb{1}$ is equal to a positive power of a unique primitive element γ_0. Define the integer $j(\gamma)$ by $\gamma = \gamma_0^{j(\gamma)}$. Let T be a finite dimensional unitary representation of Γ with character χ. Denote by U the unitary representation of G induced by T. U is a direct sum of irreducible representations of G occurring with finite multiplicities. Let $\{U_j; j \geq 0\}$ be the spherical representations that occur in U, and let $n_j(\chi)$ be their multiplicities. For technical reasons, it is always assumed that U_0 denotes the trivial representation of G. Its multiplicity $n_0(\chi)$ is equal to a_0, where a_0 is the multiplicity of the trivial representation of Γ in T. Thus $n_0(\gamma)$ may be zero. We shall nevertheless include U_0 in the collection $\{U_j\}$. Each U_j is completely determined by its elementary spherical function, say ϕ_{ν_j}, with $\nu_j \in \Lambda^{\mathbb{C}}$. Since U_j is unitary, ϕ_{ν_j} is positive definite, and one knows that $< \nu_j, \nu_j > + < \varrho, \varrho > \geq 0$. From this follows that ν_j is either purely real, i.e., $\nu_j \in \Lambda$ or purely imaginary, i.e., $\nu_j \in i\Lambda$. We choose and fix ν_j so that when it is real $\nu_j(H_0) \geq 0$, and when it is purely imaginary, we have $i\nu_j(H_0) < 0$ (these eigenvalues play the rôle of the so-called "small" eigenvalues). Set $\nu_j(\chi) = \nu_j(H_0)$. $< \nu_j, \nu_j > + < \varrho, \varrho > \geq 0$ is just the negative of the Casimir operator of G operating on ϕ_{ν_j}. Thus the numbers $(\varrho_0^2 + r_j^{+2})$ are the eigenvalues of the Laplace-Beltrami operator $-\Delta_{G/K}$ in a suitable metric, and the n_j are their multiplicities. Since U_0 is the trivial representation, we have that $\nu_0 = i\varrho$. We get

Theorem 14.1 *The Selberg trace formula for compact symmetric space forms of rank one is given by*

$$\sum_{j\geq 0} n_j(\chi)h(\nu_j^+) \;=\; \chi(1)\frac{\mathcal{V}(\Gamma\backslash G)}{4\pi}\int_{\mathbb{R}}\frac{h(r)dr}{|c(ir)|^2} + \sum_{\{\gamma\}}|u_{\gamma_0}|C(h(\gamma))g(u_\gamma) \;, \qquad (14.25)$$

$$\frac{1}{c(ir)} \;=\; \frac{\Gamma(\frac{m_\alpha+m_\beta}{2})\Gamma(ir+m_\alpha/2)\Gamma(ir/2+m_\alpha/4+m_\beta/2)}{\Gamma(m_\alpha+m_\beta)\Gamma(ir)\Gamma(ir/2+m_\alpha/4)} \;, \qquad (14.26)$$

$$C(h(\gamma)) \;=\; \epsilon_R^A(h(\gamma))\xi(h_{\mathfrak{p}}(\gamma))\prod_{\alpha\in P^+}\frac{1}{1-1/\xi_\alpha(h(\gamma))} \;. \qquad (14.27)$$

$c(z)$ is the Harish-Chandra c-function. Further, for any α, ξ_α stands for the character of $A = A_{\mathfrak{k}}A_{\mathfrak{p}}$ defined by $\xi_\alpha(h) = e^{\alpha(\log(h))}$, and $\epsilon_R^A(h)$ is, for $h \in A$, equal to the sign of $\prod_{\alpha\in\Phi_R^+}(1-1/\xi_\alpha(h))$, Φ_R^+ being the set of roots of $(\mathfrak{g}^{\mathbb{C}}, \mathfrak{a}^{\mathbb{C}})$, i.e., those that are real on $\mathfrak{a}$. $C(h)$ is a positive function on A. The test function $h(r)$ must fulfill the requirements

1. $h(r)$ is holomorphic in the strip $|\Im(r)| \leq \varrho_0 + \epsilon$, $\epsilon > 0$.

2. $h(r)$ has to decrease faster than $|r|^{-2}$ for $r \to \pm\infty$.

3. $g(u) = \pi^{-1} \int_0^\infty h(r) \cos(\pi r) dp$.

We now introduce the Selberg zeta-function as follows (c.f. Chavel [102], Gangoli [173], Gangoli and Warner [174], and Takase [492])

$$Z(s) = \prod_{\{\gamma_0\}} \prod_{\lambda \in L} \left\{ \det\left[\left(\mathbb{1} - T(\gamma_0)\xi_\lambda(h(\gamma_0)) \right) \right]^{-1} e^{-su_{\gamma_0}} \right\}^{m_\lambda}, \qquad \Re(s) > 2\varrho_0 . \quad (14.28)$$

Here, L is the semi-lattice of linear forms on $\mathfrak{a}$ of the form $\sum_{i=1}^t m_i\alpha_i$, $\alpha_1, \ldots, \alpha_t$ being the elements of P_+ and $m_1, \ldots, m_t$ being non-negative integers, m_λ is the number of distinct t-tuples $(m_1, \ldots, m_t)$ of non-negative integers such that $\lambda = \sum_{i=1}^t m_i\alpha_i$, and ξ_λ is the character of $\mathfrak{a}$ corresponding to λ.

Among other things, it is possible to derive the following properties of $Z(s)$ [173, 174]:

1. $Z(s)$ is holomorphic in the half-plane $\Re(s) > 2\varrho_0$.

2. $Z(s)$ has a meromorphic continuation to the whole complex plane.

3. $Z(s)$ satisfies the functional equation

$$Z(2\varrho_0 - 2) = Z(s) \exp\left(\chi(1)\mathcal{V}(\Gamma\backslash G) \int_0^{s-\varrho_0} \frac{dt}{|c(it)|^2} \right) . \quad (14.29)$$

4. The nontrivial (or spectral) zeros of $Z(s)$ corresponding to the eigenvalues of the Laplacian $-\Delta_{G/K}$ lie on the line $\Re(s) = \varrho_0$ except for a finite number of indices j (the small eigenvalues). These zeros are located in the interval $s \in [0, 2\varrho_0]$.

5. The trivial (or topological) zeros or poles of $Z(s)$ exist only when $\dim(G/K)$ is even, i.e., when the Euler-Poincaré characteristic of M is non-zero. These zeros are determined by the pole of the function $[c(z)c(-z)]^{-1}$, c.f. [173, 174] for details.

6. When poles do not exist, $Z(s)$ is an entire function of order equal to $\dim(G/K)$.

7. The test function $h(r)$ to derive the logarithmic derivative of $Z(s)$ and its Fourier-transform $g(u)$ are given by

$$h(p, s, b) = \frac{1}{(s - \varrho_0)^2 + p^2} - \frac{1}{(b - \varrho_0)^2 + p^2} , \quad (14.30)$$

for $\Re(s), \Re(b) > 1$, and

$$g(u, s, b) = \frac{e^{-(s-\varrho_0)|u|}}{2(s - \varrho_0)} - \frac{e^{-(b-\varrho_0)|u|}}{2(b - \varrho_0)} . \quad (14.31)$$

The theory of the Selberg trace formula and the Selberg zeta-function can be extended to include also non-compact Riemannian manifolds M, i.e., one includes elliptic and parabolic elements into the group Γ. This causes, of course, additional terms in the trace formula and in the functional equation for $Z(s)$, respectively, and gives rise to additional zeros and poles in $Z(s)$, c.f. Elstrodt et al. [152], Gangoli and Warner [174], Sarnak [457] and Venkov [506].

The Selberg Trace Formula on D-Dimensional Hyperbolic Space.

Let us shortly specify the case of the Selberg trace formula on D-dimensional Riemannian manifolds, i.e., where the relevant hyperbolic space is the D-dimensional hyperboloid $\mathcal{H}^{(D)} = \mathrm{SO}_0(D,1)/\mathrm{SO}(D)$ (c.f. Chavel [102], Gangoli [173], Gangoli and Warner [174], and Venkov [505, 506]). We take into account only hyperbolic conjugacy classes from a discrete subgroup Γ of the isometries on $\mathcal{H}^{(D)}$; the winding numbers are implicitly contained, then we have for the trace formula

$$\sum_{n=0}^{\infty} h(p_n) = 2\mathcal{V} \int_0^{\infty} dp\, h(p)\Phi_D(p) + 2\sum_{\{\gamma\}} \frac{l_\gamma g(l_\gamma)}{N_\gamma^{(D-1)/2}|\det(\mathbb{1} - S^{-1}K^{-1})|} \ , \quad (14.32)$$

$$\Phi_D(p) = \frac{\Omega(D)}{(2\pi)^2}\left|\frac{\Gamma(\mathrm{i}p + \frac{D-1}{2})}{\Gamma(\mathrm{i}p)}\right|^2 \ . \quad (14.33)$$

The Harish-Chandra function in this case is relatively easy and gives a polynomial of degree $\frac{D-1}{2}$ in p^2 if D is odd, and a polynomial of degree $\frac{D-2}{2}$ in p^2 times $p \tanh \pi p$, if D is even. The matrices $K \in \mathrm{O}(D-1)$ and S are a rotation and a dilation which arise in the evaluation of the trace formula which emerge from the conjugation procedure in order to obtain a convenient fundamental domain $\Gamma\backslash\mathcal{H}^{(D-1)}$. $\mathcal{V}$ is the volume of this fundamental domain. For $D = 2$ the usual Selberg trace formula for Riemann surfaces is recovered. For $D = 3$ one has (c.f. Elstrodt [151], Elstrodt, Grunewald and Mennicke [152], Sarnak [457], Szmidt [488], and Tomaschitz [497])

$$\sum_{n=0}^{\infty} h(p_n) = \frac{\mathcal{V}}{4\pi^2} \int_{\mathbb{R}} dp\, p^2 h(p) + \sum_{\{\gamma\}} \sum_{k=1}^{\infty} \frac{l_\gamma g(kl_\gamma)}{m_\gamma |a_\gamma - 1/a_\gamma|^2} \ . \quad (14.34)$$

Here denotes $|a_\gamma|^2 = N_\gamma$ and m_γ is the order of the minimal rotation about the axis of γ. Some care is needed because in the summation over the hyperbolic conjugacy classes the l_γ can be complex, i.e., they can have a phase, and the numbers a_γ may be complex (the so-called loxodromic elements). The corresponding Selberg zeta-function $Z(s)$ (note $\varrho_0 = 1$, in comparison to Elstrodt et al. [152] we consider a shifted version)

$$Z(s) = \prod_{\{\gamma\}} \left(1 - a_\gamma^{-2k} a_\gamma^{*-2l} \mathrm{e}^{-sl_\gamma}\right) \ , \qquad \Re(s) > 2 \ , \quad (14.35)$$

is an entire function of order $3 = \mathrm{rank}(\Gamma\backslash\mathcal{H}^{(3)})$, and has as zeros the eigenvalues s_n, $n \in \mathbb{N}_0$, of the corresponding Laplacian on $\Gamma\backslash\mathcal{H}^{(3)}$ with the critical line $s = 1$.

This means that all zeros of $Z(s)$ are located at the critical line with the exception of "small eigenvalues" in the interval $s \in [0, 2]$. They may be located in the entire interval [450]. There exists no trivial zeros. This feature is typical for odd dimensions. Due to the zero-mode of the Laplacian there is a simple zero at $s = 2$. In the discussion following the next section, we see that the Selberg zeta-function in two dimensions, i.e., on Riemann surfaces, has "trivial zeros". A numerical study of eigenvalues in domains with one cusp, e.g., $\Gamma = \mathrm{PSL}(2, \mathbb{Z}[\mathrm{i}])$ is due to Grunewald and Huntebrinker [254], see also Aurich and Marklof [21].

Cosmology.

In cosmology the Selberg trace formula including its D-dimensional generalization makes it possible to evaluate the one loop effective potential [92, 93, 109, 149]. Here a, say, scalar interacting field theory is considered and one wants to obtain the so-called effective potential. One starts with the Lagrangian in this field theory, where an additional potential $V(\phi)$ may be included. Solving the corresponding classical equations of motion yields a background field $\hat{\phi}$ about which the theory is expanded. In a scalar field theory the corresponding fluctuations $\phi' = \phi - \hat{\phi}$ obey in lowest order the Klein-Gordon equation. The concept of the effective potential enables one to approximate the original action of the field theory in terms of this background field and its fluctuations by additional terms in the potential yielding $V_{eff}(\hat{\phi}) = V(\phi) + \hbar V_T^{(1)}(\hat{\phi}) + O(\hbar^2)$. The correction $V_T^{(1)} = T \ln \det(l^2 \square)/2$ is then determined by the zeta-function regularization method of the Klein-Gordon operator $\square$ with potential $V''(\hat{\phi})$. Furthermore the quantities T for the temperature, and l for a length renormalization have been introduced. The corresponding correction term in the effective action is denoted by $\Gamma^{(1)}$, and is also called the one-loop potential. In hyperbolic space-times one exploits the Selberg trace formula which gives an explicit expression for the determinant in terms of the Selberg zeta-function. The method can be applied to arbitrary hyperbolic dimensional space-times. Let us note that the consideration of the volume of three-manifolds played a role in cosmology in the search for a smallest universe [158]. However, in the sequel I will not dwell on this topic any further.

Trace Formulæ on Spheres.

There are also trace formulæ for the quantum motion on spaces of constant positive curvature, i.e., on spheres.

1. All trace formulæ are generalizations of the *Poisson* summation formula

$$\sum_{k \in \mathbb{Z}} \delta(p - 2\pi k) = \sum_{k \in \mathbb{Z}} e^{ikp} \ . \tag{14.36}$$

The characteristic feature is that on the left hand side stands the summation of the spectral momentum density.

2. One of the simplest trace formula is the trace formula for the quantum motion on the entire sphere $S^{(2)}$. From the spectral momentum density $d(p) = \sum_{k \in \mathbb{N}_0} (2k + 1)\delta(p - k - \frac{1}{2})$ one derives the trace formula valid for $p \geq 0$

$$\sum_{\mathbb{N}} (2k + 1)\delta(p - k - \tfrac{1}{2}) = 2p + 4p \sum_{k \in \mathbb{N}} (-1)^k \cos(2\pi k p) \; . \tag{14.37}$$

3. A similar approach of calculating determinants of Laplacians on surfaces of constant negative curvature is also possible for surfaces of constant positive curvature. Here, the calculation of determinants of Laplacians by means of trace formulæ have become popular, c.f. [14, 104, 136]. These results can be used to derive via a zeta-function regularization argument explicitly the vacuum energy (Casimir energy) of a free scalar field theory on orbifold factors of spheres (for instance in the orbifold space-time $\mathbb{R} \times S^{(2)}/\Gamma$, where Γ is a finite subgroup of $O(3)$ acting with fixed points), i.e., one considers the tesselations of spheres by some group actions. For instance, one has the following trace formula for the heat-kernel [104]

$$K_{S^{(2)}/\Gamma}(\tau) = \frac{2\pi}{|\Gamma|} \frac{\mathrm{e}^{\tau/4}}{(\pi\tau)^{3/2}} P \int_0^\infty d\alpha \frac{e^{-\alpha^2/\tau}}{\sin\alpha} \left[\alpha + \frac{\tau}{2} \sum_{\{\gamma\}} n_q(\cot\alpha - q\cot q\alpha) \right] \; .$$

$$\tag{14.38}$$

Here $S^{(2)}/\Gamma$ is the fundamental domain of Γ (an elliptic triangle), q is the generic order of the rotation such that for each primitive $\gamma \in \Gamma$ we have $\gamma^q = \mathbb{1}$, and n_q is the number of conjugate q-fold axes. The summation of all primitive conjugacy classes of elements $\gamma \in \Gamma$ is denoted by $\sum_{\{\gamma\}}$. The quantity $|\Gamma|$ is defined via $|\Gamma| \int_{S^{(2)}/\Gamma} d\mathbf{s} = 4\pi$ and describes the order of Γ. The formalism can also be extended to higher dimensional spheres.

4. Whereas the distribution of the normalized fluctuations $f(\alpha)$ (if it exists) for elliptic triangles does not seem to be known yet, it is possible to derive this distribution for so-called Zoll-surfaces (c.f. e.g. [59]), where the sphere $S^{(2)}$ is a special case. Here $f(\alpha)$ is a box function centered about $\alpha = 0$ of width and height one [461].

14.3   The Selberg Trace Formula on Riemann Surfaces

Let us start with some essentials about hyperbolic geometry which we cite from [72]. According to the uniformization theorem of Klein and Poincaré any compact Riemann surface M is conformally equivalent to some constant curvature surface $U\backslash\Gamma$, where U is the universal covering of M and Γ is some lattice group, isomorphic to the first homotopy group of M. In particular, the universal covering U is given by either $U = \widehat{\mathbb{C}}, \mathbb{C}$, or $\mathcal{H}$, and Γ is a discrete, fixed point-free subgroup of the conformal automorphisms of U. For either $g \geq 2$, or non-compact Riemann surfaces, the relevant universal covering is the Poincaré upper half-plane $\mathcal{H}$. Since M possesses a complex structure, one may change to complex coordinates $z = x + iy$ and $\bar{z} = x - iy$ which implies $ds^2 = 2g_{z\bar{z}}dzd\bar{z}$ ($g_{zz} = g_{\bar{z}\bar{z}} = 0, g_{z\bar{z}} = (g^{z\bar{z}})^{-1} = \frac{1}{2}e^{2\varphi}$), and traceless tensors of weight n may be represented as

$$T^n := \{f(z)dz^n | f(z)dz^n = f'(z')dz'^n\} , \qquad n \in \mathbb{N}_0 . \tag{14.39}$$

A prime denotes quantities with respect to new coordinates $z' = z'(z)$. If one fixes for a genus $g \geq 2$ one out of the possible 2^{2g} spin structures on M, n will be allowed to take half-integer values: $T^{1/2}$ denotes the space of spinors on M. We then define covariant derivative operators on M

$$\nabla_n^z : T^n \mapsto T^{n-1} , \qquad \nabla_n^z[f(z)dz^n] := [g^{z\bar{z}}\partial_z f(z)]dz^{n-1} , \tag{14.40}$$

$$\nabla_z^n : T^n \mapsto T^{n+1} , \qquad \nabla_z^n[f(z)dz^n] := [(g_{z\bar{z}})^n \partial_z (g^{z\bar{z}})^n f(z))]dz^{n+1} . \tag{14.41}$$

By means of these derivative operators we can define invariant, second order differential operators which we call Laplace-like operators

$$\Delta_n^+ : T^n \mapsto T^n , \qquad \Delta_n^+ := -2\nabla_{n+1}^z \nabla_z^n , \tag{14.42}$$

$$\Delta_n^- : T^n \mapsto T^n , \qquad \Delta_n^- := -2\nabla_z^{n-1}\nabla_n^z . \tag{14.43}$$

This implies that $\Delta_0^+ = \Delta_0^- = -\Delta = -y^2(\partial_x^2 + \partial_y^2)$ which is the usual Laplacian on $\mathcal{H}$.

All the operators $\Delta_n^\pm$ are non-negative and self-adjoint provided one introduces a scalar product on T^n

$$\begin{aligned} f_1, f_2 \in T^n : \quad < f_1, f_2 >_{T^n} &:= \int_M d^2 z \sqrt{g}(g^{z\bar{z}})^n f_1^*(z)f_2(z) \\ &= 2^{n-1} \int_M d^2 z y^{2n-2} f_1^*(z)f_2(z) . \end{aligned} \tag{14.44}$$

One then can define spinor fields in the following way: We introduce $\bar{\Gamma} \subset \mathrm{PSL}(2,\mathbb{R})$, $-\mathbb{1} \in \bar{\Gamma}$, with $\bar{\Gamma}\backslash\{\pm\mathbb{1}\} = \Gamma$. Elements $\gamma \in \Gamma$ act on $z \in \mathcal{H}$ via the Möbius transformation

$$z \mapsto \gamma z = \frac{az+b}{cz+d} . \tag{14.45}$$

We further introduce a character χ such that $\chi : \bar{\Gamma} \mapsto \{\pm\mathbb{1}\}$, $\chi(-1) = -1$. We define the space $S(2n)$ of automorphic forms of weight n on M by

$$S(2n) := \left\{ f : \mathcal{H} \mapsto \mathbb{C} \middle| f(\gamma z) = \chi^{2n}(\gamma)\left(\frac{cz+d}{|cz+d|}\right)^{2n} f(z); \gamma = \begin{pmatrix} a & b \\ c & d \end{pmatrix} \in \bar{\Gamma} \right\} . \tag{14.46}$$

On $S(2n)$ the scalar product has the following form

$$f_1, f_2 \in T^n : \quad < f_1, f_2 >_{T^n} = 2^{n-1} \int_M d^2 z y^{-2} f_1^*(z)f_2(z) . \tag{14.47}$$

We then find that one can introduce self-adjoint second order differential operators on spinor-fields in the following way

$$D_{2n} := -\Delta + 2iny\partial_x = -y^2(\partial_x^2 + \partial_y^2) + 2iny\partial_x \ , \tag{14.48}$$

and the operators $D_{2n} + n(n \pm 1)$ and $\Delta_n^\pm$ are conjugate under the isometry $I : T^n \mapsto S(2n)$, $f(z)dz^n \mapsto y^n f(z)$.

In the following I formulate the Selberg trace formula for the Maass Laplacians D_{2n}. Later on we can calculate determinants of these Laplacians on Riemann surfaces by means of the zeta-function regularization and express these determinants in terms of the Selberg zeta-function. In the sequel we will do this for closed Riemann surfaces and for bordered Riemann surfaces.

Now, let g denote the genus of a Riemann surface. Let us denote by s the number of inequivalent elliptic fixed points and by κ the number of inequivalent cusps. ν_j denotes the order of the generators of the elliptic subgroups $R_j \subset \Gamma$ ($1 \leq j \leq s$); this means that $R_j^{\nu_j} = \mathbb{1}$ for $1 \leq j \leq s$ ($1 < \nu_j < \infty$). The non-Euclidean area of the Riemann surface is given by

$$\mathcal{A}(\mathcal{F}) = 2\pi \left[2(g-1) + \kappa + \sum_{j=1}^s \left(1 - \frac{1}{\nu_j}\right) \right] \ . \tag{14.49}$$

If Γ contains no elliptic elements, i.e., no $\gamma \in \Gamma$ has a fixed point on $\mathcal{H}$, $\Gamma\backslash\mathcal{H}$ is a regular surface. If elliptic elements are present in Γ, $\Gamma\backslash\mathcal{H}$ will only have a manifold structure outside the respective fixed points. Including these turns $\Gamma/\backslash\mathcal{H}$ into an orbifold. Despite this slight complication any $\Gamma\backslash\mathcal{H}$ will in the following be called a hyperbolic surface, irrespective of a possible existence of orbifold points.

On a regular hyperbolic surface the conjugacy classes $\{\gamma\}_\Gamma$ of any hyperbolic $\gamma \in \Gamma$ are in one-to-one correspondence with the closed geodesics on the surface and we denote by l_γ the *length* of the closed geodesic on the surface related to $\{\gamma\}_\Gamma$. The *norm* N_γ of a hyperbolic element $\gamma \in \Gamma$ and the length l_γ are related by $N_\gamma = e^{l_\gamma}$. The norm of a hyperbolic element is given by $N_\gamma = a + d$. This one-to-one correspondence is no longer true if elliptic elements are present. However, the norms N_γ of conjugacy classes in Γ are still properly defined, and we use sometimes the notion "lengths of closed orbits" and "norms of conjugacy classes" irrespective of a possible existence of such orbifold points, keeping in mind that "norms of conjugacy classes" is the more correct one.

We define ($m \in \mathbb{Z}$)

$$j(\gamma, z) = \left(\frac{cz+d}{c\bar{z}+d}\right)^{m/2} = \left(\frac{cz+d}{|cz+d|}\right)^m \ , \qquad \gamma = \begin{pmatrix} a & b \\ c & d \end{pmatrix} \in \bar{\Gamma} \ , \qquad z \in \mathcal{H} \ . \tag{14.50}$$

Obviously we have

$$\begin{aligned} |j(\gamma, z)| &= 1 \ , \\ j(\gamma\sigma, z) &= j(\gamma, \sigma z)j(\sigma, z) \ , \qquad \forall \gamma, \sigma \in \bar{\Gamma} \ . \end{aligned} \tag{14.51}$$

We define automorphic forms $f(z)$ of weight m to be $\mathbb{C}$-valued functions on $\mathcal{H}$ having the property

$$f(\gamma z) = \chi_\gamma^m j(\gamma, z) f(z) , \qquad \forall \gamma \in \bar{\Gamma} . \tag{14.52}$$

The set of such differentiable automorphic forms will be denoted by $C^\infty(\chi, m)$, and $\mathcal{L}^2(\chi, m)$ is the space of square integrable automorphic forms, i.e.,

$$\int_{\Gamma \backslash \mathcal{H}} dV(z) |f(z)|^2 < \infty, \qquad dV(z) = \frac{dxdy}{y^2} . \tag{14.53}$$

We consider the operator $D_m = -y^2(\partial_x^2 + \partial_y^2) + imy\partial_x$ acting on $C^\infty(\chi, m)$. There exists a (unique) self-adjoint extension of this operator on $\mathcal{L}^2(\chi, m)$ which we also denote by D_m and which is known as the Maass-Laplacian. For a general Fuchsian group of the first kind D_m has a discrete and a continuous spectrum. If $\Gamma \backslash \mathcal{H}$ is compact, however, there is only a discrete spectrum. In case of a non-cocompact Fuchsian group it is not known in general whether Maass-Laplacians have infinitely many eigenvalues. Only for arithmetic Fuchsian groups it is known that the discrete spectrum is infinite, see e.g., [509].

We introduce the point pair invariant

$$
\begin{aligned}
k(z, w) &= \left(\frac{w - \bar{z}}{z - \bar{w}}\right)^{m/2} \Phi\left(\frac{|z - w|^2}{\Im(z)\Im(w)}\right) , \tag{14.54} \\
&= i^m \left(\frac{w - \bar{z}}{|w - \bar{z}|}\right)^m \Phi\left(\frac{|z - w|^2}{\Im(z)\Im(w)}\right) , \tag{14.55}
\end{aligned}
$$

for some $\Phi \in C_c^2(\mathbb{R})$ for $z, w \in \mathcal{H}$. $k(z, w)$ has the properties

$$
\begin{aligned}
k(z, w) &= \overline{k(w, z)} , \\
k(\gamma z, \gamma w) &= j(\gamma, z) k(z, w) j^{-1}(\gamma, w) ,
\end{aligned}
\tag{14.56}
$$

for all $\gamma \in \bar{\Gamma}$. $k(z, w)$ is the integral operator corresponding to an operator valued function $h(D_m)$ such that

$$< \Psi | h(D_m) | z > = \int_{\mathcal{H}} k(z, w) \Psi(w) dV(w) . \tag{14.57}$$

We introduce the automorphic kernel by

$$K(z, w) = \frac{1}{2} \sum_{\gamma \in \bar{\Gamma}} \chi_\gamma^m j(\gamma, w) k(z, \gamma w) , \tag{14.58}$$

and the factor "1/2" is included because $\gamma \in \bar{\Gamma}$ runs through γ and $-\gamma$, respectively. The automorphic kernel has the properties

$$
\begin{aligned}
K(z, w) &= \overline{K(w, z)} , \\
K(\gamma z, \sigma w) &= \chi_\gamma^m j(\gamma, z) K(z, w) j^{-1}(\sigma, w) \chi_\sigma^{-m} ,
\end{aligned}
\tag{14.59}
$$

for all $\gamma, \sigma \in \bar{\Gamma}$. This construction of the automorphic kernel is valid for arbitrary Fuchsian groups.

In the case that Γ is strictly hyperbolic, i.e., except the identity it contains only hyperbolic elements, the trace of the Selberg operator L is given by

$$\text{tr}(L) = \int_{\mathcal{F}} dV(z)K(z,z) = \sum_{n=0}^{\infty} h(p_n) \ , \tag{14.60}$$

where $\Lambda(\lambda_n) = \Lambda(p_n^2 + \frac{1}{4}) = h(p_n)$. One now obtains the following

Theorem 14.2 *The Selberg trace formula for automorphic forms of weight $m \in \mathbb{Z}$ on compact Riemann surfaces has the form*

$$\sum_{n=0}^{\infty} h(p_n) = -\frac{A}{8\pi^2}\int_0^{\infty} \frac{\cosh \frac{m}{2}u}{\sinh \frac{u}{2}}g'(u)du + \sum_{\{\gamma\}_p}\sum_{k=1}^{\infty} \frac{\chi_\gamma^{mk}l_\gamma g(kl_\gamma)}{2\sinh \frac{kl_\gamma}{2}} \ . \tag{14.61}$$

Here on the left hand side n labels all eigenvalues $\lambda_n = \frac{1}{4} + p_n^2$ of D_m, where only one root p_n is counted. $h(p)$ denotes an even function in p and has the properties

1. *$h(p)$ is holomorphic in the strip $|\Im(p)| \leq \frac{1}{2} + \epsilon$, $\epsilon > 0$.*

2. *$h(p)$ has to decrease faster than $|p|^{-2}$ for $p \to \pm\infty$.*

3. *$g(u) = \pi^{-1}\int_0^{\infty} h(p)\cos(\pi p)dp$.*

The first requirement causes a growing condition on the Fourier transform $g(u)$ of $h(p)$ such that the exponential proliferation in the number of the norms of the hyperbolic conjugacy classes is matched in order that the sum of the hyperbolic conjugacy classes is absolutely convergent.

Let us note that the first term on the right hand side in the trace formula of Theorem 14.2 can also be written as

$$-\frac{A}{8\pi^2}\int_0^{\infty} \frac{\cosh \frac{m}{2}u}{\sinh \frac{u}{2}}g'(u)du = \frac{A}{4\pi}\left\{\int_{\mathbb{R}} dp\, p h(p)\frac{\sinh(2\pi p)}{\cosh(2\pi p) + \cos(\pi m)}\right.$$
$$\left. + \sum_{n=0}^{[(m-1)/2]} (m - 2n - 1)h[\tfrac{i}{2}(m - 2n - 1)]\right\} \ , \tag{14.62}$$

where $[x]$ is the integer part of x.

$\Phi(x)$ is the kernel function of the operator valued function $h(\sqrt{D_m - 1/4})$. As usual, the l_γ's denote the *lengths* of closed geodesics on Σ, where we use the fact that the conjugacy classes $\{\gamma\}_\Gamma$ of hyperbolic $\gamma \in \Gamma$ are in one-to-one correspondence with the closed geodesics on Σ. The *norm* N_γ of an element $\gamma \in \Gamma$ and the length l_γ are related by $N_\gamma = e^{l_\gamma}$. $\Phi(x)$ and $g(u)$ are connected through

$$g(u) = (-1)^{m/2} \int_{\mathbb{R}} d\zeta \left(\frac{\zeta + 2i \cosh \frac{u}{2}}{\zeta - 2i \cosh \frac{u}{2}} \right)^{m/2} \Phi(\zeta^2 + 4 \sinh^2 \tfrac{u}{2}) , \qquad (14.63)$$

$$\Phi(x) = -\frac{1}{\pi} \int_{\mathbb{R}} Q'(x + t^2) \left(\frac{\sqrt{x + 4 + t^2} \, t}{\sqrt{x + 4 + t^2} + t} \right)^{m/2} dt, \qquad (14.64)$$

with $Q(w) = g(u)$, where $w = 4 \sinh^2 \frac{u}{2}$. $Q(w)$ can also be expanded according to $(w \geq 0)$

$$Q(w) = \int_w^\infty \frac{du \Phi(u)}{\sqrt{u - w}} (u + 4)^{-m/2} \left[\sum_{l=0}^{m/2} \binom{m}{2l} (u - w)^l i^{2l} (w + 4)^{-l+m/2} \right] . \qquad (14.65)$$

Let κ be the number of inequivalent cusps z_α ($\alpha = 1, \ldots, \kappa$) of Γ, i.e., the number of zero interior angles of the fundamental polygon $\mathcal{F}$. To each cusp there is associated an *Eisenstein series*

$$e(z, s, \alpha) = \sum_{\gamma \in \Gamma_\alpha \backslash \Gamma} y^s(\gamma z) , \qquad (14.66)$$

$z \in \mathcal{H}$, $\Re(s) > 1$, $\alpha = 1, \ldots, \kappa$, with Γ_α being the stabilizer of the cusp α. In the spectral decomposition of D_m on $\mathcal{L}^2(\chi, m)$ these Eisenstein series span the continuous spectrum. For each z_α we consider the maximal subgroup $\Gamma_\alpha \subset \Gamma$ which stabilizes it. The subgroup Γ_α is generated by a single parabolic element S_α. For each $\alpha = 1, \ldots, \kappa$ there exists a transformation $g_\alpha \in \mathrm{PSL}(2, \mathbb{R})$ such that $g_\alpha \infty = z_\alpha$, $g_\alpha^{-1} S_\alpha g_\alpha z = S_\infty z = z + 1$ ($z \in \mathcal{H}$). Let V be an h-dimensional complex vector space, $V = \mathbb{C}^h$. Let U be a representation of Γ which acts in the space V and is unitary with respect to the inner product in V. For each $\alpha = 1, \ldots, \kappa$ we have a subspace $V_\alpha \subset V$ of the operator $U(S_\alpha)$, i.e., $V_\alpha = \{v \in V | U(s_j)v = v\}$. Let $k_\alpha = \dim V_\alpha$ and $\kappa_0 = \sum_{\alpha=1}^\kappa k_\alpha$. k_α denotes the degree of singularity of the representation U relative to the generator S_α of $\Gamma_\alpha \subset \Gamma$ and κ_0 denotes the degree of singularity of Γ relative to U. For each α ($\alpha = 1, \ldots, \kappa$) one chooses a basis $v_1(\alpha), \ldots, v_h(\alpha)$ for V such that

$$U(S_\alpha)(1_V - P_\alpha)v_l(\alpha) = \nu_{l\alpha} v_l(\alpha), \qquad (14.67)$$

with P_α the projector on the subspace V_α, and where it is supposed that we have the alternatives

$$\nu_{l\alpha} v_l(\alpha) = \begin{cases} 0 \\ e^{2\pi i \vartheta_{l\alpha}}, \end{cases} \qquad (14.68)$$

with the numbers $\vartheta_{l\alpha} \in (0, 1)$. For $\kappa_0 \geq 1$ the contributions corresponding to the parabolic conjugacy classes must be regularized, i.e., the continuous spectrum of the automorphic kernel must be subtracted form the Selberg trace formula. In order to do this one considers the Eisenstein series

$$e(z, s, \alpha, v, \Gamma, U) = e(z, s, \alpha, v) = \sum_{\gamma \in \Gamma_\alpha \backslash \Gamma} y^s(g_\alpha^{-1} \gamma z) u^*(\gamma) v , \qquad (14.69)$$

where $\alpha = 1, \ldots, \kappa$, $v \in V_\alpha \subset V$, $g_\alpha \in \mathrm{PSL}(2, \mathbb{R})$, and $z \in \mathcal{H}$. These Eisenstein series span the continuous spectrum of the Laplacian on the Riemann surface. A Fourier expansion of the Eisenstein series (14.66) then yields

$$e_\beta(z, s, \alpha, v_l(\alpha)) = P_\beta e(z, s, \alpha, v_l(\alpha)) = \sum_{m \in \mathbb{Z}} a_m(y, s) e^{2\pi i m x} , \qquad (14.70)$$

with the coefficients

$$\left.\begin{aligned} a_0(y, s) &= \delta_{\alpha\beta} v_l(\alpha) y^s + \sqrt{\pi} \frac{\Gamma(s - \tfrac{1}{2})}{\Gamma(s)} y^{1-s} \eta_0(s) , \\ a_m(y, s) &= \frac{2\pi^s}{\Gamma(s)} |m|^{s-1/2} \sqrt{y}\, K_{s-\frac{1}{2}} \eta_m(s) , \qquad (m \neq 0) , \end{aligned}\right\} \qquad (14.71)$$

with (see [509] for proper definitions and more details)

$$\eta_m(s) = \sum_{\gamma \in \Gamma_\alpha \backslash \Gamma / \Gamma_\beta} \frac{P_\beta U^*(\gamma) v_l(\alpha)}{|c(g_\alpha^{-1} \gamma g_\beta)|^{2s}} \exp\left[2\pi i m \frac{d(g_\alpha^{-1} \gamma g_\beta)}{c(g_\alpha^{-1} \gamma g_\beta)} \right] , \qquad (14.72)$$

where $c > 0$, $d \bmod c$, $\begin{pmatrix} * & * \\ c & d \end{pmatrix} \in g_\beta^{-1} \Gamma g_\alpha$. The $(\kappa_0 \times \kappa_0)$-matrix

$$\mathfrak{S}_{\alpha l, \beta k} = \sqrt{\pi} \frac{\Gamma(s - \tfrac{1}{2})}{\Gamma(s)} \eta_0(s) \qquad (14.73)$$

is called scattering matrix and has the properties

1. $e(z, s, \alpha, v, \Gamma, U) = \mathfrak{S}(s) e(z, 1 - s, \alpha, v, \Gamma, U)$.

2. $\mathfrak{S}(s) \mathfrak{S}(1 - s) = 1$ which is evidently true for $\kappa = 1$ and is explicitly given in matrix notation by

$$\sum_{\beta=1}^{\kappa} \sum_{k=1}^{k_\beta} \mathfrak{S}_{\alpha l, \beta k}(s) \mathfrak{S}_{\beta k, \gamma m}(1 - s) = \delta_{\alpha\gamma} \delta_{lm} . \qquad (14.74)$$

Let us denote $\Delta(s) = \det(\mathfrak{S}_{\alpha l, \beta k})$. The function $\Delta(s)$ has the following properties [509]:

1. For any $p \in \mathbb{R}$, $\Delta(\tfrac{1}{2} + ip) \neq 0$.

2. It satisfies the functional equation $\Delta(s)\Delta(1 - s) = 1$, $\Delta^*(s^*) = \Delta(s)$, and in particular $|\Delta(\tfrac{1}{2} + ip)| = 1$.

3. It is regular in the half-plane $\Re(s) > \tfrac{1}{2}$ except for a finite number of poles on the interval of the real axis $s \in (\tfrac{1}{2}, 1]$ denoted by $\sigma_1, \sigma_2, \ldots, \sigma_M$, which give due to the functional relation zeros, symmetric with respect to $s = \tfrac{1}{2}$ in the interval $s \in [0, \tfrac{1}{2})$.

4. In the half-plane $\Re(s) < \frac{1}{2}$, $\Delta(s)$ has poles at $\varrho = \beta + i\gamma$ ($\beta < \frac{1}{2}$) and the logarithmic derivative of $\Delta(\frac{1}{2} + ip)$ can be represented as

$$-\frac{\Delta'(\frac{1}{2} + ip)}{\Delta(\frac{1}{2} + ip)} = \sum_\varrho \frac{1 - 2\varrho}{(\beta - \frac{1}{2})^2 + (p - \gamma)^2} + \text{const.} + O\left(\frac{1}{p^2}\right), \qquad (p \to \infty, p \in \mathbb{R})$$

(14.75)

with the summation over all poles $\varrho = \beta + i\gamma$ of $\Delta(s)$ in the half-plane $\Re(s) < \frac{1}{2}$. Note that $\varrho^* = \beta - i\gamma$ are poles as well, and due to the functional relation we have at $s = 1 - \varrho$ and $s = 1 - \varrho^*$, respectively, zeros for $\Delta(s)$.

We can write down the Selberg trace formula as follows (Hejhal [270] and Venkov [509, p.76])

Theorem 14.3 *The Selberg trace formula for arbitrary Riemann surfaces has the following form*

$$\sum_{n=0}^{\infty} h(p_n) = \frac{A}{2\pi}\dim V \int_{\mathbb{R}} p \tanh \pi p \, h(p) dp + \sum_{\{\gamma\}} \sum_{k=1}^{\infty} \frac{\text{tr}_V[U^k(\gamma)]l_\gamma}{\sinh \frac{kl_\gamma}{2}} g(kl_\gamma)$$

$$+ \sum_{\{R\}} \sum_{k=1}^{\nu-1} \frac{\text{tr}_V[U^k(R)]}{2\nu \sin(k\pi/\nu)} \int_{\mathbb{R}} h(p) \frac{\cosh[\pi(1 - 2k/\nu)p]}{\cosh \pi p} dp$$

$$+ \frac{1}{2\pi} \int_{\mathbb{R}} \frac{\Delta'(\frac{1}{2} + ip)}{\Delta(\frac{1}{2} + ip)} h(p) dp - \frac{\kappa_0}{\pi} \int_{\mathbb{R}} h(p) \Psi(1 + ip) dp$$

$$+ \frac{1}{2}[\kappa_0 - \text{tr}(\mathfrak{S})]h(0) - 2\left[\kappa_0 \ln 2 + \sum_{\alpha=1}^{\nu} \sum_{l=1+k_\alpha}^{\dim V} \ln\left|1 - e^{2\pi i\vartheta_{l\alpha}}\right|\right] g(0). \quad (14.76)$$

On the left hand side n runs through the set of eigenvalues $E_n = \frac{1}{4} + p_n^2$ of the discrete spectrum of the Laplacian, where we take both values of p_n which give the same E_n. On the right side $\{\gamma\}$ runs through the set of all primitive hyperbolic conjugacy classes in Γ, and $\{R\}$ runs through the set of all primitive elliptic conjugacy classes in Γ. The test function h must satisfy the following properties

1. *$h(p)$ is an even function in p,*

2. *$h(p)$ is analytic in the strip $|\Im(p)| < \frac{1}{2} + \epsilon$ for some $\epsilon > 0$,*

3. *and $h(p)$ vanishes according to $h(p) = O[1/(1 + p^2)^{2+\epsilon}]$ for some $\epsilon > 0$ for $p \to \pm\infty$.*

Sometimes one writes $h(\frac{1}{4} + p^2) = h(E)$ instead of $h(p)$ to emphasize the dependence on the energy eigenvalues of the Laplacian. Whereas this Selberg trace formula corresponds to the scalar Laplacian, we want to go further and state the Selberg trace formula for the Maass Laplacian of weight m, $m \in \mathbb{Z}$ (actually it is even possible to state the trace formula for an arbitrary weight $m \in \mathbb{R}$ which can be done by analytic continuation [270]). Therefore we have [270, pp.403, 413]:

Theorem 14.4 *The Selberg trace formula for automorphic forms of weight m, $m \in \mathbb{Z}$ on arbitrary Riemann surfaces has the form*

$$
\sum_{n=0}^{\infty} h(p_n) = \frac{A}{4\pi^2} \dim V \int_0^{\infty} \frac{\cosh \frac{m}{2}u}{\sinh \frac{u}{2}} g'(u) du + \sum_{\{\gamma\}} \sum_{k=1}^{\infty} \frac{\operatorname{tr}_V[U^k(\gamma)]\chi_\gamma^{mk}l_\gamma}{\sinh \frac{kl_\gamma}{2}} g(kl_\gamma)
$$

$$
+ \frac{i}{2} \sum_{\{R\}} \sum_{k=1}^{\nu-1} \frac{\operatorname{tr}_V[U^k(R)]\chi_R^{mk}\mathrm{e}^{\mathrm{i}(m-1)k\pi/\nu}}{\nu \sin(k\pi/\nu)} \int_{\mathbb{R}} du\, g(u)\mathrm{e}^{(m-1)u/2} \frac{\mathrm{e}^u - \mathrm{e}^{2\mathrm{i}k\pi/\nu}}{\cosh u - \cos[\pi(1 - 2k/\nu)]}
$$

$$
+ \sum_{\alpha=1}^{\nu} \sum_{l=1+k_\alpha}^{\dim V} (\tfrac{1}{2} - \vartheta_{l\alpha}) P \int_{\mathbb{R}} du\, g(u)\mathrm{e}^{(m-1)u/2} \frac{\mathrm{e}^u - 1}{\cosh u - 1}
$$

$$
+ \frac{1}{2\pi} \int_{\mathbb{R}} \frac{\Delta'(\frac{1}{2} + \mathrm{i}p)}{\Delta(\frac{1}{2} + \mathrm{i}p)} h(p) dp - \frac{\kappa_0}{\pi} \int_{\mathbb{R}} h(p) \Psi(1 + \mathrm{i}p) dp
$$

$$
+ \frac{1}{2}[\kappa_0 - \operatorname{tr}(\mathfrak{S})]h(0) - 2\left[\kappa_0 \ln 2 + \sum_{\alpha=1}^{\nu} \sum_{l=1+k_\alpha}^{\dim V} \ln\left|1 - \mathrm{e}^{2\pi\mathrm{i}\vartheta_{l\alpha}}\right|\right] g(0)
$$

$$
+ \kappa_0 \int_0^{\infty} du\, g(u) \frac{1 - \cosh \frac{m}{2}u}{\sinh \frac{u}{2}} \ . \tag{14.77}
$$

On the left hand side n runs through the set of eigenvalues $E_n = \frac{1}{4} + p_n^2$ of the discrete spectrum of the Laplacian, where we take both values of p_n which give the same E_n. On the right side $\{\gamma\}$ runs through the set of all primitive hyperbolic conjugacy classes in Γ, and $\{R\}$ runs through the set of all primitive elliptic conjugacy classes in Γ. The test function h must fulfill the same conditions as in the previous theorem.

14.3.1 The Selberg Zeta-Function

The Selberg zeta-function is defined as

$$
Z_\nu(s) = \prod_{\{\gamma\}} \prod_{k=0}^{\infty} \left[1 - \chi^\nu(\gamma)\mathrm{e}^{-(k+s)l_\gamma}\right] \ , \qquad \nu = 0, 1 \ . \tag{14.78}
$$

Here $\{\gamma\}$ runs over all primitive conjugacy classes in Γ (if $\nu = 0$) or $\bar{\Gamma}$ (if $\nu = 1$). In order to study the analytic properties of this function we must choose a proper test function in the trace formula. An appropriate function is

$$
h(p^2 + \tfrac{1}{4}, s, a) = \frac{1}{(s - \frac{1}{2})^2 + p^2} - \frac{1}{(a - \frac{1}{2})^2 + p^2} \ . \tag{14.79}
$$

The corresponding Fourier transform has the form

$$
g(u, s, a) = \frac{\mathrm{e}^{-(s-1/2)|u|}}{2s - 1} - \frac{\mathrm{e}^{-(a-1/2)|u|}}{2a - 1} \ . \tag{14.80}
$$

We only give the case of $m = 0$. The case $m \in \mathbb{Z}$ is very similar, and will actually be cited in the case of bordered Riemann surfaces (see below). We therefore obtain (Hejhal [270, pp.435] and Venkov [509, pp.81]):

Theorem 14.5 *Suppose that $s \in \mathbb{C}$, $\Re(s) > 1$, and let $a \in \mathbb{R}$, $a > 1$, be fixed. The Selberg trace formula for the Selberg zeta-function has the form:*

$$\frac{Z'(s)}{Z(s)} = -(s - \tfrac{1}{2})\frac{\mathcal{A}\dim V}{\pi} \sum_{k=0}^{\infty} \left(\frac{1}{s+k} - \frac{1}{a+k} \right)$$

$$- \sum_{\{R\}} \sum_{k=1}^{\nu-1} \frac{\mathrm{tr}_V[U^k(R)]}{\nu \sin(\pi k/\nu)} \frac{\pi e^{-2\pi i k(s-1/2)/\nu}}{1 - e^{-2\pi i s}} + i(s - \tfrac{1}{2}) \sum_{l=1}^{\infty} \frac{e^{-2\pi i k(l-1/2)/\nu}}{(s - \tfrac{1}{2})^2 - (l - \tfrac{1}{2})^2}$$

$$- \frac{1}{2}\frac{\Delta'(s)}{\Delta(s)} + (s - \tfrac{1}{2}) \sum_{\varrho,\beta < \frac{1}{2}} \left(\frac{1}{(s - \tfrac{1}{2})^2 - (\varrho - \tfrac{1}{2})^2} - \frac{1}{(a - \tfrac{1}{2})^2 - (\varrho - \tfrac{1}{2})^2} \right)$$

$$- (s - \tfrac{1}{2}) \sum_{l=1}^{\mathcal{M}} \frac{1}{(s - \tfrac{1}{2})^2 - (\sigma_l - \tfrac{1}{2})^2} + \kappa_0 \Psi(\tfrac{3}{2} - s)$$

$$- 2\kappa_0(s - \tfrac{1}{2}) \sum_{k=1}^{\infty} \frac{1}{(s - \tfrac{1}{2})^2 - k^2} - \frac{1}{2s - 1}\left[\kappa_0 - \mathrm{tr}(\mathfrak{S}(\tfrac{1}{2})) \right]$$

$$+ (s - \tfrac{1}{2}) \sum_{j} \left(\frac{1}{(s - \tfrac{1}{2})^2 + p_j^2} - \frac{1}{(a - \tfrac{1}{2})^2 + p_j^2} \right) + \mathrm{const}_1 + \mathrm{const}_2(s - \tfrac{1}{2}) \ . \quad (14.81)$$

with some constants $\mathrm{const}_{1,2}$.

The zero- and pole-structure of the Selberg zeta-function $Z(s)$ can be read off:

Theorem 14.6 *The Selberg zeta-function $Z(s)$ is a meromorphic function of $s \in \mathbb{C}$ of order equal to two. The zeros of the function $Z(s)$ are at the following points*

1. *Nontrivial zeros:*

   (a) *On the line $\Re(s) = 1/2$, symmetric relative to the point $s = 1/2$ and on the interval $[0,1]$, symmetric to the point $s = 1/2$. We call these zeros s_j. Each s_j has a multiplicity equal twice the multiplicity of the corresponding eigenvalue E_j, $E_j = \frac{1}{4} + p_j^2$, and E_j runs through the discrete spectrum of $-\Delta$.*

   (b) *At the points s_j on the interval $\frac{1}{2} < s \leq \frac{1}{2}$ of the real axis; s_j has multiplicity no greater than κ_0, the total degree of singularity of the representation χ relative to the group Γ; every such zero s_j is of the form $s_j = \sigma_j$, $j = 1, \ldots, \mathcal{M}$, and is connected with an eigenvalue E_j of the discrete spectrum of the Laplacian in the interval $[0, \frac{1}{4}]$.*

   (c) *At the poles ϱ of the function $\Delta(s)$ which lie in the half-plane $\Re(s) < \frac{1}{2}$ and have the same multiplicity as the poles of $\Delta(s)$.*

2. *Trivial zeros: At the points $s = -l$, $l \in \mathbb{N}_0$, with multiplicity $\#N_{-l}$ given by*

$$\#N_{-l} = \frac{\mathcal{A}\dim V}{\pi}(l + \tfrac{1}{2}) - \sum_{\{R\}_p} \sum_{k=1}^{\nu-1} \sin\left(\frac{k\pi(2l+1)}{\nu} \right) \frac{\mathrm{tr}_V[U^k(R)]}{\nu \sin(k\pi/\nu)} \ . \quad (14.82)$$

$Z(s)$ has poles at the points

1. $s = \frac{1}{2}$ *with multiplicity* $\frac{1}{2}(\kappa_0 - \mathrm{tr}\mathfrak{S}(\frac{1}{2}))$,

2. $s = -l + \frac{1}{2}$, $l \in \mathbb{N}$, *with multiplicity* κ_0. $s = l$, $l \in \mathbb{Z}$, *with multiplicity* $\#P_l = 2\kappa_0$.

Furthermore we have:

Theorem 14.7 *The functional equation has the form*

$$Z(1-s) = Z(s)\Delta(s)\psi(s) \tag{14.83}$$

with the function $\psi(s)$ given by

$$
\begin{aligned}
\psi(s) = {}& \left(\frac{\Gamma(\frac{3}{2} - s)}{\Gamma(s + \frac{1}{2})}\right)^{\kappa_0} \exp\Bigg\{ -\mathcal{A}\dim V \int_0^{s-1/2} t \tanh \pi t \, dt \\
& + \pi \sum_{\{R\}} \sum_{k=1}^{\nu-1} \frac{\mathrm{tr}_V[U^k(R)]}{\nu \sin(k\pi/\nu)} \int_0^{s-1/2} \left(\frac{e^{-2\pi ikt/\nu}}{1 + e^{-2\pi it}} + \frac{e^{-2\pi ikt/\nu}}{1 + e^{2\pi it}}\right) \\
& + (1 - 2s)\left(\kappa_0 \ln 2 + \sum_{\alpha=1}^{\nu} \sum_{l=1+k_\alpha}^{\dim V} \ln\left|1 - e^{2\pi i\vartheta_{l\alpha}}\right|\right) - i \arg\Delta(\tfrac{1}{2})\Bigg\} .
\end{aligned}
\tag{14.84}
$$

14.3.2   Determinants of Maass-Laplacians

We want to calculate determinants of Maass Laplacians on Riemann surfaces. This is done by the zeta-function regularization method. We assume that an operator A is non-negative and self-adjoint on a compact manifold, thus possesses a discrete spectrum with a complete set of eigenvectors. Denote by $0 = \lambda_0 \leq \lambda_1 \leq \lambda_2 \leq \ldots \leq \lambda_j \nearrow \infty$ the eigenvalues of A. One then uses the zeta-function of Minakshisundaram and Pleijel [407] (MP-zeta-function) to regularize the functional determinant of A in the following way:

$$\zeta_A(s) := \mathrm{tr}'(A^{-s}) = \sum_{n=1}^{\infty} \lambda_n^{-s} , \qquad \Re(s) > 1 . \tag{14.85}$$

A prime denotes the omission of possible zero-modes of A. The convergence domain $\Re(s) > 1$ follows from the property of the eigenvalues $\lambda_n = O(n)$, $n \to \infty$. This method of regularization of determinants by zeta-functions was introduced by Ray and Singer [451] in differential geometry and Hawking [266] in field theory.

The functional determinant of the operator A is defined by

$$\det{}' A := \exp\left[-\frac{d}{ds}\zeta_A(s)\Big|_{s=0}\right] . \tag{14.86}$$

Let us consider the trace of the heat kernel of the operator A

$$\Theta_A(t) := \text{tr}(e^{-At}) = \sum_{n=0}^{\infty} e^{-\lambda_n t} , \qquad t > 0 . \tag{14.87}$$

By means of $\Theta_A(t)$ we can rewrite the MP-zeta-function as follows

$$\zeta_A(s) = \frac{1}{\Gamma(s)} \int_0^{\infty} dt \, t^{-s} [\Theta_A(s) - d_0] , \qquad \Re(s) > 1 . \tag{14.88}$$

Here $d_0 = \dim(\ker A)$ is the number of zero modes of the operator A. We will make extensive use of this important equation.

For operators A (i.e., Laplacians) on surfaces, the heat kernels have a small-t asymptotic according to $\Theta_A(t) = \frac{a}{t} + O(1)$ $(t \to 0)$. Therefore the leading term of the asymptotic eigenvalue distribution of A is given by Weyl's law

$$\Theta_A(t) = \int_0^{\infty} d\lambda \frac{dN(\lambda)}{d\lambda} e^{-\lambda t} , \qquad N(\lambda) := \#\{\lambda_n | \lambda_n \leq \lambda\} , \tag{14.89}$$

$$\Rightarrow \quad N(\lambda) \sim a\lambda , \qquad \lambda \to \infty . \tag{14.90}$$

The last line implies $\lambda_n \sim \lambda/\mathcal{A}$ for $n \to \infty$. The constant a is essentially given by the *area* of the compact surface, i.e., $a = \mathcal{A}/4\pi$. If the area is non-compact but still finite the same asymptotic behaviour is valid, for instance in the modular domain [507]. If the area is infinite as, e.g., in the hyperbola billiard, these considerations must be modified and the small t behaviour is considerably altered [481].

Let us sketch the evaluation of the determinant of the scalar Laplacian $-\Delta$ [479]. We calculate the determinant of the scalar Laplacian by considering the following trace

$$\zeta_\Delta(s) = \text{tr}'(-\Delta)^s = \sum_{n \in \mathbb{N}} \lambda^{-s} . \tag{14.91}$$

Due to $Z(1) = 0$ we must consider the following limit

$$\zeta_\Delta(s) = \lim_{\sigma \to 1+} \left\{ k_\Delta(s; \sigma) - \frac{1}{[\sigma(1-\sigma)]^s} \right\} , \tag{14.92}$$

$$k_\Delta(s; \sigma) = \text{tr}'(-\Delta + \sigma(1-\sigma))^s = \sum_{n \in \mathbb{N}} [\lambda + \sigma(1-\sigma)]^{-s} \tag{14.93}$$

$$= \frac{1}{\Gamma(s)} \int_0^{\infty} dt \, t^{s-1} e^{-\sigma(1-\sigma)t} \text{tr}(e^{\Delta t}) . \tag{14.94}$$

Performing in a proper way the limiting procedure yields

$$\det'(-\Delta) = e^{-2(g-1)C} Z'(1) , \tag{14.95}$$

$$C = -\frac{\pi}{2} \int_0^{\infty} dr \frac{r^2 + \frac{1}{4}}{\cosh^2 \pi r} [1 - \ln(r^2 + \frac{1}{4})] = \frac{1}{4} - \frac{1}{2} \ln 2\pi - 2\zeta'(-1) . \tag{14.96}$$

A rather tedious manipulation enables one to derive the following relation between the functional determinant $D_\Delta(z) = \det'(-\Delta + z)$ and the Selberg zeta-function which reads as follows [479]

$$Z(s) = s(s-1)D_\Delta(s(s-1))\left[(2\pi)^{1-s}e^{C+s(s-1)}G(s)G(s+1)\right]^{2(g-1)} , \qquad (14.97)$$

with $G(s)$ the Barnes' double Γ-function (γ = Euler's constant) [187, p.661]

$$G(z+1) = (2\pi)^{z/2}e^{-z/2(\gamma+1)/2}\prod_{n=1}^{\infty}\left[\left(1+\frac{z}{n}\right)^n e^{-z+z^2/2n}\right] . \qquad (14.98)$$

In order to apply this technique to the calculation of determinants of Maass Laplacians on Riemann surfaces, we first observe that for $n \geq 2$ the spectrum of D_{2n} on the entire $\mathcal{H}$ has both a continuous (c) and a discrete (d) contribution, e.g., c.f. [197]. Therefore the determinant factorizes into two contributions corresponding to these two parts, i.e.,

$$\det'(c\Delta_n^+) = \det'(c\Delta_n^+)_c\det'(c\Delta_n^+)_d . \qquad (14.99)$$

We have to distinguish two cases: $n \in \mathbb{N}$ and $n \in \mathbb{N}_0 + \frac{1}{2}$. We obtain

$$\det'(c\Delta_n^+)_c = c^{\zeta_n(0)}\det(-\Delta + n(n+1)) , \qquad n \in \mathbb{N} , \qquad (14.100)$$

$$\det'(c\Delta_n^+)_c = c^{\tilde{\zeta}_n(0)}\det(D_1 + n(n+1)) , \qquad n \in \mathbb{N} + \tfrac{1}{2} , \qquad (14.101)$$

where we have introduced

$$\zeta_n(s) = \text{tr}(D_0 + n(n+1))^{-s} , \qquad \tilde{\zeta}_n(s) = \text{tr}(D_1 + n(n+1))^{-s} . \qquad (14.102)$$

The latter two determinants are explicitly known and given by [72, 458, 479, 515]

$$\det(-\Delta + s(s-1)) = s(s-1)\mathcal{D}_{D_0}(s(s-1))$$

$$= Z_0(s)\left(\frac{(2\pi)^s e^{-1/4-\frac{1}{2}\ln 2\pi + 2\zeta'(-1)-s(s-1)}}{G(s)G(s+1)}\right)^{2(g-1)} , \qquad (14.103)$$

$$\det(D_1 + s(s-1)) = \mathcal{D}_{D_1}(s(s-1))$$

$$= Z_1(s)\left(\frac{(2\pi)^s e^{-1/4-\frac{1}{2}\ln 2\pi + 2\zeta'(-1)-s(s-1)}}{G^2(s+\frac{1}{2})}\right)^{2(g-1)} . \qquad (14.104)$$

Therefore we need the small t behaviour of the operators D_0 and D_1 to finish the calculation. For D_0 one has [479]

$$\Theta_{D_0}(t) = \frac{g-1}{t}\sum_{n=0}^{N}b_n t^n + O(t^N) , \qquad (14.105)$$

$$b_0 = 1, \quad b_n = \frac{(-1)^n}{2^{2n}n!}\left[1+2\sum_{k=1}^{n}\binom{n}{k}(2^{2k-1}-1)|B_{2k}|\right] , \quad n \geq 1 , \qquad (14.106)$$

$$\Theta_{D_0}(t) = (g-1)\left(\frac{1}{t}-\frac{1}{3}+O(t)\right) , \qquad (14.107)$$

and the B_{2k} are the Bernoulli numbers. Similarly one obtains for the operator D_1 [72]

$$\Theta_{D_1}(t) = \frac{g-1}{t} \sum_{n=0}^{N} a_n t^n + O(t^N) \ , \tag{14.108}$$

$$a_n = \frac{(-1)^n}{2^{2n} n!} \sum_{k=1}^{n} \binom{n}{k} (-1)^k 2^{2k} B_{2k} \ , \qquad \text{for } n \geq 0 \ , \tag{14.109}$$

$$\Theta_{D_1}(t) = (g-1)\Big(\frac{1}{t} - \frac{1}{12} + O(t)\Big) \ . \tag{14.110}$$

This gives finally for the determinants of Maass-Laplacians on Riemann surfaces for $m \in \mathbb{Z}$

$$\det(c\Delta_n^+) = Z_{2(n-[n])}(1+n)$$
$$\times \exp\Big\{(g-1)\Big[-\Big(\frac{1}{3}-\frac{1}{2}(n-[n])+n(n+1)\Big)\ln c + (2n+1)\ln(2\pi)$$
$$+ 4\zeta'(-1) - 2(n+\tfrac{1}{2})^2 - 4 \sum_{3\leq k\leq n+1} \ln\Gamma(k) + 4\Big(n-\frac{1}{2}-[n]\Big)\ln\Gamma(n+1)$$
$$+ \sum_{0\leq m\leq n-\frac{1}{2}} (2n-2m-1)\ln[c(2nm+2n-m^2-m)]\Big]\Big\} \ . \tag{14.111}$$

Here empty sums are understood as to be ignored and $[x]$ is the integer part of x. $\zeta'(-1)$ denotes the derivative of the Riemann-zeta-function $\zeta(s)$ at $s=-1$. The cases $m=0,2$, respectively, reduce to $(c=1)$

$$\det(\Delta_0^+) = Z_0'(1)\exp\Big\{(g-1)\Big[-\frac{1}{2}+\ln(2\pi)+4\zeta'(-1)\Big]\Big\} \ , \tag{14.112}$$

$$\det(\Delta_1^+) = Z_1(2)\exp\Big\{(g-1)\Big[-\frac{9}{2}+3\ln(2\pi)+4\zeta'(-1)\Big]\Big\} \ . \tag{14.113}$$

14.4 The Selberg Trace Formula on Bordered Riemann Surfaces

Let $\tilde{\Sigma}$ be a Riemann surface of genus g, and $d_1,\ldots,d_m$ conformal, non-overlapping discs on $\tilde{\Sigma}$. Then $\Sigma := \tilde{\Sigma}\setminus\{d_1,\ldots,d_m\}$ is a bordered Riemann surface with signature (g,n). $c_i = \partial d_i$ $(i=1,\ldots,n)$ are the n components of $\partial\Sigma$. Now one takes a copy $\mathcal{I}\Sigma$ of Σ, a mirror image, and glues both surfaces together along $\partial\Sigma$ and $\partial\mathcal{I}\Sigma$, yielding the doubled surface $\hat{\Sigma} := \Sigma\cup\mathcal{I}\Sigma$. Furthermore $\Sigma = \hat{\Sigma}/\mathcal{I}$, and $\hat{\Sigma}$ is a closed Riemann surface of genus $\hat{g} = 2g+n-1$. The uniformization theorem for Riemann surfaces states that $\hat{\Sigma}$ is conformally equivalent to $\Gamma\backslash U$ with the universal covering $U = \hat{\mathbb{C}}, \mathbb{C}$, or $\mathcal{H}$, and Γ a discrete, fixed point-free subgroup of the conformal automorphisms of U. For either $\hat{g}\geq 2$, or non-compact Riemann surfaces $\hat{\Sigma}$, the relevant universal covering is the Poincaré upper half-plane $\mathcal{H}$. Hence, $\hat{\Sigma}$ may be represented as $\hat{\Sigma} \cong \hat{\Gamma} \backslash \mathcal{H}$, where $\hat{\Gamma}$ is a Fuchsian group, i.e., a discrete subgroup of $\mathrm{PSL}(2, \mathbb{R})$. Σ and $\hat{\Sigma}$ may be represented as fundamental polygons $\mathcal{F}$ and $\hat{\mathcal{F}}$, respectively, tesselating the entire Poincaré upper half-plane by means of the group action.

In order to construct a convenient fundamental domain and representation of the involution $\mathcal{I}$ on it, one takes, according to Sibner [466] and Venkov [509], $\hat{\Sigma}$ as a symmetric Riemann surface with reflection symmetry $\mathcal{I}$. Then $\hat{\mathcal{F}}$ may be chosen as the interior of a fundamental polygon in $\mathcal{H}$ with $4\hat{g} + 2n - 2$ edges, and area $\mathcal{A}(\hat{\mathcal{F}})$. The fundamental polygon $\hat{\mathcal{F}}$ is symmetric with respect to the imaginary axis. That is, we can translate the polygon $\hat{\mathcal{F}}$ in such a way that the side across which $\mathcal{I}$ is a reflection, say the boundary curve c_n, runs along the y-axis in the Poincaré upper half-plane. Here $\mathcal{I}$ takes on the form

$$\mathcal{I} : z \to -\bar{z} \ . \tag{14.114}$$

The other sides are among the edges of the fundamental polygon. This choice of $\hat{\mathcal{F}}$ is adopted for convenience and in no way reduces the generality of the considerations.

Now let $\hat{\Gamma}$ be a Fuchsian group for the doubled surface $\hat{\Sigma}$, and $\bar{\Gamma} \subset \mathrm{SL}(2, \mathbb{R})$ such that $\hat{\Gamma} = \bar{\Gamma}/\{\pm 1\}$. Let $\chi : \bar{\Gamma} \to \{\pm 1\}$ be a multiplier system with $\chi(-1) = -1$. $\chi(\gamma)$ will also be denoted by χ_γ. We define $(m \in \mathbb{Z})$ even and odd automorphic forms, respectively, by having the property $[f \in \mathcal{L}^2(\chi, m)]$

$$f(\mathcal{I}z) = \chi(\mathcal{I})f(z) \ , \tag{14.115}$$

where we have extended the multiplier system χ from $\bar{\Gamma}$ to $\bar{\Gamma} \cup \bar{\Gamma}\mathcal{I}$ by setting: $\chi(\gamma\mathcal{I}) = \chi(\gamma)\chi(\mathcal{I})$ for $\gamma \in \bar{\Gamma}$. We have $\chi(\mathcal{I}) = +1$ for Neumann, and $\chi(\mathcal{I}) = -1$ for Dirichlet boundary conditions. $\Lambda(\lambda)$ only depends on Φ, m and λ. On the doubled Riemann surface $\hat{\Sigma}$, i.e., concerning the Fuchsian group $\hat{\Gamma}$, we are lead to a natural definition of the Selberg integral operator $\hat{L}_\pm$ acting on even and odd $f \in \mathcal{L}^2(\chi, m)$, respectively, as follows

$$\begin{aligned}(\hat{L}_\pm f)(z) &= \int_{\hat{\mathcal{F}}} dV(w)\hat{K}_\pm(z, w)f(w) \\ &= \frac{1}{2}\int_{\hat{\mathcal{F}}} dV(w)K(z, w)f(w) \pm \frac{1}{2}\int_{\hat{\mathcal{F}}} dV(w)K(z, -\bar{w})f(w) \ , \end{aligned} \tag{14.116}$$

with the $\pm$-sign for Dirichlet and Neumann boundary conditions on $\partial\Sigma$, respectively. Therefore we obtain for the automorphic kernel the expression

$$\hat{K}_\pm(z, w) = \frac{1}{4}\sum_{\gamma \in \bar{\Gamma}} chi_\gamma^m j(\gamma, w)k(z, \gamma w) \pm \frac{1}{4}\sum_{\gamma \in \bar{\Gamma}} \chi_\gamma^m j(\gamma, -\bar{w})k[z, \gamma(-\bar{w})] \ . \tag{14.117}$$

Let us now restrict ourselves to Dirichlet boundary conditions. In this case only the odd automorphic forms survive in the spectral expansion of the automorphic kernel. A glance at the continuous spectrum shows that the Eisenstein series $e(z, s, \alpha)$ drop out, according to a result of Venkov [509]. In the case of Dirichlet boundary conditions we are thus left with the spectral expansion of the automorphic kernel in *odd discrete* eigenfunctions Ψ_n on $\mathcal{H}$

$$\hat{K}_D(z, w) = \sum_n h(p_n)\Psi_n(z)\Psi_n(w) \ . \tag{14.118}$$

In the case of Neumann boundary conditions both the discrete and continuous spectrum contribute to the spectral expansion of the automorphic kernel. Using even eigenfunctions Φ_n and Eisenstein series $e(z, s, \alpha)$, respectively, we get

$$\hat{K}_N(z, w) = \sum_n h(p_n)\Phi_n(z)\Phi_n(w) + \frac{1}{4\pi}\int_{\mathbb{R}} dp\, h(p) \sum_{\alpha=1}^{\kappa} e(z, \tfrac{1}{2} + ip, \alpha)\overline{e(w, \tfrac{1}{2} + ip, \alpha)} \ . \tag{14.119}$$

Let us denote the composition of a $\gamma \in \bar{\Gamma}$ and $\mathcal{I}$ by $\rho = \gamma\mathcal{I}$. In order to investigate the various conjugacy classes for the formulation of the Selberg trace-formula for bordered Riemann surfaces, we have to distinguish the original conjugacy classes which appear already for closed Riemann surfaces from the additional conjugacy classes of the $\gamma\mathcal{I}$. The new conjugacy classes can be characterized by their traces. We consider first compact Riemann surfaces, i.e., compact polygons as fundamental domains. The case of closed Riemann surfaces gives us hyperbolic and elliptic conjugacy classes which correspond to $|\text{tr}(\gamma)| > 2$ and $|\text{tr}(\gamma)| < 2$, respectively. Let us denote by $g \in \text{PSL}(2, \mathbb{R})$ some arbitrary element. As it turns out we have to consider two cases for conjugacy classes of the ρ's. The first is for $\text{tr}(\rho) \neq 0$. The relative centralizer Γ_ρ of ρ then is of the form

$$\begin{pmatrix} b & 0 \\ 0 & -b^{-1} \end{pmatrix} \ , \qquad (\text{mod} \pm 1) \ , \tag{14.120}$$

where we define $\Gamma_\rho := \{\gamma \in \hat{\Gamma}|\gamma^{-1}\rho\gamma = \rho\}$, and the relative conjugacy classes by $\{\rho\}_{\hat{\Gamma}} := \{\rho' \in \hat{\Gamma}\mathcal{I}|\rho' = \gamma^{-1}\rho\gamma, \gamma \in \hat{\Gamma}\}$. Γ_ρ consists of hyperbolic elements and the identity, and since $\hat{\Gamma}$ is discrete of a single hyperbolic element. The second case is given by $\text{tr}(\rho) = 0$. The relative centralizers consist of elements of the form

$$\rho_1 = \begin{pmatrix} c & 0 \\ 0 & c^{-1} \end{pmatrix} \ , \qquad \rho_2 = \begin{pmatrix} 0 & d \\ -d^{-1} & 0 \end{pmatrix} \ , \qquad (\text{mod} \pm 1) \ . \tag{14.121}$$

ρ_2 is an elliptic element of order two. Thus Γ_ρ consists of hyperbolic, elliptic and the identity elements. However, due to the construction $\rho_1^n\rho_2$ ($n \in \mathbb{Z}$) we see that we can generate infinitely many elliptic conjugacy classes, which is impossible, since $\hat{\Gamma}$ is discrete. Therefore the relative centralizer of ρ with $\text{tr}(\rho) = 0$ consists either of hyperbolic elements and the identity or by a single elliptic generator of order two. The explicit computation reveals that for cocompact groups only the former case is possible, the latter leads to a divergence.

The conjugacy classes of $\rho \in \hat{\Gamma}\mathcal{I}$ can therefore be distinguished in two ways [61, 74] according to their squares $\rho^2 \in \hat{\Gamma}$. Let $\rho \in \hat{\Gamma}$ be primitive, that is, it is not a positive power of any other element of $\hat{\Gamma}\mathcal{I}$. Then

1. $\rho = \rho_i$, $\rho_i^2 \in \{C_i\}_{\hat{\Gamma}}$, $i = 1, \ldots, n$. The $\{C_i\}_{\hat{\Gamma}}$ are the conjugacy classes of the C_i in $\hat{\Gamma}$ which correspond to the closed geodesics c_i on $\hat{\Sigma}$.

2. $\rho = \rho_p$, ρ_p^2 being a primitive element in $\hat{\Gamma}$ and $\rho_p^2 \neq \{C_i\}_{\hat{\Gamma}}$.

Thus the sum over conjugacy classes for $\rho \in \hat{\Gamma}\mathcal{I}$ is divided into first the conjugacy classes of the C_i in $\hat{\Gamma}$, which correspond to the closed geodesics c_i on $\hat{\Sigma}$, and second into conjugacy classes such that for all $\rho \in \hat{\Gamma}\mathcal{I}$ there is a unique description $\gamma = k^{-1}\rho^{2n-1}k$ ($n \in \mathbb{N}$), for $\rho \in \hat{\Gamma}\mathcal{I}$ inconjugate and primitive, and $k \in \Gamma_{\rho^2}\backslash\hat{\Gamma}$. In the notation of Venkov [509] the relative conjugacy classes with $\mathrm{tr}(\rho) = 0$ correspond to the case 1), and the relative classes with $\mathrm{tr}(\rho) \neq 0$ correspond to 2). In this case $P(\rho) = \rho^2$ generates the relative centralizer Γ_ρ, whereas this is generated by $P(\rho) = P(\gamma\mathcal{I}) = \gamma$ in the former case. Also, for any γ under consideration with $\mathrm{tr}(\gamma\mathcal{I}) \neq 0$, $(\gamma\mathcal{I}\gamma\mathcal{I})$ is hyperbolic.

In addition, we call a relative conjugacy class $\{\rho\}_{\hat{\Gamma}}$ primitive if it is not an odd power of any other relative class $\{\rho'\}_{\hat{\Gamma}}$.

Let us continue by considering a non-compact fundamental polygon with corresponding non-cocompact Fuchsian group $\hat{\Gamma}$. Besides the already known relative conjugacy classes of $\rho \in \hat{\Gamma}\mathcal{I}$ there appear additional classes with $\mathrm{tr}(\rho) = 0$ for which the relative centralizers Γ_ρ are generated by single elliptic elements of order two. These have been excluded in the cocompact case. For each such $\rho = \gamma\mathcal{I}$ there exists an element $g \in \mathrm{PSL}(2, \mathbb{R})$ having the properties

$$g\gamma\mathcal{I}g^{-1} = \mathcal{I} \tag{14.122}$$

$$g\Gamma_{\gamma\mathcal{I}}g^{-1} = \left\{\mathbb{1}_2, \begin{pmatrix} 0 & a \\ -1/a & 0 \end{pmatrix} (\mathrm{mod} \pm 1)\right\}, \tag{14.123}$$

where $a \geq 1$. These classes play the role of the parabolic conjugacy classes in the classical Selberg trace formula.

We return with our discussion to the Selberg operator with the automorphic kernel (14.117). We consider the second term with an odd automorphic function ϕ which has the property

$$\frac{1}{2}\int_{\hat{\mathcal{F}}} dV(w)K(z, \mathcal{I}w)\phi(w) = \frac{1}{4}\sum_{\gamma \in \hat{\Gamma}}\int_{\mathcal{H}} dV(w)\overline{k(-\bar{z}, \gamma w)}\,\overline{\phi(w)} = \frac{1}{2}\overline{(L\phi)(-\bar{z})} \tag{14.124}$$

due to $k(z, \mathcal{I}w) = \overline{k(\mathcal{I}z, w)}$. Therefore we have obtained for an odd automorphic function the following identity

$$(\hat{L}_-\phi)(z) = \frac{1}{2}(L\phi)(z) - \frac{1}{2}\overline{(L\phi)(-\bar{z})}. \tag{14.125}$$

Compact Fundamental Domain.

For convenience we set $\rho = \gamma\mathcal{I}$ and use the classification of the inverse-hyperbolic transformations according to $\rho \in \bar{\Gamma}\mathcal{I}$, respectively, $\rho^2 \in \bar{\Gamma}$. We obtain ($m = 2n \in \mathbb{N}_0$)

$$\sum_{\gamma \in \bar{\Gamma}} \chi_\gamma^m \int_{\hat{\mathcal{F}}} dV(z)j(\gamma, -\bar{z})k[z, \gamma(-\bar{z})]$$

$$= \sum_{\rho \in \bar{\Gamma}\mathcal{I}} \chi^m_{\rho\mathcal{I}} \int_{\widehat{\mathcal{F}}} dV(z) j(\rho, z) k(z, \rho z) =: \sum_{\rho \in \bar{\Gamma}\mathcal{I}} A(\rho)$$

$$= \sum_{\rho_p} \sum_{k=0}^{\infty} A(\rho_p^{2k+1}) + \sum_{i=1}^{n} \sum_{\rho_i} \sum_{k=0}^{\infty} A(\rho_i^{2k+1}), \qquad (14.126)$$

where

$$\sum_{\rho} \sum_{k=0}^{\infty} A(\rho^{2k+1}) = \sum_{\{\rho\}} \sum_{k=0}^{\infty} \chi^{m(2k+1)}_{\rho} \chi^m_{\mathcal{I}} \int_{\Gamma_{\rho^2} \backslash \mathcal{H}} dV(z) j(\rho^{2k+1}, z) k(z, \rho^{2k+1} z)$$

$$= \sum_{\{\rho\}} \sum_{k=0}^{\infty} \chi^{m(2k+1)}_{\rho} \chi^m_{\mathcal{I}} B_k(\rho) . \qquad (14.127)$$

By an overall conjugation we can arrange that $\rho^2 z = Nz$, therefore $\rho z = \sqrt{N}(-\bar{z})$. This yields

$$B_k(\rho) = \int_{\Gamma_{\rho^2} \backslash \mathcal{H}} dV(z) j(\rho^{2k+1}, z) k(z, \rho^{2k+1} z) = \int_1^N \frac{dy}{y^2} \int_{\mathbb{R}} dx \, k[z, N^{k+1/2}(-\bar{z})]$$

$$= (-1)^{m/2} \int_1^N \frac{dy}{y^2} \int_{\mathbb{R}} dx \left(\frac{\bar{z}}{z}\right)^{m/2} \Phi\left(\frac{|z + N^{k+\frac{1}{2}} \bar{z}|^2}{N^{k+1/2} y^2}\right)$$

$$= \frac{(-1)^{m/2} \ln N}{2 \cosh \frac{u}{2}} \int_{\mathbb{R}} d\zeta \left(\frac{\zeta + 2i \cosh \frac{u}{2}}{\zeta - 2i \cosh \frac{u}{2}}\right)^{m/2} \Phi(\zeta^2 + 4 \sinh^2 \frac{u}{2}) = \frac{l_0 g(u)}{2 \cosh \frac{u}{2}} , \qquad (14.128)$$

with the abbreviation $u = (2k + 1) \ln \sqrt{N} = (k + \frac{1}{2}) l_{\rho^2}$ and $g(u)$ as in (14.63). Considering all relevant contributions we can derive the following Selberg trace formula (Bolte and Grosche [70], Bolte and Steiner [74], and Venkov [509]).

Theorem 14.8 *The Selberg trace formula on compact bordered Riemann surfaces for automorphic forms of weight m is given by*

$$\sum_{n=1}^{\infty} h(p_n) = -\frac{\mathcal{A}(\widehat{\mathcal{F}})}{16\pi^2} \int_0^{\infty} \frac{\cosh \frac{mu}{2}}{\sinh \frac{u}{2}} g'(u) du + \frac{1}{4} \sum_{\{\gamma\}_p} \sum_{k=1}^{\infty} \frac{\chi^{mk}_{\gamma} l_{\gamma} g(kl_{\gamma})}{\sinh \frac{kl_{\gamma}}{2}}$$

$$+ \frac{i}{4} \sum_{\{R\}_p} \sum_{k=1}^{\nu-1} \frac{\chi^{mk}_{R} e^{i(m-1)k\pi/\nu}}{\nu \sin(k\pi/\nu)} \int_{\mathbb{R}} du \, g(u) \frac{e^{(m-1)u/2}(e^u - e^{2ik\pi/\nu})}{\cosh u + \cos[\pi - 2(k\pi/\nu)]}$$

$$- \frac{1}{4} \sum_{\{\rho\}_p} \sum_{k=0}^{\infty} \frac{\chi^{m(2k+1)}_{\rho} \chi^m_{\mathcal{I}} l_{\rho^2} g[(k + \frac{1}{2}) l_{\rho^2}]}{\cosh \frac{1}{2}[(k + \frac{1}{2}) l_{\rho^2}]} - \frac{1}{2} \sum_{i=1}^{n} \sum_{k=1}^{\infty} \frac{\chi^{mk}_{C_i} l_{C_i} g(kl_{C_i})}{\cosh \frac{kl_{C_i}}{2}} - \frac{L}{4} g(0) . \qquad (14.129)$$

Here we have abbreviated $\sum_{i=1}^{n} l_{C_i} = L$. $h(p)$ denotes an even function in p with the corresponding $g(u)$ given in (14.63) and has the same properties as in Theorem 14.2. Note that for Neumann boundary conditions the last three terms change their signs.

Non-compact Fundamental Domain.

To evaluate the trace formula in the case when also parabolic conjugacy classes are present we recall the enumeration before Theorem 14.8. In order that the regularization of the terms which correspond to the parabolic conjugacy classes is actually possible we require the following property of the multiplier system

$$\kappa_0 := \sum_{\{S\}} \chi_S^m = \sum_{\substack{\{\rho\};\hat{\Gamma}_{\rho,ell} \\ \mathrm{tr}(\rho)=0}} \chi_\rho^m \ . \tag{14.130}$$

We include all relevant conjugacy classes. There are the hyperbolic ones $\{\gamma\}_{\hat{\Gamma}}$, the inverse hyperbolic ones $\{\gamma\mathcal{I}\}_{\hat{\Gamma}}$, $\mathrm{tr}(\gamma\mathcal{I}) \neq 0$, the elliptic ones $\{R\}_{\hat{\Gamma}}$, the parabolic ones $\{S\}_{\hat{\Gamma}}$, $\mathrm{tr}(S) = 2$, and the inverse elliptic ones $\{\gamma\mathcal{I}\}_{\hat{\Gamma}}$, $\mathrm{tr}(\gamma\mathcal{I}) = 0$. Following Venkov [509] we hence have to consider

$$\mathrm{tr}(L) = \frac{1}{2} \int_{\hat{\mathcal{F}}} \sum_{\{\gamma\}} \Big[k(z,\gamma z) - k(z,\rho z) \Big] dV(z)$$

$$= \frac{1}{2} \mathcal{A}(\hat{\mathcal{F}})\, \Phi(0) + \frac{1}{2} \sum_{\{\gamma\}} \chi_\gamma^m \int_{\hat{\mathcal{F}}(\gamma)} dV(z) k(z,\gamma z) + \frac{1}{2} \sum_{\{R\}} \chi_R^m \int_{\hat{\mathcal{F}}(R)} dV(z) k(z,Rz)$$

$$- \frac{1}{2} \sum_{\substack{\{\rho\} \\ \mathrm{tr}(\rho)\neq 0}} \chi_\rho^m \int_{\hat{\mathcal{F}}(\rho)} dV(z) k(z,\rho z) - \frac{1}{2} \sum_{\substack{\{\rho\};\hat{\Gamma}_{\rho,hyp} \\ \mathrm{tr}(\rho)=0}} \chi_\rho^m \int_{\hat{\mathcal{F}}(\rho)} dV(z) k(z,\rho z)$$

$$+ \frac{1}{2} \lim_{Y\to\infty} \int_{\hat{\mathcal{F}}_Y} dV(z)$$

$$\times \left\{ \sum_{\{S\}} \chi_S^m \sum_{\gamma'\in\hat{\Gamma}_S\backslash\Gamma} k(z,\gamma'^{-1}S\gamma'z) - \sum_{\substack{\{\rho\};\hat{\Gamma}_{\rho,ell} \\ \mathrm{tr}(\rho)=0}} \chi_\rho^m \sum_{\gamma'\in\hat{\Gamma}_\rho\backslash\Gamma} k(z,\gamma'^{-1}\rho\gamma'z) \right\} \ , \tag{14.131}$$

with some properly defined compact domain $\hat{\mathcal{F}}_Y$ depending on a large parameter Y, and where the sum is taken over all hyperbolic conjugacy classes $\{\gamma\}$, elliptic conjugacy classes $\{R\}$ and parabolic conjugacy classes $\{S\}$ in $\bar{\Gamma}$, over all relative non-degenerate classes $\{\gamma\mathcal{I}\}$ with $\mathrm{tr}(\gamma\mathcal{I}) \neq 0$, and all relative conjugacy classes $\{\gamma\mathcal{I}\}$ with $\mathrm{tr}(\gamma\mathcal{I}) = 0$. Parabolic transformations have the form

$$S = \begin{pmatrix} 1 & n \\ 0 & 1 \end{pmatrix} \ , \qquad (n \in \mathbb{Z} \setminus \{0\}) \ , \tag{14.132}$$

as we already know from section 10.2. In the case of bordered Riemann surfaces we have the same term corresponding to these transformations, which must be regularized by the elliptic conjugacy classes, $\mathrm{tr}(\rho) = 0$. There may be some $\rho \in \hat{\Gamma}\mathcal{I}$, $\mathrm{tr}\rho = 0$, such that the relative centralizer Γ_ρ is generated by an elliptic element of order two. To deal with these ρ one assumes that in (14.123) $g = \mathbb{1}_2$, that is

$$\gamma_a = \begin{pmatrix} 0 & a \\ -a^{-1} & 0 \end{pmatrix} \ , \qquad (\mathrm{mod} \pm 1) \ , \tag{14.133}$$

with some $a \geq 1$, is the elliptic generator of order two of Γ_ρ. We therefore have to consider

$$\int\limits_{\bigcup_{\gamma'\widehat{\mathcal{F}}_Y, \gamma'\in\bar{\Gamma}}} k(z, \rho z) = |\Gamma_\rho| \int\limits_{\bigcup_{\gamma'\widehat{\mathcal{F}}_Y, \gamma'\in\bar{\Gamma}\mathcal{I}\backslash\bar{\Gamma}}} k(z, \rho z) , \tag{14.134}$$

where $|\Gamma_\rho| = \text{order}[\Gamma_\rho] = 2$, which yields an additional factor $1/2$.

For a properly defined asymptotic expansion of the corresponding integral we remove from $\mathcal{H}$ two regions, denoted by $B_1(Y) = \{z \in \mathcal{H} | x \geq Y\}$ and $B_2(Y) = \gamma_a B_1(Y)$, respectively, i.e., we consider $B(Y) = \mathcal{H} - B_1(Y) - B_2(Y)$. Since we have the entire domain $\widehat{\mathcal{F}}$ taken into account, we must in the sequel consider the domain $B(Y) \cup \mathcal{I}B(Y)$. Finally one then considers the union

$$\int\limits_{\bigcup_{g\gamma\widehat{\mathcal{F}}_Y, \{\gamma\}}} k(z, \rho z) , \tag{14.135}$$

for some appropriate $g \in SL(2, \mathbb{R})$ (see (14.123)). According to Venkov [509], this has the consequence that the asymptotic behaviour of all relevant expressions in the limit $Y \to \infty$ is changed such that one considers the variable νY instead of Y. The fixed number ν is denoted by $\nu(\rho)$. Similarly, a is denoted by $a(\rho)$. Therefore we must multiply the results by $q(\widehat{\mathcal{F}})$ which denotes the number of classes $\{\rho\}_{\bar{\Gamma}}$ having the property $\text{tr}(\gamma\mathcal{I}) = 0$ and Γ_ρ being generated by an elliptic element of order two. Because we know that all terms in the trace formula must be finite we then find that $q(\widehat{\mathcal{F}}) = 4\kappa_0$. This gives the following result [70]

Theorem 14.9 *The Selberg trace formula on arbitrary bordered Riemann surfaces for automorphic forms of weight m, $m \in \mathbb{N}_0$, is given by*

$$\sum_{n=1}^{\infty} h(p_n) = -\frac{\mathcal{A}(\widehat{\mathcal{F}})}{4\pi} \int_0^{\infty} \frac{\cosh\frac{mu}{2}}{\sinh\frac{u}{2}} g'(u)du + \frac{1}{4} \sum_{\{\gamma\}_p} \sum_{k=1}^{\infty} \frac{\chi_\gamma^{mk} l_\gamma g(kl_\gamma)}{\sinh\frac{kl_\gamma}{2}}$$

$$+ \frac{i}{4} \sum_{\{R\}_p} \sum_{k=1}^{\nu-1} \chi_R^{mk} \frac{e^{i(m-1)k\pi/\nu}}{\nu\sin(k\pi/\nu)} \int_{\mathbb{R}} du\, g(u) \frac{e^{(m-1)u/2}(e^u - e^{2ik\pi/\nu})}{\cosh u + \cos[\pi - 2(k\pi/\nu)]}$$

$$- \frac{1}{4} \sum_{\{\rho^2\}_p} \sum_{k=0}^{\infty} \frac{\chi_\rho^{m(2k+1)} \chi_{\mathcal{I}}^m l_{\rho^2} g[(k+\frac{1}{2})l_{\rho^2}]}{\cosh[\frac{1}{2}(k+\frac{1}{2})l_{\rho^2}]} - \frac{1}{2} \sum_{i=1}^{n} \sum_{k=1}^{\infty} \frac{\chi_{C_i}^{mk} l_{C_i} g(kl_{C_i})}{\cosh\frac{kl_{C_i}}{2}}$$

$$+ \frac{g(0)}{2}\left[\frac{1}{4} \sum_{\substack{\{\rho\};\bar{\Gamma}_{\rho,ell}\\ \text{tr}(\rho)=0}} \chi_\rho^m \ln\left(\frac{a(\rho)}{\nu(\rho)}\right) - \kappa_0 \ln 2 - \frac{L}{2}\right] + \frac{\kappa_0}{8} h(0)$$

$$- \frac{\kappa_0}{4\pi} \int_{\mathbb{R}} h(p)\Psi(\tfrac{1}{2} + ip)dp + \frac{\kappa_0}{4} \int_0^{\infty} \frac{g(u)}{\sinh\frac{u}{2}}\left(1 - \cosh\frac{um}{2}\right)du . \tag{14.136}$$

$h(p)$ *denotes an even function in p with the corresponding $g(u)$ given in (14.63) and has the same properties as in Theorem 14.2.*

Note that for Neumann boundary conditions the inverse-hyperbolic terms change their signs. In this case, however, the parabolic terms are quite different, due to the additional presence of the continuous spectrum represented by the Eisenstein-series.

14.4.1   The Selberg Zeta-Function

We have introduced the Selberg zeta-function for bordered Riemann surfaces according to [70, 74, 509] as

$$
Z(s) = \prod_{\{\gamma\}_p} \prod_{k=0}^{\infty} \left[1 - \chi_\gamma^m e^{-l_\gamma(s+k)} \right] \times \prod_{\{\rho\}_p} \prod_{k=0}^{\infty} \left(\frac{1 + \chi_\rho^m e^{-l_\rho(s+k)}}{1 - \chi_\rho^m e^{-l_\rho(s+k)}} \right)^{(-1)^k \chi_{\mathcal{I}}^m}
$$

$$
\times \prod_{i=1}^{n} \prod_{k=0}^{\infty} \left(\frac{1}{1 - \chi_{C_i}^m e^{-l_{C_i}(s+k)}} \right)^{2(-1)^k} . \tag{14.137}
$$

We treat the general case covered by (14.136) and use the test-function

$$
h(p, s, b) = \frac{1}{(s - \frac{1}{2})^2 + p^2} - \frac{1}{(b - \frac{1}{2})^2 + p^2} , \tag{14.138}
$$

which fulfills the requirements of Theorem 14.9 if $\Re(s), \Re(b) > 1$. One finds

$$
g(u, s, b) = \frac{e^{-(s-1/2)|u|}}{2(s - \frac{1}{2})} - \frac{e^{-(b-1/2)|u|}}{2(b - \frac{1}{2})} . \tag{14.139}
$$

We consider the definition of the Selberg zeta-function on bordered Riemann surfaces (14.137), which generalizes the definition of Venkov [507] for $m \neq 0$. Then one derives [70]:

Theorem 14.10 *Suppose that $s \in \mathbb{C}$, $\Re(s) > 1$, and let $a \in \mathbb{R}$, $a > 1$, be fixed. The Selberg trace formula for the Selberg zeta-function has the form:*

$$
\frac{Z'(s)}{Z(s)} = (s - \tfrac{1}{2}) \frac{2\mathcal{A}(\widehat{\mathcal{F}})}{\pi} \left[\Psi\left(s + \frac{m}{2}\right) + \Psi\left(s - \frac{m}{2}\right) - \Psi\left(b + \frac{m}{2}\right) - \Psi\left(b - \frac{m}{2}\right) \right]
$$

$$
- i \sum_{\{R\}_p} \sum_{k=1}^{\nu-1} \frac{1}{\nu \sin(k\pi/\nu)} \sum_{l=0}^{\infty} \left[\frac{e^{-2i(k\pi/\nu)(l+1/2-m/2)}}{s+l-m/2} - \frac{e^{2i(k\pi/\nu)(l+1/2+m/2)}}{s+l+m/2} \right]
$$

$$
+ 4(s - \tfrac{1}{2}) \sum_{j} \left(\frac{1}{(s - \frac{1}{2})^2 + p_j^2} - \frac{1}{(b - \frac{1}{2})^2 + p_j^2} \right) + \mathrm{const}_1 + \mathrm{const}_2(s - \tfrac{1}{2})
$$

$$
+ 2\kappa_0 \Psi(1 - s) - 4\kappa_0(s - \tfrac{1}{2}) \sum_{k=1}^{\infty} \frac{1}{(s - \frac{1}{2})^2 - (k + \frac{1}{2})^2}
$$

$$
+ \kappa_0 \left[2\Psi(s) - \Psi\left(s + \frac{m}{2}\right) - \Psi\left(s - \frac{m}{2}\right) \right] , \tag{14.140}
$$

with some constants $\mathrm{const}_{1,2}$*.*

The zero- and pole-structure can be read off [70]

Theorem 14.11 *$Z(s)$ is a meromorphic function of $s \in \mathbb{C}$ of order equal to two. The zeros of the function $Z(s)$ are at the following points*

1. *Nontrivial zeros: on the line $\Re(s) = 1/2$, symmetric relative to the point $s = 1/2$, and on the interval $[0, 1]$, symmetric to the point $s = 1/2$. We call these zeros s_j. Each s_j has multiplicity equal twice the multiplicity of the corresponding eigenvalue E_j of the operator D_m of the corresponding Dirichlet boundary value problem, $E_j = \frac{1}{4} + p_j^2$ and E_j runs through the entire spectrum of D_m.*

2. *Trivial zeros:*

   (a) at the points $s = -l + m/2$, $l \in \mathbb{N}_0$, with multiplicity $\#N_{-l}$ given by

   $$\#N_{-l} = \frac{2\mathcal{A}(\widehat{\mathcal{F}})}{\pi}\left(l - \frac{m-1}{2}\right) + \kappa_0 + \mathrm{i} \sum_{\{R\}_p} \sum_{k=1}^{\nu-1} \frac{\mathrm{e}^{2\mathrm{i}(k\pi/\nu)(l+1/2+m/2)}}{\nu \sin(k\pi/\nu)} \quad . \quad (14.141)$$

   (b) at the points $s = -l - m/2$, $l \in \mathbb{N}_0$, with multiplicity $\#N_{-l}$ given by

   $$\#N_{-l} = \frac{2\mathcal{A}(\widehat{\mathcal{F}})}{\pi}\left(l + \frac{m+1}{2}\right) + \kappa_0 - \mathrm{i} \sum_{\{R\}_p} \sum_{k=1}^{\nu-1} \frac{\mathrm{e}^{2\mathrm{i}(k\pi/\nu)(l+1/2-m/2)}}{\nu \sin(k\pi/\nu)} \quad . \quad (14.142)$$

   (c) at the points $s = l \in \mathbb{N}$ with multiplicity $\#N_l = 2\kappa$.

3. *$Z(s)$ has poles at the points*

   (a) $s = l$, $l \in \mathbb{Z}$, with multiplicity $\#P_l = 2\kappa_0$.

   (b) $s = -l$, $l = 0, 1, 2, \ldots$, with multiplicity $\#P_l = 2\kappa_0$.

4. *Note that for $m = 0$ the zeros of 2 and the poles of 3 combine into zeros with multiplicity*

   $$\#N_{-l} = \frac{4\mathcal{A}(\widehat{\mathcal{F}})}{\pi}\left(l + \tfrac{1}{2}\right) - 2 \sum_{\{R\}_p} \sum_{k=1}^{\nu-1} \frac{\sin[k\pi(2l+1)/\nu]}{\nu \sin(k\pi/\nu)} \quad . \quad (14.143)$$

The full picture of the zeros and poles emerges by combining 2 and 3, and there remain only poles at $s = -l$, $l \in \mathbb{N}_0$. Note, in particular that if no elliptic and parabolic terms are present, $Z(s)$ has, of course, no poles, and no zero at $s = 1$; this stands in contrast to the Selberg zeta-function on a closed Riemann surface, where the zero at $s = 1$ stems from the one-fold zero-mode of the Maass-Laplacian.

Furthermore we have [70]

Theorem 14.12 *The functional equation has the form*

$$Z(1 - s) = Z(s)\psi(s) \tag{14.144}$$

with the function $\psi(s)$ given by

$$
\begin{aligned}
\psi(s) = {}& \left[\frac{\Gamma(1-s)}{\Gamma(s)}\right]^{2\kappa} \exp\Bigg\{ -4\mathcal{A}(\widehat{\mathcal{F}}) \int_0^{s-1/2} t \left(\begin{array}{c} \tan \pi t \\ \cot \pi t \end{array}\right) dt + 4c_{\widehat{\mathcal{F}}}(s - \tfrac{1}{2}) \\
& + i \sum_{\{R\}_p} \sum_{k=1}^{\nu-1} \frac{1}{\nu \sin(k\pi/\nu)} \int_0^{s-1/2} \left[\frac{e^{-2i(k\pi/\nu)(l+1/2-m/2)}}{s+l-m/2} + \frac{e^{-2i(k\pi/\nu)(l+1/2-m/2)}}{s-l+(m-3)/2}\right. \\
& \left. - \frac{e^{-2i(k\pi/\nu)(l+1/2+m/2)}}{s+l+(m-1)/2} - \frac{e^{-2i(k\pi/\nu)(l+1/2+m/2)}}{s-l-(m-3)/2}\right] dt\Bigg\} \, ,
\end{aligned}
\tag{14.145}
$$

where the $\tan \pi t$-, respectively the $\cot \pi t$-term must be taken whether m is even or odd, and the constant $c_{\widehat{\mathcal{F}}}$ is given by

$$
c_{\widehat{\mathcal{F}}} = \frac{1}{4} \sum_{\substack{\{\rho\}; \hat{\Gamma}_{\rho,ell} \\ tr(\rho)=0}} \chi_\rho^m \ln\left(\frac{a(\rho)}{\mu(\rho)}\right) - \frac{\kappa_0}{4} \ln 2 - \frac{L}{8} \, .
\tag{14.146}
$$

14.4.2   Determinants of Maass-Laplacians

Similarly as in the case of closed Riemann surfaces we are going to calculate determinants of Maass-Laplacians on bordered Riemann surfaces. Some examples of calculations of the scalar determinant are due to Bolte and Steiner [74] and Blau et al. [61, 62]. In particular, we calculate the determinant of the operator $\Delta_m^{(\pm)} = D_m + m(m \pm 1)$, because $\Delta_m^{(+)}$ is the relevant operator in string theory. First we only consider the case where the Fuchsian group $\hat{\Gamma}$ is strictly hyperbolic, since then it is known that the discrete spectrum of D_m (and of $\Delta_m^{(\pm)}$) is infinite and no continuous spectrum appears. We denote the omission of zero-modes by primes and define the determinants by zeta-function regularization, i.e., we set

$$
\det{}'(\Delta_m^{(\pm)}) := \exp\left(-\frac{d}{ds}\zeta_m^{(\pm)}(s)\right) \, ,
\tag{14.147}
$$

$$
\zeta_m^{(\pm)}(s) := tr'(\Delta_m^{(\pm)})^{-s} = \frac{1}{\Gamma(s)} \int_0^\infty dt\, t^{s-1} \left[e^{-tm(m\pm1)} tr(e^{-tD_m}) - N_m^{(\pm)}\right] \, .
\tag{14.148}
$$

Here $N_m^{(\pm)}$ denote the numbers of respective zero-modes. In the case of Dirichlet boundary conditions no zero-modes are present, and we discuss this case first. With the heat-kernel function $h(p) = e^{-t(p^2+\frac{1}{4})}$ we determine $g(u) = e^{-u^2/4t-t/4}/\sqrt{4\pi t}$ and consider the trace formula (14.136) for this test function. We split $\zeta_m(s)$ up into a term $\zeta_I(s)$ corresponding to the first summand on the r.h.s. of (14.136), a term

$\zeta_\Gamma(s)$ corresponding to the second, fourth and fifth summand, and a term $\zeta_{g(0)}(s)$ corresponding to the last summand, respectively. We then find $\zeta'_\Gamma(0) = -\frac{1}{2}\ln Z(n+1)$, and therefore obtain for the determinant of the operator Δ^+_n for $m = 2n \in \mathbb{N}_0$ [70]

$$
\det(\Delta^+_n) = \sqrt{Z(1+n)}
$$
$$
\times \exp\Bigg[-\frac{L}{8}(2n+1) + \frac{\mathcal{A}(\widehat{\mathcal{F}})}{8\pi}\Big\{ (2n+1)\ln(2\pi) + 4\zeta'(-1) - \frac{1}{2}(2n+1)^2
$$
$$
-4\sum_{3\leq k\leq n+1}\ln\Gamma(k) + 4(n - \tfrac{1}{2} - [n])\ln\Gamma(n+1)
$$
$$
+ \sum_{0\leq k\leq(2n-1)/2}(2n-2k-1)\ln(2nk + 2n - k^2 - k)\Big\}\Bigg] . \tag{14.149}
$$

Here empty sums are understood as to be ignored and $[x]$ is the integer part of x. $\zeta'(-1)$ denotes the derivative of the Riemann zeta-function $\zeta(s)$ at $s = -1$. The cases $m = 0, 2$ are given by

$$
\det(\Delta^+_0) = \sqrt{Z(1)}\exp\Bigg\{ -\frac{L}{8} + \frac{\mathcal{A}(\widehat{\mathcal{F}})}{8\pi}\Big[-\frac{1}{2} + \ln(2\pi) + 4\zeta'(-1)\Big]\Bigg\}, \tag{14.150}
$$
$$
\det(\Delta^+_1) = \sqrt{Z(2)}\exp\Bigg\{ -\frac{3L}{8} + \frac{\mathcal{A}(\widehat{\mathcal{F}})}{8\pi}\Big[-\frac{9}{2} + 3\ln(2\pi) + 4\zeta'(-1)\Big]\Bigg\}. \tag{14.151}
$$

Note that the construction of the Selberg zeta-function on bordered Riemann surfaces guarantees that in the case of Neumann boundary conditions the powers of the terms involving the inverse-hyperbolic conjugacy classes change their signs. The product of the determinants of Dirichlet and Neumann boundary conditions, denoted by superscripts D and N, respectively, just gives in a natural way the determinant of the corresponding operator on the Riemann surface $\widehat{\Sigma}$, denoted by the superscript $\widehat{\Sigma}$, i.e.,

$$
\det(\Delta^{(D,+)}_n) \times \det'(\Delta^{(N,+)}_n) = \det'(\Delta^{(\widehat{\Sigma},+)}_n) \propto \begin{cases} Z'_{\widehat{\Sigma}}(1) , & \text{for } n = 0 , \\ Z_{\widehat{\Sigma}}(n+1) , & \text{for } n \neq 0 , \end{cases}
$$
$$
\tag{14.152}
$$

where $Z_{\widehat{\Sigma}}$ is the Selberg zeta-function on the entire Riemann surface $\widehat{\Sigma}$.

Chapter 15

The Selberg Super-Trace Formula

15.1 Automorphisms on Super-Riemann Surfaces

We sketch some important facts about super-Riemann surfaces. The theory of super-Riemann surfaces has been developed by Batchelor et al. [47, 48], DeWitt [123], [199, 203, 213, 215, 221], Moore, Nelson and Polchinski [413], Ninnemann [422], Rabin and Crane [449], and Rogers [455]. Let us start with a $(1|1)$ (complex)-dimensional (not necessarily) flat super-space, parameterized by even coordinates $Z \in \mathbb{C}_c$ and odd (Grassmann) coordinates $\theta \in \mathbb{C}_a$, respectively. Let Λ_∞ be the infinite dimensional vector space generated by elements ζ_a ($a = 1, 2, \ldots$) with basis $1, \zeta_a, \zeta_a\zeta_b, \ldots$ ($a < b$) and the anticommuting relation $\zeta_a\zeta_b = -\zeta_b\zeta_a$, $\forall_{a,b}$. Every $Z \in \Lambda_\infty$ can be decomposed as $Z = Z_B + Z_S$ with $Z_B \in \mathbb{C}_c \equiv \mathbb{C}$, $Z_S = \sum_n \frac{1}{n!} c_{a_1,\ldots,a_n} \zeta^{a_n} \ldots \zeta^{a_1}$, with the $c_{a_1,\ldots,a_n} \in \mathbb{C}_a$ totally antisymmetric. Z_B and Z_S are called the *body* (sometimes denoted by $Z_B = Z_{red}$) and *soul* of the super-number Z, respectively. The notion of super-space and super-manifolds enables one to represent super-symmetry transformations as pure geometric transformations in the coordinates $Z = (z, \theta) \in \mathbb{C}_c \times \mathbb{C}_a$. As it is well-known, a usual complex manifold of complex dimension equal to one is already a Riemann surface. The definition of a super-Riemann surface, however, requires the introduction of a super-conformal structure. Let us consider the operator $D = \theta\partial_z + \partial_\theta$ (note $D^2 = \partial_z$). Further, we consider a general superanalytic coordinate transformation $\tilde{z} = \tilde{z}(z, \theta)$, $\tilde{\theta} = \tilde{\theta}(z, \theta)$. A superanalytic coordinate transformation is called super-conformal, iff the $(0|1)$-dimensional subspace of the tangential space generated by the action of D is invariant under such a coordinate transformation, i.e., $D = (D\tilde{\theta})\tilde{D}$. This means that a coordinate transformation is super-conformal iff $Dz' = \theta'D\theta'$.

To study super-symmetric field theories one needs even and odd super-fields. Here the definition of DeWitt [123] of super-Riemann manifolds conveniently comes into play. The infinite dimensional algebra Λ_∞ supplies all the required quantities. Domains in $\mathbb{C}^{(1|1)}$ with coordinates (z, θ) are constructed in such a way that the entire Grassmann algebra is attached to the usual complex coordinates. If one considers the universal family of DeWitt super-Riemann manifolds with genus g, then only $2g - 2$

parameters of Λ_∞ are required, the remaining ones are redundant.

An important property we need in our investigations is, when a super-manifold is split. This means that for a coordinate transformation $Z \to Z'$ ($Z, Z' \in \Lambda_\infty$) the coefficient functions do not mix with each other. Let z be usual local coordinates, and $\zeta \in \Lambda_\infty$ local Grassmann coordinates. When a super-manifold is split there is a global isomorphism such that the coefficient functions y and η of a super-function $F(z, \zeta)$ transform according to

$$
\begin{aligned}
y &= a_0(z) + a_{ij}(z)\zeta^i\zeta^j + \ldots, \\
\eta &= b_{1,i}(z)\zeta^i + b_{3,ijk}(z)\zeta^i\zeta^j\zeta^k + \ldots,
\end{aligned}
\quad \to \quad
\begin{aligned}
&a_0'(z') + a_{ij}'(z')\zeta'^i\zeta'^j + \ldots, \\
&b_{1,i}'(z')\zeta'^i + b_{3,ijk}'(z')\zeta'^i\zeta'^j\zeta'^k + \ldots,
\end{aligned}
$$
$$(15.1)$$

for $Z \to Z'$. Due to a theorem of Batchelor [47] every differentiable super-manifold is split, in particular every complex super-manifold of dimension $(d|1)$. The super-Riemann surfaces in question can be seen as a complex $(1|1)$-dimensional super-manifold, or a real $(2|2)$-dimensional manifold, respectively, where the coordinate transformations are super-conformal mappings [449].

To generalize the uniformization theorem for Riemann surfaces to super-Riemann surfaces $\mathcal{M}$, one shows that unique generalizations $\widehat{\mathbb{C}}^{(1|1)}$, $\mathbb{C}^{(1|1)}$ and $\mathcal{H}^{(1|1)} := \{(z, \theta) \in \mathbb{C}^{(1|1)} | \Im(z) > 0\}$ of simple connected Riemann surfaces exist, and endow $U = \widehat{\mathbb{C}}^{(1|1)}$, $\mathbb{C}^{(1|1)}$ and $\mathcal{H}^{(1|1)}$ with a super-conformal structure, such that the local coordinate transformations are super-conformal mappings [449].

In the case of non-Euclidean harmonic analysis in the context of super-Riemann surfaces we consider the group $\mathrm{OSp}(2, \mathbb{C})$ of super-conformal automorphisms on super-Riemann surfaces as a natural generalization of Möbius transformations. They have the form

$$
\mathrm{OSp}(2, 1; \mathbb{C}_c^2 \times \mathbb{C}_a) := \left\{ \gamma = \begin{pmatrix} a & b & \chi_\gamma(b\alpha - a\beta) \\ c & d & \chi_\gamma(d\alpha - c\beta) \\ \alpha & \beta & \chi_\gamma(1 - \alpha\beta) \end{pmatrix} \right|
$$

$$
a, b, c, d \in \mathbb{C}_c; \alpha, \beta \in \mathbb{C}_a; ad - bc = 1 + \alpha\beta; \mathrm{sdet}\gamma = \chi_\gamma \in \{\pm 1\} \Bigg\} \quad (15.2)
$$

(α, β real, with the complex conjugate rules $\overline{f + g} = \bar{f} + \bar{g}$, and $\overline{f \cdot g} = \bar{f} \cdot \bar{g}$). Its generators are the operators L_0, L_1, L_{-1}, $G_{1/2}$ and $G_{-1/2}$ of the Neveu-Schwarz sector of the super-Virasoro algebra of the fermionic string. Elements $\gamma \in \mathrm{OSp}(2, 1; \mathbb{C}_c^2 \times \mathbb{C}_a)$ act on elements $x = (z_1, z_2, \xi) \in \mathbb{C}_c^2 \times \mathbb{C}_a \setminus \{0\}$ by matrix multiplication, i.e., $x' = \gamma x$. By means of a local coordinate system $(z, \theta) = (z_1/z_2, \xi/z_2)$ and the requirements of super-conformal transformation the local coordinate transformations are fixed and the super-Möbius transformations explicitly have the form [42, 203, 422, 449, 499]

$$
z' = \frac{az + b}{cz + d} + \theta\frac{\alpha z + \beta}{(cz + d)^2}, \qquad \theta' = \frac{\alpha + \beta z}{cz + d} + \frac{\chi_\gamma \theta}{cz + d}. \quad (15.3)
$$

The χ_γ with $\chi_\gamma = \pm 1$ lead to the description of spin structures on a super-Riemann surface. The transformation factor of the D operator yields to

$$
F_\gamma := (D\theta')^{-1} = \chi_\gamma(cz + d + \delta\theta), \quad (15.4)
$$

with $\delta = \chi_\gamma \sqrt{1 + \alpha\beta}\,(\alpha d + \beta c)$. This general super-Möbius transformation does mix the coefficient functions of super-functions $F \in \Lambda_\infty$. Since we have required that the super-Riemann surfaces in question are split, the odd quantities α, β are not necessary and can be omitted. It is sufficient to consider transformations $\gamma \in \mathrm{OSp}(2,1)$ with $\alpha = \beta = 0$ and characters χ_γ which describe spin structures. Furthermore γ and $-\gamma$ describe the same transformation. Thus the automorphisms on $\mathcal{H}^{(1|1)}$ are given by (note $\mathcal{H}^{(1|1)}{}_{red} \equiv \mathcal{H}^{(2)} = \mathcal{H} \cong \Lambda^{(2)}$)

$$\mathrm{Aut}\,\mathcal{H}^{(1|1)} = \frac{\mathrm{OSp}(2|1,\mathbb{R})}{\{\pm\mathbb{1}\}} \ , \tag{15.5}$$

and a super-Fuchsian group Γ denotes a discrete subgroup of $\mathrm{Aut}\,\mathcal{H}^{(1|1)}$. Therefore we obtain for the transformations $z \to z'$ and $\theta \to \theta'$ [42, 422]

$$z' = \frac{az + b}{cz + d} \ , \qquad \theta' = \frac{\chi_\gamma \theta}{cz + d} \ , \tag{15.6}$$

[here $F_\gamma = \chi_\gamma(cz + d)$]. $M_{\xi=0}$ corresponds to the usual Riemann surface M_{red} with some spin-structure, since a $\gamma \in \mathrm{Aut}\,\mathcal{H}^{(1|1)}$ is fixed by a $\mathrm{PSL}(2,\mathbb{R})$ transformation and a character $\chi_\gamma = \pm 1$. The properties of the odd coordinates are determined by the properties of M_{red} and θ is the cut of a spinor-bundle.

We need some further ingredients. Let us introduce the quantities N_γ and l_γ

$$2 \cosh \frac{l_\gamma}{2} = N_\gamma^{1/2} + N_\gamma^{-1/2} = a + d + \chi_\gamma \alpha\beta \ . \tag{15.7}$$

N_γ is called *norm* of a hyperbolic $\gamma \in \Gamma$ in a super-Fuchsian group, and N_{γ_0} will denote the norm of a primitive hyperbolic $\gamma \in \Gamma$, and $l_\gamma = \ln N_\gamma$ denotes the *length* corresponding to a $\gamma \in \Gamma$ and all notions from the bosonic case are interpreted in a straightforward way in terms of their super generalizations. Each element $\gamma \in \Gamma/\{\pm\mathbb{1}\}$ is uniquely described as $\gamma = k^{-1}\gamma_0 k$ for some primitive γ_0, $n \in \mathbb{N}$ and $k \in \Gamma/\Gamma_{\gamma_0}$. For $\mathrm{OSp}(2,\mathbb{R})/\{\pm\mathbb{1}\}$ in homogeneous coordinates a hyperbolic transformation is always conjugate to the transformation $z' = N_\gamma z$, $\theta' = \chi_\gamma \sqrt{N_\gamma}\,\theta$, or in matrix representation

$$\text{hyperbolic } \gamma \in \Gamma \text{ conjugate to} \qquad \begin{pmatrix} N_\gamma^{1/2} & 0 & 0 \\ 0 & N_\gamma^{-1/2} & 0 \\ 0 & 0 & \chi_\gamma \end{pmatrix} \ . \tag{15.8}$$

In analogy with the usual Selberg theory hyperbolic transformations are called dilations.

The generators of a particular super-Fuchsian group of a super-Riemann surface with genus g obey the constraint

$$(\gamma_0 \gamma_1^{-1} \dots \gamma_{2g-2} \gamma_{2g-1}^{-1})(\gamma_0^{-1} \gamma_1 \dots \gamma_{2g-2}^{-1} \gamma_{2g-1}) = \mathbb{1}_{2|1} \ . \tag{15.9}$$

In order to construct explicitly a metric on $\mathcal{H}^{(1|1)}$ one starts with the super-Vierbeins in flat super-space and performs a super-Weyl transformation [280] to obtain the metric $ds^2 = dq^a{}_a g_b dq^b$ in $\mathcal{H}^{(1|1)}$ [499]. The scalar product has the form

$$(\Phi_1, \Phi_2) = \int_{\mathcal{H}^{(1|1)}} \frac{dz d\bar{z} d\theta d\bar{\theta}}{2Y} \Phi_1(Z)\bar{\Phi}_2(Z) \equiv \int_{\mathcal{H}^{(1|1)}} dV(Z)\Phi_1(Z)\bar{\Phi}_2(Z) \; , \qquad (15.10)$$

for super-functions $\Phi_1, \Phi_2 \in L^2(\mathcal{H}^{(1|1)})$ and $Y = y + \mathrm{i}\theta\bar{\theta}/2 = y + \theta_1\theta_2$ $(\theta = \theta_1 + \mathrm{i}\theta_2)$. We have one even and one odd point pair invariant given by

$$R(Z,W) = \frac{|z - w - \theta\nu|^2}{YV} \; , \qquad (15.11)$$

$$
\begin{aligned}
r(Z,W) &= \mathrm{i}\frac{2\theta\bar{\theta} + (\nu + \bar{\nu})(\theta - \bar{\theta})}{4Y} + \mathrm{i}\frac{2\nu\bar{\nu} + (\theta + \bar{\theta})(\nu - \bar{\nu})}{4V} \\
&\qquad + \frac{(\nu - \bar{\nu})(\theta - \bar{\theta})\Re(z - w - \theta\nu)}{4YV} \qquad (15.12) \\
&= \frac{(\theta_1 - \nu_1)\theta_2}{y} + \frac{(\nu_1 - \theta_1)\nu_2}{v} + \frac{\theta_2\nu_2\Re(z - w - \theta\nu)}{4YV} \qquad (15.13)
\end{aligned}
$$

$(Z, W \in \mathcal{H}^{(1|1)}$, $W = (w, \nu) = (u + \mathrm{i}v, \nu_1 + \mathrm{i}\nu_2)$, $V = v + \mathrm{i}\nu\bar{\nu}/2)$ as derived from classical mechanics on the Poincaré super-upper half-plane. We introduce the Dirac(-Laplace) operators $\square_m$ and $\hat{\square}_m$, respectively [8, 42]

$$\square_m = 2YD\bar{D} + \mathrm{i}m(\bar{\theta} - \theta)\bar{D} \; , \qquad \hat{\square}_m = 2YD\bar{D} + \frac{\mathrm{i}m}{2}(\bar{\theta} - \theta)(D + \bar{D}) \; , \qquad (15.14)$$

and $\square_m$ and $\hat{\square}_m$ are related by a linear isomorphism $\square_m = Y^{-m/2}(\hat{\square}_m + \mathrm{i}m/2)Y^{m/2}$. Particularly we have for $m = 0$

$$\hat{\square}_0 = \square_0 \equiv \square = 2Y(\partial_\theta\partial_{\bar{\theta}} + \theta\bar{\theta}\partial_z\partial_{\bar{z}} + \theta\partial_{\bar{\theta}}\partial_z - \bar{\theta}\partial_\theta\partial_{\bar{z}}) \; . \qquad (15.15)$$

With the notation $-\Delta_m = -4y^2\partial_z\partial_{\bar{z}} + \mathrm{i}my\partial_x = -y^2(\partial_x^2 + \partial_y^2) + \mathrm{i}my\partial_x$ we obtain for a super-function

$$\Psi(Z, \bar{Z}) = A(z, \bar{z}) + \frac{\theta\bar{\theta}}{y}B(z, \bar{z}) + \frac{1}{\sqrt{y}}\left(\theta\chi(z, \bar{z}) + \bar{\theta}\tilde{\chi}(z, \bar{z})\right) \qquad (15.16)$$

the following equivalence

$$
\hat{\square}_m\Psi(Z, \bar{Z}) = s\Psi(Z, \bar{Z})
$$
$$
\Longleftrightarrow \quad
\begin{cases}
-\Delta_m A(z, \bar{z}) = s(s + \mathrm{i})A(z, \bar{z}) \; , \\
B(z, \bar{z}) = \frac{s}{2}A(z, \bar{z}) \; , \\
\left(s - \frac{\mathrm{i}m}{2}\right)\tilde{\chi}(z, \bar{z}) = -2y\partial_{\bar{z}}\chi(z, \bar{z}) + \frac{\mathrm{i}}{2}(m + 1)\chi(z, \bar{z}) \; , \\
-\Delta_{(m+1)}\chi(z, \bar{z}) = (\frac{1}{4} + s^2)\chi(z, \bar{z}) \; .
\end{cases}
\qquad (15.17)
$$

An explicit solution of (15.17) for $m = 0$ on the entire $\mathcal{H}^{(1|1)}$ is given by [397]

$$\Phi_{p,k}(z, \bar{z}, \theta, \bar{\theta}) = \sqrt{\frac{2\mathrm{i}\sinh\pi p}{\pi^3}}\left(1 - \mathrm{i}\frac{1 + 2\mathrm{i}p}{4y}\theta\bar{\theta}\right)\sqrt{y}\,\mathrm{e}^{\mathrm{i}kx}K_{\mathrm{i}p}(|k|y) \ , \qquad (15.18)$$

$$\phi_{p,k}(z, \bar{z}, \theta, \bar{\theta}) = \sqrt{\frac{\cos[\pi(c + \mathrm{i}p)]}{2\pi^2(c + \mathrm{i}p)^{\sigma_k - 1}}}\frac{\mathrm{e}^{\mathrm{i}kx}}{\sqrt{y}}$$
$$\times\left[\theta W_{\sigma_k/2, c+\mathrm{i}p}(2|k|y) + \mathrm{i}(c + \mathrm{i}p)^{\sigma_k}\bar{\theta}W_{-\sigma_k/2, c+\mathrm{i}p}(2|k|y)\right] \ , \qquad (15.19)$$

with $s = -\mathrm{i}(\frac{1}{2} + \mathrm{i}p)$, $\sigma_k = \mathrm{sign}(k)$, $(k \neq 0)$, and $c \in \mathbb{R}$, $|c| \leq \frac{1}{2}$, $\Psi = \Phi + \phi$. K_ν and $W_{\mu,\nu}$ denote modified Bessel- and Whittaker-functions, respectively. Due to the particular form of the differential equation for $\Phi(Z, \bar{Z})$ we see that the solutions can be characterized by their parity with respect to the coordinate x, i.e., they can have even and odd parity with respect to x.

I have proposed in [213] a decomposition of an appropriate $T \in \Gamma$ into a hyperbolic, elliptic and parabolic contribution as follows

$$(T \in \Gamma \text{ conjugate to}) \qquad \gamma \times R \times S = \begin{pmatrix} N_\gamma^{1/2} & 0 & 0 \\ 0 & N_\gamma^{-1/2} & 0 \\ 0 & 0 & \chi_\gamma \end{pmatrix}$$
$$\times \begin{pmatrix} \cos\varphi & -\sin\varphi & 0 \\ \sin\varphi & \cos\varphi & 0 \\ 0 & 0 & \chi_R \end{pmatrix} \cdot \begin{pmatrix} 1 & n & 0 \\ 0 & 1 & 0 \\ 0 & 0 & \chi_S \end{pmatrix} \ , \qquad (15.20)$$

with $n \in \mathbb{N}$ and $0 < \varphi < \pi$, and γ, R and S, respectively, denote hyperbolic, elliptic and parabolic transformations, acting by matrix multiplication. The corresponding fundamental domains $\mathcal{F}^{(1|1)}$ are chosen appropriately. The body $\mathcal{F}$ of a fundamental domain $\mathcal{F}^{(1|1)}$ has according to [269] $4g + 2s + 2\kappa$ sides, the boundaries being geodesics, of course. We also maintain the notion of χ_T irrespective of $T \in \Gamma$ hyperbolic, elliptic or parabolic, and we choose χ_T according to the spin structure of the super-Riemann surface in question. For a super-Riemann surface of genus g there are obviously $2^{(\#generators)} = 2^{(2g+s+\kappa)}$ possible spin structures.

The constraint (15.9) is altered due to the presence of elliptic fixed points and cusps according to [270, 508, 509]

$$(\gamma_0\gamma_1^{-1}\ldots\gamma_{2g-2}\gamma_{2g-1}^{-1})(\gamma_0^{-1}\gamma_1\ldots\gamma_{2g-2}^{-1}\gamma_{2g-1})R_1\ldots R_s S_1\ldots S_\kappa = \mathbb{1}_{2|1} \ . \qquad (15.21)$$

15.1.1 Closed Super-Riemann Surfaces

Turning to the Selberg super-trace formula, let us introduce the Selberg super-operator L by [41]–[43, 203]

$$(L\phi)(Z) = \int_{\mathcal{H}^{(1|1)}} dV(W)k_m(Z, W)\phi(W) \ , \qquad (15.22)$$

$$k_m(Z, W) = J^m(Z, W)\{\Phi[R(Z, W)] + r(Z, W)\Psi[R(Z, W)]\} \ , \qquad (15.23)$$

$$J^m(Z, W) = \left(\frac{z - \bar{w} - \theta \bar{\nu}}{\bar{z} - w - \bar{\theta}\nu} \right)^{m/2} . \tag{15.24}$$

$k_m(Z, W)$ is the integral kernel of an operator valued function of the Dirac-Laplace operator $\Box_m$ (respectively $\hat{\Box}_m$), and Φ and Ψ are sufficiently decreasing functions at infinity. Note $J^m(\gamma Z, \gamma W) = j(\gamma, Z) J^m(Z, W) j^{-1}(\gamma, W)$ with $j(\gamma, Z)$ given by $j(\gamma, Z) = (F_\gamma / |F_\gamma|)^m$, where $F_\gamma = D\theta'$ [203, 422]. We have $j(\gamma \sigma, Z) = j(\gamma, \sigma Z) j(\sigma, Z)$ ($\forall \gamma, \sigma \in \Gamma$ and $Z \in \mathcal{H}^{(1|1)}$). A super-automorphic form $f(Z)$ is then defined by [41, 203] $f(\gamma Z) = j(\gamma, Z) f(Z)$ ($\forall \gamma \in \Gamma$). The super-automorphic kernel is defined as

$$K(Z, W) = \frac{1}{2} \sum_{\{\gamma\}} k_m(Z, \gamma W) j(\gamma, W) , \tag{15.25}$$

("$\frac{1}{2}$" because both γ and $-\gamma$ have to be included in the sum) i.e., $(L\phi)(z) = [h(\Box_m)](z)$. L is acting on super-automorphic functions $f(Z)$.

15.1.2   Compact Fundamental Domain

Let f be a super-automorphic function with $f(\gamma Z) = j(\gamma, Z) f(Z)$ and $g = Lf$. Let $\mathcal{F}^{(1|1)}(\gamma)$ be a fundamental domain of $\gamma \in \Gamma$ whose body equals $\mathcal{F}^{(1|1)}_{red} = \mathcal{F}$ (and is constructed in the same sense as the generalization $\mathcal{H}^{(1|1)}$ of $\mathcal{H}$). The expansion into hyperbolic conjugacy classes yields

$$
\begin{aligned}
\operatorname{str}(L) &= \int_{\mathcal{F}^{(1|1)}(\gamma)} dV(Z) K(Z, Z) \\
&= \int_{\mathcal{F}^{(1|1)}(\gamma)} \sum_{\gamma \in \Gamma} k_m(Z, \gamma Z) dV(Z) = \frac{\mathrm{i}^m}{2} \mathcal{A} \, \Phi(0) + \sum_{\substack{\{\gamma\} \\ \operatorname{str}(\gamma) + \chi_\gamma > 2}} \chi_\gamma^m A(\gamma) .
\end{aligned} \tag{15.26}
$$

Here I have assumed without loss of generality $a + d \geq 0$ for a $\gamma \in \hat{\Gamma}$, since $\operatorname{Aut}\mathcal{H}^{(1|1)} = \mathrm{OSp}(2|1, \mathbb{R})/\{\pm \mathbb{1}\}$. The first term corresponds to the identity transformation (zero-length term) and the second, $A(\gamma)$, is given by

$$A(\gamma) = \chi_\gamma^{-m} \int_{\mathcal{F}^{(1|1)}(\gamma)} k_m(Z, \gamma Z) j(\gamma, W) dV(Z) . \tag{15.27}$$

In [42, 203] these two terms corresponding to the identity transformation and hyperbolic conjugacy classes, respectively, were calculated, and I have obtained the Selberg super-trace formula on super-Riemann surfaces with hyperbolic conjugacy classes (c.f. [41]–[43, 203, 213]).

Theorem 15.1 *The Selberg super-trace formula for m-weighted Dirac-Laplace operators on closed super-Riemann surfaces for hyperbolic conjugacy classes is given by:*

$$\sum_{n=0}^{\infty} \left[h\left(\frac{1+m}{2} + \mathrm{i}p_n^{(B)} \right) - h\left(\frac{1+m}{2} + \mathrm{i}p_n^{(F)} \right) \right]$$

$$= -\frac{\mathcal{A}(\mathcal{F})}{4\pi} \int_0^\infty \frac{g(u) - g(-u)}{\sinh \frac{u}{2}} \cosh\left(\frac{um}{2}\right) du$$

$$+ \sum_{\{\gamma\}} \sum_{k=1}^\infty \frac{l_\gamma \chi_\gamma^{mk}}{2 \sinh \frac{kl_\gamma}{2}} \left[g(kl_\gamma) + g(-kl_\gamma) - \chi_\gamma^k \left(g(kl_\gamma) e^{-kl_\gamma/2} + g(-kl_\gamma) e^{kl_\gamma/2} \right) \right] . \tag{15.28}$$

The test function h is required to have the following properties

1. $h(\frac{1+m}{2} + \mathrm{i}p) \in C^\infty(\mathbb{R})$.

2. $h(p)$ vanishes faster than $1/|p|$ for $p \to \pm\infty$.

3. $h(\frac{1+m}{2} + \mathrm{i}p)$ is holomorphic in the strip $|\Im(p)| \leq \frac{1}{2} + \epsilon$, $\epsilon > 0$, to guarantee absolute convergence in the summation over $\{\gamma\}$.

The above Selberg super-trace formula (15.28) is valid for discrete hyperbolic conjugacy classes and in this case the noneuclidean area of the ("bosonic") fundamental domain is $\mathcal{A} = 4\pi(g - 1)$. The Fourier transform g of h is given by

$$\begin{aligned}
g(u) &= \frac{1}{2\pi} \int_\mathbb{R} h\left(\frac{1+m}{2} + \mathrm{i}p\right) e^{-\mathrm{i}up} dp \\
&= \frac{1}{4} \int_{4\sinh^2 \frac{u}{2}}^\infty \frac{dx}{(x+4)^{m/2}} \left\{ \frac{\Psi(x) + 2(e^u - 1)\Phi'(x)}{\sqrt{x - 4\sinh^2 \frac{u}{2}}} [\alpha_+^m(x, u) + \alpha_-^m(x, u)] \right. \\
&\qquad\qquad \left. - \mathrm{i}m e^{u/2} \Phi(x) \frac{\alpha_+^m(x, u) - \alpha_-^m(x, u)}{x + 4} \right\} , \tag{15.29}
\end{aligned}$$

where $\alpha_\pm^m(x, u) = \left(\pm\sqrt{x - 4\sinh^2(u/2)} - 2\mathrm{i}\cosh(u/2) \right)^{m/2}$. Specific trace formulæ, in particular for the heat kernel were considered by Aoki [8], Oshima [427], Yasui [396, 397] and Uehara and Yasui [499], as well as an explicit evaluation of the energy dependent resolvent kernel for the operator $\hat{\Box}^2$ [8, 427]. From (15.29) an explicit formula for $\Phi(x)$ can be derived [203] which has the form

$$\mathrm{i}^m \Phi(x) = \frac{1}{\pi\sqrt{x+4}} \int_x^\infty \frac{dy}{\sqrt{y+4}} \int_\mathbb{R} Q_1'(y + t^2) \left(\frac{\sqrt{y + t^2 + 4} - t}{\sqrt{y + t^2 + 4} + t} \right)^{m/2} dt , \tag{15.30}$$

with $Q_1(w) = 2\coth\frac{u}{2}[g(u) - g(-u)]$, $w = 4\sinh^2\frac{u}{2}$. Let us consider the combination

$$g(u)e^{-u/2} - g(-u)e^{u/2} = \frac{\mathrm{i}^m}{2} \sinh\frac{u}{2} \int_\mathbb{R} d\xi \left(\frac{\sqrt{w+4} + \mathrm{i}\xi}{\sqrt{w+4} - \mathrm{i}\xi} \right)^{m/2} \left[4\Phi'(w+\xi^2) - \Psi(w+\xi^2) \right] . \tag{15.31}$$

We define $Q_3(w) = 2[g(u)e^{-u/2} - g(-u)e^{u/2}]/\sinh\frac{u}{2}$ and obtain the general inversion formula for $\Psi(x)$

$$\mathrm{i}^m \Psi(x) = 4\mathrm{i}^m \Phi'(x) + \frac{1}{\pi} \int_\mathbb{R} Q_3'(x + t^2) \left(\frac{\sqrt{x + 4 + t^2} - t}{\sqrt{x + 4 + t^2} - t} \right)^{m/2} dt . \tag{15.32}$$

Alternatively, this can be rewritten as

$$
\mathrm{i}^m \Psi(x) = -\frac{\mathrm{i}^m \Phi(x)}{2(x+4)} + \frac{1}{\pi} \int_{\mathbb{R}} \left(\frac{\sqrt{x+4+t^2} - t}{\sqrt{x+4+t^2} - t} \right)^{m/2} \left[Q_3'(x+t^2) - \frac{Q_1'(x+t^2)}{x+4} \right] dt \ .
$$
(15.33)

For $m = 0$ we obtain simple inversion formulæ for $\Phi(t)$ and $\Psi(t)$, respectively

$$
\Phi(t) = -\frac{1}{\pi} \int_t^\infty \frac{Q_1(w) dw}{(w+4)\sqrt{w-t}} \ , \qquad \Psi(t) = -\frac{1}{2\pi} \int_t^\infty \frac{Q_2'(w) dw}{\sqrt{w-t}} \ ,
$$
(15.34)

with $Q_2(w) = 2[g(u)\mathrm{e}^{-u/2} + g(-u)\mathrm{e}^{u/2}]/\cosh \frac{u}{2}$.

15.1.3   Non-Compact Fundamental Domain

In this paragraph I include not only the case of parabolic conjugacy classes, respectively the case of a non-compact fundamental domain, but also the elliptic conjugacy classes. In the paper [213] I have dealt with the $m = 0$ case for both conjugacy classes.

In order to discuss the incorporation of elliptic conjugacy classes elements into the Selberg super-trace formula I have proposed in [213] for elliptic elements $\gamma \in \Gamma$ the representation

$$
\text{(elliptic } \gamma \in \Gamma \text{ conjugate to)} \qquad \begin{pmatrix} \cos\varphi & -\sin\varphi & 0 \\ \sin\varphi & \cos\varphi & 0 \\ 0 & 0 & \chi_R \end{pmatrix} \equiv R \ .
$$
(15.35)

We have $0 < \varphi < \pi$ and $\varphi = \pi/\nu_j$, $j = 1, \dots, s$. Therefore the effect of an elliptic transformation on super-coordinates $Z = (z, \theta)$ is as follows

$$
w = z' = \frac{z \cos\varphi - \sin\varphi}{z \sin\varphi + \cos\varphi} \ , \qquad \nu = \theta' = \frac{\chi_R \theta}{z \sin\varphi + \cos\varphi} \ .
$$
(15.36)

This yields for the even and odd point-pair invariants, respectively

$$
R(Z, W) = \sin^2\varphi \left[\frac{(1+x^2)^2}{y^2} + 2(1+x^2) + y^2 - 4 \right] \left(1 - \frac{\theta\bar\theta}{y} \right) \ ,
$$
(15.37)

$$
r(Z, W) = \frac{\theta\bar\theta}{y}(1 - \chi_R \cos\varphi) \ .
$$
(15.38)

Due to $D\theta' = \chi_R/(z \sin\varphi + \cos\varphi)$ we obtain furthermore

$$
j_R^m(Z) J^m(Z, RZ) = \chi_R^m \left(\frac{\sin\varphi(1+|z|^2) + 2\mathrm{i}y \cos\varphi}{\sin\varphi(1+|z|^2) - 2\mathrm{i}y \cos\varphi} \right)^{m/2}
$$
$$
\times \left(1 + \frac{\mathrm{i}m(1+|z|^2)\chi_R \sin\varphi}{(1+|z|^2)^2 \sin^2\varphi + 4y^2 \cos^2\varphi} \theta\bar\theta \right) \ .
$$
(15.39)

Restricting myself to hyperbolic and elliptic conjugacy classes gives for the Selberg super-trace formula

$$\text{str}(L) = \sum_{\{\gamma\}} \chi_\gamma^m A(\gamma) = \frac{i^m}{2}\mathcal{A}\dim V\,\Phi(0)$$

$$+ \sum_{\substack{\{\gamma\}\\ \text{str}(\gamma)+\chi_\gamma>2}} \chi_\gamma^m \text{str}_V[U(\gamma)]A(\gamma) + \sum_{\substack{\{R\}\\ \text{str}(R)+\chi_R<2}} \chi_R^m \text{str}_V[U(R)]A(R) \ . \quad (15.40)$$

Evaluating the relevant contributions we get [221]

$$\sum_{\substack{\{R\}\\ \text{str}(R)+\chi_R<2}} \text{str}_V[U(R)]A(R)$$

$$= \frac{i}{2}\sum_{\{R\}}\frac{\text{str}_V[U(R)]}{\nu\cos(\pi/\nu)}\left\{\frac{1-\chi_R\cos(\pi/\nu)}{\sin(\pi/\nu)}\int_\mathbb{R}\frac{\sinh[(p-\frac{i}{2})(\pi-2\pi/\nu)]}{\cosh\pi p}h(\tfrac{1}{2}+ip)dp\right.$$

$$\left. - \int_\mathbb{R}\frac{\sinh[(\pi-2\pi/\nu)p]}{\cosh\pi p}h(\tfrac{1}{2}+ip)dp\right\}$$

$$= \sum_{\{R\}}\frac{\text{str}_V[U(R)]}{\nu}\left\{\left(1-\chi_R\cos\frac{\pi}{\nu}\right)\int_0^\infty\frac{g(u)e^{-u/2}+g(-u)e^{u/2}}{\cosh u - \cos(2\pi/\nu)}du\right.$$

$$\left. + \int_0^\infty\frac{g(u)-g(-u)}{\cosh u - \cos(2\pi/\nu)}\sinh\frac{u}{2}du\right\}$$

$$= \sum_{\{R\}}\sum_{k=1}^{\nu-1}\frac{\text{str}_V[U^k(R)]}{\nu}\left\{\left(1-\chi_R^k\cos\frac{k\pi}{\nu}\right)\int_0^\infty\frac{g(u)e^{-u/2}+g(-u)e^{u/2}}{\cosh u - \cos(2k\pi\nu)}du\right.$$

$$\left. + \int_0^\infty\frac{g(u)-g(-u)}{\cosh u - \cos(2k\pi/\nu)}\sinh\frac{u}{2}du\right\} \ , \quad (15.41)$$

and I have displayed the result in three alternative ways.

Let us first assume that there is only one cusp. We propose similarly as for the hyperbolic and elliptic conjugacy classes for parabolic conjugacy classes $\gamma\in\Gamma$:

$$\text{(parabolic } \gamma\in\Gamma \text{ conjugate to)} \qquad \begin{pmatrix} 1 & n & 0 \\ 0 & 1 & 0 \\ 0 & 0 & \chi_S \end{pmatrix} \equiv S \ . \quad (15.42)$$

We have for the super-trace formula by including all conjugacy classes

$$\text{str}(L) = \sum_{\{\gamma\}}\text{str}[U(\gamma)]\chi_\gamma^m A(\gamma) = \frac{i^m}{2}\mathcal{A}\dim V\,\Phi(0) + \sum_{\substack{\{\gamma\}\\ \text{str}(\gamma)+\chi_\gamma>2}} \chi_\gamma^m\text{str}_V[U(\gamma)]A(\gamma)$$

$$+ \sum_{\substack{\{R\}\\ \text{str}(R)+\chi_R<2}} \chi_R^m\text{str}_V[U(R)]A(R) + \sum_{\substack{\{S\}\\ \text{str}(S)+\chi_S=2}} \text{str}_V[U(S)]\chi_S^m A(S) \ , \quad (15.43)$$

and we must investigate the fourth term. This gives for S acting on super-coordinates $Z = (z, \theta)$

$$w = z' = z + n \ , \qquad \nu = \theta' = \chi_S \theta \ . \tag{15.44}$$

For the even and odd point-pair invariants $R(Z, W)$ and $r(Z, W)$, respectively, this yields

$$R(Z, W) = \frac{n^2}{y^2}\left(1 - \frac{\theta\bar{\theta}}{y}\right) \ , \qquad r(Z, W) = \frac{\theta\bar{\theta}}{y}(1 - \chi_S) \ . \tag{15.45}$$

Furthermore $j_S^m(Z) = \chi_S^m$, and

$$J_S^m(Z, W) = \left(\frac{n - 2\mathrm{i}y}{n + 2\mathrm{i}y}\right)^{m/2}\left(1 - \frac{\mathrm{i}nm\chi_S\theta\bar{\theta}}{n^2 + 4y^2}\right) \ . \tag{15.46}$$

In order to be on the safe side we choose the fundamental domain for a parabolic transformation in such a way, that we consider the domain $[0, 1] \times (0, \infty)$ of integration for the x- and y-integrations, respectively, and truncate it to $[0, 1] \times (0, y_M)$ and take finally the limit $y_M \to \infty$. Therefore with $A(S) = \lim_{y_M \to \infty} A_{y_M}(S)$

$$A_{y_M}(S) = \chi_S^{-m} \int_0^1 dx \int_0^{y_M} dy \int \frac{d\theta d\bar{\theta}}{Y} \sum_{n \neq 0} j_S^m(Z) k_m(Z, S^n Z)$$

$$= \sum_{n \neq 0} \frac{1}{n} \int_{n/y_M}^\infty du \left(\frac{u - 2\mathrm{i}}{u + 2\mathrm{i}}\right)^{m/2}$$

$$\times \left[\frac{1}{2}\Phi(u^2) + u^2\Phi'(u^2) + (1 - \chi_S)\Psi(u^2) + \frac{\mathrm{i}um\chi_S}{u^2 + 4}\Phi(u^2)\right] \ . \tag{15.47}$$

We see clearly that this expression is divergent for $y_M \to \infty$. If $\kappa = 0$ we have to consider by including $\mathrm{str}[U(S)]$

$$A_{y_M}(S) = \sum_{n \neq 0}^\infty \frac{\mathrm{e}^{2\pi \mathrm{i}n\theta}}{n} \int_{n/y_M}^\infty du \left(\frac{u - 2\mathrm{i}}{u + 2\mathrm{i}}\right)^{m/2}$$

$$\times \left[\frac{1}{2}\Phi(u^2) + u^2\Phi'(u^2) + (1 - \chi_S)\Psi(u^2) + \frac{\mathrm{i}um\chi_S}{u^2 + 4}\Phi(u^2)\right] \tag{15.48}$$

(in the notation for one cusp) and the summation is convergent. However, in the general case we must find a regularization procedure. This will be done along the lines of the usual Selberg trace formula on non-compact Riemann surfaces. Therefore I have proposed for each parabolic conjugacy class, $j = 1, \ldots, \kappa$, the following continuum regularization for the Selberg super-trace formula in the presence of cusps and $\kappa_0 \geq 1$

$$\mathrm{str}(L)|_{\mathrm{cusp}_j} = \int_{\mathcal{D}^{(1|1)}} K(Z, SZ) dV(Z)$$

$$- c_{S_j} \frac{1 - \chi_{S_j}}{\pi} \int_\infty^\infty dp \, h(\mathrm{i}p + \tfrac{1}{2}) \int_{\mathcal{H}^{(1|1)}} dV(Z) E_j(Z, \mathrm{i}p + \tfrac{1}{2}) E_j^*(Z, \mathrm{i} + \tfrac{1}{2}) \tag{15.49}$$

with some normalization constants c_{S_j}. Let us define the super-Eisenstein series for one cusp and $\kappa_0 = 1$

$$E(Z, s) := \sum_{S \in \Gamma_0 \backslash \Gamma} [Y(SZ)]^s \ , \tag{15.50}$$

with Γ_0 in the stabilizer of Γ and with elements of the form of γ_{ell} of (15.42). This definition is completely analogous as in, e.g., [269], Kubota [362]. Note that $Y(\gamma Z)$ is understood as

$$Y \left[\begin{pmatrix} a & b & 0 \\ c & d & 0 \\ 0 & 0 & \chi s \end{pmatrix} Z \right] = \frac{Y}{|cz + d|^2} \ . \tag{15.51}$$

We analyze $E(Z, s)$ by a Fourier transformation, i.e.,

$$E(Z, s) = \sum_{m \in \mathbb{Z}} a_m(Z, s) \, e^{2\pi i m x} \ . \tag{15.52}$$

This yields for the coefficients $a_m(Z, s)$

$$a_m(Z, s) = \int_0^1 E(Z, s) e^{-2\pi i m x} \, dx = \int_0^1 \sum_{S \in \Gamma_0 \backslash \Gamma} Y(SZ)^s e^{-2\pi i m x} \, dx$$

$$= Y^s + \int_{\mathbb{R}} \sum_{S \in \Gamma_0 \backslash \Gamma} Y(SZ)^s e^{-2\pi i m x} \, dx \ , \qquad S = \begin{pmatrix} a & b & 0 \\ c & d & 0 \\ 0 & 0 & \chi s \end{pmatrix} \in \Gamma_0 \backslash \Gamma / \Gamma_0, c \neq 0 \ ,$$

$$= Y^s + Y^s y^{1-2s} \sum_c \frac{1}{|c|^{2s}} \left(\sum_d e^{2\pi i m d/c} \right) \int_{\mathbb{R}} \frac{e^{-2\pi i m y t}}{(1 + t^2)^s} dt \ . \tag{15.53}$$

The decomposition $\Gamma_0 \backslash \Gamma / \Gamma_0$ guarantees that there are no $\gamma \in \Gamma$ left containing parabolic transformations. We obtain for $E(Z, s)$ the expansion

$$E(Z, s) = \left(1 + \frac{s}{2y} \theta\bar{\theta} \right) \left[y^s + \phi_0(s) y^{1-s} + \sum_{m \neq 0} \phi_m(s) \sqrt{y} \, e^{2\pi i m x} K_{s-\frac{1}{2}}(2\pi |m| y) \right] \ , \tag{15.54}$$

and $\phi_0(s) \equiv \phi(s), \phi_m(s)$ respectively, are given by

$$\phi(s) \quad = \quad \sum_c \frac{1}{|c|^{2s}} \left(\sum_d e^{2\pi i m d/c} \right) \sqrt{\pi} \, \frac{\Gamma(s - \frac{1}{2})}{\Gamma(s)} \ , \tag{15.55}$$

$$\phi_m(s) \quad = \quad \sum_c \frac{1}{|c|^{2s}} \left(\sum_d e^{2\pi i m d/c} \right) \frac{2\pi^s}{\Gamma(s)} |m|^{s-\frac{1}{2}} \ , \tag{15.56}$$

with $c > 0$, $d \bmod c$, $\begin{pmatrix} * & * \\ c & d \end{pmatrix} \in S^{-1} \Gamma S$. Equation (15.54) shows the general structure of an even super-function which is an eigenfunction of the Laplace-Dirac operator $\square$ on the Poincaré super-upper half-plane, c.f. [42, 203, 397]. The set of super-Eisenstein

$E(Z, s)$ series therefore span the continuous spectrum of the Dirac-Laplace operator $\square$ on the super-Riemann surface, similarly as the Eisenstein series $e(z, s)$ span the continuous spectrum of the Laplacian Δ on the Riemann surface. Now consider the general case of the presence of several cusps and some numbers $\vartheta_{l\alpha}$. Of course, we must only regularize the super-automorphic kernel by the incorporation of $E(Z, s)$ whenever $\chi_{S_j} = -1$ ($j \in \{1, \ldots, \kappa\}$). We consider

$$E(Z, s, \alpha, v) = \sum_{\gamma \in \Gamma_\alpha \backslash \Gamma} Y^s(g_\alpha^{-1} \gamma Z) U^*(\gamma) v \ . \tag{15.57}$$

We assume a Fourier expansion according to

$$E_\beta(g_\beta Z, s, \alpha, v_l(\alpha)) = P_\beta E(g_\beta Z, s, \alpha, v_l(\alpha)) = \sum_{m \in \mathbb{Z}} a_m(Y, s) e^{2\pi i m x} \ , \tag{15.58}$$

with the coefficients $a_m(Y, s)$

$$\begin{aligned}
a_m(Y, s) &= \int_0^1 \sum_{\gamma \in \Gamma_\alpha \backslash \Gamma} Y^s(g_\alpha^{-1} \gamma g_\beta Z) P_\beta U^*(\gamma) v_l(\alpha) e^{-2\pi i m x} dx \\
&= \sum_{\gamma \in \Gamma_\alpha \backslash \Gamma / \Gamma_\beta}^{\cdot} P_\beta U^*(\gamma) v_l(\alpha) \int_{\mathbb{R}} Y^s(g_\alpha^{-1} \gamma g_\beta Z) e^{-2\pi i m x} dx \\
&= Y^s + Y^s y^{1-2s} \frac{P_\beta U^*(\gamma) v_l(\alpha)}{|c(g_\alpha^{-1} \gamma g_\beta)|^{2s}} \exp\left[2\pi i m \frac{d(g_\alpha^{-1} \gamma g_\beta)}{c(g_\alpha^{-1} \gamma g_\beta)}\right] \int_{\mathbb{R}} \frac{e^{-2\pi i m y t}}{(1 + t^2)^s} dt \\
&= \left(1 + \frac{s}{2y} \theta \bar\theta\right) e_\beta(g_\beta z, s, \alpha, v_l(\alpha)) \ , \tag{15.59}
\end{aligned}$$

with $e_\beta(g_\beta Z, s, \alpha, v_l(\alpha))$ as in (14.70) and all the results deduced from the properties of the usual Eisenstein series can be used appropriately in the Selberg super-trace formula. Thus we obtain [213]

Theorem 15.2 *The Selberg super-trace formula for Dirac-Laplace operators on closed super-Riemann surfaces for hyperbolic, elliptic and parabolic conjugacy classes is given by:*

$$\begin{aligned}
\sum_{n=0}^\infty & \left[h(\tfrac{1}{2} + i p_n^B) - h(\tfrac{1}{2} + i p_n^F)\right] = i \cdot \dim V \frac{\mathcal{A}}{4\pi} \int_{\mathbb{R}} h(ip + \tfrac{1}{2}) \tanh \pi p \, dp \\
& + \sum_{\{\gamma\}} \sum_{k=1}^\infty \frac{l_\gamma \mathrm{str}[U^k(\gamma)]}{2 \sinh \frac{k l_\gamma}{2}} \left[g(k l_\gamma) + g(-k l_\gamma) - \chi_\gamma^k \left(g(k l_\gamma) e^{-k l_\gamma / 2} + g(-k l_\gamma) e^{k l_\gamma / 2}\right)\right] \\
& + \sum_{\{R\}} \sum_{k=1}^{\nu-1} \frac{\mathrm{str}[U^k(R)]}{\nu} \left\{ \left(1 - \chi_R^k \cos \frac{k\pi}{\nu}\right) \int_0^\infty \frac{g(u) e^{-u/2} + g(-u) e^{u/2}}{\cosh u - \cos(2k\pi/\nu)} du \right. \\
& \left. \hspace{4cm} + \int_0^\infty \frac{g(u) - g(-u)}{\cosh u - \cos(2k\pi/\nu)} \sinh \frac{u}{2} du \right\}
\end{aligned}$$

$$-2\Big[\tilde{\kappa}_0 \ln 2 + \kappa_- \ln|\mathrm{sdet}(1-U(S))|\Big]g(0) - \frac{\kappa_-}{2}h(\tfrac{1}{2})\mathrm{str}[\mathfrak{S}(\tfrac{1}{2})] + \kappa_- \int_0^\infty g(-u)du$$

$$+ \frac{\kappa_-}{2\pi}\int_{\mathbb{R}} h(\mathrm{i}p + \tfrac{1}{2})\frac{\Delta'(\tfrac{1}{2}+\mathrm{i}p)}{\Delta(\tfrac{1}{2}+\mathrm{i}p)}dp + \frac{\tilde{\kappa}_0}{2\pi}\int_{\mathbb{R}} h(\mathrm{i}p + \tfrac{1}{2})[\Psi(1+\mathrm{i}p) + \Psi(1-\mathrm{i}p)]dp$$

$$+ \frac{\kappa_0}{2}\int_0^\infty [g(u) - g(-u)]du \ , \tag{15.60}$$

where $\tilde{\kappa}_0 = \sum_{\{S_j\}} \kappa_{S_j}(1 - \chi_{S_j})$ *whenever* $\chi_{S_j} = -1$ *and the other terms similarly interpreted. In particular the p-integral over* $\Delta'(\tfrac{1}{2}+\mathrm{i}p)/\Delta(\tfrac{1}{2}+\mathrm{i}p)$ *is only present if* $\kappa_0 \neq 0$. *Of course,* $g(u)$ *and* $h(\tfrac{1}{2}+\mathrm{i}p)$ *can be replaced by each other through their corresponding Fourier transforms.*

The test function h *is required to have the following properties*

1. $h(\tfrac{1}{2} + \mathrm{i}p) \in C^\infty(\mathbb{R})$.

2. $h(p)$ *vanishes faster than* $1/|p|$ *for* $p \to \pm\infty$.

3. $h(\tfrac{1}{2}+\mathrm{i}p)$ *is holomorphic in the strip* $|\Im(p)| \leq \tfrac{1}{2} + \epsilon$, $\epsilon > 0$, *to guarantee absolute convergence in the summation over* $\{\gamma\}$.

15.2 Selberg Super-Zeta-Functions

Let us consider the two Selberg super-zeta-functions Z_0 and Z_1, respectively, defined by [42, 203]

$$Z_0(s) = \prod_{\{\gamma\}}\prod_{k=0}^\infty \mathrm{sdet}\Big[1_V - U(\gamma)\mathrm{e}^{-(s+k)l_\gamma}\Big] \ , \qquad \Re(s) > 1 \ , \tag{15.61}$$

$$Z_1(s) = \prod_{\{\gamma\}}\prod_{k=0}^\infty \mathrm{sdet}\Big[1_V - U(\gamma)\chi_\gamma\mathrm{e}^{-(s+k)l_\gamma}\Big] \ , \qquad \Re(s) > 1 \ . \tag{15.62}$$

For convenience we will consider the functions

$$R_0(s) = \frac{Z_0(s)}{Z_0(s+1)} \ , \qquad R_1(s) = \frac{Z_1(s)}{Z_1(s+1)} \ , \qquad \Re(s) > 1 \ , \tag{15.63}$$

and the analytic properties of the $Z_{0,1}$ functions can be easily derived from the $R_{0,1}$ functions. As we shall see, only functional relations for the $R_{0,1}$ functions can be derived, but not for the $Z_{0,1}$ functions.

15.2.1 The Selberg Super-Zeta-Function Z_0

Let us turn to the discussion of the Selberg super-zeta-function R_0. We consider the testfunction $(\Re(s,a) > 1)$

$$h_0(\mathrm{i}p + \tfrac{1}{2}, s, a) = 2(\tfrac{1}{2} + \mathrm{i}p)\left(\frac{1}{s^2 - (\tfrac{1}{2}+\mathrm{i}p)^2} - \frac{1}{a^2 - (\tfrac{1}{2}+\mathrm{i}p)^2}\right) \ , \tag{15.64}$$

with the Fourier-transform $g_0(u, s, a)$ given by

$$g_0(u, s, a) = \text{sign}(u) e^{u/2} \left(e^{-s|u|} - e^{-a|u|} \right) . \tag{15.65}$$

A regularization term is needed to match the requirements of a valid test function for the trace formula. We obtain the Selberg super-trace formula for the test function $h_0(ip + \frac{1}{2}, s, a)$ as follows

$$\frac{R_0'(s)}{R_0(s)} - \frac{R_0'(a)}{R_0(a)}$$

$$= 2 \sum_{n=1}^{\infty} \left[\frac{\lambda_n^B}{s^2 - (\lambda_n^B)^2} - \frac{\lambda_n^B}{a^2 - (\lambda_n^B)^2} - \frac{\lambda_n^F}{s^2 - (\lambda_n^F)^2} + \frac{\lambda_n^F}{a^2 - (\lambda_n^F)^2} \right]$$

$$- \sum_{\{R\}} \sum_{k=1}^{\nu-1} \frac{\text{str}[U^k(R)]}{\nu \sin(2k\pi/\nu)} \sum_{l=1}^{\infty} \sin\left(\frac{2lk\pi}{\nu}\right) \left[\frac{1}{s+l-1} - \frac{1}{s+l+1} - \frac{1}{a+l-1} + \frac{1}{a+l+1} \right]$$

$$- \frac{A \dim V}{2\pi} \left[\Psi(s) + \Psi(s+1) - \Psi(a) - \Psi(a+1) \right]$$

$$- \frac{\kappa_0}{2} \left(\frac{1}{s - \frac{1}{2}} + \frac{1}{s + \frac{1}{2}} - \frac{1}{a - \frac{1}{2}} - \frac{1}{a + \frac{1}{2}} \right)$$

$$+ \frac{1}{2} \text{str}[\mathfrak{S}(\tfrac{1}{2})] \left(\frac{1}{s^2 - \frac{1}{4}} - \frac{1}{a^2 - \frac{1}{4}} \right) + \kappa_- \frac{\Delta'(s+1)}{\Delta(s+1)} - \kappa_- \frac{\Delta'(a+1)}{\Delta(a+1)}$$

$$+ \kappa_- \sum_{\rho, \beta < \frac{1}{2}} \left[\frac{1}{s - \rho} - \frac{1}{s + \rho} - \frac{1}{a - \rho} + \frac{1}{a + \rho} \right]$$

$$- \kappa_- \sum_{j=1}^{\mathcal{M}} \left[\frac{1}{s + (\sigma_j - 1)} - \frac{1}{s - (\sigma_j - 1)} - \frac{1}{a + (\sigma_j - 1)} + \frac{1}{a - (\sigma_j - 1)} \right] . \tag{15.66}$$

Thus we obtain [203, 213]

Theorem 15.3 *The Selberg super-zeta-function $R_0(s)$ is a meromorphic function on Λ_∞ and has furthermore the following properties:*

1. *The Selberg super-zeta-function $R_0(s)$ has "trivial" zeros at the following points and nowhere else*

   (a) *First note that*

$$\frac{1}{\sin(2k\pi/\nu)} \sum_{l=1}^{\infty} \sin\left(\frac{2lk\pi}{\nu}\right) \left(\frac{1}{s+l-1} - \frac{1}{s+l+1} \right)$$

$$= \frac{1}{s} + \frac{\cos(\frac{2k\pi}{\nu})}{s+1} + 2 \sum_{l=2}^{\infty} \frac{\cos(\frac{2lk\pi}{\nu})}{s+l} . \tag{15.67}$$

Therefore:
$s = 0$ *with multiplicity*

$$\# N_0 = \frac{\mathcal{A}\dim V}{2\pi} - \sum_{\{R\}}\sum_{k=1}^{\nu-1}\frac{\mathrm{str}[U^k(R)]}{\nu} \quad . \tag{15.68}$$

$s = -1$ *with multiplicity*

$$\# N_1 = \frac{\mathcal{A}\dim V}{\pi} - 2\sum_{\{R\}}\sum_{k=1}^{\nu-1}\frac{\mathrm{str}[U^k(R)]}{\nu}\cos\left(\frac{2k\pi}{\nu}\right) \quad . \tag{15.69}$$

$s = -n, n = 2, 3, \ldots,$ *with multiplicity*

$$\# N_n = \frac{\mathcal{A}\dim V}{\pi} - 2\sum_{\{R\}}\sum_{k=2}^{\nu-1}\frac{\mathrm{str}[U^k(R)]}{\nu}\sum_{l=2}^{\infty}\cos\left(\frac{2lk\pi}{\nu}\right) \quad . \tag{15.70}$$

Note that if $\# N_n < 0$, we have poles instead of zeros.

(b) $s = \frac{1}{2}$ *with multiplicity* $\# N_{\frac{1}{2}} = \kappa_-\mathrm{str}[\mathfrak{S}(\frac{1}{2})]$.

(c) $s = -\sigma_j$, $j = 1, \ldots, M$, *with the same multiplicity as the poles σ_j of $\Delta(s)$.*

(d) $s = \rho$ *with the multiplicity as the pole ρ of $\Delta(s)$ in the half-plane $\Re(s) < \frac{1}{2}$.*

2. *The Selberg super-zeta-function $R_0(s)$ has "trivial" poles at the following points and nowhere else*

   (a) $s = \pm\frac{1}{2}$ *with multiplicity $\kappa_0/2$.*

   (b) $s = -\frac{1}{2}$ *with multiplicity* $\# N_{-\frac{1}{2}} = \kappa_-\mathrm{str}[\mathfrak{S}(\frac{1}{2})]$.

   (c) $s = 1 - \sigma_j$, $j = 1, \ldots, M$, *with the multiplicity of the pole σ_j of the function $\Delta(s)$.*

   (d) $s = \rho - 1$ *with κ_- times the same multiplicity as the pole ρ of $\Delta(s)$ in the half-plane $\Re(s) < \frac{1}{2}$.*

   (e) *The items 1b–1d and 2a–2d are only present if $\kappa_0 \neq 0$.*

3. *The Selberg super-zeta-function $R_0(s)$ has "non-trivial" zeros and poles at the following points and nowhere else*

   (a) $s = \mathrm{i}p_n^{B(F)} - \frac{1}{2}$: *there are zeros (poles) of the same multiplicity as the corresponding Eigenvalue of $\Box$.*

   (b) $s = -\mathrm{i}p_n^{B(F)} - \frac{1}{2}$: *reversed situation for poles and zeros.*

   (c) $s = \lambda_n^{B(F)}$ *there are zeros (poles), and*

(d) $s = -\lambda_n^{B(F)}$ there are poles of the same multiplicity as the corresponding Eigenvalue of $\square$, respectively. The last two cases describe so-called small Eigenvalues of the operator $\square$.

Of course, (15.66) can be extended meromorphically to all $s \in \Lambda_\infty$.

The test function $h_0(ip + \frac{1}{2}, s, a)$ is symmetric with respect to $s \to -s$. Therefore subtracting the trace formulæ of $h_0(ip + \frac{1}{2}, s, a)$ and $h_0(ip + \frac{1}{2}, -s, a)$ from each other yields the functional equation for the R_0 function in differential form

$$
\frac{\mathrm{d}}{\mathrm{d}s} \ln \left[R_0(s) R_0(-s) \right]
$$
$$
= \frac{\mathcal{A}\dim V}{\pi} \frac{\mathrm{d}}{\mathrm{d}s} \ln(\sin \pi s) + \left[\frac{\Delta'(s)}{\Delta(s)} - \frac{\Delta'(1+s)}{\Delta(1+s)} \right]^{\kappa_-} + \kappa_0 \left(\frac{1}{s - \frac{1}{2}} + \frac{1}{s + \frac{1}{2}} \right)
$$
$$
- \sum_{\{R\}} \sum_{k=1}^{\nu-1} \frac{\mathrm{str}[U^k(R)]}{\nu \sin(2k\pi/\nu)} \sum_{l=1}^{\infty} \sin\left(\frac{2lk\pi}{\nu} \right)
$$
$$
\times \left[\frac{1}{s + l - 1} + \frac{1}{s - (l-1)} - \frac{1}{s + l + 1} - \frac{1}{s - (l+1)} \right] . \qquad (15.71)
$$

In integrated form, this gives the functional equation

$$
R_0(s) R_0(-s) = const.(\sin \pi s)^{\mathcal{A}\dim V/\pi} \left(\frac{\Delta(s+1)}{\Delta(s)} \right)^{\kappa_-} \ln\left(s^2 - \frac{1}{4} \right)^{-\kappa_0} \Psi_0(s) , \quad (15.72)
$$

with the function $\Psi_0(s)$ given by

$$
\Psi_0(s) = \exp\left\{ - \sum_{\{R\}} \sum_{k=1}^{\nu-1} \frac{\mathrm{str}[U^k(R)]}{\nu \sin(2k\pi/\nu)} \sum_{l=1}^{\infty} \sin\left(\frac{2lk\pi}{\nu} \right) \ln\left| \frac{(s^2 - (l-1)^2)}{(s^2 - (l+1)^2)} \right| \right\}. \qquad (15.73)
$$

We check easily the consistency of the functional equation with respect to the analytical properties of the Selberg super-zeta function R_0. Note the similarity of the corresponding relation (15.73) for the classical Selberg zeta-function.

15.2.2   The Selberg Super-Zeta-Function Z_1

We discuss the function $Z_1(s)$. In order to do this we choose the test function $(\Re(s, a) > 1)$

$$
h_1(\tfrac{1}{2} + ip, s, a) = 2ip\left(\frac{1}{s^2 + p^2} - \frac{1}{a^2 + p^2} \right) , \qquad (15.74)
$$

with the Fourier-transform function $g_1(u)$ given by

$$
g_1(u, s, a) = \mathrm{sign}(u)\left(\mathrm{e}^{-s|u|} - \mathrm{e}^{-a|u|} \right) . \qquad (15.75)
$$

Again a regularization term is needed to match the requirements of a valid test function for the trace formula. We obtain the Selberg super-trace formula for the test function $h_1(ip + \frac{1}{2}, s, a)$ as follows

$$
\frac{R_1'(s)}{R_1(s)} - \frac{R_1'(a)}{R_1(a)}
$$

$$
= 2 \sum_{n=1}^{\infty} \left[\frac{\lambda_n^B - \frac{1}{2}}{s^2 - (\lambda_n^B - \frac{1}{2})^2} - \frac{\lambda_n^B - \frac{1}{2}}{a^2 - (\lambda_n^B - \frac{1}{2})^2} - \frac{\lambda_n^F - \frac{1}{2}}{s^2 - (\lambda_n^F - \frac{1}{2})^2} + \frac{\lambda_n^F - \frac{1}{2}}{a^2 - (\lambda_n^F - \frac{1}{2})^2} \right]
$$

$$
- \Delta n_0^{(0)} \left(\frac{1}{s^2 - \frac{1}{4}} - \frac{1}{a^2 - \frac{1}{4}} \right)
$$

$$
- 2 \sum_{\{R\}} \sum_{k=1}^{\nu-1} \frac{\mathrm{str}[U^k(R)]}{\nu} \chi_R^k \sum_{l=0}^{\infty} \cos \left[(2l+1) \frac{k\pi}{\nu} \right] \left(\frac{1}{s+l+\frac{1}{2}} - \frac{1}{a+l+\frac{1}{2}} \right)
$$

$$
= -\frac{\mathcal{A} \dim V}{\pi} \left[\Psi(s + \tfrac{1}{2}) - \Psi(a + \tfrac{1}{2}) \right] + \kappa_- \left(\frac{1}{s} - \frac{1}{a} \right) - \kappa_0 \left(\frac{1}{s} - \frac{1}{a} \right)
$$

$$
+ \kappa_- \frac{\Delta'(s + \frac{1}{2})}{\Delta(s + \frac{1}{2})} - \kappa_- \frac{\Delta'(a + \frac{1}{2})}{\Delta(a + \frac{1}{2})}
$$

$$
+ \kappa_- \sum_{j=1}^{\mathcal{M}} \left[\frac{1}{s - (\sigma_j - \frac{1}{2})} - \frac{1}{s + (\sigma_j - \frac{1}{2})} - \frac{1}{a - (\sigma_j - \frac{1}{2})} + \frac{1}{a + (\sigma_j - \frac{1}{2})} \right]
$$

$$
+ \kappa_- \sum_{\rho, \beta < \frac{1}{2}} \left[\frac{1}{s - (\rho - \frac{1}{2})} - \frac{1}{s + (\rho - \frac{1}{2})} - \frac{1}{a - (\rho - \frac{1}{2})} + \frac{1}{a + (\rho - \frac{1}{2})} \right] . \quad (15.76)
$$

$\Delta n_0^{(0)} = n_0^B - n_0^F$ denotes the difference between the number of even- and odd zero-modes of the Dirac-Laplace operator $\square$.

According to [422] $\Delta n_0^{(0)} = 1 - 2q$ with $q = \dim(\ker \bar{\partial}_1)$ and $\bar{\partial}_p^\dagger = -y^2 \partial_z + \frac{1}{2} p y$. We find [213]

Theorem 15.4 *The Selberg super-zeta-function $R_1(s)$ is a meromorphic function on Λ_∞ and has furthermore the following properties:*

1. *The Selberg super-zeta-function $R_1(s)$ has "trivial" zeros at the following points and nowhere else*

   (a) $s = -\frac{1}{2} - l$, $l = 0, 1, 2, \ldots$, *and the multiplicity of these zeros is given by*

$$
\#N_l = \frac{\mathcal{A} \dim V}{\pi} - 2 \sum_{\{R\}} \sum_{k=1}^{\nu-1} \frac{\mathrm{str}[U^k(R)]}{\nu} \chi_R^k \sum_{l=0}^{\infty} \cos \left[(2l+1) \frac{k\pi}{\nu} \right] . \quad (15.77)
$$

   Note that for $l = 0$ there is an additional $\Delta n_0^{(0)}$ term coming from the super-trace of $h_1(ip + \frac{1}{2}, s, a)$ for $\lambda = 0$. Note also that if $\#N_l < 0$, we have poles instead of zeros.

(b) $s = 0$ with multiplicity $\kappa_- - \kappa_0$. If $\kappa_- - \kappa_0 < 0$, $s = 0$ is a pole. Note that the contributions from the zeros (poles) of $\Delta(s)$ and the poles (zeros) of the summation over j and ρ, respectively, cancel each other.

2. The Selberg super-zeta-function $R_1(s)$ has "non-trivial" zeros and poles at the following points and nowhere else

   (a) $s = \mathrm{i}p_n^{B(F)}$: there are zeros (poles) of the same multiplicity as the corresponding Eigenvalue of $\square$.

   (b) $s = -\mathrm{i}p_n^{B(F)}$: reversed situation for poles and zeros.

   (c) $s = \lambda_n^{B(F)} - \frac{1}{2}$ there are zeros (poles), and

   (d) $s = -(\lambda_n^F - \frac{1}{2})$ there are poles of the same multiplicity as the corresponding Eigenvalue of $\square$, respectively. The last two cases describe so-called small Eigenvalues of the operator $\square$.

Of course, (15.76) can be extended meromorphically to all $s \in \Lambda_\infty$.

The test function $h_1(\mathrm{i}p + \frac{1}{2}, s, a)$ is symmetric by the interchange $s \to -s$. Subtracting the trace formula for $h_1(\mathrm{i}p + \frac{1}{2}, s, a)$ and $h_1(\mathrm{i}p + \frac{1}{2}, -s, a)$ yields the functional equation for R_1 in differential form

$$\frac{\mathrm{d}}{\mathrm{d}s} \ln[R_1(s)R_1(-s)] = -\frac{\mathcal{A}\dim V}{\pi}\pi\tan\pi s + \frac{2(\kappa_- - \kappa_0)}{s}$$
$$- 2\sum_{\{R\}}\sum_{k=1}^{\nu-1}\frac{\mathrm{str}[U^k(R)]}{\nu}\chi_R^k\sum_{l=0}^{\infty}\cos\left[(2l+1)\frac{k\pi}{\nu}\right]\left(\frac{1}{s+l+\frac{1}{2}} + \frac{1}{s-(l+\frac{1}{2})}\right) \quad (15.78)$$

(note $\Psi(\frac{1}{2} + s) = \Psi(\frac{1}{2} - s) + \pi\tan\pi s$). The integrated functional equation therefore has the form

$$R_1(s)R_1(-s) = const.(\cos\pi s)^{\mathcal{A}\dim V/\pi}s^{2(\kappa_- - \kappa_0)}\Psi_1(s) \;, \quad (15.79)$$

with the function $\Psi_1(s)$ given by

$$\Psi_1(s) = \exp\left\{-2\sum_{\{R\}}\sum_{k=1}^{\nu-1}\frac{\mathrm{str}[U^k(R)]}{\nu}\chi_R^k\sum_{l=0}^{\infty}\cos\left[(2l+1)\frac{k\pi}{\nu}\right]\ln\left|s^2 - (l+\frac{1}{2})^2\right|\right\} \;.$$
$$(15.80)$$

We check easily the consistency of the functional equation with respect to the analytical properties of the Selberg super-zeta function R_1.

15.2.3 The Selberg Super-Zeta-Function Z_S

Following [396] we can also introduce the Selberg super-zeta function $Z_S(s)$ defined by

$$Z_S(s) = \prod_{\{\gamma\}} \prod_{k=0}^{\infty} \mathrm{sdet}\left[1_V - U(\gamma)\mathrm{diag}(1, \mathrm{e}^{-l_\gamma}, \chi_\gamma \mathrm{e}^{-l_\gamma/2}, \chi_\gamma \mathrm{e}^{-l_\gamma/2})\mathrm{e}^{-(s+k)l_\gamma}\right] \quad (15.81)$$

$$= \prod_{\{\gamma\}} \prod_{k=0}^{\infty} \frac{\mathrm{sdet}\left[1_V - U(\gamma)\mathrm{e}^{-(s+k)l_\gamma}\right]\mathrm{sdet}\left[1_V - U(\gamma)\mathrm{e}^{-(s+k+1)l_\gamma}\right]}{\left(\mathrm{sdet}\left[1_V - U(\gamma)\chi_\gamma \mathrm{e}^{-(s+k+1/2)l_\gamma}\right]\right)^2} \quad (15.82)$$

$$= \frac{Z_0(s)Z_0(s+1)}{Z_1^2(s+\frac{1}{2})} . \quad (15.83)$$

The appropriate test function is ($\Re(s) > 1$)

$$h_S(p, s) = \frac{1}{s^2 - \lambda^2}\Big|_{\lambda = \frac{1}{2}+ip} = \frac{1}{(s^2 - \frac{1}{4}) - ip + p^2} . \quad (15.84)$$

The corresponding Fourier-transform g_S has the form

$$g_S(u, s) = \frac{1}{2s}\mathrm{e}^{u/2 - s|u|} . \quad (15.85)$$

The evaluation of the various terms in the Selberg super-trace formula is straightforward and we obtain similarly to the previous two cases

$$\frac{1}{2s}\frac{Z_S'(s)}{Z_S(s)} = \frac{1}{2s}\frac{\mathrm{d}}{\mathrm{d}s}\ln\left[\frac{Z_0(s)Z_0(s+1)}{Z_1^2(s+\frac{1}{2})}\right]$$

$$= \sum_{n=1}^{\infty}\left[\frac{1}{s^2 - (\lambda_n^B)^2} - \frac{1}{s^2 - (\lambda_n^F)^2}\right] + \left(\Delta_0^{(0)} + \frac{\mathcal{A}\dim V}{4\pi}\right)\frac{1}{s^2}$$

$$- \frac{1}{2s}\sum_{\{R\}}\sum_{k=1}^{\nu-1}\frac{\mathrm{str}[U^k(R)]}{\nu\sin(2k\pi/\nu)}\sum_{l=1}^{\infty}\sin\left(\frac{2lk\pi}{\nu}\right)$$

$$\times\left[\frac{4(1 - \chi_R^k\cos(\frac{k\pi}{\nu}))}{s+l} + \frac{1}{s+l-1} + \frac{1}{s+l+1} - \frac{2}{s+l}\right]$$

$$+ \frac{1}{s}\left[\tilde{\kappa}_0 + \kappa_-\ln|\mathrm{sdet}(1 - U(S))|\right] + \frac{\kappa_-}{2}\frac{\mathrm{str}[\mathfrak{S}(\frac{1}{2})]}{s^2 - \frac{1}{4}} - \frac{\kappa_-}{2s}\frac{1}{s+\frac{1}{2}} + \frac{\kappa_0}{4s}\left(\frac{1}{s+\frac{1}{2}} - \frac{1}{s-\frac{1}{2}}\right)$$

$$+ \frac{\tilde{\kappa}_0}{2s}\left[\Psi\left(s+\frac{1}{2}\right) + \Psi\left(s+\frac{3}{2}\right)\right]$$

$$- \frac{\kappa_-}{2s}\frac{\Delta'(1+s)}{\Delta(1+s)} - \kappa_-\sum_{j=1}^{M}\frac{1}{s^2 - (\sigma_j - 1)^2} + \kappa_-\sum_{\rho,\beta<\frac{1}{2}}\frac{1}{s^2 - \rho^2} . \quad (15.86)$$

We therefore find [213]

Theorem 15.5 *The Selberg super-zeta-function Z_S is a meromorphic function on Λ_∞ and has furthermore the following properties:*

1. *The Selberg super-zeta-function $Z_S(s)$ has "trivial" zeros at the following points and nowhere else*

    (a) $s = 0$ *with multiplicity*

    $$\#N_0 = 2\left(\Delta_0^{(0)} + \frac{\mathcal{A}\dim V}{4\pi}\right) - \sum_{\{R\}}\sum_{k=1}^{\nu-1}\frac{\mathrm{str}[U^k(R)]}{\nu} . \tag{15.87}$$

    $s = -1$ *with multiplicity*

    $$\#N_1 = -2\sum_{\{R\}}\sum_{k=1}^{\nu-1}\frac{\mathrm{str}[U^k(R)]}{\nu}\left[1 - 2\chi_R\cos\left(\frac{k\pi}{\nu}\right) + \cos\left(\frac{k\pi}{\nu}\right)\right] . \tag{15.88}$$

    $s = -n$, $n = -2, 3, 4, \ldots$, *with multiplicity*

    $$\#N_n = 4\sum_{\{R\}}\sum_{k=2}^{\nu-1}\frac{\mathrm{str}[U^k(R)]}{\nu\sin(2k\pi/\nu)}\left[\sin^2\left(\frac{k\pi}{\nu}\right) - \left(1 - \chi_R\cos\frac{k\pi}{\nu}\right)\right]\sin\left(\frac{2lk\pi}{\nu}\right) . \tag{15.89}$$

    (b) $s = \pm\frac{1}{2}$ *with multiplicity* $\#N_{\pm\frac{1}{2}} = \kappa_-\mathrm{str}[\mathfrak{S}(\frac{1}{2})]/2$.

    (c) $s = \rho$ *with* κ_- *times the same multiplicity as the pole ρ of $\Delta(s)$ in the half-plane $\Re(s) < \frac{1}{2}$.*

    (d) $s = \rho - 1$ *with* κ_- *times the same multiplicity as the pole ρ of $\Delta(s)$ in the half-plane $\Re(s) < \frac{1}{2}$.*

2. *The Selberg super-zeta-function $Z_S(s)$ has "trivial" poles at the following points and nowhere else*

    (a) $s = \frac{1}{2}$ *with multiplicity* $\kappa_0/2$.

    (b) $s = -\frac{1}{2}$ *with multiplicity* $2\kappa_- - \kappa_0/2$.

    (c) $s = -\frac{1}{2} - l$ $(l = 1, \ldots)$ *with multiplicity* $\#N_l = 2\tilde{\kappa}_0$.

    (d) $s = -\sigma_j$, $j = 1, \ldots, \mathcal{M}$, *with* κ_- *times the multiplicity of the pole σ_j of the function $\Delta(s)$.*

    (e) *The items 1b–1d and 2a–2d are only present if $\kappa_0 \neq 0$.*

3. *The Selberg super-zeta-function $Z_S(s)$ has "non-trivial" zeros and poles at the following points and nowhere else*

    (a) $s = \pm(\frac{1}{2} + ip_n^B)$ *there are zeros (poles) and (b)* $s = \pm(\frac{1}{2} + ip_n^F)$ *there are poles (zeros),*

with the same multiplicity as the corresponding Eigenvalue of $\square$, respectively.

Of course, (15.86) can be extended meromorphically to all $s \in \Lambda_\infty$.

The test function $h_S(ip + \frac{1}{2}, s)$ is symmetric with respect to $s \to -s$ and therefore we can deduce the functional relation

$$\frac{Z_S(s)}{Z_S(-s)} = const.\, e^{4[\kappa_0 + \ln |sdet(1 - U(S))|]} \frac{1}{\Delta(s)\Delta(s+1)}$$

$$\times \left(\frac{s - \frac{1}{2}}{s + \frac{1}{2}}\right)^{\kappa_- - \kappa_0} \left(\frac{\Gamma(s + \frac{1}{2})\Gamma(s + \frac{3}{2})}{\Gamma(\frac{1}{2} - s)\Gamma(\frac{3}{2} - s)}\right)^{\tilde{\kappa}_0} \Psi_S(s) \,, \tag{15.90}$$

with the function $\Psi_S(s)$ given by

$$\Psi_S(s) = \exp\left\{-2 \sum_{\{R\}} \sum_{k=1}^{\nu-1} \frac{str[U^k(R)]}{\nu \sin(2k\pi/\nu)}\right.$$

$$\times \sum_{l=1}^{\infty} \sin\left(\frac{2lk\pi}{\nu}\right) \left[2\left(1 - 2\chi_R^k \cos\frac{k\pi}{\nu}\right) \ln\left|\frac{s+l}{s-l}\right| + \ln\left|\frac{(s+l-1)(s+l+1)}{(s-l+1)(s-l-1)}\right|\right]\right\} .$$

$$\tag{15.91}$$

We check easily the consistency of the functional equation with respect to the analytical properties of the Selberg super-zeta-function Z_S. In the case, where only hyperbolic conjugacy classes are present in the super-Fuchsian group, (15.90) reduces to the simple functional equation $Z_S(s) = Z_S(-s)$. Let us note that the relation

$$\frac{d}{ds} \ln\left[\frac{Z_0(s)Z_0(s+1)}{Z_1(s+\frac{1}{2})}\right] - \frac{d}{ds} \ln\left[\frac{Z_0(s+1)Z_0(s+2)}{Z_1(s+\frac{3}{2})}\right]$$

$$= \frac{R_0'(s)}{R_0(s)} + \frac{R_0'(s+1)}{R_0(s+1)} - 2\frac{R_1'(s+\frac{1}{2})}{R_1(s+\frac{1}{2})} \tag{15.92}$$

provides a consistency check for the zeta-functions R_0, R_1 and Z_S, respectively.

15.3 Super-Determinants of Dirac Operators

Since $\square_m^2$ is not a positive definite operator the super-determinant of the operator $c^2 - \square_m^2$ for $\Re(c) > m$ is calculated and analytically continued in c. Let $m \in \mathbb{N}_0$. The super-determinant is defined using the zeta-function regularization as

$$sdet(c^2 - \square_m^2) = \exp\left[-\frac{\partial}{\partial s}\zeta_m(s;c)\Big|_{s=0}\right] \tag{15.93}$$

$$\zeta_m(s;c) = str[(c^2 - \square_m^2)^{-s}]$$

$$= \frac{1}{\Gamma(s)} \int_0^\infty dt\, t^{s-1} str\{\exp[-t(c^2 - \square_m^2)]\} . \tag{15.94}$$

For the calculation one uses the heatkernel function in the super-trace formula, i.e.

$$h_{hk}(s) = e^{t[(s+\frac{m}{2})^2 - c^2]} \; . \tag{15.95}$$

Performing the limit $c \to \epsilon$ for $|\epsilon| \ll 1$:

$$\mathrm{sdet}(-\square_0^2) = \frac{1}{(2g-2)!} \cdot \frac{Z_0(1)Z_0^{(2g-2)}(0)}{[\tilde{Z}_1(\frac{1}{2})]^2} \epsilon^{2\Delta n_0^{(0)}} \; . \tag{15.96}$$

By $\tilde{Z}_1(\frac{1}{2})$ the appropriate derivative or residuum of Z_1 at $s = \frac{1}{2}$ is denoted, depending on whether $\Delta n_0^{(0)} \leq 0$ or $\Delta n_0^{(0)} > 0$, respectively. To make this quantity well-defined I subtract from $\mathrm{sdet}(-\square_0^2)$ the zero-mode which I denote by priming the sdet. Using further the functional relations for Z_0 and Z_1 this yields [203]

$$\mathrm{sdet}'(-\square_0^2) = \left[\pi^{g-1}\frac{Z_0(1)}{\tilde{Z}_1(\frac{1}{2})}\right]^2 \frac{Z_1(0)}{Z_1(1)} \; . \tag{15.97}$$

For calculating the super-determinant for m even and $m \geq 2$ a subtraction of zero- or trivial-modes is not necessary. Proceeding similarly as for $m = 0$ we get for $m = 2, 4, \ldots$ [203]:

$$\mathrm{sdet}(-\square_m^2) = \left[\left(\frac{\pi}{m!}\right)^{g-1}\frac{Z_0(1+\frac{m}{2})}{Z_1(\frac{m+1}{2})}\right]^2 \frac{Z_1(0)}{Z_1(1)} \; . \tag{15.98}$$

Similarly for $m = 2, 4, \ldots$:

$$\mathrm{sdet}(-\square_{-m}^2) = \left[\left(\frac{(m-2)!}{\pi}\right)^{g-1}\frac{Z_0(\frac{m}{2})}{Z_1(\frac{m+1}{2})}\right]^2 \frac{Z_1(1)}{Z_1(0)} \; . \tag{15.99}$$

For $m = 1, 3, \ldots$:

$$\mathrm{sdet}(-\square_m^2) = \left[\left(\frac{\pi}{\mathrm{i}\,m!}\right)^{g-1}\frac{Z_1(1+\frac{m}{2})}{Z_0(\frac{m+1}{2})}\right]^2 \frac{Z_1(0)}{Z_1(1)} \; . \tag{15.100}$$

For $m = 3, 5, \ldots$:

$$\mathrm{sdet}(-\square_{-m}^2) = \left[\left(\frac{(m-2)!\mathrm{i}}{\pi}\right)^{g-1}\frac{Z_1(\frac{m}{2})}{Z_0(\frac{m+1}{2})}\right]^2 \frac{Z_1(1)}{Z_1(0)} \; . \tag{15.101}$$

The case of $\square_{-1}^2$ must be treated separately because of the appearance of zero-modes which must be subtracted. Therefore denoting the omission of zero-modes by priming the super-determinant one gets

$$\mathrm{sdet}'(-\square_{-1}^2) = \left[\pi^{1-g}\frac{\tilde{Z}_1(\frac{1}{2})}{Z_0(1)}\right]^2 \frac{Z_1(1)}{Z_1(0)} \; . \tag{15.102}$$

From the introductory remarks in Chapter 10 we know that the relevant string integrand is given by $\text{sdet}'(-\Box_0^2)$ and $\text{sdet}(-\Box_2^2)$. Hence, we get for the partition function

$$
\begin{aligned}
Z_g &= \int_{s\mathcal{M}_g} d(SWP)[\text{sdet}'(-\Box_0^2)]^{-5/2}[\text{sdet}(-\Box_2^2)]^{1/2} \\
&= \left(\frac{1}{2\pi^4}\right)^{g-1} \int_{s\mathcal{M}_g} d(SWP)\left(\frac{Z_0(1)}{\tilde{Z}_1(\frac{1}{2})}\right)^{-5} \frac{Z_0(2)}{Z_1(\frac{3}{2})}\left(\frac{Z_1(1)}{Z_1(0)}\right)^2 .
\end{aligned}
\tag{15.103}
$$

15.4 The Selberg Super-Trace Formula on Bordered Super-Riemann Surfaces

Because it is sufficient to consider super-Riemann surfaces without odd parameters I have proposed a construction of a bordered super-Riemann surface. In order to do this we take the construction of a usual bordered Riemann surface and endow it with the Grassmann algebra Λ_∞. Because we know how to define a closed super-Riemann surface, we take $\hat{\Sigma}$ and enlarge it to $\hat{\Sigma}^{(1|1)}$ together with its corresponding super-Fuchsian group $\hat{\Gamma}^{(1|1)}$ constructed from $\hat{\Gamma}$ and the super fundamental domain $\hat{\mathcal{F}}^{(1|1)}$. A convenient way to introduce the super-analogue of the involution $\mathcal{I}$ turns out to be the *super-involution*

$$
\left.\begin{aligned}
\mathcal{I}Z &= \mathcal{I}(z,\theta) = (-\bar{z}, -i\bar{\theta}) , \\
\mathcal{I}\bar{Z} &= \mathcal{I}(\bar{z},\bar{\theta}) = (-z, -i\theta) ,
\end{aligned}\right\}
\tag{15.104}
$$

respectively $\mathcal{I}(z,\theta_1,\theta_2) = (-\bar{z}, -i\theta_1, i\theta_2)$. It has the properties

$$
\mathcal{I}D = i\bar{D} , \qquad \mathcal{I}\bar{D} = iD .
\tag{15.105}
$$

Note that $\mathcal{I}^4 Z = Z$ and $\mathcal{I}^4 D = D$. Furthermore for the Dirac-Laplace operator $\hat{\Box}_m$ we have

$$
\mathcal{I}\hat{\Box}_m = \hat{\Box}_{-m} = \overline{\hat{\Box}_m} .
\tag{15.106}
$$

Similarly as for the usual bordered Riemann surface where $\Sigma = \hat{\Sigma}\backslash\mathcal{I}$, we then define the bordered super-Riemann surface $\Sigma^{(1|1)}$ as $\Sigma^{(1|1)} = \hat{\Sigma}^{(1|1)}\backslash\mathcal{I}$. The corresponding discs $d_1^{(1|1)}, \ldots, d_n^{(1|1)}$ are super-conformal non-overlapping super-discs seen as usual conformal non-overlapping discs endowed with the Grassmann algebra Λ_∞. The particular form of the involution (15.104) enables us to work directly on the fundamental domains $\hat{\mathcal{F}}^{(1|1)}$. The super-Fuchsian group $\hat{\Gamma}$ is consequently a symmetric super-Fuchsian group.

For the point pair invariants we find for the action of $\mathcal{I}$

$$
\begin{aligned}
R(Z, \mathcal{I}W) &= \overline{R(\mathcal{I}Z, W)} \tag{15.107} \\
r(Z, \mathcal{I}W) &= \overline{r(\mathcal{I}Z, W)} , \tag{15.108}
\end{aligned}
$$

furthermore $J(Z, \mathcal{I}W) = \overline{J(\mathcal{I}Z, W)}$, and due to the construction of k_m

$$
k_m(Z, \mathcal{I}W) = \overline{k_m(\mathcal{I}Z, W)} .
\tag{15.109}
$$

Let us consider the super-automorphic Selberg operator with Dirichlet boundary conditions

$$
\begin{aligned}
(\hat{L}f)(Z) &= \frac{1}{4}\int_{\mathcal{H}^{(1|1)}} dV(W)[k_m(Z,W) - k_m(Z,\mathcal{I}W)]f(W) \\
&= \frac{1}{4}\sum_{\{\gamma\}}\int_{\gamma\widehat{\mathcal{F}}^{(1|1)}(\gamma)} dV(W)[k_m(Z,W) - k_m(Z,\mathcal{I}W)]f(W) \\
&= \frac{1}{2}\int_{\widehat{\mathcal{F}}^{(1|1)}(\gamma)} dV(W)K(Z,W)f(W) ,
\end{aligned}
\tag{15.110}
$$

where

$$
K(Z,W) = \frac{1}{2}\sum_{\{\gamma\}}[k_m(Z,\gamma W) - k_m(Z,\gamma\mathcal{I}W)]
\tag{15.111}
$$

is the super-automorphic kernel on bordered super-Riemann surfaces. Now we have for a super-function ϕ which is odd with respect to x

$$
\begin{aligned}
&\frac{1}{2}\int_{\widehat{\mathcal{F}}^{(1|1)}(\gamma)} dV(W)K(Z,\mathcal{I}W)\phi(W) \\
&= \frac{1}{2}\int_{\mathcal{H}^{(1|1)}} dV(W)k_m(Z,\mathcal{I}W)\phi(W) \\
&= \frac{1}{2}\int_{\mathcal{H}^{(1|1)}} dV(W)\overline{k(\mathcal{I}Z,W)}\phi(W) = \frac{1}{2}\overline{(L\bar{\phi})(\mathcal{I}Z)} ,
\end{aligned}
\tag{15.112}
$$

due to the properties of the super-Selberg operator. Let now Φ be an eigenfunction of $\hat{\Box}_m$ which is odd with respect to x, i.e., $\hat{\Box}_m\Phi = s\Phi$. Then $\overline{s\Phi} = \bar{s}\bar{\Phi} = \hat{\Box}_{-m}\bar{\Phi}$ and $\bar{\Phi}$ is an odd eigenfunction of $\hat{\Box}_{-m}$ with eigenvalue $\bar{s}$. Denote by $\hat{L}$ the Selberg super-operator on the super-Riemann surface $\hat{\Sigma}$; let $(L\phi)(Z) = \Lambda(s)\phi(Z)$ and $\overline{(L\bar{\phi})(\mathcal{I}Z)} = \overline{\Lambda'(\bar{s})}\phi(\mathcal{I}Z)$ on Σ and $\mathcal{I}\Sigma$, respectively. Then

$$
\begin{aligned}
(\hat{L}\phi)(Z) &= \frac{1}{2}(L\phi)(Z) - \frac{1}{2}\overline{(L\bar{\phi})(\mathcal{I}Z)} \\
&= \frac{1}{2}\Lambda(s)\phi(Z) - \frac{1}{2}\overline{\Lambda'(\bar{s})}\phi(\mathcal{I}Z) = \frac{1}{2}\big[\Lambda(s) + \Lambda'(s)\big]\phi(Z) .
\end{aligned}
\tag{15.113}
$$

The equivalence relation (15.17) shows that the eigenvalue problem of the operator $\hat{\Box}_m$ is closely related to the eigenvalue problem of the operator $-\Delta_m$, both for eigenfunctions which are even or odd with respect to x. Now, an odd eigenfunction of $-\Delta_m$ is also an odd eigenfunction of $-\Delta_{-m}$, and the solution of the corresponding differential equations depends only on m^2 but not on m ([269, pp.266–268], [150, pp.203–205]), hence, the spectrum depends only on $|m|$ (compare also [70] and the discussion before concerning the Selberg trace formula on bordered Riemann surfaces). Therefore we conclude that a with-respect-to-x odd eigenfunction of $\Box_m$ is also a with-respect-to-x odd eigenfunction of $\mathcal{I}\Box_m$ with the eigenvalue $\bar{s}$, furthermore $\Lambda = \Lambda'$ [70], and we can infer [together with the usual identification $h(p) = \Lambda(\frac{1}{2} + ip)$]

$$
(\hat{L}\phi)(Z) = h(p)\phi(z) .
\tag{15.114}
$$

Let $Z_\Gamma(\gamma)$ be the centralizer of a $\gamma \in \Gamma$. For $\mathrm{str}(\hat{L})$ we obtain on the one hand

$$\mathrm{str}(\hat{L}) = \sum_n \left[h(p_n^{(B)}) - h(p_n^{(F)}) \right] , \tag{15.115}$$

where $s_n^{(B,F)} = \frac{1}{2} + ip_n^{(B,F)}$ are the bosonic and fermionic eigenvalues, respectively, of $\square_m$. [According to (15.17) we should consequently write $s = -i(\frac{1}{2} + ip)$, which looks, however, somewhat artificial and is therefore not adopted.] On the other hand we have

$$\begin{aligned}
\mathrm{str}(\hat{L}) &= \frac{1}{2} \int_{\widehat{\mathcal{F}}^{(1|1)}(\gamma)} dV(W) K(Z, Z) \\
&= \frac{1}{4} \sum_{\{\gamma\}} \int_{\widehat{\mathcal{F}}^{(1|1)}(\gamma)} [k_m(Z, \gamma Z) - k_m(Z, \gamma \mathcal{I} Z)] dV(Z) ,
\end{aligned} \tag{15.116}$$

where $\widehat{\mathcal{F}}^{(1|1)}(\gamma)$ denotes the fundamental region for the super-Fuchsian group $Z_\Gamma(\gamma)$, the centralizer of $\gamma \in \Gamma$.

15.4.1 Compact Fundamental Domain

For convenience we set $\rho = \gamma \mathcal{I}$ and use the classification of the inverse-hyperbolic transformations according to $\rho \in \bar{\Gamma} \mathcal{I}$, respectively, $\rho^2 \in \bar{\Gamma}$. We generalize the result of the conjugacy classes for the usual case of bordered Riemann surfaces and consider the two cases of the conjugacy classes in $\gamma \mathcal{I}$. The expansion into the conjugacy classes yields for the Selberg super-operator for Dirichlet boundary conditions

$$\begin{aligned}
\mathrm{str}(\hat{L}) &= \frac{1}{2} \int_{\widehat{\mathcal{F}}^{(1|1)}(\gamma)} \sum_{\{\gamma\}} \left[k_m(Z, \gamma Z) - k_m(Z, \gamma \mathcal{I} Z) \right] dV(Z) \\
&= \frac{\widehat{\mathcal{A}}}{4} \Phi(0) + \frac{1}{2} \sum_{\{\gamma\}} \int_{\widehat{\mathcal{F}}^{(1|1)}(\gamma)} dV(Z) k_m(Z, \gamma Z) - \frac{1}{2} \sum_{\{\rho\}; \widehat{\Gamma}_{\rho, hyp}} \int_{\widehat{\mathcal{F}}^{(1|1)}(\rho)} dV(Z) k_m(Z, \rho Z) .
\end{aligned} \tag{15.117}$$

Let us consider the involution term. We obtain $\nu_2 = \rho\theta_2 = -\chi_\rho N^{1/4} \theta_2$. Similarly as in the usual hyperbolic case we find for the two-point invariants $[M = N^{k+1/2}]$

$$R(Z, \rho Z) = \frac{|z + M\bar{z}|^2}{My^2} \left(1 - \frac{2\theta_1 \theta_2}{y} \right) , \tag{15.118}$$

$$r(Z, \rho Z) = \frac{\theta_1 \theta_2}{y} \left[2 + \chi(M^{1/2} + M^{-1/2}) \right] . \tag{15.119}$$

Furthermore $j(\rho^{2k+1}, Z) = \chi_\rho^{(2k+1)m}$ and

$$J^m(Z, \rho^{2k+1} Z) = \left(\frac{\zeta + 2i \cosh \frac{u}{2}}{\zeta - 2i \cosh \frac{u}{2}} \right)^{m/2} \left(1 - \frac{2i \, m \chi_\rho^{2k+1} \zeta \theta_1 \theta_2}{y(\zeta^2 + 4 \cosh^2 \frac{u}{2})} \right) , \tag{15.120}$$

where $\zeta = 2x \cosh \frac{u}{2}/y$ and $u = (2k+1)\ln\sqrt{M} = (k+1/2)l_{\rho^2}$. The evaluation of the conjugacy classes $\{\rho\}$ is straightforward and similar to the usual hyperbolic case. Evaluating the relevant terms we obtain [215, 221]

Theorem 15.6 *The Selberg super-trace formula for m-weighted Dirac-Laplace operators $\Box_m$ on compact bordered super-Riemann surfaces with Dirichlet boundary conditions is given by:*

$$\sum_{n=1}^{\infty} \left[h(p_n^{(B)}) - h(p_n^{(F)}) \right] = -\frac{\hat{\mathcal{A}}}{4\pi} \int_0^\infty \frac{g(u)-g(-u)}{\sinh\frac{u}{2}} \cosh\left(\frac{um}{2}\right) du$$

$$+ \frac{1}{4} \sum_{\{\gamma\}} \sum_{k=1}^{\infty} \frac{\chi_\gamma^{km} l_\gamma}{\sinh\frac{kl_\gamma}{2}} \left[g(kl_\gamma) + g(-kl_\gamma) - \chi_\gamma^k \left(g(kl_\gamma)e^{-kl_\gamma/2} + g(-kl_\gamma)e^{kl_\gamma/2} \right) \right]$$

$$- \frac{1}{4} \sum_{\{\rho^2\}} \sum_{k=0}^{\infty} \frac{\chi_{\rho^2}^{(k+1/2)m} l_{\rho^2}}{\cosh\left[\frac{1}{2}(k+\frac{1}{2})l_{\rho^2}\right]} \left\{ g[(k+\tfrac{1}{2})l_{\rho^2}] + g[-(k+\tfrac{1}{2})l_{\rho^2}] \right.$$

$$\left. - \chi_{\rho^2}^{k+1/2}\left(g[(k+\tfrac{1}{2})l_{\rho^2}]e^{-\frac{1}{2}(k+1/2)l_{\rho^2}} + g[-(k+\tfrac{1}{2})l_{\rho^2}]e^{(k+1/2)l_{\rho^2}/2} \right) \right\}$$

$$- \frac{1}{2} \sum_{i=1}^{n} \sum_{k=1}^{\infty} \frac{\chi_{C_i}^{km} l_{C_i}}{\cosh\frac{kl_{C_i}}{2}}$$

$$\times \left[g(kl_{C_i}) + g(-kl_{C_i}) - \chi_{C_i}^k \left(g(kl_{C_i})e^{-kl_{C_i}/2} + g(-kl_{C_i})e^{kl_{C_i}/2} \right) \right] , \qquad (15.121)$$

where $\lambda_n^{(B,F)} = \frac{1}{2} + ip_n^{(B,F)}$ on the left runs through the set of all eigenvalues of this Dirichlet problem, and the summation on the right is taken over all primitive conjugacy classes $\{\gamma\}_{\hat{\Gamma}}$, $\mathrm{str}(\gamma) + \chi_\gamma > 2$, and $\{\rho\}_{\hat{\Gamma}}$, ρ hyperbolic.
The test function h is required to have the following properties

1. *$h(p) \equiv h(\frac{1+m}{2} + ip) \in C^\infty(\mathbb{R})$,*

2. *$h(p)$ vanishes faster than $1/|p|$ for $p \to \pm\infty$.*

3. *$h(p)$ is holomorphic in the strip $|\Im(p)| \leq \frac{1}{2} + \epsilon$, $\epsilon > 0$, to guarantee absolute convergence in the summation over $\{\gamma\}$ and $\{\rho\}$.*

Note that there is no $k = 0$ contribution from the last summand. $g(u)$ is given by (15.29).

Note that in the case of Neumann boundary conditions the last two terms just change their signs.

15.4.2 Non-Compact Fundamental Domain

I consider only the case $m = 0$. We now include all relevant conjugacy classes and get

$$\mathrm{str}(\hat{L}) = \frac{1}{2} \int_{\widehat{\mathcal{F}}^{(1|1)}(T)} \sum_{\{T\}} \Big[k(Z, TZ) - k(Z, T\mathcal{I}Z) \Big] dV(Z)$$

$$= \frac{1}{4} \hat{\mathcal{A}} \, \Phi(0) + \frac{1}{2} \sum_{\substack{\{\gamma\} \\ \mathrm{str}(\gamma) + \chi_\gamma > 2}} \int_{\widehat{\mathcal{F}}^{(1|1)}(\gamma)} dV(Z) k(Z, \gamma Z)$$

$$- \frac{1}{2} \sum_{\{\rho\}; \widehat{\Gamma}_{\rho, hyp}} \int_{\widehat{\mathcal{F}}^{(1|1)}(\rho)} dV(Z) k(Z, \rho Z) + \frac{1}{2} \sum_{\substack{R \in \widehat{\Gamma} \\ \mathrm{str}(R) + \chi_R < 2}} \int_{\widehat{\mathcal{F}}^{(1|1)}(R)} dV(Z) k(Z, RZ)$$

$$+ \frac{1}{2} \lim_{y_m \to \infty} \int_{\widehat{\mathcal{F}}^{(1|1)}_{y_M}} dV(Z)$$

$$\times \left\{ \sum_{\{S\}} \sum_{\gamma' \in \widehat{\Gamma}_S \backslash \Gamma} k(Z, \gamma'^{-1} S \gamma' Z) - \sum_{\substack{\{\rho\}; \widehat{\Gamma}_{\rho, ell} \\ \mathrm{str}(\rho) + \chi_\rho = 0}} \sum_{\gamma' \in \widehat{\Gamma}_S \backslash \Gamma} k(Z, \gamma'^{-1} \rho \gamma' Z) \right\} , \quad (15.122)$$

with some properly defined compact domain $\widehat{\mathcal{F}}^{(1|1)}_{y_M}$ depending on a large parameter y_M, and where the sum is taken over all hyperbolic conjugacy classes $\{\gamma\}$, elliptic conjugacy classes $\{R\}$ and parabolic conjugacy classes $\{S\}$ in $\bar{\Gamma}$ with representatives γ, R and S, respectively, over all relative non-degenerate classes $\{\rho\}$, ρ hyperbolic, and over the relative conjugacy classes $\{\rho\}$ with $\mathrm{str}(\rho) + \chi_\rho = 0$, ρ elliptic.

As we know from the discussion of the usual Selberg case, the conjugacy class with $\mathrm{tr}(\rho) = 0$, ρ elliptic, contains an element of order two. In the super-case this is generalized to

$$\gamma_a = \begin{pmatrix} 0 & a & 0 \\ -a^{-1} & 0 & 0 \\ 0 & 0 & \chi_{\gamma_a} \end{pmatrix} , \quad (\mathrm{mod} \pm 1) , \quad (15.123)$$

with some $a \geq 1$. Because γ_a is an elliptic element of order two we have to consider

$$\int_{\bigcup_{\gamma' \widehat{\mathcal{F}}^{(1|1)}_{y_M}, \gamma' \in \widehat{\Gamma}}} dV(Z) k(Z, \rho Z) = |\widehat{\Gamma}(\rho)| \int_{\bigcup_{\gamma' \widehat{\mathcal{F}}^{(1|1)}_{y_M}, \gamma' \in \widehat{\Gamma}\mathcal{I} \backslash \widehat{\Gamma}}} dV(Z) k(Z, \rho Z) \quad (15.124)$$

and $|\widehat{\Gamma}(\rho)| = \mathrm{order}[\widehat{\Gamma}(\rho)] = 2$ which yields an additional factor $\frac{1}{2}$ in the second term of the parabolic contribution of (15.122).

For a proper asymptotic expansion [507] of the corresponding integral we remove from $\mathcal{H}^{(1|1)}$ two regions, denoted by $B_1^{(1|1)} = \{Z \in \mathcal{H}^{(1|1)} | x \geq y_M\}$ and $B_2^{(1|1)} = \gamma_a B_1$, respectively, i.e., we consider

$$B^{(1|1)} = \mathcal{H}^{(1|1)} - B_1^{(1|1)} - B_2^{(1|1)} . \quad (15.125)$$

Therefore we can derive [221]

Theorem 15.7 *The Selberg super-trace formula for the Dirac-Laplace operator □ on bordered super-Riemann surfaces with hyperbolic, elliptic and parabolic conjugacy classes with Dirichlet boundary conditions is given by:*

$$
\sum_{n=1}^{\infty} \left[h(p_n^{(B)}) - h(p_n^{(F)}) \right] = i\frac{\widehat{\mathcal{A}}}{4\pi} \int_{\mathbb{R}} h(p) \tanh \pi p dp
$$

$$
+ \frac{1}{4} \sum_{\{\gamma\}} \sum_{k=1}^{\infty} \frac{l_\gamma}{\sinh \frac{kl_\gamma}{2}} \left[g(kl_\gamma) + g(-kl_\gamma) - \chi_\gamma^k \left(g(kl_\gamma)e^{-kl_\gamma/2} + g(-kl_\gamma)e^{kl_\gamma/2} \right) \right]
$$

$$
- \frac{1}{4} \sum_{\{\rho^2\}} \sum_{k=0}^{\infty} \frac{l_{\rho^2}}{\cosh\left[\frac{1}{2}(k+\frac{1}{2})l_{\rho^2}\right]} \left\{ g[(k+\tfrac{1}{2})l_{\rho^2}] + g[-(k+\tfrac{1}{2})l_{\rho^2}] \right.
$$

$$
\left. - \chi_{\rho^2}^{k+1/2} \left(g[(k+\tfrac{1}{2})l_{\rho^2}]e^{-\frac{1}{2}(k+1/2)l_{\rho^2}} + g[-(k+\tfrac{1}{2})l_{\rho^2}]e^{(k+\frac{1}{2})l_{\rho^2}/2} \right) \right\}
$$

$$
- \frac{1}{2} \sum_{i=1}^{n} \sum_{k=1}^{\infty} \frac{l_{C_i}}{\cosh \frac{kl_{C_i}}{2}} \left\{ g(kl_{C_i}) + g(-kl_{C_i}) - \chi_{C_i}^k \left(g(kl_{C_i})e^{-kl_{C_i}/2} + g(-kl_{C_i})e^{kl_{C_i}/2} \right) \right\}
$$

$$
+ \frac{1}{2} \sum_{\{R\}} \sum_{k=1}^{\nu-1} \frac{1}{\nu} \left\{ \left(1 - \chi_R^k \cos \frac{k\pi}{\nu} \right) \right.
$$

$$
\times \int_0^\infty \frac{g(u)e^{-u/2} + g(-u)e^{u/2}}{\cosh u - \cos(2k\pi/\nu)} du + \int_0^\infty \frac{g(u) - g(-u)}{\cosh u - \cos(2k\pi/\nu)} \sinh \frac{u}{2} du \left. \right\}
$$

$$
+ (\kappa_S + \kappa_-)g(0) + \frac{\kappa_-}{2} \int_0^\infty g(-u)du
$$

$$
- \frac{\kappa_-}{2} \int_0^\infty \ln(1 - e^{-u}) \left\{ \frac{d}{du}\left[g(u) + g(-u) \right] \right\} du + \frac{\kappa}{4} \int_0^\infty \left[g(u) - g(-u) \right] du
$$

$$
- \frac{\kappa_-}{4} \int_0^\infty \tanh \frac{u}{4}\left[g(u) + g(-u) \right] du - \frac{\kappa_+}{4} \int_0^\infty \tanh \frac{u}{4} \tanh \frac{u}{2}\left[g(u) - g(-u) \right] du ,
$$

$$
(15.126)
$$

where $\lambda_n^{(B,F)} = \frac{1}{2} + ip_n^{(B,F)}$ *on the left runs through the set of all eigenvalues of this Dirichlet problem, and the summation on the right is taken over all primitive conjugacy classes* $\{\gamma\}_{\widehat{\Gamma}}$, $\mathrm{str}(\gamma) + \chi_\gamma > 2$, $\{R\}_{\widehat{\Gamma}}$, $\mathrm{str}(R) + \chi_R < 2$, $\{\rho\}_{\widehat{\Gamma}}$, ρ *hyperbolic,* $\{S\}_{\widehat{\Gamma}}$, $\mathrm{str}(S) + \chi_S = 2$, *and* $\{\rho\}_{\widehat{\Gamma}}$, $\mathrm{str}(\rho) + \chi_\rho = 0$, ρ *elliptic.*
The test function h is required to have the following properties

1. $h(p) \in C^\infty(\mathbb{R})$,

2. $h(p)$ *vanishes faster than* $1/|p|$ *for* $p \to \pm\infty$.

3. $h(p)$ *is holomorphic in the strip* $|\Im(p)| \leq \frac{1}{2} + \epsilon$, $\epsilon > 0$, *to guarantee absolute convergence in the summation over* $\{\gamma\}$ *and* $\{\rho\}$.

In the case of Neumann boundary conditions the regularization procedure is similar to the treatment in the previous subsection which is due to the fact that in this case the continuous spectrum does not drop out and must be taken into account. The full picture emerges from a proper combination of the results of Theorem 15.2 and Theorem 15.7. In particular, the sum of the Selberg super-trace formula for Dirichlet and Neumann boundary conditions, respectively, must yield the result of Theorem 15.2, and the Selberg super-trace formula for Neumann boundary conditions follows straightforward by a subtraction.

15.5   Selberg Super-Zeta-Functions

For the case of bordered super-Riemann surfaces we will consider the *modified Selberg super-zeta-functions* on bordered super-Riemann surfaces as follows

$$
\hat{Z}_0(s) = \prod_{\{\gamma\}}\prod_{k=0}^{\infty}\left[1 - e^{-(s+k)l_\gamma}\right] \times \prod_{\substack{\{\rho\}\\ \mathrm{str}(\rho)+\chi_\rho\neq 0}}\prod_{k=0}^{\infty}\left(\frac{1 + e^{-(s+k)l_\rho}}{1 - e^{-(s+k)l_\rho}}\right)^{(-1)^k}
$$
$$
\times \prod_{i=1}^{n}\prod_{k=0}^{\infty}\left(\frac{1}{1 - e^{-l_{C_i}(s+k)}}\right)^{2(-1)^k}, \tag{15.127}
$$

$$
\hat{Z}_1(s) = \prod_{\{\gamma\}}\prod_{k=0}^{\infty}\left[1 - \chi_\gamma e^{-(s+k)l_\gamma}\right] \times \prod_{\substack{\{\rho\}\\ \mathrm{str}(\rho)+\chi_\rho\neq 0}}\prod_{k=0}^{\infty}\left(\frac{1 + \chi_\rho e^{-(s+k)l_\rho}}{1 - \chi_\rho e^{-(s+k)l_\rho}}\right)^{(-1)^k}
$$
$$
\times \prod_{i=1}^{n}\prod_{k=0}^{\infty}\left(\frac{1}{1 - \chi_{C_i} e^{-l_{C_i}(s+k)}}\right)^{2(-1)^k} \tag{15.128}
$$

for $\Re(s) > 1$. For convenience we will consider the functions

$$
\hat{R}_0(s) := \frac{\hat{Z}_0(s)}{\hat{Z}_0(s+1)} = \prod_{\{\gamma\}}\left(1 - e^{-sl_\gamma}\right) \times \prod_{\substack{\{\rho\}\\ \mathrm{str}(\rho)+\chi_\rho\neq 0}}\prod_{k=0}^{\infty}\left(\frac{1 + e^{-(s+k)l_\rho}}{1 - e^{-(s+k)l_\rho}}\right)^{\alpha_k(-1)^k}
$$
$$
\times \prod_{i=1}^{n}\prod_{k=0}^{\infty}\left(\frac{1}{1 - e^{-l_{C_i}(s+k)}}\right)^{2\alpha_k(-1)^k}, \tag{15.129}
$$

$$
\hat{R}_1(s) := \frac{\hat{Z}_1(s)}{\hat{Z}_1(s+1)} = \prod_{\{\gamma\}}\left(1 - \chi_\gamma e^{-sl_\gamma}\right) \times \prod_{\substack{\{\rho\}\\ \mathrm{str}(\rho)+\chi_\rho\neq 0}}\prod_{k=0}^{\infty}\left(\frac{1 + \chi_\rho e^{-(s+k)l_\rho}}{1 - \chi_\rho e^{-(s+k)l_\rho}}\right)^{\alpha_k(-1)^k}
$$
$$
\times \prod_{i=1}^{n}\prod_{k=0}^{\infty}\left(\frac{1}{1 - \chi_{C_i} e^{-l_{C_i}(s+k)}}\right)^{2\alpha_k(-1)^k} \tag{15.130}
$$

with $\alpha_k = 1\,(m = 0)$, $\alpha_k = 2\,(k \in \mathbb{N})$ for $\Re(s) > 1$. As we shall see, only functional relations for the $\hat{R}_{0,1}$ functions can be derived, but not for the $\hat{Z}_{0,1}$ functions.

15.5.1   The Selberg Super-Zeta-Function $\hat{R}_0$

We discuss the analytic properties of the Selberg super-zeta-function $\hat{R}_0$ by means of the test-function $h_0(p, s, a)$ from (15.64) and have the following Selberg super-trace formula for $\hat{R}_0$

$$
\frac{\hat{R}_0'(s)}{\hat{R}_0(s)} - \frac{\hat{R}_0'(a)}{\hat{R}_0(a)}
$$

$$
= 4 \sum_{n=1}^{\infty} \left[\frac{\lambda_n^{(B)}}{s^2 - (\lambda_n^{(B)})^2} - \frac{\lambda_n^{(B)}}{a^2 - (\lambda_n^{(B)})^2} - \frac{\lambda_n^{(F)}}{s^2 - (\lambda_n^{(F)})^2} + \frac{\lambda_n^{(F)}}{a^2 - (\lambda_n^{(F)})^2} \right]
$$

$$
- \sum_{\{R\}} \sum_{k=1}^{\nu-1} \frac{1}{\nu \sin(2k\pi/\nu)} \sum_{l=1}^{\infty} \sin\left(\frac{2lk\pi}{\nu}\right)
$$

$$
\times \left[\frac{1}{s+l-1} - \frac{1}{s+l+1} - \frac{1}{a+l-1} + \frac{1}{a+l+1} \right]
$$

$$
- \frac{\hat{\mathcal{A}}}{4\pi} \Big[\Psi(s) + \Psi(s+1) - \Psi(a) - \Psi(a+1) \Big]
$$

$$
+ \frac{\kappa}{2} \left(\frac{1}{s-\frac{1}{2}} + \frac{1}{s+\frac{1}{2}} - \frac{1}{a-\frac{1}{2}} - \frac{1}{a+\frac{1}{2}} - \frac{4}{s} + \frac{4}{a} \right) . \tag{15.131}
$$

Thus we obtain [221]

Theorem 15.8 *The Selberg super-zeta-function $\hat{R}_0(s)$ is a meromorphic function on Λ_∞ and has furthermore the following properties:*

1. *The Selberg super-zeta-function $\hat{R}_0(s)$ has "trivial" zeros at the following points and nowhere else.*

   (a) $s = 0$ with multiplicity

   $$
   \#N_0 = \frac{\hat{\mathcal{A}}}{4\pi} - 2\kappa - \sum_{\{R\}} \frac{\nu - 1}{\nu} . \tag{15.132}
   $$

   $s = -1$ with multiplicity

   $$
   \#N_1 = \frac{\hat{\mathcal{A}}}{2\pi} - \sum_{\{R\}} \sum_{k=1}^{\nu-1} \frac{1}{\nu} \cos\left(\frac{2k\pi}{\nu}\right) . \tag{15.133}
   $$

   $s = -n$, $n = 2, 3, \ldots$, with multiplicity

   $$
   \#N_n = \frac{\hat{\mathcal{A}}}{2\pi} - 2 \sum_{\{R\}} \sum_{k=2}^{\nu-1} \frac{1}{\nu} \sum_{l=2}^{\infty} \cos\left(\frac{2lk\pi}{\nu}\right) . \tag{15.134}
   $$

   Note that if $\#N_n < 0$, we have poles instead of zeros.

(b) $s = -\frac{1}{2}$ with multiplicity $\#N_{-1/2} = \kappa/2$.

(c) $s = \frac{1}{2}$ with multiplicity $\#N_{\frac{1}{2}} = \kappa/2$.

2. The Selberg super-zeta-function $\hat{R}_0(s)$ has "non-trivial" zeros and poles at the following points and nowhere else

(a) $s = ip_n^{(B,F)} + \frac{1}{2}$: there are zeros (poles) with twice the multiplicity as the corresponding eigenvalue of $\square$.

(b) $s = -ip_n^{(B,F)} - \frac{1}{2}$: reversed situation for poles and zeros.

(c) $s = \lambda_n^{(B,F)}$ there are zeros (poles), and

(d) $s = -\lambda_n^{(B,F)}$ there are poles with twice the multiplicity y as the corresponding eigenvalue of $\square$, respectively. The last two cases describe so-called small eigenvalues of the operator $\square$.

Of course, (15.131) can be extended meromorphically to all $s \in \Lambda_\infty$.

The test function $h_0(p, s, a)$ is symmetric with respect to $s \to -s$. Therefore subtracting the trace formulæ of $h_0(p, s, a)$ and $h_0(p, -s, a)$ from each other yields the functional equation for the $\hat{R}_0$ function in differential form

$$
\frac{\mathrm{d}}{\mathrm{d}s} \ln\left[\hat{R}_0(s)\hat{R}_0(-s)\right] = \frac{\hat{A}}{2\pi} \frac{\mathrm{d}}{\mathrm{d}s} \ln(\sin \pi s) + \kappa\left(\frac{1}{s+\frac{1}{2}} + \frac{1}{s-\frac{1}{2}} - \frac{4}{s}\right)
$$
$$
- \sum_{\{R\}} \sum_{k=1}^{\nu-1} \frac{1}{\nu \sin(2k\pi/\nu)} \sum_{l=1}^\infty \sin\left(\frac{2lk\pi}{\nu}\right)
$$
$$
\times \left[\frac{1}{s+l-1} + \frac{1}{s-(l-1)} - \frac{1}{s+l+1} - \frac{1}{s-(l+1)}\right] . \quad (15.135)
$$

In integrated form, this gives the functional equation

$$
\hat{R}_0(s)\hat{R}_0(-s) = const. \, (\sin \pi s)^{\hat{A}/2\pi} \left(\frac{s^2 - \frac{1}{4}}{s^4}\right)^\kappa \hat{\Psi}_0(s) , \quad (15.136)
$$

with the function $\hat{\Psi}_0(s)$ given by

$$
\hat{\Psi}_0(s) = \exp\left\{-\sum_{\{R\}} \sum_{k=1}^{\nu-1} \frac{1}{\nu \sin(2k\pi/\nu)} \sum_{l=1}^\infty \sin\left(\frac{2lk\pi}{\nu}\right) \ln\left|\frac{(s^2 - (l-1)^2)}{(s^2 - (l+1)^2)}\right|\right\} . \quad (15.137)
$$

15.5.2   The Selberg Super-Zeta-Function $\hat{R}_1$

We discuss the analytic properties of the Selberg super-zeta-function $\hat{R}_1$ by means of the test-function $h_1(p, s, a)$ from (15.74) and have the following Selberg super-trace formula for $\hat{R}_1$

$$\frac{\hat{R}_1'(s)}{\hat{R}_1(s)} - \frac{\hat{R}_1'(a)}{\hat{R}_1(a)}$$

$$= 4 \sum_{n=1}^{\infty} \left[\frac{\lambda_n^{(B)} - \frac{1}{2}}{s^2 - (\lambda_n^{(B)} - \frac{1}{2})^2} - \frac{\lambda_n^{(B)} - \frac{1}{2}}{a^2 - (\lambda_n^{(B)} - \frac{1}{2})^2} - \frac{\lambda_n^{(F)} - \frac{1}{2}}{s^2 - (\lambda_n^{(F)} - \frac{1}{2})^2} + \frac{\lambda_n^{(F)} - \frac{1}{2}}{a^2 - (\lambda_n^{(F)} - \frac{1}{2})^2} \right]$$

$$- 2 \sum_{\{R\}} \sum_{k=1}^{\nu-1} \frac{\chi_R^k}{\nu} \sum_{l=0}^{\infty} \cos\left[(2l+1)\frac{k\pi}{\nu} \right] \left(\frac{1}{s+l+\frac{1}{2}} - \frac{1}{a+l+\frac{1}{2}} \right)$$

$$- \frac{\widehat{\mathcal{A}}}{2\pi} \left[\Psi(s+\tfrac{1}{2}) - \Psi(a+\tfrac{1}{2}) \right] + \kappa\left(\frac{1}{s} - \frac{1}{a} \right)$$

$$- \kappa_+ \left[\Psi\left(\frac{s}{2} + \frac{3}{4}\right) - \Psi\left(\frac{s}{2} + \frac{1}{4}\right) - \Psi\left(\frac{a}{2} + \frac{3}{4}\right) + \Psi\left(\frac{a}{2} + \frac{1}{4}\right) \right] . \qquad (15.138)$$

Thus we obtain [221]

Theorem 15.9 *The Selberg super-zeta-function $\hat{R}_1(s)$ is a meromorphic function on Λ_∞ and has furthermore the following properties:*

1. *The Selberg super-zeta-function $\hat{R}_1(s)$ has "trivial" zeros at the following points and nowhere else*

   (a) *$s = -\frac{1}{2} - l$, $l = 0, 1, 2, \ldots$, and the multiplicity of these zeros is given by*

   $$\#N_l = \frac{\widehat{\mathcal{A}}}{2\pi} - 2 \sum_{\{R\}} \sum_{k=1}^{\nu-1} \frac{\chi_R^k}{\nu} \sum_{l=0}^{\infty} \cos\left[(2l+1)\frac{k\pi}{\nu} \right] . \qquad (15.139)$$

   Note that if $\#N_l < 0$, we have poles instead of zeros.

   (b) *$s = 0$ with the multiplicity given by $\#N_0 = \kappa$.*

   (c) *$s = -\frac{3}{2} - 2l$, $l = 0, 1, 2, \ldots$, with the multiplicity given by $\#N_l = 2\kappa_+$.*

2. *The Selberg super-zeta-function $\hat{R}_1(s)$ has "trivial" poles at the following points and nowhere else*

   (a) *$s = -\frac{1}{2} - 2l$, $l = 0, -1, -2, \ldots$, with the multiplicity given by $\#P_l = 2\kappa_+$.*

3. *The Selberg super-zeta-function $\hat{R}_1(s)$ has "non-trivial" zeros and poles at the following points and nowhere else*

(a) $s = \mathrm{i}p_n^{(B,F)}$: there are zeros (poles) with twice the multiplicity as the corresponding eigenvalue of $\square$.

(b) $s = -\mathrm{i}p_n^{(B,F)}$: reversed situation for poles and zeros.

(c) $s = \lambda_n^{(B,F)} - \frac{1}{2}$ there are zeros (poles), and

(d) $s = -(\lambda_n^{(B,F)} - \frac{1}{2})$ there are poles with twice the multiplicity as the corresponding eigenvalue of $\square$, respectively. The last two cases describe so-called small eigenvalues of the operator $\square$.

Of course, (15.138) can be extended meromorphically to all $s \in \Lambda_\infty$.

The test function $h_1(p, s, a)$ is symmetric by the interchange $s \to -s$. Therefore subtracting the trace formula for $h_1(p, s, a)$ and $h_1(p, -s, a)$ yields the functional equation for $\hat{R}_1$ in differential form

$$\frac{\mathrm{d}}{\mathrm{d}s} \ln\left[\hat{R}_1(s)\hat{R}_1(-s)\right]$$

$$= -\frac{\hat{A}}{2}\tan\pi s - \kappa_+\left[\Psi\left(\frac{s}{2} + \frac{3}{4}\right) - \Psi\left(\frac{3}{4} - \frac{s}{2}\right) - \Psi\left(\frac{1}{4} + \frac{s}{2}\right) + \Psi\left(\frac{1}{4} - \frac{s}{2}\right)\right]$$

$$+ \frac{2\kappa}{s} - 2\sum_{\{R\}}\sum_{k=1}^{\nu-1}\frac{\chi_R^k}{\nu}\sum_{l=0}^{\infty}\cos\left[(2l+1)\frac{k\pi}{\nu}\right]\left(\frac{1}{s+l+\frac{1}{2}} + \frac{1}{s-(l+\frac{1}{2})}\right) \quad (15.140)$$

[note $\Psi(\frac{1}{2} + s) = \Psi(\frac{1}{2} - s) + \pi\tan\pi s$]. The integrated functional equation therefore has the form

$$\hat{R}_1(s)\hat{R}_1(-s) = const.\,(\cos\pi s)^{\hat{A}/2\pi}s^{2\kappa}\left[\frac{\Gamma(\frac{1}{4} + \frac{s}{2})\Gamma(\frac{1}{4} - \frac{s}{2})}{\Gamma(\frac{3}{4} - \frac{s}{2})\Gamma(\frac{3}{4} + \frac{s}{2})}\right]^{2\kappa_+}\hat{\Psi}_1(s)\,, \quad (15.141)$$

with the function $\hat{\Psi}_1(s)$ given by

$$\hat{\Psi}_1(s) = \exp\left\{-2\sum_{\{R\}}\sum_{k=1}^{\nu-1}\frac{\chi_R^k}{\nu}\sum_{l=0}^{\infty}\cos\left[(2l+1)\frac{k\pi}{\nu}\right]\ln\left|s^2 - (l+\frac{1}{2})^2\right|\right\}\,. \quad (15.142)$$

15.5.3 The Selberg Super-Zeta-Function $\hat{Z}_S$

Following [203, 198, 396] we can also introduce the Selberg super-zeta-function $\hat{Z}_S(s)$ defined by

$$\hat{Z}_S(s) = \frac{\hat{Z}_0(s)\hat{Z}_0(s+1)}{\hat{Z}_1^2(s+\frac{1}{2})}\,. \quad (15.143)$$

The appropriate test function is $h_S(p, s)$ of (15.84). Therefore we obtain

$$\frac{1}{2s}\frac{\hat{Z}_S'(s)}{\hat{Z}_S(s)} = \frac{1}{2s}\frac{\mathrm{d}}{\mathrm{d}s}\ln\left[\frac{\hat{Z}_0(s)\hat{Z}_0(s+1)}{\hat{Z}_1^2(s+\frac{1}{2})}\right]$$

$$= 2 \sum_{n=1}^{\infty} \left[\frac{1}{s^2 - (\lambda_n^{(B)})^2} - \frac{1}{s^2 - (\lambda_n^{(F)})^2} \right] + \frac{\hat{\mathcal{A}}}{8\pi} \frac{1}{s^2}$$

$$- \frac{1}{2s} \sum_{\{R\}} \sum_{k=1}^{\nu-1} \sum_{l=1}^{\infty} \frac{\sin(2lk\pi/\nu)}{\nu \sin(2k\pi/\nu)} \left[\frac{4(1 - \chi_R^k \cos(\frac{k\pi}{\nu}))}{s+l} + \frac{1}{s+l-1} + \frac{1}{s+l+1} - \frac{2}{s+l} \right]$$

$$+ \frac{C\kappa_- - \kappa_s - \kappa_-}{s} + \frac{\kappa}{4s} \left(\frac{4}{s} + \frac{1}{s-\frac{1}{2}} - \frac{1}{s+\frac{1}{2}} \right) + \frac{\kappa_-}{s} \Psi(s) + \frac{\kappa_+}{2s} \left[\Psi\left(\frac{s}{2}\right) - \Psi\left(\frac{s+1}{2}\right) \right] .$$

$$(15.144)$$

Thus we get [221]

Theorem 15.10 *The Selberg super-zeta-function $\hat{Z}_S$ is a meromorphic function on Λ_∞ and has furthermore the following properties:*

1. *The Selberg super-zeta-function $\hat{Z}_S(s)$ has "trivial" zeros at the following points and nowhere else*

   (a) $s = 0$ with multiplicity

   $$\#N_0 = \frac{\hat{\mathcal{A}}}{4\pi} - \sum_{\{R\}} \frac{\nu-1}{\nu} + 2\kappa . \qquad (15.145)$$

   $s = -1$ *with multiplicity*

   $$\#N_1 = -2 \sum_{\{R\}} \sum_{k=1}^{\nu-1} \frac{1}{\nu} \left[1 - 2\chi_R \cos\left(\frac{k\pi}{\nu}\right) + \cos\left(\frac{k\pi}{\nu}\right) \right] - 4\kappa . \qquad (15.146)$$

   $s = -n$ $(n = -2, 3, 4, \ldots)$ *with multiplicity*

   $$\#N_n = \sum_{\{R\}} \sum_{k=2}^{\nu-1} \frac{4}{\nu \sin(2k\pi/\nu)} \left[\sin^2\left(\frac{k\pi}{\nu}\right) - \left(1 - \chi_R \cos\frac{k\pi}{\nu}\right) \right] \sin\left(\frac{2lk\pi}{\nu}\right) - 4\kappa .$$

   $$(15.147)$$

   (b) $s = \frac{1}{2}$ with multiplicity $\#N_{1/2} = \kappa/2$.

   (c) $s = -1 - 2l$, $l = 0, 1, 2, \ldots$, with multiplicity $\#N_l = 2\kappa_+$.

   Note that if $\#N_l < 0$, we have poles instead of zeros.

2. *The Selberg super-zeta-function $\hat{Z}_S(s)$ has "trivial" poles at the following points and nowhere else*

   (a) $s = -\frac{1}{2}$ with multiplicity $\#P_{-1/2} = \kappa/2$.

3. *The Selberg super-zeta-function $\hat{Z}_S(s)$ has "non-trivial" zeros and poles at the following points and nowhere else [203, 396]*

(a) $s = \pm(\frac{1}{2} + ip_n^{(B)})$ there are zeros and

(b) $s = \pm(\frac{1}{2} + ip_n^{(F)})$ there are poles, with twice the multiplicity as the corresponding eigenvalue of $\Box$, respectively.

Of course, (15.144) can be extended meromorphically to all $s \in \Lambda_\infty$.

The test function $h_S(p, s)$ is symmetric with respect to $s \to -s$ and therefore we can deduce the functional relation

$$
\frac{\hat{Z}_S(s)}{\hat{Z}_S(-s)} = const. \, e^{4s(C\kappa_- - \kappa_S - \kappa_-)}
$$
$$
\times \left(\frac{s - \frac{1}{2}}{s + \frac{1}{2}}\right)^\kappa \left(\frac{\Gamma(s)}{\Gamma(-s)}\right)^{2\kappa_-} \left(\frac{\Gamma(\frac{s}{2})\Gamma(\frac{1}{2} - \frac{s}{2})}{\Gamma(\frac{1}{2} - \frac{s}{2})\Gamma(-\frac{s}{2})}\right)^{2\kappa_+} \hat{\Psi}_S(s) , \qquad (15.148)
$$

with the function $\hat{\Psi}_S(s)$ given by

$$
\hat{\Psi}_S(s) = \exp\left\{ -2 \sum_{\{R\}} \sum_{k=1}^{\nu-1} \frac{1}{\nu \sin(2k\pi/\nu)} \sum_{l=1}^\infty \sin\left(\frac{2lk\pi}{\nu}\right) \right.
$$
$$
\left. \times \left[2\left(1 - 2\chi_R^k \cos\frac{k\pi}{\nu}\right) \ln\left|\frac{s+l}{s-l}\right| + \ln\left|\frac{(s+l-1)(s+l+1)}{(s-l+1)(s-l-1)}\right| \right] \right\} . \qquad (15.149)
$$

We can check the consistency of the functional equation with respect to the analytical properties of the Selberg super-zeta function $\hat{Z}_S$. In the case, where only hyperbolic conjugacy classes are present in the super-Fuchsian group, (15.148) reduces to the simple functional equation [203] $\hat{Z}_S(s) = \hat{Z}_S(-s)$. Let us note that in the case of Neumann boundary conditions the Selberg super-zeta-functions must be differently defined due to the changed signs of the $\gamma\mathcal{I}$-terms, i.e., the power of the corresponding terms in the super-zeta-functions is reversed.

15.6 Super-Determinants of Dirac Operators

Finally, I have evaluated super-determinants of Dirac-Laplace operators on bordered super-Riemann surfaces. They are obtained in a straightforward way as before by means of the zeta-function regularization method. We obtain [221]

$$
\text{sdet}(-\Box_m^2) = \left(\frac{\pi}{m!}\right)^{\hat{A}/4\pi} \frac{\hat{Z}_0(1 + \frac{m}{2})}{\hat{Z}_1(\frac{m+1}{2})} \sqrt{\frac{\hat{Z}_1(0)}{\hat{Z}_1(1)}} , \qquad m = 2, 4, \ldots , \quad (15.150)
$$

$$
\text{sdet}(-\Box_{-m}^2) = \left(\frac{(m-2)!}{\pi}\right)^{\hat{A}/4\pi} \frac{\hat{Z}_0(\frac{m}{2})}{\hat{Z}_1(\frac{m+1}{2})} \sqrt{\frac{\hat{Z}_1(1)}{\hat{Z}_1(0)}} , \qquad m = 2, 4, \ldots , \quad (15.151)
$$

$$\text{sdet}(-\Box_m^2) = \left(\frac{\pi}{im!}\right)^{\widehat{\mathcal{A}}/4\pi} \frac{\hat{Z}_1(1+\frac{m}{2})}{\hat{Z}_0(\frac{m+1}{2})} \sqrt{\frac{\hat{Z}_1(0)}{\hat{Z}_1(1)}} \, , \qquad m = 1, 3, \ldots \, , \quad (15.152)$$

$$\text{sdet}(-\Box_{-m}^2) = \left(\frac{i(m-2)!}{\pi}\right)^{\widehat{\mathcal{A}}/4\pi} \frac{\hat{Z}_1(\frac{m}{2})}{\hat{Z}_0(\frac{m+1}{2})} \sqrt{\frac{\hat{Z}_1(1)}{\hat{Z}_1(0)}} \, , \qquad m = 3, 5, \ldots \quad (15.153)$$

(note for instance the relation $\text{sdet}(-\Box_0^2) \cdot \text{sdet}(-\Box_{-1}^2) = 1$). Here, of course, use has been made of the functional relations for the modified Selberg super-zeta-functions. In particular we get

$$[\text{sdet}(-\Box_0^2)]^{-5/2}[\text{sdet}(-\Box_2^2)]^{1/2} = \frac{\pi^{-\widehat{\mathcal{A}}/2\pi}}{\sqrt{2}}\left(\frac{\hat{Z}_0(1)}{\hat{Z}_1(\frac{1}{2})}\right)^{-5/2}\left(\frac{\hat{Z}_0(2)}{\hat{Z}_1(\frac{3}{2})}\right)^{1/2}\frac{\hat{Z}_1(0)}{\hat{Z}_1(1)} \, . \quad (15.154)$$

These determinants are for Dirac-Laplace operators with Dirichlet boundary conditions on bordered super-Riemann surfaces. In order to distinguish them from those with Neumann boundary conditions, $\text{sdet}_\Sigma^{(N)\prime}(-\Box_0^2)$, I denote $\text{sdet}(-\Box_0^2) \equiv \text{sdet}_\Sigma^{(D)}(-\Box_0^2)$. Now I know that the Selberg super-zeta-functions have concerning the $\gamma\mathcal{I}$-length product the reverse power behaviour, denoted by an index "(N)", i.e., $Z^{(N)}(s)$. Furthermore I have to take into account that instead of bosonic and fermionic eigenfunctions which are odd with respect to x of $\Box_0$, we have bosonic and fermionic eigenfunctions which are even with respect to x, i.e., we have for instance

$$\text{sdet}_\Sigma^{(N)\prime}(-\Box_0^2) = (-1)^{1-2q}\pi^{\widehat{\mathcal{A}}/4\pi}\frac{\hat{Z}_0^{(N)}(1)}{\tilde{Z}_1^{(N)}(\frac{1}{2})}\sqrt{\frac{\hat{Z}_0^{(N)}(0)}{\hat{Z}_0^{(N)}(1)}} \, . \quad (15.155)$$

Here by $\tilde{Z}_1^{(N)}(\frac{1}{2})$ the order of $\hat{Z}_1^{(N)}$ at $s = \frac{1}{2}$ is denoted, depending on whether $\Delta n_0^{(0)} \leq 0$ or $\Delta n_0^{(0)} > 0$, respectively. $\Delta n_0^{(0)} = n_0^B - n_0^F$ denotes the difference between the number of even bosonic- and fermionic zero-modes of the Dirac-Laplace operator $\Box_0$. According to [422] $\Delta n_0^{(0)} = 1 - 2q$ with $q = \dim(\ker\bar{\partial}_1)$ and $\bar{\partial}_p^\dagger = -y^2\partial_z + \frac{1}{2}py$. From the corresponding expressions for $\text{sdet}_\Sigma^{(D)}(-\Box_m^2)$ and $\text{sdet}_\Sigma^{(N)\prime}(-\Box_m^2)$, respectively, now follows

$$\text{sdet}_\Sigma^{(D)}(-\Box_m^2) \cdot \text{sdet}_\Sigma^{(N)\prime}(-\Box_m^2) = \text{sdet}_{\hat{\Sigma}}^\prime(-\Box_m^2) \, , \quad (15.156)$$

where $\hat{\Sigma}$ denotes the closed double of the bordered super-Riemann surface.

15.7 Asymptotic Distributions on Super-Riemann Surfaces

Similarly as in the usual case, we can evaluate asymptotic distributions on super Riemann surfaces, i.e. analogues of Huber's law and of Wely's law. We find [229]:

1. Huber's law for compact super-Riemann surfaces

$$\bar{N}(L) \propto \Delta n_0^{(1)} \left[\mathrm{Ei}(L) - \frac{1}{2}\mathrm{Ei}\left(\frac{L}{2}\right) \right] + \sum_{n=1}^{\infty} \left[\mathrm{Ei}(s_n^{(B)}L) - \mathrm{Ei}(s_n^{(F)}L) \right] \ , \qquad L \to \infty \ .$$
$$(15.157)$$

2. Huber's law for non-compact super-Riemann surfaces

$$\bar{N}(L) \propto \mathrm{Ei}(\Delta n_0^{(1)}L) \propto \frac{e^{\Delta n_0^{(1)}L}}{\Delta n_0^{(1)}L} \ , L \to \infty \ . \qquad (15.158)$$

3. Weyl's law for compact super-Riemann surfaces (only hyperbolic conjugacy classes)

$$\bar{N}^{(B)}(p) - \bar{N}^{(F)}(p) = -\mathrm{dim}V\frac{\mathcal{A}}{4\pi} \ . \qquad (15.159)$$

4. Weyl's law for compact super-Riemann surfaces (elliptic an hyperbolic conjugacy classes)

$$\bar{N}^{(B)}(p) - \bar{N}^{(F)}(p) = -\mathrm{dim}V\frac{\mathcal{A}}{4\pi} + \sum_{\{R\}}\sum_{k=1}^{\nu-1}\frac{1}{\nu}\left(1 - \chi_R^k\cos\frac{k\pi}{\nu}\right)\frac{2}{1 - \cos(2k\pi/\nu)} \ . \qquad (15.160)$$

5. Weyl's law for non-compact super-Riemann surfaces

$$\bar{N}^{(B)}(p) - \bar{N}^{(F)}(p) - \frac{\kappa_-}{2\pi}\int_{-p}^{p}\frac{\Delta'(\frac{1}{2} + iq)}{\Delta(\frac{1}{2} + iq)}dq$$
$$= -\mathrm{dim}V\frac{\mathcal{A}}{4\pi} + 2\sum_{\{R\}}\sum_{k=1}^{\nu-1}\frac{1 - \chi_R^k\cos(k\pi/\nu)}{\nu[1 - \cos(2k\pi/\nu)]} - \kappa_-\left(\frac{1}{2}\mathrm{tr}[\mathfrak{S}(\frac{1}{2})] + 1\right) + \frac{\kappa_0}{2}$$
$$- \frac{4p}{\pi}\left[\kappa_0\ln 2 + \kappa_-\ln|\mathrm{sdet}(1 - U(S))| + \frac{\kappa_0}{2}\right] - \frac{2\kappa_0}{\pi}p\ln p + O(p^{-1}). \qquad (15.161)$$

The additional states due to cusps increase this expression to $-\infty$, in contrast to the compact case, where it is approaching a constant.

Chapter 16

Summary and Discussion

16.1 Results on Path Integrals

16.1.1 General Results

Let us start with the results on path integration. I have archived for two important topics new results on exactly solvable path integrals: Path integrals on pseudo-Euclidean spaces and several path integral identities involving parametric coordinate systems. The purpose of this endeavour was to present as many as possible explicit path integral representations in spaces of constant (zero, positive or negative) curvature. The project has been started in [223], including some earlier contributions [202, 210, 247]. I could implement many of these earlier results in my listings which are now almost complete as far as path integral representations for one-parametric coordinate systems are concerned.

The path integral solutions given in Chapter 4 were entirely new results in the first edition of this monograph. In the second edition, new path integral representations have been added in Chapters 8 to 10. Several techniques have been used to get the various path integral solutions. The (pseudo-) cartesian path integral solutions have been obtained in a straightforward way from the usual free particle path integral solution, only the indefinite metric had to be taken properly into account. In the case of (pseudo-) polar coordinates I have exploited the two- and three-dimensional versions of the expansion (4.12) to get the corresponding radial path integral solution. The path integrals corresponding to parabolic coordinates have been calculated by space-time transformations and results from the polar coordinate systems. Some of the archived path integrals and path integral identities are conjectural, e.g., (4.17, 4.47), due to the lack of proper expansion theorems. The same is true for the path integral representations in some of the spheroidal coordinates which could only be constructed in a heuristic way.

For the path integrals formulated in elliptic, spheroidal, and hyperbolic coordinates new interbasis expansions have been constructed and used to perform the (group) path integration explicitly in these specific coordinate space representations.

Therefore I have obtained explicit path integral representations for all one-parametric two- and three-dimensional coordinate systems in pseudo-Euclidean space. However, explicit path integral representations for two-parametric coordinate systems could not be found.

16.1.2 Higher Dimensions

Naturally one asks about path integral representations in higher dimensional pseudo-Euclidean spaces. From the literature it is known that in four-dimensional Euclidean space there are 42 coordinate systems, and in four-dimensional pseudo-Euclidean space 261 coordinate systems, respectively [322]. From the corresponding path integral representations 35 out of the 42 and 182 out of the 261 are explicitly solvable. However, only a few of them are of any interest. Let us for instance consider the parabolic and spherical systems in four-dimensional Euclidean space for which we obtain (c.f. table 16.1 for the definition of the coordinates).

Parabolic, $\xi, \eta > 0, \mathbf{\Omega} \in S^{(2)}$:

$$
\int_{\xi(t')=\xi'}^{\xi(t'')=\xi''} \mathcal{D}\xi(t) \int_{\eta(t')=\eta'}^{\eta(t'')=\eta''} \mathcal{D}\eta(t)(\xi^2 + \eta^2)\xi^2\eta^2 \int_{\mathbf{\Omega}(t')=\mathbf{\Omega}'}^{\mathbf{\Omega}(t'')=\mathbf{\Omega}''} \mathcal{D}\mathbf{\Omega}(t)
$$

$$
\times \exp\left\{ \frac{i}{\hbar} \int_{t'}^{t''} \left[\frac{m}{2}\left((\xi^2 + \eta^2)(\dot{\xi}^2 + \dot{\eta}^2) + \xi^2\eta^2\dot{\mathbf{\Omega}}^2 \right) - \frac{\Delta V(\mathbf{\Omega})}{\xi^2\eta^2} \right] dt \right\}
$$

$$
= \sum_{l=0}^{\infty} \sum_{n=-l}^{l} \Psi_{ln}^{S^{(2)}}(\mathbf{\Omega}'')\Psi_{ln}^{S^{(2)}\,*}(\mathbf{\Omega}') \int_{\mathbb{R}} d\zeta \int_0^{\infty} \frac{dp}{p} \frac{|\Gamma[\frac{1}{2}(l+3/2) + \frac{i\zeta}{2p}]|^4}{4\pi^2(\xi'\xi''\eta'\eta'')^{3/2}\Gamma^4(l+3/2)} e^{-i\hbar p^2 T/2m}
$$

$$
\times M_{-i\zeta/2p,(l+1/2)/2}(-ip\xi''^2)M_{i\zeta/2p,(l+1/2)/2}(ip\xi'^2)
$$

$$
\times M_{i\zeta/2p,(l+1/2)/2}(-ip\eta''^2)M_{-i\zeta/2p,(l+1/2)/2}(ip\eta'^2) \ .
$$
(16.1)

Spherical, $r > 0, \mathbf{\Omega} \in S^{(3)}$:

$$
\int_{r(t')=r'}^{r(t'')=r''} \mathcal{D}r(t)r^3 \int_{\mathbf{\Omega}(t')=\mathbf{\Omega}'}^{\mathbf{\Omega}(t'')=\mathbf{\Omega}''} \mathcal{D}\mathbf{\Omega}(t)
$$

$$
\times \exp\left\{ \frac{i}{\hbar} \int_{t'}^{t''} \left[\frac{m}{2}\left(\dot{r}^2 + r^2\dot{\mathbf{\Omega}}^2 \right) + \frac{1}{r^2}\left(\frac{3\hbar^2}{2m} - \Delta V(\mathbf{\Omega}) \right) \right] dt \right\}
$$

$$
= \frac{m}{2\pi^2 i\hbar T r'r''} \exp\left[-\frac{m}{2i\hbar T}(r'^2 + r''^2) \right] \sum_{l=0}^{\infty} (l+1)C_l^1(\cos\psi_{S^{(3)}})I_{l+1}\left(\frac{mr'r''}{i\hbar T} \right)
$$
(16.2)

$$
= \frac{m}{i\hbar T r'r''} \exp\left[-\frac{m}{2i\hbar T}(r'^2 + r''^2) \right] \sum_{l=0}^{\infty} \sum_{m_1,m_2} \Psi_{l,m_1,m_2}^{S^{(3)}}(\mathbf{\Omega}'')\Psi_{l,m_1,m_2}^{S^{(3)}\,*}(\mathbf{\Omega}')I_{l+1}\left(\frac{mr'r''}{i\hbar T} \right) \ .
$$
(16.3)

In the path integral representations in parabolic coordinates, $\Psi_{ln}^{S^{(2)}}(\mathbf{s})$ denote any of the wave-functions in the two coordinate systems on the sphere $S^{(2)}$, and in the polar systems, $\Psi^{S^{(3)}}_{l,m_1,m_2}(\mathbf{s})$ denote any of the wave-functions in the six coordinate systems on the sphere $S^{(3)}$. Therefore we have two parabolic and six polar path integral representations. We see that the only difference in comparison to the three-dimensional case is a shift in the angular momentum number from $l \to l + \frac{1}{2}$ in the indices. The same feature is, of course, observed in the spheroidal systems where a $S^{(2)}$-coordinate system is attached.

Similarly we have in four-dimensional pseudo-Euclidean space for the path integral representations in polar coordinates ($r > 0, \mathbf{u} = (u_0, \mathbf{u}) \in \Lambda^{(3)}$)

$$
\int_{r(t')=r'}^{r(t'')=r''} \mathcal{D}r(t) r^3 \int_{\mathbf{u}(t')=\mathbf{u}'}^{\mathbf{u}(t'')=\mathbf{u}''} \frac{\mathcal{D}\mathbf{u}(t)}{u_0} \exp\left[\frac{\mathrm{i}}{\hbar}\int_{t'}^{t''}\left(\frac{m}{2}\dot{r}^2 + r^2\dot{\mathrm{u}}^2 - \frac{1}{r^2}\left(\frac{3\hbar^2}{8m} + \Delta V(\mathrm{u})\right)\right)dt\right]
$$
$$
= \frac{1}{r'r''}\int d\omega_1 \int d\omega_2 \int_0^\infty dk\, \Psi^{\Lambda^{(3)}}_{\omega_1,\omega_2,k}(\mathrm{u}'')\Psi^{\Lambda^{(3)}\,*}_{\omega_1,\omega_2,k}(\mathrm{u}')
$$
$$
\times \int_0^\infty \frac{p\,dp}{\pi^2} K_{\mathrm{i}k}(-\mathrm{i}pr'') K_{\mathrm{i}k}(\mathrm{i}pr')\, \mathrm{e}^{-\mathrm{i}\hbar p^2 T/2m} \ . \tag{16.4}
$$

Here, the $\Psi^{\Lambda^{(3)}}_{\omega_1,\omega_2,k}(\mathrm{u})$ denote the wavefunctions in any of the 34 coordinate systems on the three-dimensional pseudosphere $\Lambda^{(3)}$, and therefore there are 34 polar coordinate systems in four-dimensional pseudo-Euclidean space. In comparison to the three-dimensional pseudo-Euclidean space the only difference is in the wavefunctions of the corresponding angular coordinate subsystem. The same feature is observed in the path integral representations in parabolic or spheroidal coordinate systems, where a two-dimensional (pseudo-) sphere is attached in the description of the coordinates.

Therefore we can conclude that a consideration of separable coordinate systems in higher dimensional Euclidean and pseudo-Euclidean spaces gives nothing new as far as genuine new path integral representations are concerned. The most important cases are two and three dimensions. This limitation is due to our limited knowledge of special functions. The range of known special functions, for instance the Bessel and Whittaker functions, the Legendre and hypergeometric functions, or the spheroidal functions allows only a sufficient number of indices to cover the lower dimensional cases. However, let us consider the prolate spheroidal coordinate system in four dimensional Euclidean space (c.f. table 16.1 and [322, 117] for the definition of the coordinates)

$$
\int_{\mu(t')=\mu'}^{\mu(t'')=\mu''} \mathcal{D}\mu(t) \int_{\nu(t')=\nu'}^{\nu(t'')=\nu''} \mathcal{D}\nu(t) \frac{d^4}{4}(\sinh^2\mu + \sin^2\nu)\sinh 2\mu \sin 2\nu \prod_{i=1,2} \int_{\varphi_i(t')=\varphi_i'}^{\varphi_i(t'')=\varphi_i''} \mathcal{D}\varphi_i(t)
$$
$$
\times \exp\left\{\frac{\mathrm{i}}{\hbar}\int_{t'}^{t''}\left[\frac{m}{2}d^2\left((\sinh^2\mu + \sin^2\nu)(\dot\mu^2 + \dot\nu^2) + \sinh^2\mu\sin^2\nu\dot\varphi_1^2 + \cosh^2\mu\cos^2\nu\dot\varphi_2^2\right)\right.\right.
$$
$$
\left.\left. +\frac{\hbar^2}{2m(\sinh^2\mu + \sin^2\nu)}\left(\frac{1}{\sin^2\nu\cos^2\nu} + \frac{1}{\sinh^2\mu\cosh^2\mu}\right)\right]dt\right\}
$$

$$= \frac{d^2}{4}(\sinh 2\mu' \sinh 2\mu'' \sin 2\nu' \sin 2\nu' \sin 2\nu'')^{-1/2}$$

$$\times \sum_{m_{1,2}\in\mathbb{N}} \frac{e^{im_1(\varphi_1''-\varphi_1')+im_2(\varphi_2''-\varphi_2')}}{4\pi^2} \int_{\mathbb{R}} \frac{dE}{2\pi\hbar} e^{-iET/\hbar} \int_0^\infty ds''$$

$$\times \int_{\mu(t')=\mu'}^{\mu(t'')=\mu''} \mathcal{D}\mu(t) \int_{\nu(t')=\nu'}^{\nu(t'')=\nu''} \mathcal{D}\nu(t) \exp\left\{\frac{i}{\hbar}\int_0^{s''}\left[\frac{m}{2}(\dot\mu^2+\dot\nu^2) + Ed^2(\sinh^2\mu + \sin^2\nu)\right.\right.$$

$$\left.\left. -\frac{\hbar^2}{2m}\left(\frac{m_1^2-\frac14}{\sin^2\nu} + \frac{m_2^2-\frac14}{\cos^2\nu} + \frac{m_1^2-\frac14}{\sinh^2\mu} - \frac{m_2^2-\frac14}{\cosh^2\mu}\right)\right]ds\right\} . \qquad (16.5)$$

Table 16.1: Coordinates in Four-Dimensional Euclidean Space

Coordinate System	Coordinates
I. Cartesian	$x = x' \qquad y = y'$ $z = z' \qquad w = w'$
II.–XI. Cylindrical $E(3)$	$(x,y,z) = \mathbf{x}_{\mathbb{R}^3}$ $w = w'$
XII.–XXIII. Cylindrical $[\mathbb{R}^2]^2$	$(x,y) = \mathbf{x}_{\mathbb{R}^2}$ $(z,w) = \mathbf{x}_{\mathbb{R}^2}$
XXIV., XXV. Parabolic $S^{(3)}$ $\xi, \eta > 0$	$(x,y,z) = \xi\eta \cdot \mathbf{s}_{S^{(3)}}$ $w = \frac12(\xi^2 + \eta^2)$
XXVI., XXVII. Prolate Spheroidal $S^{(3)}$ $\mu > 0, \nu \in (0,\pi)$	$(x,y,z) = d\sinh\mu\sin\nu \cdot \mathbf{s}_{S^{(3)}}$ $w = d\cosh\mu\cos\nu$
XXVIII., XXIX. Oblate Spheroidal $S^{(3)}$ $\xi > 0, \nu \in (0,\pi)$	$(x,y,z) = d\cosh\xi\sin\nu \cdot \mathbf{s}_{S^{(3)}}$ $w = d\sinh\xi\cos\nu$
XXX.–XXXV. Spherical, $r > 0$	$\mathbf{x} = r\mathbf{s}_{S^{(3)}}$
XXXVI. Prolate Spheroidal $\mu > 0, \nu \in (0,\pi/2)$ $\varphi_{1,2} \in [0,2\pi)$	$y = d\sinh\mu\sin\nu\sin\varphi_1$ $x = d\sinh\mu\sin\nu\cos\varphi_1$ $z = d\cosh\mu\cos\nu\cos\varphi_2$ $w = d\cosh\mu\cos\nu\sin\varphi_2$
XXXVII. Oblate Spheroidal $\xi > 0, \nu \in (0,\pi/2)$ $\varphi_{1,2} \in [0,2\pi)$	$y = d\cosh\xi\sin\nu\sin\varphi_1$ $x = d\cosh\xi\sin\nu\cos\varphi_1$ $z = d\sinh\xi\cos\nu\cos\varphi_2$ $w = d\sinh\xi\cos\nu\sin\varphi_2$

It is obvious that the corresponding path integral solution in terms of the wavefunction expansion is clearly a generalization of the three-dimensional case. Whereas in three dimensions the spheroidal wavefunctions yield for $d = 0$ Legendre functions, the spheroidal wavefunctions in four dimensions have (modified) Pöschl-Teller wavefunctions as their degenerations [364]. I did not find explicit representations of the corresponding wavefunctions in the literature. However, we can propose heuristically such wavefunctions by taking into account the theory of [399]. Looking at the prolate spheroidal coordinate system in table 16.1, we see that in the limit $d \rightarrow 0$ the spherical system must emerge with the cylindrical coordinate system on $S^{(3)}$ as the proper subsystem. Therefore it follows that the wavefunctions in the variable ν must be generalization of the Pöschl-Teller wavefunctions $\Phi_n^{(\alpha,\beta)}(z)$ of section 2.9.3, and the wavefunctions in the variable μ must be again generalizations of modified Bessel functions $I_\nu(z)$. The proper quantum numbers which must be taken into account are $l \in \mathbb{N}_0$ and $m_1, m_2 \in \mathbb{Z}$. Hence we propose spheroidal wavefunctions $\mathrm{ps}_l^{(m_1,m_2)}(\cos \nu; p^2 d^2)$ and $S_l^{(m_1,m_2),(1)}(\cosh \mu; pd)$ together with the following limiting correspondence ($\gamma = pd$)

$$\mathrm{ps}_l^{(m_1,m_2)}(\cos \nu; \gamma^2) \overset{\gamma \rightarrow 0}{\simeq} (\sin \nu)^{|m_1|}(\cos \nu)^{|m_1|} P_l^{(m_1,m_2)}(\cos 2\nu) , \quad (16.6)$$

$$S_l^{(m_1,m_2),(1)}(\cosh \mu; \gamma) \overset{\gamma \rightarrow 0}{\simeq} \frac{2\pi}{pr} J_{l+1}(pr) . \quad (16.7)$$

In analogy to (5.27) we propose consequently the interbasis expansion (up to phase factors and redefinition of functions)

$$\exp\left[ipd(\sinh \mu \sin \vartheta \sin \nu \cos \alpha + \cosh \mu \cos \vartheta \cos \nu \cos \beta)\right]$$
$$= \sum_{l=0}^{\infty} \sum_{m_1,m_2 \in \mathbb{Z}} 2(|m_1|! + |m_2|! + 2l + 1)\frac{l!\Gamma(|m_1| + |m_2| + l + 1)}{\Gamma(|m_1| + l + 1)\Gamma(|m_2| + l + 1)}$$
$$\times e^{im_1\alpha + im_2\beta} S_l^{(m_1,m_2),(1)}(\cosh \mu; \gamma)\mathrm{ps}_l^{(m_1,m_2)}(\cos \vartheta; \gamma^2)\mathrm{ps}_l^{(m_1,m_2)}(\cos \nu; \gamma^2) . \quad (16.8)$$

By means of this expansion the path integration can be done in a completely analogous way as for (5.22) with the conjectural result

$$K^{(XXXVI)}(a'', a', \nu'', \nu', \varphi_1'', \varphi_1', \varphi_2'', \varphi_2'; T)$$
$$= \sum_{l=0}^{\infty} \sum_{m_1,m_2 \in \mathbb{Z}} \frac{e^{im_1(\varphi_1'' - \varphi_1') + im_2(\varphi_2'' - \varphi_2')}}{4\pi^2}$$
$$\times 2(|m_1|! + |m_2|! + 2l + 1)\frac{l!\Gamma(|m_1| + |m_2| + l + 1)}{\Gamma(|m_1| + l + 1)\Gamma(|m_2| + l + 1)}$$
$$\times \frac{2}{\pi} \int_0^{\infty} p^3 dp S_l^{(m_1,m_2),(1)}(\cosh \mu''; pd)S_l^{(m_1,m_2),(1)*}(\cosh \mu'; pd)$$
$$\times \mathrm{ps}_l^{(m_1,m_2)*}(\cos \nu'; p^2 d^2)\mathrm{ps}_l^{(m_1,m_2)}(\cos \nu''; p^2 d^2)e^{-i\hbar p^2 T/2m} . \quad (16.9)$$

Table 16.2: Parametric Coordinates in Four-Dimensional Euclidean Space

Coordinate System	Coordinates
XXXVIII. Circular Ellipsoidal 1 $\alpha \in [iK', K + iK'], \beta \in [K, K + 2iK']$ $\gamma \in [0, 4K], \varphi \in [0, 2\pi]$	$x = k^2\sqrt{a^2 - c^2}\,\mathrm{sn}\alpha\mathrm{sn}\beta\mathrm{sn}\gamma$ $y = -(k^2/k')\sqrt{a^2 - c^2}\,\mathrm{cn}\alpha\mathrm{cn}\beta\mathrm{cn}\gamma\cos\varphi$ $z = -(k^2/k')\sqrt{a^2 - c^2}\,\mathrm{cn}\alpha\mathrm{cn}\beta\mathrm{cn}\gamma\sin\varphi$ $w = (\mathrm{i}/k')\sqrt{a^2 - c^2}\,\mathrm{dn}\alpha\mathrm{dn}\beta\mathrm{dn}\gamma$
XXXIX. Circular Ellipsoidal 2 $\alpha \in [iK', K + iK'], \beta \in [K, K + 2iK']$ $\gamma \in [0, 4K], \varphi \in [0, 2\pi]$	$x = k^2\sqrt{a^2 - c^2}\,\mathrm{sn}\alpha\mathrm{sn}\beta\mathrm{sn}\gamma$ $y = -(k^2/k')\sqrt{a^2 - c^2}\,\mathrm{cn}\alpha\mathrm{cn}\beta\mathrm{cn}\gamma$ $z = (\mathrm{i}/k')\sqrt{a^2 - c^2}\,\mathrm{dn}\alpha\mathrm{dn}\beta\mathrm{dn}\gamma\cos\varphi$ $w = (\mathrm{i}/k')\sqrt{a^2 - c^2}\,\mathrm{dn}\alpha\mathrm{dn}\beta\mathrm{dn}\gamma\sin\varphi$
XL. Circular Paraboloidal $\alpha, \gamma > 0, \beta \in (0, \pi)$	$x = 2d\cosh\alpha\cos\beta\sinh\gamma\cos\varphi$ $y = 2d\cosh\alpha\cos\beta\sinh\gamma\sin\varphi$ $z = 2d\sinh\alpha\sin\beta\cosh\gamma$ $w = d(\cosh^2\alpha + \cos^2\beta - \cosh^2\gamma)$
XLI. Ellipsoidal $0 < a_1 < \varrho_1 < a_2 < \varrho_2$ $\quad < a_3 < \varrho_3 < a_4 < \varrho_4$	$x^2 = \dfrac{(\varrho_1 - a_1)(\varrho_2 - a_1)(\varrho_3 - a_1)(\varrho_4 - a_1)}{(a_4 - a_1)(a_3 - a_1)(a_2 - a_1)}$ $y^2 = \dfrac{(\varrho_1 - a_2)(\varrho_2 - a_2)(\varrho_3 - a_2)(\varrho_4 - a_2)}{(a_4 - a_2)(a_3 - a_2)(a_1 - a_2)}$ $z^2 = \dfrac{(\varrho_1 - a_3)(\varrho_2 - a_3)(\varrho_3 - a_3)(\varrho_4 - a_3)}{(a_4 - a_3)(a_2 - a_3)(a_1 - a_3)}$ $w^2 = \dfrac{(\varrho_1 - a_4)(\varrho_2 - a_4)(\varrho_3 - a_4)(\varrho_4 - a_4)}{(a_3 - a_4)(a_2 - a_4)(a_1 - a_4)}$
XLII. Paraboloidal $0 < \eta_1 < c < \eta_2 < b < \eta_3 < a < \eta_4$	$x = \frac{1}{2}(\eta_1 + \eta_2 + \eta_3 + \eta_4 - a - b - c)$ $y^2 = \dfrac{(\eta_1 - a)(\eta_2 - a)(\eta_3 - a)(\eta_4 - a)}{(c - a)(a - b)}$ $z^2 = \dfrac{(\eta_1 - b)(\eta_2 - b)(\eta_3 - b)(\eta_4 - b)}{(c - b)(b - a)}$ $w^2 = \dfrac{(\eta_1 - c)(\eta_2 - c)(\eta_3 - c)(\eta_4 - c)}{(b - c)(c - a)}$

A proper definition of these functions is involved. For instance, the function $\mathrm{ps}_\mu^{(m_1,m_2)}$ $(z; \gamma^2)$ must be expanded in terms of $(1 - z^2)^{|m_1|/2}(z)^{|m_1|} P_\mu^{(m_1,m_2)}(2z^2 - 1)$ with expansion coefficients $a_{\mu,\kappa}^{m_1,m_2}(\gamma^2)$, $\kappa \in \mathbb{Z}$. These coefficients then satisfy some recurrence relations which characterize the spheroidal functions. The case of oblate spheroidal coordinates is similar. A detailed study of these mathematical issues are beyond the scope of our work and are not discussed here any further, c.f. [243]. Of course, similar considerations can also be made for path integral formulations in four-dimensional pseudo-Euclidean space.

In tables 16.1 and 16.2 I have listed the 42 possible coordinate system in four-dimensional Euclidean space. The interested reader is invited to construct the corresponding path integral representations. A far more extensive table exists for the four-dimensional pseudo-Euclidean space which is omitted here, c.f. [322] for further information.

Table 16.3: Separation of Variables for the Super-Integrable Potentials on D_{I}

No.	Potential	Separating Coordinate System
V_1	$V_1(u,v) = \dfrac{1}{2u}\left[\dfrac{m}{2}\omega^2(4u^2+v^2) + \kappa + \dfrac{\lambda^2 - \frac{1}{4}}{2mv^2}\right]$	$\underline{(u,v)\text{-System}}$ $\underline{\text{Parabolic}}$
V_2	$V_2(u,v) = \dfrac{1}{2u}\left[\dfrac{m}{2}\omega^2(u^2+v^2) + \kappa_1 + \kappa_2 v\right]$	$\underline{(u,v)\text{-System}}$ $\underline{(r,q)\text{-System}}$
V_3	$V_3(u,v) = \dfrac{1}{2u}\dfrac{\hbar^2 v_0^2}{2m}$	$\underline{(u,v)\text{-System}}$ $\underline{(r,q)\text{-System}}$ $\underline{\text{Parabolic}}$

16.1.3 Super-Integrable Potentials in Spaces of Non-Constant Curvature

Darboux Spaces.

The aim of this monograph is not the topic of super-integrable potentials in spaces of constant and non-constant curvature. However, let us make some remarks. Similarly as in spaces of constant curvature (i.e. flat spaces, spheres or hyperboloids) one can introduce the notion of super-integrability in spaces of non-constant curvature, i.e., in the present case of the Darboux spaces [305]. In each of the four Darboux spaces several (maximally) super-integrable potentials can be found and their corresponding quantum motion can be studied. In the Darboux space D_{I} three potentials, in the Darboux space D_{II} four potentials, in the Darboux space D_{III} five potentials, and in the Darboux space D_{IV} four potentials are found [244]. In the tables 16.3–16.6 we list the explicit form of these potentials, where the coordinate systems where an explicit path integral solution is possible are underlined. For more details I refer to our publications [244].

An analogous study in three dimensions has not been performed yet and is scheduled for future investigations [246].

Koenig Spaces.

Even more complicated spaces are the so-called Koenig spaces [355]. The construction of such a space is simple. One takes a two-dimensional flat Hamiltonian, $\mathcal{H}$, including some potential V, and divides $\mathcal{H}$ by a potential $f(x,y)$ $(x,y \in \mathbb{R}^2)$ such that this potential takes on the form of a metric:

$$\mathcal{H}_{\text{Koenigs}} = \frac{\mathcal{H}}{f(x,y)}. \tag{16.10}$$

Table 16.4: Separation of Variables for the Super-Integrable Potentials on $D_{\rm II}$

No.	Potential	Separating Coordinate System
V_1	$V_1(u,v) = \dfrac{bu^2 - a}{u^2}\left[\dfrac{m}{2}\omega^2(u^2 + 4v^2) + k_1 v + \dfrac{\hbar^2}{2m}\dfrac{k_2^2 - \frac{1}{4}}{u^2}\right]$	$\underline{(u,v)\text{-System}}$ $\underline{\text{Parabolic}}$
V_2	$V_2(u,v) = \dfrac{bu^2 - a}{u^2}\left[\dfrac{m}{2}\omega^2(u^2 + v^2) + \dfrac{\hbar^2}{2m}\left(\dfrac{k_1^2 - \frac{1}{4}}{u^2} + \dfrac{k_2^2 - \frac{1}{4}}{v^2}\right)\right]$	$\underline{(u,v)\text{-System}}$ $\underline{\text{Polar}}$ Elliptic
V_3	$V_3(u,v) = \dfrac{bu^2 - a}{u^2}\dfrac{2m}{\sqrt{u^2 + v^2}}$ $\times\left[-\alpha + \dfrac{\hbar^2}{2m}\left(\dfrac{k_1^2 - \frac{1}{4}}{\sqrt{u^2 + v^2} + v} + \dfrac{k_2^2 - \frac{1}{4}}{\sqrt{u^2 + v^2} - v}\right)\right]$	$\underline{\text{Polar}}$ $\underline{\text{Parabolic}}$ Displaced Elliptic
V_4	$V_4(u,v) = \dfrac{bu^2 - a}{u^2}\dfrac{\hbar^2}{2m}v_0^2$	$\underline{(u,v)\text{-System}}$ $\underline{\text{Polar}}$ Parabolic Elliptic

Such a construction leads to a very rich structure, and attempts to classify such systems are e.g. due to Kalnins et al. [302, 303]. Indeed, simpler examples of such spaces are the *Darboux spaces*, where one chooses the potential $f(x,y)$ in such a way that it depends only on one variable [305]. Another choice consists whether one chooses for $f(x,y)$ some arbitrary potential (or some superintegrable potential) and taking into account that the Poisson bracket structure of the observables makes up a reasonable simple algebra [305]. In two dimensions the construction of Koenig spaces is possible, where one takes for $f(x,y)$ the i) isotropic singular oscillator, ii) the Holt potential, iii) and the Coulomb potential, yielding the Koenig spaces $K_{\rm I}, K_{\rm II}, K_{\rm III}$, respectively. Whereas a path integral evaluation is straightforward in principle, the determination of the, say, bound states turns out to be quite cumbersome.

Let us for instance discuss briefly the Koenig space $K_{\rm I}$. We consider the case, where we take for the metric term

$$\mathrm{d}s^2 \;=\; f_I(x,y)(\mathrm{d}x^2 + \mathrm{d}y^2) \;, \tag{16.11}$$

$$f_I(x,y) \;=\; \alpha(x^2 + y^2) + \frac{\beta}{x^2} + \frac{\gamma}{y^2} + \delta \;, \tag{16.12}$$

and $\alpha, \beta, \gamma, \delta$ are constants.

Table 16.5: Separation of Variables for the Super-Integrable Potentials on $D_{\rm III}$

No.	Potential	Separating Coordinate System
V_1	$V_1(u,v) = \dfrac{2k_1\,{\rm e}^{-u}\cos\frac{v}{2} + 2k_2\,{\rm e}^{-u}\sin\frac{v}{2} + k_3}{a + \frac{b}{4}\,{\rm e}^{-u}}$	Parabolic
		Translated Parabolic $(\xi, \eta \to \xi\eta \pm c)$
V_2	$V_2(u,v) = \dfrac{1}{a + b\,{\rm e}^{-u}}\left[-\alpha + {\rm e}^u\,\dfrac{\hbar^2}{8m}\left(\dfrac{k_1^2 - \frac{1}{4}}{\cos^2\frac{v}{2}} + \dfrac{k_1^2 - \frac{1}{4}}{\cos^2\frac{v}{2}}\right)\right]$	(u,v)-System
		Polar Parabolic
V_3	$V_3(u,v) = \dfrac{1}{a + b\,{\rm e}^{-u}}\left[-\alpha + \dfrac{\hbar^2}{2m}4{\rm e}^u\left(c_1^2\,{\rm e}^{-iv} - 2c_2\,{\rm e}^{-2iv}\right)\right]$	Polar
		Hyperbolic
V_4	$V_4(\mu,\nu) = \dfrac{1}{(a + \frac{b}{2}(\mu-\nu))(\mu+\nu)}\left[d_1\mu + d_2\nu + \dfrac{m}{2}\omega^2(\mu^2-\nu^2)\right]$	Hyperbolic
		Elliptic
V_5	$V_5(u,v) = \dfrac{1}{a + b\,{\rm e}^{-u}}\dfrac{\hbar^2 v_0^2}{2m}$	(u,v)-System
		Polar
		Parabolic
		Elliptic
		Hyperbolic

$K_{\rm I}$ is constructed by considering

$$\mathcal{H}_{K_{\rm I}} = \frac{\mathcal{H}}{f_{\rm I}(x,y)} \ , \tag{16.13}$$

hence for the Lagrangian (with potential)

$$\mathcal{L}_{K_{\rm I}} = \frac{m}{2}f_{\rm I}(x,y)(\dot{x}^2 + \dot{y}^2) - \frac{1}{f_{\rm I}(x,y)}\left[\frac{m}{2}\omega^2(x^2+y^2) + \frac{\hbar^2}{2m}\left(\frac{k_x^2 - \frac{1}{4}}{x^2} + \frac{k_y^2 - \frac{1}{4}}{y^2}\right)\right] \ . \tag{16.14}$$

Setting the potential in square-brackets equal to zero yields the Lagrangian for the free motion in $K_{\rm I}$. With this information we obtain the Green function $G(E)$ in terms of polar coordinates ($\tilde{k}_x^2 = k_x^2 - 2m\beta E/\hbar^2$, $\tilde{k}_y^2 = k_y^2 - 2m\gamma E/\hbar^2$, $\tilde{\omega}^2 = \omega^2 - 2\alpha E/m$, and $\Phi_{n_\varphi}^{(\tilde{k}_y, \tilde{k}_x)}(\varphi)$ are the usual Pöschl-Teller wave-functions):

Table 16.6: Separation of Variables for the Super-Integrable Potentials on D_{IV}

No.	Potential	Separating Coordinate System
V_1	$V_1(u,v) = \left(\dfrac{a_+}{\sin^2 u} + \dfrac{a_-}{\cos^2 u}\right)^{-1}$ $\times \left[\dfrac{\hbar^2}{2m}\left(\dfrac{k^2-\frac{1}{4}}{\cos^2 u} + \dfrac{k^2-\frac{1}{4}}{\sin^2 u}\right) - 4\alpha e^{2v} + 8m\omega^2 e^{4v}\right]$	(u,v)-System _____ Horospherical _____ Elliptic
V_2	$V_2(u,v) = \left(\dfrac{a_+}{\sin^2 u} + \dfrac{a_-}{\cos^2 u}\right)^{-1}$ $\times \left[\dfrac{\hbar^2}{2m}\left(\dfrac{k_1^2-\frac{1}{4}}{\sinh^2 v} - \dfrac{k_2^2-\frac{1}{4}}{\cosh^2 v}\right) - \dfrac{\alpha}{4}\left(\dfrac{1}{\sin^2 u} + \dfrac{1}{\cos^2 u}\right)\right]$	(u,v)-System _____ Degenerate _____ Elliptic I
V_3	$V_3(\tilde{\omega},\tilde{\varphi}) = \dfrac{\hbar^2}{2m}\left(\dfrac{a_+}{\sinh^2 \tilde{\omega}} - \dfrac{a_+}{\cosh^2 \tilde{\omega}} + \dfrac{a_-}{\sin^2 \tilde{\varphi}} + \dfrac{a_-}{\cos^2 \tilde{\varphi}}\right)^{-1}$ $\times \left[\dfrac{c_3}{\sin^2 \tilde{\varphi}} + \dfrac{c_2}{\cos^2 \tilde{\varphi}} - \dfrac{c_3}{\sinh^2 \tilde{\omega}} + \dfrac{c_2}{\cosh^2 \tilde{\omega}}\right]$	Degenerate _____ Elliptic I & II
V_4	$V_4(\mu,\nu) = \left(\dfrac{a_+}{\nu^2} + \dfrac{a_-}{\mu^2}\right)^{-1}\dfrac{\hbar^2}{2m}(k_0^2-\frac{1}{4})\left(\dfrac{1}{\mu^2} + \dfrac{1}{\nu^2}\right)$	(u,v)-System _____ Horospherical _____ Elliptic

$$G^{(K_I)}(r'',r',\varphi'',\varphi';E) = \sum_{n_\varphi} \Phi_{n_\varphi}^{(\tilde{k}_y,\tilde{k}_x)}(\varphi'')\Phi_{n_\varphi}^{(\tilde{k}_y,\tilde{k}_x)}(\varphi')$$

$$\times \frac{\Gamma[\frac{1}{2}(1+\lambda-\delta E/\hbar\tilde{\omega})]}{\hbar\tilde{\omega}\sqrt{r'r''}\,\Gamma(1+\lambda)} W_{\delta E/2\tilde{\omega},\lambda/2}\left(\frac{m\tilde{\omega}}{\hbar}r_>^2\right) M_{\delta E/2\tilde{\omega},\lambda/2}\left(\frac{m\tilde{\omega}}{\hbar}r_<^2\right) , \quad (16.15)$$

and $r_<, r_>$ is the smaller/larger of r', r''. The poles of the Γ-function give the energy-levels of the bound states: $\frac{1}{2}(1+\lambda-\delta E/\hbar\tilde{\omega}) = -n_r$, which is equivalent to ($N = n_r + n_\varphi + 1 = 1,2,\ldots$):

$$\delta E = \hbar\sqrt{\omega^2 - \frac{2\alpha}{m}E}\left(2N + \sqrt{k_x^2 - \frac{2m\beta}{\hbar^2}E} + \sqrt{k_y^2 - \frac{2m\gamma}{\hbar^2}E}\right) . \quad (16.16)$$

In general, this quantization condition is an equation of eighth order in E. In fact, this is typical feature, i.e., the equations for the energy E turn out to be of higher order, which are difficult to tackle and no general expression exist. However, limiting cases can be studied featuring Coulomb- and oscillator-like behavior [237].

The construction of three-dimensional Koenig spaces is quite similar [237]. Also, one finds limiting cases such that Darboux spaces are recovered, as well as spaces of constant curvature.

Table 16.7: Number of Path Integral Solutions

Space	Path Integral Solution for Free Motion	... for Super-Integrable Potentials
$\mathbb{R}^{(1,1)}$	11	not yet given
$\mathbb{R}^{(1,2)}$	14	not yet given
$\mathbb{R}^2$	4	7 [241]
$\mathbb{R}^3$	11	33 [241]
$E(2, \mathbb{C})$	7 [238, 244, 329]	10 [244]
$E(3, \mathbb{C})$	11 (not complete) [238, 304]	33 [245]
$S^{(2)}$	2	2 [242]
$S^{(3)}$	6	12 [242]
$\Lambda^{(2)}$	9	10 [243]
$\Lambda^{(3)}$	24	35 [243]
S_{2C}	5	13 [244, 329]
S_{3C}	17	not yet given
HH(2)	6	not yet given
D_{I}	2	5 [244]
D_{II}	4	9 [244]
D_{III}	4	11 [244]
D_{IV}	4	9 [244]
$D_{3d-\mathrm{I}}$	5	[246]
$D_{3d-\mathrm{II}}$	8	[246]
K_{I}–K_{III}	6	6
$K_{\mathrm{I}}^{(3)}$–$K_{\mathrm{II}}^{(3)}$	12	12
Total	**157**	**167**

16.1.4 Listing the Path Integral Representations

I have derived several new path integral identities arising from the discussion of the path integral representations in pseudo-Euclidean space. Among them are the path integral identity for a singular attractive $1/r^2$-potential (4.17) (valid in the distributional sense). From the horicyclic coordinate space path integral representation of the quantum motion on $SU(1, 1)$ I could derive the path integral solution of the inverted Liouville problem (2.85). Some other path integral identities have emerged from the discussion of the path integral representations in Euclidean space, on the sphere and on the pseudosphere. Some achievements concerning path integral representations for one-parametric coordinate systems could be made. The path integral solutions for the elliptic coordinate system in two-dimensional Euclidean space (5.6) and for the spheroidal coordinate systems (5.22, 5.23) in three-dimensional Euclidean space have already been stated in [223], however without proofs. This has completed by means of interbasis expansions connecting the wavefunctions in cartesian coordinates and the corresponding one-parametric ones.

In three-dimensional Euclidean space I have been able to discuss also the cases of the path integral representations in ellipsoidal (5.24) and paraboloidal coordinates (5.25). These representations, however, are on a very formal level. Whereas I have found in both cases discussions of the corresponding wavefunctions and interbasis expansions, the theory doesn't seem to be very well developed and explicit which is not surprising due to the very complicated structure. What is desirable, are more "handy" versions of the relevant formulæ which in turn have not been found or completed yet, but are in preparation.

The new path integral representations for one-parametric coordinate systems on spheres and pseudospheres have also been found by means of interbasis expansions and in comparison to the interbasis expansions in flat space these interbasis expansions are more difficult. Whereas in flat space the expansion coefficients are explicitly known and take on the form of Mathieu- and spheroidal functions, the expansion coefficients in the curved spaces are only implicitly known, c.f. for the spheres (6.10, 8.34), and c.f. for the pseudospheres (7.12–7.14, 7.62–7.64). In the latter case not all one-parametric interbasis expansion seem to be explicitly known in the literature.

In the case of the two-dimensional sphere I could present the path integral representation in elliptic coordinates. As in flat space, the path integral representations corresponding to two-parametric coordinate systems are even more complicated. Whereas we will be able to present a discussion of the corresponding case on the three-dimensional sphere [240], the cases of the three-dimensional pseudosphere are still unsolved.

Summarizing, the following results in the theory of path integrals have been achieved:

1. Systematic formulation of path integrals on curved manifolds [196, 248].

2. Systematic formulation of (explicitly time-dependent) space-time transformations in the path integral [213, 248].

3. Systematic formulation of separation of variables in the path integral [204, 223].

4. Incorporation of point interactions and boundary conditions in the path integral [205, 216, 217, 222, 226].

5. Tabulation of path integral representations in spaces of constant curvature including

   (a) the pseudo-Euclidean spaces $\mathbb{R}^{(1,1)}, \mathbb{R}^{(1,2)}$,

   (b) the Euclidean spaces $\mathbb{R}^2, \mathbb{R}^3$ [223],

   (c) the spheres $S^{(2)}, S^{(3)}$ [223, 248],

   (d) the pseudospheres $\Lambda^{(2)}, \Lambda^{(3)}, \Lambda^{(D-1)}$ [202, 210, 223, 247, 249],

   (e) the complex spheres S_{2C}, S_{3C} [236],

(f) the hermitian hyperbolic space HH(2),

(g) hyperbolic rank-one spaces [210],

(h) the hyperbolic spaces $SU(n)$ and $SU(n-1,1)$ [208].

6. Tabulation of path integral representations in spaces of non-constant curvature including

   (a) the two-dimensional Darboux spaces $D_{\mathrm{I}}, D_{\mathrm{II}}, D_{\mathrm{III}}, D_{\mathrm{IV}}$ [233],

   (b) the three-dimensional Darboux spaces $D_{3d-\mathrm{I}}, D_{3d-\mathrm{II}}$ [234],

   (c) the two-dimensional Koenig spaces $K_{\mathrm{I}}, K_{\mathrm{II}}, K_{\mathrm{III}}$ [237],

   (d) the three-dimensional Koenig spaces $K_{\mathrm{I}}^{(3)}, K_{\mathrm{II}}^{(3)}, K_{\mathrm{III}}^{(3)}, K_{\mathrm{IV}}^{(3)}, K_{\mathrm{V}}^{(3)}$ [235].

7. Tabulation of about 350 exactly solvable path integrals including (published in Springer-Verlag 1996 [252])

   (a) The general quadratic Lagrangians,

   (b) Path integrals related to the radial harmonic oscillator [197, 200, 204],

   (c) Path integrals related to the Pöschl-Teller potential [198, 219],

   (d) Path integrals related to the modified Pöschl-Teller potential [198, 200, 201, 219],

   (e) Smorodinsky-Winternitz potentials [241]–[243],

   (f) Coulombian potentials [195, 209, 212, 214],

   (g) Natanzon potential path integrals [227, 228],

   (h) Path integrals on group spaces and homogeneous spaces [202, 208, 210, 220, 223, 240, 247],

   (i) Monopole and axion path integrals [206, 207],

   (j) Path integrals with explicitly time-dependent potentials [218],

   (k) Path integrals with point interactions [205, 216, 217, 222, 226],

   (l) Path integrals with boundary conditions, i.e., path integrals in half-spaces, boxes, radial segments, rings and with discontinuities [216, 217, 226].

8. Tabulation of path integral representations of super-integrable potentials in spaces of constant curvature including

   (a) in the Euclidean spaces $\mathbb{R}^2, \mathbb{R}^3$ [241],

   (b) on the spheres $S^{(2)}, S^{(3)}$ [242],

   (c) on the pseudospheres $\Lambda^{(2)}, \Lambda^{(3)}$ [243].

9. Tabulation of path integral representations of super-integrable potentials in spaces of non- constant curvature including

(a) the two-dimensional Darboux spaces $D_{\rm I}, D_{\rm II}, D_{\rm III}, D_{\rm IV}$ [244],

(b) the three-dimensional Darboux spaces $D_{3d-{\rm I}}, D_{3d-{\rm II}}$ [246].

From table 16.7 we see that more than 300 path integral representations have been achieved (or referred to) in this monograph. Some of them have been already listed in [252] (in particular free motion in $\mathbb{R}^2$, $\mathbb{R}^3$, and $\Lambda^{(2)}$). The path integral representations which have not been given in this monograph are noted by an additional reference. Unfortunately, we have not yet completed a study of the free motion on the O(2,2) hyperboloid, which would provide comprehensive path integral representations for single-sheeted hyperboloids [239].

16.2   Results on Trace Formulæ

In Chapter 14 I have studied a particular integrable billiard system in the hyperbolic plane. It has been analyzed by means of the semiclassical periodic orbit analysis of Berry, and the conjecture of Steiner et al. concerning the energy level statistics has been checked. The billiard system has exhibited all the relevant features as predicted by the theory.

In Chapter 15 I have presented results concerning the theory of the Selberg trace formula. I have reviewed some of the classical results due to Hejhal and Venkov. This has included the formulation of the Selberg trace formula for automorphic forms, the Selberg zeta-function and its application in the calculation of determinants of Laplacians on Riemann surfaces. In the section about the Selberg trace formula on bordered Riemann surfaces, I have outlined the results which I have achieved in joint work with Jens Bolte [70]. We have formulated the Selberg trace formula for automorphic forms of weight $m \in \mathbb{Z}$ on bordered Riemann surfaces. The trace formula has been formulated for arbitrary Fuchsian groups of the first kind with reflection symmetry which included hyperbolic, elliptic and parabolic conjugacy classes. In the case of compact bordered Riemann surfaces we have formulated the corresponding Selberg zeta-functions and have discussed their analytic properties. We have explicitly evaluated determinants of Maass-Laplacians for both Dirichlet and Neumann boundary conditions. The determinants have been expressed by means of the Selberg zeta-functions. All these results have been reported in Chapters 14 and 15 in the form of theorems. In Chapter 14 I have dealt with the usual Selberg trace formula, whereas in Chapter 15 with the Selberg super-trace formula.

It was not the purpose of this work to deal with string theory. However, one of the reasons to study the Selberg trace formula and its super generalization comes from (bosonic-, fermionic- or super-) string theory. Inserting the expressions for the determinants into the partition function have yielded in all cases well-defined results. The growing behaviour of the string integrand, which depends via the scalar- and vector-Laplacian determinants on the Selberg zeta-functions, in our case $\sqrt{Z(1)}$ and $\sqrt{Z(2)}$, respectively, have been kept under control in such a way, that at most an exponential behaviour appeared. The blowing up of the bosonic perturbative string theory, both for closed and open string, is eventually due to the factorial growths of the volume of the moduli space for increasing genus. This result originally obtained for closed strings holds also true in the case of open strings.

Summarizing, I have achieved the following results in the theory of the Selberg trace formula and its super generalization

1. The formulation of the Selberg super-trace formula on super-Riemann surfaces for super-automorphic forms with integer weight (as already done in [199, 203] but reported here for completeness).

2. Determination of the analytic properties of the Selberg super-zeta-functions Z_0, Z_1 (as already done in [203] but reported here for completeness).

3. The calculation of super-determinants of Laplacians on super-Riemann surfaces (as already done in [203] but reported here for completeness).

4. The formulation of the Selberg super-trace formula for elliptic and parabolic conjugacy classes [213].

5. Discussion of the analytic properties of the Selberg super-zeta-functions Z_0, Z_1 corresponding to elliptic and parabolic conjugacy classes in the Selberg super-trace formula [213].

6. Introduction of bordered super-Riemann surfaces [215, 221].

7. Formulation of the Selberg super-trace formula on bordered super-Riemann surfaces with hyperbolic, elliptic and parabolic conjugacy classes [221].

8. Determination of the modified Selberg super-zeta-functions $\hat{Z}_0, \hat{Z}_1$ corresponding to bordered super-Riemann surfaces [221].

9. Calculation of super-determinants of Laplace operators on bordered super-Riemann surfaces [221].

10. Calculation of determinants of Laplacians on hyperbolic space forms of rank one (see below).

16.3 Miscellaneous Results, Final Remarks, and Outlook

It is possible to derive from the theory of the Selberg trace formula some further results which I am going to publish in the near future.

Determinants of Laplacians.

The first of these results is concerned with hyperbolic space forms of rank one, c.f. section 14.2. By similar techniques as in section 14.3.2 we can derive an explicit formula for the determinant of the scalar Laplacian on hyperbolic space forms of rank one. One obtains

$$\det'(-\Delta_{\Gamma\backslash G/K}) = e^{-\chi(1)\mathcal{V}(\Gamma\backslash G/K)\tilde{f}(1)} Z'_{\Gamma\backslash G/K}(2\varrho_0) , \qquad (16.17)$$

$$\tilde{f}(1) = \frac{1}{4\pi}\frac{\partial}{\partial s}\int_{\mathbb{R}}\frac{(r^2+\varrho_0^2)^{-s}}{|c(ir)|^2}dr\bigg|_{s=0} . \qquad (16.18)$$

Special cases are the determinant of the scalar Laplacian on Riemann surfaces, c.f. (14.112) and the scalar determinant on three-dimensional Riemannian spaces, i.e.

$$\det'(-\Delta_{\Gamma\backslash\mathcal{H}^{(3)}}) = e^{-\mathcal{V}(\Gamma\backslash\mathcal{H}^{(3)})/2\pi} Z'_{\Gamma\backslash\mathcal{H}^{(3)}}(2) . \qquad (16.19)$$

Weyl's Law.

Of particular importance in some applications in the theory of the Selberg trace formula is Weyl's law. I have discussed some of these aspects in Chapter 14. In numerical investigations of the asymptotic distribution of the eigenvalues on Riemann surfaces or in Riemannian spaces like the fundamental domain of the Picard group, respectively, Weyl's law is indispensable in checking the number of numerically obtained eigenvalues. The leading term in D dimensions has the form (c.f. Gangoli [173] and Brown [87])

$$\bar{N}(E) \propto \frac{\mathcal{V}}{(4\pi)^{D/2}\Gamma(1+D/2)}E^{D/2} . \qquad (16.20)$$

This gives, e.g., for Riemann surfaces $\bar{N}(E) \propto \mathcal{A}E/4\pi$. For compact Riemann surfaces where only hyperbolic conjugacy classes are present, i.e., when there are no elliptic conjugacy classes and no cusps, this simple formula is already (up to a constant) the complete asymptotic distribution of the eigenvalues. In the case of Riemann surfaces with periodic boundary conditions no "surface term" is present. Of course, the feature of Weyl's law is considerably altered if all terms up to the constant in the presence of elliptic conjugacy classes and cusps are taken into account (c.f. Hejhal [268]–[270], McKean [383], Matthies [394], and Venkov [509]).

On compact super-Riemann surfaces the corresponding result has the form [229]

$$N^{(B)}(p) - N^{(F)}(p) \propto \Delta n_{(0)} , \qquad (16.21)$$

thus displaying exact super-symmetry for all energy levels with the exception of the ground state for the quantum motion on super-Riemann surfaces [528].

Huber's Law.

The other important law in the theory of the Selberg trace formula is the so-called Huber's law. It describes the asymptotic proliferation of the length of the periodic orbits (or more precisely, the asymptotic law for the number of the norms of the hyperbolic conjugacy classes via $L = \log N$). It has the form [281]

$$\bar{N}_\Gamma \propto \mathrm{Ei}(L) \propto \mathrm{e}^L/L, \qquad (L \to \infty) \ . \tag{16.22}$$

The simple form of this law is only valid for fundamental domains in the hyperbolic plane which are compact. All these laws are also known as "prime geodesic theorems" (c.f. Buser [90], Chavel [102], Elstrodt et al. [152], Guillope [256], Huber [281], Iwaniec [291], Szmidt [488], Venkov [509] and references therein). Actually very little is known if one takes into account fundamental domains with cusps, i.e., non-compact Riemann surfaces. This failure to improve Huber's law is mainly due to the fact that the scattering matrix cannot be determined in general. Only in the case of congruence and related groups [359], e.g., $\mathrm{SL}(2, \mathbb{Z})$, where the scattering matrix is explicitly known, more detailed statements can be made. The line of reasoning translates to super-Riemann surfaces, and indeed we find that Huber's law for the asymptotic distribution of the hyperbolic conjugacy classes has the form [229, 396]

$$\bar{N}_\Gamma \propto \mathrm{Ei}(L) \propto \mathrm{e}^L/L, \qquad (L \to \infty) \ , \tag{16.23}$$

which is in complete analogy with the classical case.

Résumé.

Not all open questions and problems in path integration and the theory of the Selberg trace formula could be addressed in this work. However, it is fair to say that I was able to give a good account of the state of the art in path integration in quantum mechanics, and to present a report of my developments in the theory of the Selberg super-trace formula. In path integration I succeeded in evaluating almost all path integrals in spaces of constant curvature (with the exception of the single-sheeted hyperboloid). We have given a list of Basic Path Integrals and Master Formulæ in order to deal with general problems in quantum mechanics [251, 252].

An important part of this work dealt with the explicit representation of path integrals in spaces of constant curvature. This included the two- and three-dimensional Euclidean and Minkowski space, and the two- and three-dimensional spheres and hyperboloids. All two-dimensional cases could be evaluated. In the three-dimensional cases the two-parametric coordinate system path integrals were of considerable complexity which makes them almost all impossible to solve (the exceptions being the case of $S^{(3)}$ and $\mathbb{R}^3$). Here even the usual theory of differential equations is poorly developed and it cannot be expected that this would be better for path integrals. I have discussed in some detail with several examples the importance of having more than one coordinate system representation of a physical problem at hand, and it is therefore desirable to continue studies in this field.

In the theory of the Selberg super-trace formula some technical problems are still open. For instance, how can one calculate the contributions for automorphic forms of weight m in the presence of elliptic and parabolic conjugacy classes? Such calculations seem to be very tedious. More important is to find *physics* in the Selberg super-trace formula. Of course, it enabled us to find expressions for Laplacians on super-Riemann surfaces, which in turn appear in the Polyakov approach to the fermionic- and super-string theory. But there seems little use in them concerning more hard facts about string theory. What could be shown is that all these expressions make sense and can be expressed by means of the Selberg super-zeta-functions. However, it would be fruitful to find an application of spin $\frac{1}{2}$ trace formulæ in models of mesoscopic systems like the leaking tori model of Gutzwiller [259], something which seems to be lacking. Also, the extensive numerical studies of chaotic motion on Riemann surfaces have not been translated to the super case yet.

Of course, it is still interesting to develop more trace formulæ in mathematical physics. Here one could think of a Selberg super-trace formula for extended super-symmetry, e.g., [400], or a Selberg super-trace formula on analogues of higher dimensional hyperboloids. Hyperbolic spaces are well suited in cosmology to model non-compact but finite universes, e.g. [20], and spin and super-symmetry can be taken into account in an obvious way. Considering higher-dimensional spaces, which are actually needed in the real world, there seem to be interesting investigations and physical applications ahead.

Bibliography

[1] M.Abramowitz and I.A.Stegun (eds.): Pocketbook of Mathematical Functions (*Harry Deutsch*, Frankfurt/Main, 1984).

[2] S.Albeverio, P.Blanchard and R.Høegh-Krohn: Stationary Phase for the Feynman Integral and Trace Formula. In: "Functional Integration: Theory and Applications", Louvain-la-Neuve, Belgium, November 1979, 23–41. Eds.: J.-P.Antoine and E.Tirapequi (*Plenum Press*, New York, 1980). Feynman Path Integrals, the Poisson Formula and the Theta Function for the Schrödinger Operators. In: "Trends in Applications of Pure Mathematics to Mechanics", Vol.III, 1-22. Ed.: R.J.Knops (*Pitman*, New York, 1981). Feynman Path Integrals and the Trace Formula for the Schrödinger Operators. *Commun.Math.Phys.* **83** (1982) 49–76.

[3] S.Albeverio, P.Blanchard and R.Høegh-Krohn: Some Applications of Functional Integration. *Lecture Notes in Physics* **153** (*Springer*, Berlin, 1982).

[4] S.Albeverio, F.Gesztesy, R.J.Høegh-Krohn and H.Holden: Solvable Models in Quantum Mechanics (*Springer*, Berlin-Heidelberg, 1988).

[5] S.Albeverio and R.J.Høegh-Krohn: Mathematical Theory of Feynman Path Integrals. *Lecture Notes in Mathematics* **523** (*Springer*, Berlin, 1976).

[6] L.Alvarez-Gaumé, G.Moore and C.Vafa: Theta Functions, Modular Invariance, and Strings. *Commun.Math.Phys.* **106** (1986) 1–40.

[7] H.H.Aly and R.M.Spector: Some Solvable Potentials for the Schrödinger Equation. *Nuovo Cimento* **38** (1965) 149–152.

[8] K.Aoki: Heat Kernels and Super Determinants of Laplace Operators on Super Riemann Surfaces. *Commun.Math.Phys.* **117** (1988) 405–429.

[9] F.M.Arscott: Integral Equations for Ellipsoidal Wave Functions. *Quart.J.Math. Oxford* **8** (1957) 223–235. A New Treatment of the Ellipsoidal Wave Function. *Proc.London Math.Soc.* **9** (1959) 21–50.

[10] F.M.Arscott: Paraboloidal Co-ordinates and Laplace's Equation. *Proc.Roy.Soc.Edinburgh* **A 66** (1962) 129–139. The Whittaker-Hill Equation and the Wave Equation in Paraboloidal Co-ordinates. *Proc.Roy.Soc.Edinburgh* **A 68** (1967) 265–276.

[11] F.M.Arscott: Periodic Differential Equations (*The Macmillan Company*, New York, 1964).

[12] A.M.Arthurs: Path Integrals in Polar Coordinates. *Proc.Roy.Soc.(London)* **A 313** (1969) 445–452. Path Integrals in Curvilinear Coordinates. *Proc.Roy.Soc.(London)* **A 318** (1970) 523–529.

[13] E.Artin: Ein mechanisches System mit quasi-ergodischen Bahnen. *Abh.Math.Sem. Hamburgischen Universität* **3** (1924) 170–175.

[14] E.Aurell and P.Salomonson: On Functional Determinants of Laplacians in Polygons and Simplices. *Commun.Math.Phys.* **165** (1994) 233–259.

[15] R.Aurich, E.B.Bogomolny and F.Steiner: Periodic Orbits on the Regular Hyperbolic Octagon. *Physica* **D 48** (1991) 91–101.

[16] R.Aurich and J.Bolte: Quantization Rules for Strongly Chaotic Systems. *Mod. Phys.Lett.* **B 6** (1992) 1691–1719.

[17] R.Aurich, J.Bolte, C.Matthies, M.Sieber and F.Steiner: Crossing the Entropy Barrier of Dynamical Zeta Functions. *Physica* **D 63** (1993) 71–86.

[18] R.Aurich, J.Bolte and F.Steiner: Universal Signatures of Quantum Chaos. *Phys.Rev. Lett.* **73** (1994) 1356–1359.

[19] R.Aurich, T.Hesse and F.Steiner: On the Role of Non-Periodic Orbits in the Semi-classical Quantization of the Truncated Hyperbola Billiard. *Phys.Rev.Lett.* **74** (1995) 4408–4411.

[20] R. Aurich, S. Lustig, F. Steiner, H. Then: Can One Hear the Shape of the Universe? *Phys.Rev.Lett.* **94** (2005) 021301 [4 pages].

[21] R.Aurich and J.Marklof: Trace Formulæ for Three-Dimensional Hyperbolic Lattices and Application to a Strongly Chaotic Tetrahedral Billiard. *Physica* **D** (1996) 101–129.
J.Marklof: On Multiplicities in Length Spectra of Arithmetic Hyperbolic Three-Orbifolds. *Nonlinearity* **9** (1996) 517.

[22] R.Aurich, C.Matthies, M.Sieber and F.Steiner: Novel Rule for Quantizing Chaos. *Phys.Rev.Lett.* **68** (1992) 1629–1632.

[23] R.Aurich, F.Scheffler and F.Steiner: Subtleties of Arithmetical Quantum Chaos. *Phys.Rev.* **E 51** (1995) 4173–4202.

[24] R.Aurich, M.Sieber and F.Steiner: Quantum Chaos of the Hadamard-Gutzwiller Model. *Phys.Rev.Lett.* **61** (1988) 483–487.

[25] R.Aurich and F.Steiner: On the Periodic Orbits for a Strongly Chaotic System. *Physica* **D 32** (1988) 451–460.

[26] R.Aurich and F.Steiner: Periodic-Orbit Sum Rules for the Hadamard-Gutzwiller Model. *Physica* **D 39** (1989) 169–193.

[27] R.Aurich and F.Steiner: Energy-Level Statistics of the Hadamard-Gutzwiller Ensemble. *Physica* **D 43** (1990) 155–180.

[28] R.Aurich and F.Steiner: Exact Theory for the Quantum Eigenstates of the Hadamard-Gutzwiller Model. *Physica* **D 48** (1991) 445–470. Asymptotic Distribution of the Pseudo-Orbits and the Generalized Euler Constant γ_Δ for a Family of Strongly Chaotic Systems. *Phys.Rev.* **A 46** (1992) 771-781.

[29] R.Aurich and F.Steiner: From Classical Periodic Orbits to the Quantization of Chaos. *Proc.Roy.Soc.(London)* **A 437** (1992) 693–714.

[30] R.Aurich and F.Steiner: Staircase Functions, Spectral Rigidity and a Rule for Quantizing Chaos. *Phys.Rev.* **A 45** (1992) 583–592.

[31] R.Aurich and F.Steiner: Statistical Properties of Highly Excited Quantum Eigenstates of a Strongly Chaotic System. *Physica* **D 64** (1993) 185–214.

[32] R.Aurich and F.Steiner: Periodic–Orbit Theory of the Number Variance $\Sigma^2(L)$ of Strongly Chaotic Systems. *Physica* **D 82** (1995) 266–287.

[33] O.Babelon and M.Talon: Separation of Variables for the Classical and Neumann Model. *Nucl.Phys.* **B 379** (1992) 321–339.

[34] A.Bäcker: Privat communication.

[35] A.Bäcker, F.Steiner and P.Stifter: Spectral Statistics in the Quantized Cardioid Billiard. *Phys.Rev.* **E 52** (1995) 2463–2472.

[36] N.L.Balazs, C.S.Schmit and A.Voros: Spectral Fluctuations and Zeta Functions. *J.Stat.Phys.* **46** (1987) 1067–1090.

[37] N.L.Balazs and A.Voros: Chaos on the Pseudosphere. *Phys.Rep.* **143** (1986) 109–240.

[38] H.P.Baltes and E.R.Hilf: Spectra of Finite Systems (*Bibliographisches Institut*, Mannheim, 1976).

[39] M.Bander and C.Itzykson: Group Theory and the Hydrogen Atom (I,II). *Rev. Mod.Phys.* **38** (1968) 330–345 and 346–358

[40] A.M.Baranov, I.V.Frolov and A.S.Shvarts: Geometry of Two-Dimensional Superconformal Field Theories. *Theor.Math.Phys.* **70** (1987) 64–72.

[41] A.M.Baranov, Yu.I.Manin, I.V.Frolov and A.S.Schwarz: The Multiloop Contribution in the Fermionic String. *Sov.J.Nucl.Phys.* **43** (1986) 670–671.

[42] A.M.Baranov, Yu.I.Manin, I.V.Frolov and A.S.Schwarz: A Superanalog of the Selberg Trace Formula and Multiloop Contributions for Fermionic Strings. *Commun. Math.Phys.* **111** (1987) 373–392.

[43] A.M.Baranov and A.S.Schwarz: Baranov, A.M. and A.S.Schwarz, A.S.: Multiloop Contribution to String Theory. *JETP Lett.* **42**, 419–421 (1985) 419–421. On the Multiloop Contributions to the String Theory. *Int.J.Mod.Phys.* **A 2** (1987) 1773–1796.

[44] V.Bargmann: Zur Theorie des Wasserstoffatoms. Bemerkungen zur gleichnamigen Arbeit von V.Fock. *Zeitschr.Phys.* **99** (1936) 576–582, reprinted in [64], 411–417.

[45] A.O.Barut, A.Inomata and G.Junker: Path Integral Treatment of the Hydrogen Atom in a Curved Space of Constant Curvature. *J.Phys.A: Math.Gen.* **20** (1987) 6271–6280.

[46] A.O.Barut, A.Inomata and G.Junker: Path Integral Treatment of the Hydrogen Atom in a Curved Space of Constant Curvature: II. Hyperbolic Space. *J.Phys.A: Math.Gen.* **23** (1990) 1179–1190.

[47] M.Batchelor: The Structure of Supermanifolds. *Trans.Amer.Math.Soc.* **253** (1979) 329–338. Two Approaches to Supermanifolds. *Trans.Amer.Math.Soc.* **258** (1980) 257–270.

[48] M.Batchelor and P.Bryant: Graded Riemann Surfaces. *Commun.Math.Phys.* **114** (1988) 243–255.

[49] D.Bauch: The Path Integral for a Particle Moving in a δ-Function Potential. *Nuovo Cimento* **B 85** (1985) 118–124.

[50] J.Beckers, J.Patera, M.Perroud and P.Winternitz: Subgroups of the Euclidean Group and Symmetry Breaking in Nonrelativistic Quantum Mechanics. *J.Math.Phys.* **18** (1977) 72–83.

[51] M.A.Bég and N.Ruegg: A Set of Harmonic Functions for the Group SU(3). *J.Math.Phys.* **6** (1965) 677–683.

[52] L.Bérard-Bergery: Laplacien et Géodésiques fermées sur les Formes d'Espace Hyperbolique Compactes. In: "Séminaires Boubaki", Vol.1971/72, Exposé 406, 107–122. Eds.: A Dold and B.Eckmann *Lecture Notes in Mathematics* **317** (*Springer*, Berlin, 1973).

[53] F.A.Berezin and L.D.Faddeev: A Remark on Schrödinger's Equation with a Singular Potential. *Soviet Math.Dokl.* **2** (1961) 372–374.

[54] R.Berndt and F.Steiner (eds.): Hyperbolische Geometrie und Anwendungen in der Physik. Hamburger Beiträge zur Mathematik (aus dem Mathematischen Seminar). Heft **8** (1989) 138pp.

[55] C.C.Bernido: Path Integrals and Quantum Interference Effects in a Five-Dimensional Kaluza-Klein Theory. *Phys.Lett.* **A 125** (1987) 176–180. Path Integral Evaluation of a Charged Particle Interacting with a Kaluza-Klein Monopole. *Nucl.Phys.* **B 321** (1989) 108–120.

[56] M.V.Berry: Semiclassical Theory of Spectral Rigidity. *Proc.Roy.Soc.(London)* **A 400** (1985) 229–251.

[57] M.V.Berry: Riemann's Zeta Function: A Model for Quantum Chaos? In: "Quantum Chaos and Statistical Nuclear Physics", *Lecture Notes in Physics* **263** 1–17. Eds.: T.H.Seligman and H.Nishioka (*Springer*, Berlin, 1986). Semiclassical Formula for the Number Variance of the Riemann Zeros. *Nonlinearity* **1** (1988) 399–407.

[58] M.V.Berry: Quantum Chaology. *Proc.Roy.Soc.(London)* **A 413** (1987) 183–198.

[59] A.Besse: Manifolds all of whose Geodesics are Closed. *Ergebnisse der Mathematik und ihrer Grenzgebiete* **93** (*Springer*, New York, 1978).

[60] A.Bhattacharjie and E.C.G.Sudarshan: A Class of Solvable Potentials. *Nuovo Cimento* **25** (1962) 864–879.

[61] S.K.Blau and M.Clements: Determinants of Laplacians for World Sheets with Boundaries. *Nucl.Phys.* **B 284** (1987) 118–130.

[62] S.K.Blau, M.Clements, S.Della Pietra, S.Carlip and V.Della Pietra: The String Amplitude on Surfaces with Boundaries and Crosscaps. *Nucl.Phys.* **B 301** (1988) 285–303.

[63] O.Bohigas, M.J.Giannoni and C.Schmit: Characterization of Chaotic Quantum Spectra and Universality of Level Fluctuation Laws. *Phys.Rev.Lett.* **52** (1984) 1–4.

[64] A.Bohm, Y.Ne'eman and A.O.Barut (eds.): Dynamical Groups and Spectrum Generating Algebras Vol.I (*Academic Press*, New York, 1987).

[65] M.Böhm and G.Junker: The SU(1, 1) Propagator as a Path Integral Over Noncompact Groups. *Phys.Lett.* **A 117** (1986) 375–380.

[66] M.Böhm and G.Junker: Path Integration Over Compact and Noncompact Rotation Groups. *J.Math.Phys.* **28** (1987) 1978–1994.

[67] M.Böhm and G.Junker: Path Integration Over the n-Dimensional Euclidean Group. *J.Math.Phys.* **30** (1989) 1195–1197.

[68] M.Böhm and G.Junker: Group Theoretical Approach to Path Integrations on Spheres. In "Path Summation: Achievements and Goals", Trieste 1987, 469–480. Eds.: S.Lundquist, A.Ranfagni, V.Sa-yakit and L.S.Schulman (*World Scientific*, Singapore, 1988).

[69] J.Bolte: Periodic Orbits in Arithmetical Chaos on Hyperbolic Surfaces. *Nonlinearity* **6** (1993) 935–951. Some Studies on Arithmetical Chaos in Classical and Quantum Mechanics. *Int.J.Mod.Phys.* **B 7** (1993) 4451–4553.

[70] J.Bolte and C.Grosche: Selberg Trace Formula for Bordered Riemann Surfaces: Hyperbolic, Elliptic and Parabolic Conjugacy Classes, and Determinants of Maass-Laplacians. *Commun.Math.Phys.* **163** (1994) 217–244.

[71] J.Bolte, G.Steil and F.Steiner: Arithmetical Chaos and Violation of Universality in Energy Level Statistics. *Phys.Rev.Lett.* **69** (1992) 2188–2191.

[72] J.Bolte and F.Steiner: Determinants of Laplace-Like Operators on Riemann Surfaces. *Commun.Math.Phys.* **130** (1990) 581–597.

[73] J.Bolte and F.Steiner: The On-Shell Limit of Bosonic Off-Shell String Scattering Amplitudes. *Nucl.Phys.* **B 361** (1991) 451–468.

[74] J.Bolte and F.Steiner: The Selberg Trace Formula for Bordered Riemann Surfaces. *Commun.Math.Phys.* **156** (1993) 1–16.

[75] D.Bonatsos, C.Daskaloyannis and K.Kokkotas: Quantum-Algebraic Descripition of Quantum Superintegrable Systems in Two Dimensions. *Phys.Rev.* **A 48** (1993) R3407–R3410. Dynamical Symmetries and Deformed Oscillator Algebras for Two-Dimensional Quantum Superintegrable Systems. In *Proceedings of the "International Workshop on 'Symmetry Methods in Physics' in Memory of Prof. Ya. A. Smorodinsky"*, Dubna, July 1993, 49–56. Eds.: A.N.Sissakian, G.S.Pogosyan and S.I.Vinitsky, *JINR Publications*, E2–94–347, Dubna 1994.

[76] A.K.Bose: A Class of Solvable Potentials. *Nuovo Cimento* **32** (1964) 679–688.

[77] C.P.Boyer and G.N.Fleming: Quantum Field Theory on a Seven-Dimensional Homogeneous Space of the Poincaré Group. *J.Math.Phys.* **15** (1974) 1007–1024.

[78] C.P.Boyer and E.G.Kalnins: Symmetries of the Hamilton-Jacobi Equation. *J.Math. Phys.* **18** (1977) 1032–1045.

[79] C.P.Boyer, E.G.Kalnins and W.Miller, Jr.: Lie Theory and Separation of Variables. 7. The Harmonic Oscillator in Elliptic Coordinates and Ince Polynomials. *J.Math.Phys.* **16** (1975) 512–517.

[80] C.P.Boyer, E.G.Kalnins and W.Miller, Jr.: Symmetry and Separation of Variables for the Helmholtz and Laplace Equations. *Nagoya Math.J.* **60** (1976) 35–80.

[81] C.P.Boyer, E.G.Kalnins and W.Miller, Jr.: Separable Coordinates for Four-Dimensional Riemannian Spaces. *Commun.Math.Phys.* **59** (1978) 285–302.

[82] C.P.Boyer, E.G.Kalnins and W.Miller, Jr.: Symmetry and Separation of Variables for the Hamilton-Jacobi Equation $W_t^2 - W_x^2 - W_y^2 = 0$. *J.Math.Phys.* **19** (1978) 200–211.

[83] C.P.Boyer, E.G.Kalnins and W.Miller, Jr.: Separation of Variables in Einstein Spaces. I. Two Ignorable and One Null Coordinate. *J.Phys.A: Math.Gen.* **14** (1981) 1675–1684.

[84] C.P.Boyer, E.G.Kalnins and P.Winternitz: Completely Integrable Relativistic Hamiltonian Systems and Separation of Variables in Hermitean Hyperbolic Spaces. *J.Math.Phys.* **24** (1983) 2022–2034. Separation of Variables for the Hamilton-Jacobi Equation on Complex Projective Spaces. *SIAM J.Math.Anal.* **16** (1985) 93–109.

[85] S.Brandis: Chaos in a Coulombic Muffin-Tin Potential. *Phys.Rev.* **E 51** (1995) 3023–3031. Classical and Quantum Chaotic Scattering in a Muffin Tin Potential. *DESY Report*, DESY 95–101 (Dissertation Hamburg 1995), 105pp.

[86] P.A.Brown and E.A.Solov'ev: The Stark Effect for the Hydrogen Atom in a Magnetic Field. *Sov.Phys.JETP* **59** (1984) 38–46.

[87] F.H.Brownell: An Extension of Weyl's Asymptotic Law for Eigenvalues. *Pacific J.Math.* **5** (1955) 483-499. Extended Asymptotic Eigenvalue Distribution for Bounded Domains in n-Space. *J.Math.Mech.* **6** (1957) 119–166.

[88] H.Buchholz: The Confluent Hypergeometric Function, Springer Tracts in Natural Philosophy, Vol.15 (*Springer*, Berlin, 1969).

[89] W.Bulla and F.Gesztesy: Deficiency Indices and Singular Boundary Conditions in Quantum Mechanics. *J.Math.Phys.* **26** (1985) 2520–2528.

[90] P.Buser: Geometry and Spectra of Compact Riemann Surfaces (*Birkhäuser*, Boston, 1992).

[91] J.Bystrický, F.Lehar, J.Patera and P.Winternitz: Discrete Two-Variable Expansions of Physical Scattering Amplitudes. *Phys.Rev.* **D 13** (1976) 1276–1283.

[92] A.A.Bytsenko, G.Cognola and L.Vanzo: Vacuum Energy for $3+1$ Dimensional Space-Time with Compact Hyperbolic Spatial Part. *J.Math.Phys.* **33** (1992) 3108–3111. Erratum: Ibidem **34** (1993) 1614.
A.A.Bytsenko, S.D.Odintsov and S.Zerbini: The Effective Action in Gauged Supergravity on Hyperbolic Background and Induced Cosmological Constant. *Phys.Lett.* **B 336** (1994) 355–361.

[93] A.A.Bytsenko, L.Vanzo and S.Zerbini: Massless Scalar Casimir Effect in a Class of Hyperbolic Kaluza-Klein Space-Times. *Mod.Phys.Lett.* **A 7** (1992) 397–409.

[94] R.Camporesi: Harmonic Analysis and Propagators on Homogeneous Spaces. *Phys. Rep.* **196** (1990) 1–134.

[95] Cannata, F,. Junker, J., Trost, J.: Schrödinger Operators with Complex Potential but Real Spectrum. *Phys. Lett.* **A 246** (1998) 219–226.

[96] M.V.Carpio-Bernido: Path Integral Quantization of Certain Noncentral Systems with Dynamical Symmetries. *J.Math.Phys.* **32** (1991) 1799–1807.

[97] M.V.Carpio-Bernido: Green Function for an Axially Symmetric Potential Field: A Path Integral Evaluation in Polar Coordinates. *J.Phys.A: Math.Gen.* **24** (1991) 3013–3019.

[98] M.V.Carpio-Bernido and C.C.Bernido: An Exact Solution of a Ring-Shaped Oscillator Plus a $c\sec^2\vartheta/r^2$ Potential. *Phys.Lett.* **A 134** (1989) 395–399. Algebraic Treatment of a Double Ring-Shaped Oscillator. *Phys.Lett.* **A 137** (1989) 1–3.

[99] M.V.Carpio-Bernido, C.C.Bernido and A.Inomata: Exact Path Integral Treatment of Two Classes of Axially Symmetric Potentials. In the Proceedings of the Third International Conference on "Path Integrals From meV to MeV", Bangkok 1989, 442–459. Eds.: V.Sa-yakanit, W.Sritrakool, J.-O. Berananda, M.C.Gutzwiller, A.Inomata, S.Lundqvist, J.R.Klauder and L.S.Schulman (World Scientific, Singapore, 1989).

[100] C.Casati, B.V.Chirikov and I.Guarneri: Energy-Level Statistics of Integrable Quantum Systems. Phys.Rev.Lett. **54** (1985) 1350–1353.

[101] D.P.L.Castrigiano and F.Stärk: New Aspects of the Path Integrational Treatment of the Coulomb Potential. J.Math.Phys. **30** (1989) 2785–2788.

[102] I.Chavel: Eigenvalues in Riemannian Geometry (Academic Press, Orlando, 1984).

[103] J.Chazarain: Formule de Poisson pour les Variétés Riemanniennes. Inventiones math. **24** (1974) 65–82.

[104] P.Cheng and J.S.Dowker: Vacuum Energy on Orbifold Factors of Spheres. Nucl.Phys. **B 395** (1993) 407–432.

[105] L.Chetouani, L.Guechi and T.F.Hammann: Exact Path Integral for the Ring Potential. Phys.Lett. **A 125** (1987) 277–281. Algebraic Treatment of a General Noncentral Potential. J.Math.Phys. **33** (1992) 3410–3418.

[106] L.Chetouani, L.Guechi and T.F.Hammann: Exact Path Integral Solution of the Coulomb Plus Aharonov-Bohm Potential. J.Math.Phys. **30** (1989) 655–658.

[107] L.Chetouani, L.Guechi, M.Letlout and T.F.Hammann: Exact Path Integral Solution for a Charged Particle Moving in the Field of a Dyon. Nuovo Cimento **B 105** (1990) 387–399.

[108] A.Cisneros and H.V.McIntosh: Symmetry of the Two-Dimensional Hydrogen Atom. J.Math.Phys. **10** (1969) 277–286.

[109] C.Cognola, K.Kirsten, L.Vanzo and S.Zerbini: Finite Temperature Effective Potential on Hyperbolic Spacetimes. Phys.Rev. **D 49** (1994) 5307–5312.
C.Cognola, K.Kirsten and S.Zerbini: One Loop Effective Potential on Hyperbolic Manifolds. Phys.Rev. **D 48** (1993) 790–799.
C.Cognola and L.Vanzo: Thermodynamic Potential for Scalar Fields in Space Time with Hyperbolic Spatial Part. Mod.Phys.Lett. **A 7** (1992) 3677–3688. The Trace of the Heat Kernel on a Compact Hyperbolic 3-Orbifold. J.Math.Phys. **35** (1994) 3109–3116.

[110] C.A.Coulson and A.Joseph: A Constant of the Motion for the Two-Centre Kepler Problem. Int.J.Quantum Chem. **1** (1967) 337–347.

[111] C.A.Coulson and A.Joseph: Spheroidal Wave Functions for the Hydrogen Atom. Proc.Phys.Soc. **90** (1967) 887–893.

C.A.Coulson and P.D.Robinson: Wave Functions for the Hydrogen Atom in Spheroidal-Coordinates I: The Derivation and Properties of the Functions. *Proc.Phys.Soc.* **71** (1958) 815–827.

[112] A.Csordas, R.Graham and P.Szépfalusy: Level Statistics of a Noncompact Cosmological Billiard. *Phys.Rev.* **A 44** (1991) 1491–1499.

[113] A.Csordas, R.Graham, P.Szépfalusy and G.Vattay: Transition from Poissonian to Gaussian-Orthogonal-Ensemble Level Statistics in a Modified Artin's Billiard. *Phys.Rev.* **E 49** (1994) 325–332.
R.Graham, R.Hübner, P.Szépfalusy and G.Vattay: Level Statistics of a Noncompact Integrable Billiard. *Phys.Rev.* **A 44** (1991) 7002–7015.

[114] J.Daboul, P.Slodowy and C.Daboul: The Hydrogen Algebra as Centerless Twisted Kac-Moody Algebra. *Phys.Lett.* **B 317** (1993) 321–328.

[115] G.Darboux: Leçons sur les Systéme Orthogonaux et les Coordonées Curvilignes (Paris, 1910).

[116] L.S.Davtyan, L.G.Mardoyan, G.S.Pogosyan, A.N.Sissakian and V.M.Ter-Antonyan: Generalised KS Transformation: From Five-Dimensional Hydrogen Atom to Eight-Dimensional Isotrope Oscillator. *J.Phys.A: Math.Gen.* **20** (1987) 6121–6125.

[117] Л.С.Давтян, Л.Г.Мардоян, Г.С.Погосян, А.Н.Сисакян и В.М.Тер-Антонян: Сфероидальный базис четырехмерного изотропного осциллятора. *ОИЯИ препринт* P2–87–453, Дубна 1987, 15pp.
[L.S.Davtyan, L.G.Mardoyan, G.S.Pogosyan, A.N.Sissakian and V.M.Ter-Antonyan: Spheroidal Basis of the Four-Dimensional Isotropic Oscillator. *JINR Communications*, P2-87-453, Dubna 1987, 15pp., unpublished].

[118] L.S.Davtyan, G.S.Pogosyan, A.N.Sissakian and V.M.Ter-Antonyan: Two-Dimensional Hydrogen Atom. Reciprocal Expansions of the Polar and Parabolic Bases of the Continuous Spectrum. *Theor.Math.Phys.* **66** (1986) 146–153. On the Hidden Symmetry of a One-Dimensional Hydrogen Atom: *J.Phys.A: Math.Gen.* **20** (1987) 2765–2772. Transformations Between Parabolic Bases of the Two-Dimensional Hydrogen Atom in the Continuous Spectrum. *Theor.Math.Phys.* **74** (1988) 157–161.

[119] S.De Bievre and J.Renaud: Quantization of the Nilpotent Orbits in $so(1, 2)$ and Massless Particles on (Anti-) De Sitter Space-Time. *J.Math.Phys.* **35** (1994) 3775–3793.

[120] M.A.Del Olmo, M.A.Rodríguez and P.Winternitz: The Conformal Group $SU(2, 2)$ and Integrable Systems on a Lorentzian Hyperboloid. *Centre de Recherches Mathématiques*, Montreal, CRM-2194, 1994, 37pp.

[121] Yu.N.Demkov: Symmetry Group of the Isotropic Oscillator. *Sov.Phys.JETP* **26** (1954) 757. Ibidem **36** (1959) 63–66.

[122] Yu.N.Demkov and I.V.Komarov: Hypergeometric Partial Solutions in the Problem of Two Coulomb Centers. *Theor.Math.Phys.* **38** (1979) 174–176.

[123] B.S.DeWitt: Dynamical Theory in Curved Spaces. I. A Review of the Classical and Quantum Action Principles. *Rev.Mod.Phys.* **29** (1957) 377–397.

[124] C.DeWitt-Morette: The Semiclassical Expansion. *Ann.Phys.(N.Y.)* **97** (1976) 367–399.

[125] E.D'Hoker and D.H.Phong: Loop Amplitudes for the Bosonic Polyakov String. *Nucl.Phys.* **B 269** (1986) 205–234.

[126] E.D'Hoker and D.H.Phong: On Determinants of Laplacians on Riemann Surfaces. *Commun.Math.Phys.* **104** (1986) 537–545.

[127] E.D'Hoker and D.H.Phong: Loop Amplitudes for the Fermionic String. *Nucl.Phys.* **B 278** (1986) 225–241.

[128] E.D'Hoker and D.H.Phong: The Geometry of String Perturbation Theory. *Rev.Mod. Phys.* **60** (1988) 917–1065.

[129] W.Dietz: Separable Coordinate Systems for the Hamilton-Jacobi, Klein-Gordon and Wave Equations in Curved Spaces. *J.Phys.A: Math.Gen.* **9** (1976) 519–533.

[130] P.A.M.Dirac: The Lagrangian in Quantum Mechanics. *Phys.Zeitschr.Sowjetunion* **3** (1933) 64–72, reprinted in: "Quantum Electrodynamics", 312–320. Ed.: J.Schwinger (*Dover*, New York, 1958). On the Analogy Between Classical and Quantum Mechanics. *Rev.Mod.Phys.* **17** (1945) 195–199.

[131] W.Dittrich and M.Reuter: Classical and Quantum Dynamics. From Classical Paths to Path Integrals (*Springer*, Berlin, 1992).

[132] J.C.D'Olivo and M.Torres: The Canonical Formalism and Path Integrals in Curved Spaces. *J.Phys.A: Math.Gen.* **21** (1988) 3355–3363. The Weyl Ordering and Path Integrals in Curved Spaces. In "Path Summation: Achievements and Goals", Trieste 1987, 481–497. Eds.: S.Lundquist, A.Ranfagni, V.Sa-yakanit and L.S.Schulman (*World Scientific*, Singapore, 1988).

[133] B.Dorizzi, B.Grammaticos, A.Ramani and P.Winternitz: Integrable Hamiltonian Systems with Velocity-Dependent Potentials. *J.Math.Phys.* **26** (1985) 3070–3079.

[134] J.S.Dowker: When is the 'Sum Over Classical Paths' Exact? *J.Phys.A: Gen.Phys.* **3** (1970) 451–461.

[135] J.S.Dowker: Quantum Mechanics on Group Space and Huygens' Principle. *Ann. Phys.(N.Y.)* **62** (1971) 361–382.

[136] J.S.Dowker: Functional Determinants on Spheres and Sectors. *J.Math.Phys.* **35** (1994) 4989–4999. Functional Determinants on Regions of the Plane and Sphere. *Class.Quant.Grav.* **11** (1994) 557–566.

[137] J.J.Duistermaat and V.W.Guillemin: The Spectrum of Positive Elliptic Operators and Periodic Bicharacteristics. *Inventiones math.* **29** (1975) 39–79.

[138] H.Dürr and A.Inomata: Path Integral Quantization of the Dyonium. *J.Math.Phys.* **26** (1985) 2231–2233.

[139] I.H.Duru: Path Integrals Over SU(2) Manifold and Related Potentials. *Phys.Rev.* **D 30** (1984) 2121–2127.

[140] I.H.Duru: Quantum Treatment of a Class of Time-Dependent Potentials. *J.Phys.A: Math.Gen.* **22** (1989) 4827–4833.

[141] I.H.Duru and H.Kleinert: Solution of the Path Integral for the H-Atom. *Phys.Lett.* **B 84** (1979) 185–188.

[142] I.H.Duru and H.Kleinert: Quantum Mechanics of H-Atoms From Path Integrals. *Fortschr.Phys.* **30** (1982) 401–435.

[143] F.J.Dyson and M.L.Metha: Statistical Theory of the Energy Levels of Complex Systems. IV. *J.Math.Phys.* **4** (1963) 701–712.

[144] S.F.Edwards and Y.V.Gulyaev: Path Integrals in Polar Co-ordinates. *Proc.Roy. Soc.(London)* **A 279** (1964) 229–235.

[145] I.Efrat: The Selberg Trace Formula for $\mathrm{PSL}_2(\mathbb{R})^n$. *Memoirs Amer.Math.Soc.* **65** (1987) 1–111.

[146] I.Efrat: Determinants of Laplacians on Surfaces of Finite Volume. *Commun. Math.Phys.* **119** (1988) 443–451. Erratum: Ibidem **119** (1991) 607.

[147] A.Einstein: Zum Quantensatz von Sommerfeld und Epstein. *Verh.Deutsche Phys. Ges.* **19** (1917) 82–92.

[148] L.P.Eisenhart: Enumeration of Potentials for Which One-Particle Schroedinger Equations Are Separable. *Phys.Rev.* **74** (1948) 87–89.

[149] E.Elizalde, S.D.Odintsov, A.Romeo, A.A.Bytsenko and S.Zerbini: Zeta Regularization Techniques with Applications (*World Scientific*, Singapore, 1994).

[150] J.Elstrodt: Die Resolvente zum Eigenwertproblem der automorphen Funktionen in der hyperbolischen Ebene. Teil I. *Math.Ann.* **203** (1973) 295–330.

[151] J.Elstrodt: Eisenstein Series on the Three-Dimensional Hyperbolic Space and Imaginary Quadratic Number Fields. *J.Reine Angew.Math.* **360** (1985) 160–215.

[152] J.Elstrodt, F.Grunewald and J.Mennicke: Discontinuous Groups on the Three-Dimensional Hyperbolic Space: Analytical Theory and Arithmetic Applications. *Russian Math.Surveys* **38** (1983) 137–147. The Selberg Zeta-Function for Compact Discrete Subgroups of PSL(2, $\mathbb{C}$). *Banach Center Publications* **17** (1985) 83–120. Eisenstein Series on Three-Dimensional Hyperbolic Space and Imaginary Quadratic Number Fields. *J.Reine Angew.Math.* **360** (1985) 160–215. Zeta-Functions of Binary Hermitian Forms and Special Values of Eisenstein Series on Three-Dimensional Hyperbolic Spaces. *Math.Ann.* **277** (1987) 655–708.

[153] A.Erdélyi, W.Magnus, F.Oberhettinger and F.G.Tricomi (eds.): Higher Transcendental Functions (*McGraw Hill*, New York, 1955).

[154] A.Erdélyi, W.Magnus, F.Oberhettinger and F.G.Tricomi (eds.): Tables of Integral Transforms (*McGraw Hill*, New York, 1954).

[155] N.W.Evans: Superintegrability in Classical Mechanics. *Phys.Rev.* **A 41** (1990) 5666–5676. Group Theory of the Smorodinsky-Winternitz System. *J.Math.Phys.* **32** (1991) 3369–3375.

[156] N.W.Evans: Super-Integrability of the Winternitz System. *Phys.Lett.* **A 147** (1990) 483–486.

[157] L.D.Faddeev and A.A.Slavnov: Gauge Fields. Introduction to Quantum Theory (*Benjamin/Cummings*, Reading, 1980).

[158] H.V.Fagundes: Smallest Universe of Negative Curvature. *Phys.Rev.Lett.* **70** (1993) 1579–1582.

[159] R.P.Feynman: The Principle of Least Action in Quantum Mechanics. Ph. D. Thesis, Princeton University, May 1942.

[160] R.P.Feynman: Space-Time Approach to Non-Relativistic Quantum Mechanics. *Rev. Mod.Phys.* **20** (1948) 367–387.

[161] R.P.Feynman: Mathematical Formulation of the Quantum Theory of Electromagnetic Interaction. *Phys.Rev.* **80** (1950) 440–457.

[162] R.P.Feynman: The λ-Transition in Liquid Helium. *Phys.Rev.* **90** (1953) 1116–1117. Atomic Theory of the λ Transition in Helium. *Phys.Rev.* **91** (1953) 1291–1301. Atomic Theory of Liquid Helium Near Absolute Zero. *Phys.Rev.* **91** (1953) 1301-1308. Atomic Theory of the Two-Fluid Model of Liquid Helium. *Phys.Rev.* **94** (1954) 262–277.

[163] R.P.Feynman: Slow Electrons in a Polar Crystal. *Phys.Rev.* **97** (1955) 660–665.

[164] R.P.Feynman and A.Hibbs: Quantum Mechanics and Path Integrals (*McGraw Hill*, New York, 1965).

[165] R.P.Feynman and H.Kleinert: Effective Classical Partition Function. *Phys.Rev.* **A 34** (1986) 5080–5084.

[166] J.Fischer, J.Niederle and R.Raczka: Generalized Spherical Functions for the Non-compact Rotation Groups. *J.Math.Phys.* **7** (1966) 816–821.
N.Limić, J.Niederle and R.Raczka: Discrete Degenerate Representations of Noncompact Rotation Groups. I. *J.Math.Phys.* **7** (1966) 1861–1876. Continuous Degenerate Representations of Noncompact Rotation Groups. II. *J.Math.Phys.* **7** (1966) 2026–2035. Eigenfunction Expansions Associated with the Second-Order Invariant Operator on Hyperboloids and Cones. III. *J.Math.Phys.* **8** (1967) 1079–1093.
J.Niederle: Decomposition of Discrete Most Degenerate Representations of $SO_0(p,q)$ when Restricted to Representations of $SO_0(p,q-1)$ or $SO_0(p-1,q)$. *J.Math.Phys.* **8** (1967) 1921–1930.

[167] W.Fischer, H.Leschke and P.Müller: Changing Dimension and Time: Two Well-Founded and Practical Techniques for Path Integration in Quantum Physics. *J.Phys.A: Math.Gen.* **25** (1992) 3835–3853.

[168] W.Fischer, H.Leschke and P.Müller: Path Integration in Quantum Physics by Changing the Drift of the Underlying Diffusion Process: Application of Legendre Processes. *Ann.Phys.(N.Y.)* **227** (1993) 206–221. Path Integration in Quantum Physics by Changing Drift and Time of the Underlying Diffusion Process. In "VI. International Conference for Path Integrals from meV to MeV", Tutzing, Germany 1992, 259–267. Eds.: H.Grabert, A.Inomata, L.S.Schulman and U.Weiss (*World Scientific*, Singapore, 1993).

[169] V.Fock: Zur Theorie des Wasserstoffatoms. *Zeitschr.Phys.* **98** (1935) 145–154, reprinted in [64], 400–410.

[170] H.Friedrich and D.Wintgen: The Hydrogen Atom in a Uniform Magnetic Field – An Example of Chaos. *Phys.Rep.* **183** (1989) 37–79.

[171] J.Friš, V.Mandrosov, Ya.A.Smorodinsky, M.Uhlir and P.Winternitz: On Higher Symmetries in Quantum Mechanics. *Phys.Lett.* **16** (1965) 354–356.
J.Friš, Ya.A.Smorodinskiĭ, M.Uhlíř and P.Winternitz: Symmetry Groups in Classical and Quantum Mechanics. *Sov.J.Nucl.Phys.* **4** (1967) 444–450.

[172] O.F.Gal'bert, Ya.I.Granovskii and A.S.Zhedanov: Dynamical Symmetry of Anisotropic Singular Oscillator. *Phys.Lett.* **A 153** (1991) 177–180.

[173] R.Gangoli: The Length Spectra of Some Compact Manifolds of Negative Curvature. *J.Diff.Geom.* **12** (1977) 403–424. Zeta Functions of Selberg's Type for Compact Space Forms of Symmetric Spaces of Rank One. *Illinois J.Math.* **21** (1977) 1–41.

[174] R.Gangoli and G.Warner: Zeta Functions of Selberg's Type for Some Noncompact Quotients of Symmetric Spaces of Rank One. *Nagoya Math.J.* **78** (1980) 1–44.

[175] I.M.Gelfand and A.M.Jaglom: Die Integration in Funktionenräumen und ihre Anwendung in der Quantentheorie. *Fortschr.Phys.* **5** (1957) 517–556.
I.M.Gel'fand and A.M.Yaglom: Integration in Functional Spaces and its Applications in Quantum Physics. *J.Math.Phys.* **1** (1960) 48–69.

[176] I.M.Gel'fand and M.I.Graev: Geometry of Homogeneous Spaces, Representations of Groups in Homogeneous Spaces and Related Questions. I. *Amer.Math.Soc.Transl., Ser.2* **37** (1964) 351–429.

[177] I.M.Gel'fand, M.I.Graev, and N.Ya.Vilenkin: Generalized Functions, Vol.5 (*Academic Press*, New York, 1966).

[178] C.C.Gerry: Dynamical Group for a Ring Potential. *Phys.Lett.* **A 118** (1986) 445–447.

[179] J.-L.Gervais and A.Jevicki: Point Canonical Transformations in the Path Integral. *Nucl.Phys.* **B 110** (1976) 93–112.

[180] R.Giachetti and V.Tognetti: Quantum Corrections to the Thermodynamics of Nonlinear Systems; *Phys.Rev.* **B 33** (1986) 7647–7658.
R.Giachetti, V.Tognetti and R.Vaia: An Effective Hamiltonian for Quantum Statistical Mechanics. In "Path Integrals From meV to MeV", Bangkok 1989, 195–216. Eds.: V.Sa-yakanit et al. (*World Scientific*, Singapore, 1989).

[181] G.Gilbert: String Theory Path Integral - Genus Two and Higher. *Nucl.Phys.* **B 277** (1986) 102–124.

[182] J.Glimm and A.Jaffe: Quantum Physics: A Functional Point of View (*Springer*, Berlin, 1981).

[183] M.J.Goovaerts: Path-Integral Evaluation of a Nonstationary Calogero Model. *J.Math.Phys.* **16** (1975) 720–723.

[184] M.J.Goovaerts, A.Babcenco and J.T.Devreese: A New Expansion Method in the Feynman Path Integral Formalism: Application to a One-Dimensional Delta-Function Potential. *J.Math.Phys.* **14** (1973) 554–559.

[185] M.J.Goovaerts and F.Broeckx: Analytic Treatment of a Periodic δ-Function Potential in the Path Integral Formalism. *SIAM J.Appl.Math.* **45** (1985) 479–490.

[186] M.J.Goovaerts and J.T.Devreese: Analytic Treatment of the Coulomb Potential in the Path Integral Formalism by Exact Summation of a Perturbation Expansion. *J.Math.Phys.* **13** (1972) 1070–1082. Erratum: Ibidem **14** (1973) 153.

[187] I.S.Gradshteyn and I.M.Ryzhik: Table of Integrals, Series, and Products (*Academic Press*, New York, 1980).

[188] Ya.A.Granovsky, A.S.Zhedanov and I.M.Lutzenko: Quadratic Algebra as a 'Hidden' Symmetry of the Hartmann Potential. *J.Phys.A: Math.Gen.* **24** (1991) 3887–3894.

[189] Ya.A.Granovsky, A.S.Zhedanov and I.M.Lutzenko: Quadratic Algebras and Dynamics in Curved Spaces. I. Oscillator. *Theor.Math.Phys.* **91** (1992) 474–480.

[190] Ya.A.Granovsky, A.S.Zhedanov and I.M.Lutzenko: Quadratic Algebras and Dynamics in Curved Spaces. II. The Kepler Problem. *Theor.Math.Phys.* **91** (1992) 604–612.

[191] M.B.Green: Supersymmetrical Dual String Theories and Their Field Theory Limits. *Surveys in High Energy Physics* **3** (1983) 127–160.

[192] M.B.Green and J.H.Schwarz: Supersymmetrical Dual String Theory, I-III. *Nucl.Phys.* **B 181** (1982) 502–530. Ibidem **B 198** (1982) 252–268, 441–460.

[193] M.B.Green and J.H.Schwarz: Anomaly Cancellations in Supersymmetric D=10 Gauge Theory and Superstring Theory. *Phys.Lett.* **B 149**, 117–122 (1984). Infinity Cancellations in SO(32) Superstring Theory. *Phys.Lett.* **B 151** (1985) 21–25. The Hexagon Gauge Anomaly in Type I Superstring Theory. *Nucl.Phys.* **B 255** (1985) 93–114.

[194] M.B.Green, J.H.Schwarz and E.Witten: Superstring Theory I&II (*Cambridge University Press*, Cambridge, 1988).

[195] C.Grosche: Das Coulombpotential im Pfadintegral. Diploma Thesis, Universität Hamburg, 1985, 100pp.

[196] C.Grosche: The Product Form for Path Integrals on Curved Manifolds. *Phys.Lett.* **A 128** (1988) 113–122.

[197] C.Grosche: The Path Integral on the Poincaré Upper Half-Plane with a Magnetic Field and for the Morse Potential. *Ann.Phys.(N.Y.)* **187** (1988) 110–134.

[198] C.Grosche: Path Integral Solution of a Class of Potentials Related to the Pöschl-Teller Potential. *J.Phys.A: Math.Gen.* **22** (1989) 5073–5087.

[199] C.Grosche: Analogon der Selbergschen Spurformel für Super-Riemannsche Flächen. In: "Hyperbolische Geometrie und Anwendungen in der Physik." Hamburger Beiträge zur Mathematik (aus dem Mathematischen Seminar), **8** (1989) 218–238. Eds.: R.Berndt and F.Steiner.

[200] C.Grosche: Path Integration on the Hyperbolic Plane with a Magnetic Field. *Ann.Phys.(N.Y.)* **201** (1990) 258–284.

[201] C.Grosche: The Path Integral for the Kepler Problem on the Pseudosphere. *Ann.Phys.(N.Y.)* **204** (1990) 208–222.

[202] C.Grosche: The Path Integral on the Poincaré Disc, the Poincaré Upper Half-Plane and on the Hyperbolic Strip. *Fortschr.Phys.* **38** (1990) 531–569.

[203] C.Grosche: Selberg Supertrace Formula for Super Riemann Surfaces, Analytic Properties of the Selberg Super Zeta-Functions and Multiloop Contributions to the Fermionic String. *DESY Report*, DESY 89–010 (Dissertation Hamburg 1989), 102pp., and *Commun. Math.Phys.* **133** (1990) 433–485.

[204] C.Grosche: Separation of Variables in Path Integrals and Path Integral Solution of Two Potentials on the Poincaré Upper Half-Plane. *J.Phys.A: Math.Gen.* **23** (1990) 4885–4901.

[205] C.Grosche: Path Integrals for Potential Problems with δ-Function Perturbation. *J. Phys.A: Math.Gen.* **23** (1990) 5205–5234.

[206] C.Grosche: Path Integral Solution for an Electron Moving in the Field of a Dirac Monopole. *Phys.Lett.* **A 151** (1990) 365–370.

[207] C.Grosche: Kaluza-Klein Monopole System in Parabolic Coordinates by Functional Integration. *J.Phys.A: Math.Gen.* **24** (1991) 1771–1783.
C.Grosche, G.S.Pogosyan and A.N.Sissakian: On the Interbasis Expansion of the Kaluza-Klein Monopole System. *Annalen der Physik* **6** (1997) 144–161.

[208] C.Grosche: The SU(u, v)-Group Path-Integration. *J.Math.Phys.* **32** (1991) 1984–1997.

[209] C.Grosche: Coulomb Potentials by Path-Integration. *Fortschr.Phys.* **40** (1992) 695–737.

[210] C.Grosche: Path Integration on Hyperbolic Spaces. *J.Phys.A: Math.Gen.* **25** (1992) 4211–4244.

[211] C.Grosche: Energy-Level Statistics of an Integrable Billiard System in a Rectangle in the Hyperbolic Plane. *J.Phys.A: Math.Gen.* **25** (1992) 4573–4594.

[212] C.Grosche: Path Integral Solution of a Non-Isotropic Coulomb-Like Potential. *Phys. Lett.* **A 165** (1992) 185–190.

[213] C.Grosche: Selberg Super-Trace Formula for Super Riemann Surfaces II: Elliptic and Parabolic Conjugacy Classes, and Selberg Super Zeta Functions. *Commun.Math.Phys.* **151** (1993) 1–37.

[214] C.Grosche: Path Integral Solution of Two Potentials Related to the SO($2, 1$) Dynamical Algebra. *J.Phys.A: Math.Gen.* **26** (1993) L279–L287.

[215] C.Grosche: Selberg Trace-Formulæ in Mathematical Physics. In *Conference Proceedings Vol.41 of the Workshop "From Classical to Quantum Chaos (1892-1992)"*, Trieste, July 1992, 45–55. Eds.: G. Dell'Antonio, S. Fantoni and V. R. Manfredi (*Società Italiana Di Fisica*, Bologna, 1993).

[216] C.Grosche: Path Integration via Summation of Perturbation Expansions and Application to Totally Reflecting Boundaries and Potential Steps. *Phys.Rev.Lett.* **71** (1993) 1–4.

[217] C.Grosche: δ-Function Perturbations and Boundary Problems by Path Integration. *Ann.Physik* **2** (1993) 557–589.

[218] C.Grosche: Path Integral Solution of a Class of Explicitly Time-Dependent Potentials. *Phys.Lett.* **A 182** (1993) 28–36.

[219] C.Grosche: Path Integral Solution of Scarf-Like Potentials. *Nuovo Cimento* **B 108** (1993) 1365–1376.

[220] C.Grosche: On the Path Integral in Imaginary Lobachevsky Space. *J.Phys.A: Math. Gen.* **27** (1994) 3475–3489.

[221] C.Grosche: Selberg Supertrace Formula for Super Riemann Surfaces III: Bordered Super Riemann Surfaces. *Commun.Math.Phys.* **162** (1994) 591–631.

[222] C.Grosche: Path Integrals for Two- and Three-Dimensional δ-Function Perturbations. *Ann.Physik* **3** (1994) 283–312.

[223] C.Grosche: Path Integration and Separation of Variables in Spaces of Constant Curvature in Two and Three Dimensions. *Fortschr.Phys.* **42** (1994) 509–584.

[224] C.Grosche: Towards the Classification of Exactly Solvable Feynman Path Integrals: δ-Function Perturbations and Boundary Conditions as Miscellaneous Solvable Models. In the *Proceedings of the "International Workshop on 'Symmetry Methods in Physics' in Memory of Prof. Ya. A. Smorodinsky"*, Dubna, July 1993, 129–139. Eds.: A.N.Sissakian, G.S.Pogosyan and S.I.Vinitsky, *JINR Publications*, E2–94–347, Dubna 1994.

[225] C.Grosche: δ'-Function Perturbations and Neumann Boundary-Conditions by Path Integration. *J.Phys.A: Math.Gen.* **28** (1995) L99–L105.

[226] C.Grosche: Boundary Conditions in Path Integrals. In the *Proceedings of the "Workshop on Singular Schrödinger Operators"*, SISSA, Trieste, September 1994, 20pp. *Trieste Report* ILAS/FM–16/1995.

[227] C.Grosche: Conditionally Solvable Path Integral Problems. *J.Phys.A: Math.Gen.* **28** (1995) 5889–5902.
Conditionally Solvable Path Integral Problems II: Natanzon Potentials. *J.Phys.A: Math.Gen.* **29** (1996) 365–383.

[228] C.Grosche: The General Besselian and Legendrian Path Integral. *J.Phys.A: Math.Gen.* **29** (1996) L183–L189. Path Integral Solution for Natanzon Potentials. *Proceedings of the "Barut's Memorial Conference on Group Theory in Physics"*, 21.–27. December 1995, Edirne, Turkey; Eds.: I. H. Duru (*World Scientific*, Singapore, 1996).

[229] C.Grosche: Asymptotic Distributions on Super Riemann Surfaces. *Class.Quant.Grav.* **13** (1996) 2329–2348.

[230] C.Grosche: Energy Fluctuations in Integrable Billiards in Hyperbolic Geometry. In *IMA Volumes in Mathematics and its Applications* **109**. *Emerging Applications of Number Theory*, Minneapolis, 1996. Eds.: D. A. Hejhal, J. Friedman, M. C. Gutzwiller and A. M. Odlyzko. Springer, New York, 1998, pp. 269–290.

[231] C.Grosche: On the Path Integral Treatment for an Aharonov-Bohm Field on the Hyperbolic Plane. *Int.J.Theor.Phys.* **38** (1999) 955–969.

[232] C.Grosche: Path Integration on Hermitian Hypberbolic Space. *J.Phys.A: Math. Gen.* **38** (2005) 3625–3650.

[233] C.Grosche: Path Integration on Darboux Spaces. *Phys.Part.Nucl.* **37** (2006) 368–389.

[234] C.Grosche: Path Integral Approach for Spaces of Nonconstant Curvature in Three Dimensions. *Phys.Atom.Nucl.* **70** (2007) 537–544.

[235] C.Grosche: Path Integral Approach for Quantum Motion on Spaces of Non-constant Curvature According to Koenigs: II. Three Dimensions. *DESY preprint*, DESY 07–132, August 2007. `quant-ph/arXiv:0708.3090`

[236] C.Grosche: Path Integral Representations on the Complex Sphere. *DESY preprint*, DESY 07–133, August 2007. `quant-ph/arXiv:0710.4232`

[237] C.Grosche: Path Integral Approach for Koenigs Spaces. *Phys.Atom.Nucl.* (2008) 899–904.

[238] C.Grosche: Path Integral Representations on Complex Euclidean Space. *In preparation.*

[239] C.Grosche: Path Integral Representations on the O(2,2) Hyperboloid. *In preparation.*

[240] C.Grosche, Kh.Karayan, G.S.Pogosyan and A.N.Sissakian: Free Motion on the Three-Dimensional Sphere: The Ellipso-Cylindrical Bases. *J.Phys.A: Math.Gen.* **30** (1997) 1629–1657.

[241] C.Grosche, G.S.Pogosyan and A.N.Sissakian: Path Integral Discussion for Smorodinsky-Winternitz Potentials: I. Two- and Three-Dimensional Euclidean Space. *Fortschr.Phys.* **43** (1995) 453–521.

[242] C.Grosche, G.S.Pogosyan and A.N.Sissakian: Path Integral Discussion for Smorodinsky-Winternitz Potentials: II. The Two- and Three-Dimensional Sphere. *Fortschr.Phys.* **43** (1995) 523–563.

[243] C.Grosche, G.S.Pogosyan and A.N.Sissakian: Path-Integral Approach to Superintegrable Potentials on the Two-Dimensional Hyperboloid. *Phys.Part.Nucl.* **27** (1996) 244–278.
Path Integral Discussion for Smorodinsky-Winternitz Potentials: IV. The Three-Dimensional Hyperboloid. *Phys.Part.Nucl.* **28** (1997) 486–519.

[244] C.Grosche, G. S. Pogosyan and A. N. Sissakian: Path Integral Approach for Superintegrable Potentials on Spaces of Non-constant Curvature: I. Darboux Spaces D_I and D_{II}. *Phys.Part.Nucl.* **38** (2007) 299–325.
C.Grosche, G. S. Pogosyan and A. N. Sissakian: Path Integral Approach for Superintegrable Potentials on Spaces of Non-constant Curvature: II. Darboux Spaces D_{III} and D_{IV}. *Phys.Part.Nucl.* **38** (2007) 525–563.

[245] C.Grosche and G. S. Pogosyan: Path Integral Approach for Superintegrable Potentials on Complex Euclidean Space. *In preparation.*

[246] C.Grosche and G. S. Pogosyan: Path Integral Approach for Superintegrable Potentials on Spaces of Non-constant Curvature: Three-Dimensional Darboux Spaces. *In preparation.*

[247] C.Grosche and F.Steiner: The Path Integral on the Poincaré Upper Half Plane and for Liouville Quantum Mechanics. *Phys.Lett.* **A 123** (1987) 319–328.

[248] C.Grosche and F.Steiner: Path Integrals on Curved Manifolds. *Zeitschr.Phys.* **C 36** (1987) 699–714.

[249] C.Grosche and F.Steiner: The Path Integral on the Pseudosphere. *Ann.Phys.(N.Y.)* **182** (1988) 120–156.

[250] C.Grosche and F.Steiner: Classification of Solvable Feynman Path Integrals. In the Proceedings of the IV. International Conference on "Path Integrals from meV to MeV", Tutzing, Germany 1992, 276–288. Eds.: H.Grabert, A.Inomata, L.S.Schulman and U.Weiss (*World Scientific*, Singapore, 1993).

[251] C.Grosche and F.Steiner: How to Solve Path Integrals in Quantum Mechanics. *J.Math.Phys.* **36** (1995) 2354–2385.

[252] C.Grosche and F.Steiner: Handbook of Feynman Path Integrals. *Springer Tracts in Modern Physics* **145** (*Springer-Verlag*, Berlin, Heidelberg, 1998).

[253] D.J.Gross and V.Periwal: String Perturbation Theory Diverges. *Phys.Rev.Lett.* **60** (1988) 2105–2108.

[254] F.Grunewald and W.Huntebrinker: A Numerical Study of Eigenvalues of the Hyperbolic Laplacian for Polyhedra with one Cusp. *University of Düsseldorf Preprint*, October 1994, 44pp.

[255] A.Guha and S.Mukherjee: Exact Solution of the Schrödinger Equation with Noncentral Parabolic Potentials. *J.Math.Phys.* **28** (1987) 840–843.

[256] L.Guillope: Sur la Distribution des Longueurs des Geodesique Fermées d'une Surface Compacte a Bord Totalement Geodesique. *Duke Math.J.* **53** (1986) 827–848.

[257] D.Gurarie: Quantized Neumann Problem, Separable Potentials on S^n and the Lamé Equation. *J.Math.Phys.* **36** (1995) 5355–5391.

[258] M.C.Gutzwiller: Phase-Integral Approximation in Momentum Space and the Bound States of an Atom. *J.Math.Phys.* **8** (1967) 1979–2000. Phase-Integral Approximation in Momentum Space and the Bound States of an Atom. II. *J.Math.Phys.* **10** (1969) 1004–1020. Energy Spectrum According to Classical Mechanics. *J.Math.Phys.* **11** (1970) 1791–1806. Periodic Orbits and Classical Quantization Conditions. *J.Math. Phys.* **12** (1971) 343–358.

[259] M.C.Gutzwiller: Classical Quantization of a Hamiltonian with Ergodic Behaviour. *Phys.Rev.Lett.* **45** (1980) 150–153. The Quantization of a Classically Ergodic System. *Physica* **D 5** (1982) 183–207. Stochastic Behaviour in Quantum Scattering. *Physica* **D 7** (1983) 341–355. The Geometry of Quantum Chaos. *Physica Scripta* **T 9** (1985) 184–192. Physics and Selberg's Trace Formula. *Contemporary Math.* **53** (1986) 215–251. Mechanics on a Surface of Constant Negative Curvature. In "Number Theory", *Lecture Notes in Mathematics* **1240** 230–258. Eds.: D.V.Chudnovsky, G.V.Chudnovsky, H.Cohn and M.B.Nathanson (*Springer*, Berlin, 1985).

[260] M.C.Gutzwiller: Chaos in Classical and Quantum Mechanics (*Springer*, Berlin, 1990).

[261] J.Hadamard: Les Surfaces a Coubures Opposées et Leurs Lignes Geodésiques. *J.de Math.pure et Appl.* **4** (1898) 27–87.

[262] J.Harnad and P.Winternitz: Harmonics on Hyperspheres, Separation of Variables and the Bethe Ansatz. *Lett.Math.Phys.* **33** (1995) 61–74.

[263] J.Harnad and P.Winternitz: Classical and Quantum Integrable Systems in $\tilde{\mathfrak{gl}}(2)^{+*}$ and Separation of Variables. *Centre de Recherches Mathématiques*, Montreal, CRM-1921, 1993, 27pp., *Commun.Math.Phys.*, to appear.

[264] H.Hartmann: Bewegung eines Körpers in einem ringförmigen Potentialfeld. *Theoret. Chim.Acta* **24** (1972) 201–206.

[265] M.Hashizume, K.Minemura and K.Okamoto: Harmonic Functions on Hermitian Hyperbolic Spaces. *Hiroshima Math.J.* **3** (1973) 81–108.

[266] S.Hawking: Zeta Function Regularization of Path Integrals in Curved Spacetime. *Commun.Math.Phys.* **55** (1977) 133–148.

[267] D.R.Heath-Brown: The Distribution and Moments of the Error Term in the Dirichlet Divisor Problem. *Acta Arithmetica* **40** (1992) 389–415.

[268] D.A.Hejhal: The Selberg Trace Formula and the Riemann Zeta Function. *Duke Math.J.* **43** (1976) 441–482.

[269] D.A.Hejhal: The Selberg Trace Formula for PSL(2, $\mathbb{R}$) I. *Lecture Notes in Mathematics* **548** (*Springer*, Berlin, 1976).

[270] D.A.Hejhal: The Selberg Trace Formula for PSL(2, $\mathbb{R}$) II. *Lecture Notes in Mathematics* **1001** (*Springer*, Berlin, 1981).

[271] D.A.Hejhal and B.N.Rackner: On the Topography of Maass Waveforms for PSL(2, $\mathbb{Z}$). *Exp.Math.* **1** (1992) 275–305.

[272] S.Helgason: Eigenspaces of the Laplacian. Integral Representations and Irreducibility. *J.Funct.Anal.* **17** (1974) 328–353.

[273] S.Helgason: Differential Geometry, Lie Groups, and Symmetric Spaces (*Academic Press*, New York, 1978).

[274] S.Helgason: Geometric Analysis on Symmetric Spaces. (*Mathematical Surveys and Monographs* **39** Providence, 1994).

[275] E.J.Heller: Bound-State Eigenfunctions of Classically Chaotic Hamiltonian Systems: Scars of Periodic Orbits. *Phys.Rev.Lett.* **53** (1984) 1515–1518. E.J.Heller, P.O'Connor and J.Gehlen: The Eigenfunctions of Classically Chaotic Systems. *Physica Scripta* **40** (1989) 354–359.

[276] D.R.Herrick: Symmetry of the Quadratic Zeemann Effect for Hydrogen. *Phys.Rev.* **A 26** (1982) 323–329.

[277] P.W.Higgs: Dynamical Symmetries in a Spherical Geometry. *J.Phys.A: Math.Gen.* **12** (1979) 309–323.

[278] R.Ho and A.Inomata: Exact Path Integral Treatment of the Hydrogen Atom. *Phys. Rev.Lett.* **48** (1982) 231–234.

[279] D.B.Hodge: Eigenvalues and Eigenfunctions of the Spheroidal Wave Equation. *J.Math.Phys.* **11** (1970) 2308–3312.

[280] P.S.Howe: Superspace and the Spinning String. *Phys.Lett.* **B 70** (1977) 453–456. Super Weyl Transformations in Two Dimensions. *J.Phys.A: Math.Gen.* **12** (1979) 393–402.

[281] H.Huber: Zur analytischen Theorie hyperbolischer Raumformen und Bewegungsgruppen. *Math.Ann.* **138** (1959) 1–26.

[282] T.Ichinose and H.Tamura: Imaginary-Time Path Integral for a Relativistic Spinless Particle in an Electromagnetic Field. *Commun.Math.Phys.* **105** (1986) 239–257. Path Integral Approach to Relativistic Quantum Mechanics. *Prog.Theor.Phys.Supp.* **92** (1987) 144–175. The Zitterbewegung of a Dirac Particle in Two-Dimensional Space-Time. *J.Math.Phys.* **29** (1988) 103–109.

[283] A.Inomata: Exact Path-Integration for the Two Dimensional Coulomb Problem. *Phys.Lett.* **A 87** (1982) 387–390. Alternative Exact-Path-Integral-Treatment of the Hydrogen Atom. *Phys.Lett.* **A 101** (1984) 253–257.

[284] A.Inomata: Roles of Dynamical Symmetries in Path Integration. In *Proceedings of the "International Workshop on 'Symmetry Methods in Physics' in Memory of Prof. Ya. A. Smorodinsky"*, Dubna, July 1993, 148–159. Eds.: A.N.Sissakian, G.S.Pogosyan and S.I.Vinitsky, *JINR Publications*, E2–94–347, Dubna 1994.

[285] A.Inomata: Privat communication.

[286] A.Inomata and G.Junker: Quantization of the Kaluza-Klein Monopole System by Path Integration. *Phys.Lett.* **B 234** (1990) 41–44. Path-Integral Quantization of Kaluza-Klein Monopole Systems. *Phys.Rev.* **D 43** (1991) 1235–1242.

[287] A.Inomata and G.Junker: Path Integrals and Lie Groups. In "Noncompact Lie Groups and Some of Their Applications", San Antonio, Texas, USA, 1993, 199–224. Eds.: E.A.Tanner and R.Wilson (*Kluwer Academic Publishers*, Dordrecht, 1993).

[288] A.Inomata, G.Junker and R.Wilson: Topological Charge Quantization Via Path Integration: An Application of the Kustaanheimo-Stiefel Transformation. *Found.Phys.* **23** (1993) 1073–1091.

[289] A.Inomata, H.Kuratsuji and C.C.Gerry: Path Integrals and Coherent States of $SU(2)$ and $SU(1,1)$ (*World Scientific*, Singapore, 1992).

[290] A.Inomata and R.Wilson: Path Integral Realization of a Dynamical Group. *Lecture Notes in Physics* **261**, 42–47 (*Springer*, Berlin, 1985).

[291] H.Iwaniec: Prime Geodesic Theorem. *J.Reine Angew.Math.* **349** (1984) 136–159.

[292] A.A.Izmest'ev, G.S.Pogosyan, A.N.Sissakian and P.Winternitz: Contraction of Lie Algebras and Separation of Variables. The n–Dimensional Sphere. *J. Math. Phys. 40* (1999) 1549–1574.

[293] T.Jacobson: Spinor Chain Path Integral for the Dirac Equation. *J.Phys.A: Math.Gen.* **17** (1984) 2433–2451.

[294] T.Jacobson and L.S.Schulman: Quantum Stochastics: The Passage from a Relativistic to a Non-Relativistic Path Integral. *J.Phys.A: Math.Gen.* **17** (1984) 375–383.

[295] G.Junker: Remarks on the Local Time Rescaling in Path Integration. *J.Phys.A: Math.Gen.* **23** (1990) L881–L884.

[296] M.Kac: Can One Hear the Shape of a Drum? *Amer.Math.Monthly (Part II)* **23** (1966) 1–23.

[297] W.Kallies, I.Lukáč, G.S.Pogosyan and A.N.Sissakian: Ellipsoidal Basis for Isotropic Oscillator. In *Proceedings of the "International Workshop on 'Symmetry Methods in Physics' in Memory of Prof. Ya. A. Smorodinsky"*, Dubna, July 1993, 206–217. Eds.: A.N.Sissakian, G.S.Pogosyan and S.I.Vinitsky, *JINR Publications*, E2–94–347, Dubna 1994.

[298] E.G.Kalnins: Mixed Basis Matrix Elements for the Subgroup Reductions of $SO(2,1)$. *J.Math.Phys.* **14** (1973) 654–657.

[299] E.G.Kalnins: The Relativistic Invariant Expansion of a Scalar Function on Imaginary Lobachevski Space. *J.Math.Phys.* **14** (1973) 1316–1319.

[300] E.G.Kalnins: On the Separation of Variables for the Laplace Equation $\Delta\psi + K^2\psi = 0$ in Two- and Three-Dimensional Minkowski Space. *SIAM J.Math.Anal.* **6** (1975) 340–374.

[301] E.G.Kalnins: Separation of Variables for Riemannian Spaces of Constant Curvature (*Longman Scientific & Technical*, Essex, 1986).

[302] Kalnins, E.G., Kress, J.M., Pogosyan, G., Miller, W.Jr.: Complete Sets of Invariants for Dynamical Systems that Admit a Separation of Variables. *J. Math. Phys.* **43** (2002) 3592–3609.
Infinite-Order Symmetries for Quantum Separable Systems. *Phys. Atom. Nucl.* **68** (2005) 1756-1763.

[303] Kalnins, E.G., Kress, J.M., Miller, W.Jr.: Second Order Superintegrable Systems in Conformally Flat Spaces. I. 2D Classical Structure Theory. *J. Math. Phys.* **46** (2005) 053509.
Second Order Superintegrable Systems in Conformally Flat Spaces. II. The Classical Two-Dimensional Stäckel Transform. *J. Math. Phys.* **46** (2005) 053510.

[304] Kalnins, E.G., Kress, J.M., Miller, W.Jr.: Nondegenerate Three-Dimensional Complex Euclidean Superintegrable Systems and Algebraic Varieties. *J. Math. Phys.* **48** (2007) 113518 [26 pages].

[305] Kalnins, E.G., Kress, J.M., Miller, W.Jr., Winternitz, P.: Superintegrable Systems in Darboux Spaces. *J. Math. Phys.* **44** (2003) 5811–5848.
Kalnins, E.G., Kress, J.M., Winternitz, P.: Superintegrability in a Two-Dimensional Space of Non-constant Curvature. *J. Math. Phys.* **43** (2002) 970–983.

[306] E.G.Kalnins, V.B.Kuznetsov and W.Miller, Jr.: Quadrics on Complex Riemannian Spaces of Constant Curvature, Separation of Variables, and the Gaudin Magnet. *J.Math.Phys.* **35** (1994) 1710–1731.

[307] E.G.Kalnins, R.G.McLenaghan and G.C.William: Symmetry Operators for Maxwell's Equations on Curved Space-Time. *Proc.Roy.Soc.(London)* **A 439** (1992) 103–113.

[308] E.G.Kalnins and W.Miller, Jr.: Lie Theory and Separation of Variables. 3. The Equation $f_{tt} - f_{ss} = \gamma^2 f$. *J.Math.Phys.* **15** (1974) 1025–1032. Erratum: Ibidem **16** (1975) 1531.

[309] E.G.Kalnins and W.Miller, Jr.: Lie Theory and Separation of Variables. 4. The Groups SO(2, 1) and SO(3). *J.Math.Phys.* **15** (1974) 1263–1274.

[310] E.G.Kalnins and W.Miller, Jr.: Lie Theory and Separation of Variables. 5. The Equations $iU_t + U_{xx} = 0$ and $iU_t + U_{xx} - c/x^2 U = 0$. *J.Math.Phys.* **15** (1974) 1728–1737.

[311] E.G.Kalnins and W.Miller, Jr.: Lie Theory and Separation of Variables. 6. The Equation $iU_t + \Delta_2 U = 0$. *J.Math.Phys.* **16** (1975) 499–511.

[312] E.G.Kalnins and W.Miller, Jr.: Lie Theory and Separation of Variables. 7. The Harmonic Oscillator in Elliptic Coordinates and Ince Polynomials. *J.Math.Phys.* **16** (1975) 512–517.

[313] E.G.Kalnins and W.Miller, Jr.: Lie Theory and Separation of Variables. 8. Semisubgroup Coordinates for $\Psi_{tt} - \Delta_2 \Psi$. *J.Math.Phys.* **16** (1975) 2507–2516.

[314] E.G.Kalnins and W.Miller, Jr.: Lie Theory and Separation of Variables. 9. Orthogonal R-Separable Coordinate Systems for the Wave Equation $\partial_{tt}\Psi = \Delta_2\Psi$. *J.Math.Phys.* **17** (1976) 331–355.

[315] E.G.Kalnins and W.Miller, Jr.: Lie Theory and Separation of Variables. 10. Non-orthogonal R-Separable Solutions for the Wave Equation $\Psi_{tt} - \Delta_2\Psi = 0$. *J.Math. Phys.* **17** (1976) 356–368.

[316] E.G.Kalnins and W.Miller, Jr.: Lie Theory and Separation of Variables. 11. The EPD Equation. *J.Math.Phys.* **17** (1976) 369–377.

[317] E.G.Kalnins and W.Miller, Jr.: The Wave Equation, O(2, 2), and Separation of Variables on Hyperboloids. *Proc.Roy.Soc.Edinburgh* **A 79** (1977) 227–256.

[318] E.G.Kalnins and W.Miller, Jr.: Lie Theory and the Wave Equation in Space-Time. 1. The Lorentz Group. *J.Math.Phys.* **18** (1977) 1–16.

[319] E.G.Kalnins and W.Miller, Jr.: Lie Theory and the Wave Equation in Space-Time. 3. Semigroup Coordinates. *J.Math.Phys.* **18** (1977) 271–280.

[320] E.G.Kalnins and W.Miller, Jr.: Lie Theory and the Wave Equation in Space-Time. 5. R-Separable Solutions of the Wave-Equation $\psi_{tt} - \Delta_3\psi = 0$. *J.Math.Phys.* **18** (1977) 1741–1751. *J.Math.Phys.* **19** (1978) 1247–1257 (rev.).

[321] E.G.Kalnins and W.Miller, Jr.: Lie Theory and the Wave Equation in Space-Time. 2. The Group SO(4, ℂ). *SIAM J.Math.Anal.* **9** (1978) 12–33.

[322] E.G.Kalnins and W.Miller, Jr.: Lie Theory and the Wave Equation in Space-Time. 4. The Klein-Gordon Equation and the Poincaré Group. *J.Math.Phys.* **19** (1978) 1233–1246.

[323] E.G.Kalnins and W.Miller, Jr.: Separable Coordinates for Three-Dimensional Complex Riemannian Spaces. *J.Diff.Geometry* **14** (1979) 221–236.

[324] E.G.Kalnins and W.Miller, Jr.: Non-Orthogonal Separable Coordinate Systems for the Flat 4-Space Helmholtz Equation. *J.Phys.A: Math.Gen.* **12** (1979) 1129–1147.

[325] E.G.Kalnins and W.Miller, Jr.: Separation of Variables on n-Dimensional Riemannian Manifolds I. The n-Sphere and Euclidean n-Space. *J.Math.Phys.* **27** (1986) 1721–1736.

[326] E.G.Kalnins and W.Miller, Jr.: R-Separation of Variables for the Time-Dependent Hamilton-Jacobi and Schrödinger Equations. *J.Math.Phys.* **28** (1987) 1005–1015.

[327] E.G.Kalnins and W.Miller, Jr.: Hypergeometric Expansions of Heun Polynomials. *SIAM J.Math.Anal.* **22** (1991) 1450. Addendum: Ibidem 1803.

[328] E.G.Kalnins and W.Miller, Jr.: Complete Set of Functions for Perturbations of Robertson-Walker Cosmologies and Spin 1 Equations in Robertson-Walker-Type Space-Times. *J.Math.Phys.* **32** (1991) 698–707. Complete Sets of Functions for Perturbations of Robertson-Walker Cosmologies. *J.Math.Phys.* **32** (1991) 1415–1423.

[329] Kalnins, E.G., Miller Jr., W., Pogosyan, G.S.: Completeness of Multiseparable Super-integrability on the Complex 2-Sphere. *J. Phys. A: Math. Gen.* **33** (2000) 6791–6806. E.G. Kalnins, J.M. Kress, G.S. Pogosyan, W. Miller Jr.: Completeness of Superinte-grability in Two-Dimensional Constant Curvature Spaces. *J. Phys. A: Math. Gen.* **34** (2001) 4705–4720.
Superintegrability on Two-Dimensional Complex Euclidean Space. In *Algebraic Meth-ods in Physics. A Symposium for the 60th Birthdays of JiříPatera and Pavel Winter-nitz*, pp.95–103. *CRM Series in Mathematical Physics, Eds.: Yvan Saint-Aubin, Luc Vinet.* Springer, Berlin, Heidelberg, 2001.

[330] E. G. Kalnins, W. Miller, and G. S. Pogosyan: Exact and Quasiexact Solvability of Second-Order Superintegrable Quantum Systems: I. Euclidean Space Preliminaries. *J.Math.Phys.* **47** (2006) 033502.

[331] E.G.Kalnins, W.Miller, Jr., and G.J.Reid: Separation of Variables for Complex Rie-mannian Spaces of Constant Curvature I. Orthogonal Separable Coordinates for $S_{n\mathbb{C}}$ and $E_{n\mathbb{C}}$. *Proc.Roy.Soc.(London)* **A 394** (1984) 183–206.

[332] E.G.Kalnins, W.Miller, Jr. and G.C.Williams: Matrix Operator Symmetries of the Dirac Equation and Separation of Variables. *J.Math.Phys.* **27** (1986) 1893–1900.

[333] E.G.Kalnins, W.Miller, Jr. and P.Winternitz: The Group O(4), Separation of Vari-ables and the Hydrogen Atom. *SIAM J.Appl.Math.* **30** (1976) 630–664.

[334] E.G.Kalnins, J.Patera, R.T.Sharp and P.Winternitz: Two-Variable Galilei-Group Expansions of Nonrelativistic Scattering Amplitudes. *Phys.Rev.* **D 8** (1973) 2552–2572. Potential Scattering and Galilei-Invariant Expansions of Scattering Amplitudes. *Phys.Rev.* **D 8** (1973) 3527–3538.

[335] E.G.Kalnins and G.C.William: Symmetry Operators and Separation of Variables for Spin-Wave Equations in Oblate Spheroidal Coordinates. *J.Math.Phys.* **31** (1990) 1739–1744.

[336] N.Katayama: A Note on a Quantum-Mechanical Harmonic Oscillator in a Space of Constant Curvature. *Nuovo Cimento* **B 107** (1992) 763–768.
M.Ikeda and N.Katayama: On the Generalization of Bertrand's Theorem to Spaces of Constant Curvature. *Tensor* **38** (1982) 37–48.

[337] D.C.Khandekar and S.V.Lawande: Feynman Path Integrals: Some Exact Results and Applications. *Phys.Rep.* **137** (1986) 115–229.

[338] G.A.Kerimov: Lobachevskian Quantum Field Theory in (2ν) Dimensions. *J.Math. Phys.* **27** (1986) 1358–1362. Lobachevskian Dirac Fields. *J.Math.Phys.* **31** (1990) 1745–1754.

[339] M.Kibler and C.Campigotto: On a Generalized Aharonov-Bohm Plus Oscillator Sys-tem. *Phys.Lett.* **A 181** (1993) 1–6.

[340] M.Kibler and C.Campigotto: Classical and Quantum Study of a Generalized Kepler-
       Coulomb System. *Int.J.Quantum Chem.* **45** (1993) 209–224.
       M.Kibler, G.-H.Lamot and P.Winternitz: Classical Trajectories for Two Ring-Shaped
       Potentials. *Int.J.Quantum Chem.* **43** (1992) 625–645.
       M.Kibler and P.Winternitz: Periodicity and Quasi-Periodicity for Super-Integrable
       Hamiltonian Systems. *Phys.Lett.* **147** (1990) 338–342.

[341] M.Kibler, L.G.Mardoyan and G.S.Pogosyan: On a Generalized Kepler-Coulomb Sys-
       tem: Interbasis Expansions. *Int.J.Quantum Chem.* **52** (1994) 1301–1316.

[342] M.Kibler and T.Negadi: Motion of a Particle in a Ring-Shaped Potential: An
       Approach Via a Nonbijective Canonical Transformation. *Int.J.Quantum Chem.* **26**
       (1984) 405–410.

[343] M.Kibler and T.Negadi: Motion of a Particle in a Coulomb Field Plus Aharonov-
       Bohm Potential. *Phys.Lett.* **A 124** (1987) 42–46.

[344] M.Kibler and P.Winternitz: Dynamical Invariance Algebra of the Hartmann Poten-
       tial. *J.Phys.A: Math.Gen.* **20** (1987) 4097–4108.

[345] H.M.Kiefer and D.M.Fradkin: Comments on Separability Operators, Invariance Lad-
       der Operators, and Quantization of the Kepler Problem in Prolate-Spheroidal Coor-
       dinates. *J.Math.Phys.* **9** (1968) 627–632.

[346] I.Kim: Die Spurformeln für integrable und pseudointegrable Systeme. Diploma The-
       sis, Hamburg University, 1994. 113pp. (unpublished).

[347] J.R.Klauder: The Action Option and a Feynman Quantization of Spinor Fields in
       Terms of Ordinary C-Numbers. *Ann.Phys.(N.Y.)* **11** (1960) 123–168. Path Integrals
       and Stationary-Phase Approximations. *Phys.Rev.* **D 19** (1979) 2349–2356. Construct-
       ing Measures for Spin-Variable Path Integrals. *J.Math.Phys.* **23** (1982) 1797–1805.
       New Measures for Nonrenormalizable Quantum Field Theory. *Ann. Phys.(N.Y.)* **117**
       (1979) 19–55. Quantization *Is* Geometry, After All. *Ann.Phys.(N.Y.)* **188** (1988)
       120–144.

[348] J.R.Klauder and I.Daubechies: Measures for Path Integrals. *Phys.Rev.Lett.* **48** (1982)
       117–120. Quantum Mechanical Path Integrals with Wiener Measures for all Polyno-
       mial Hamiltonians. *Phys.Rev.Lett.* **52** (1984) 1161–1164.
       J.R.Klauder and B.-S.Skagerstam: Coherent States. Applications in Physics and
       Mathematical Physics (*World Scientific*, Singapore, 1985).
       J.R.Klauder and W.Tomé: The Universal Propagator for Affine [or $SU(1,1)$] Coher-
       ent States. *J.Math.Phys.* **33** (1992) 3700–3709.
       J.R.Klauder and B.F.Whiting: Extended Coherent States and Path Integrals with
       Auxiliary Variables. *J.Phys.A: Math.Gen.* **26** (1993) 1697–1715.

[349] H.Kleinert: Path Integral for Coulomb System with Magnetic Charges. *Phys.Lett.* **A**
       116 (1986) 201–206.

[350] H.Kleinert: How to do the Time Sliced Path Integral for the H Atom. *Phys.Lett.* **A 120** (1987) 361–366.

[351] H.Kleinert: Quantum Mechanics and Path Integral in Spaces with Curvature and Torsion. *Mod.Phys.Lett.* **A 4** (1990) 2329–2337.

[352] H.Kleinert: Path Integrals in Quantum Mechanics, Statistics and Polymer Physics (*World Scientific*, Singapore, 1990).

[353] H.Kleinert and I.Mustapic: Summing the Spectral Representations of Pöschl-Teller and Rosen-Morse Fixed-Energy Amplitudes. *J.Math.Phys.* **33** (1992) 643–662.

[354] G.Kleppe: Green's Function for Anti-de Sitter Space Gravity. *Phys.Rev.* **D 50** (1994) 7335–7345.

[355] Koenigs, G.: Sur les géodésiques a intégrales quadratiques. A note appearing in "Lecons sur la théorie générale des surface". Darboux, G., Vol.4, 368–404, Chelsea Publishing, 1972.

[356] I.V.Komarov and V.B.Kuznetsov: Kowaleswki's Top on the Lie Algebra $o(4), e(3)$ and $o(3, 1)$. *J.Phys.A: Math.Gen.* **23** (1990) 841–846. Quantum Euler-Manakov Top on the 3-Sphere S_3. *J.Phys.A: Math.Gen.* **24** (1991) L737–L742.

[357] D.V.Kosygin, A.A.Minasov and Ya.G.Sinai: Statistical Properties of the Spectra of Laplace-Beltrami Operators on Liouville Surfaces. *Russ. Math.Surveys* **48** (1993) 1–142.

[358] T.H.Koornwinder: Jacobi Functions and Analysis on Noncompact Semisimple Lie Groups. In Special Functions: Group Theoretical Aspects and Applications, 1–85. Eds.: R.A.Askey, T.H.Koornwinder and W.Schempp (*Reidel*, Dordrecht, 1984).

[359] S.-Y.Koyama: Determinant Expressions of Selberg Zeta Functions. I.–III. *Trans. Amer.Math.Soc.* **324** (1991) 149–168. Ibidem **329** (1992) 755–772. *Proc.Amer. Math.Soc.* **113** (1991) 303–311.

[360] H.A.Kramers and G.P.Ittmann: Zur Quantelung des asymmetrischen Kreisels (I-III). *Zeitschr.Phys.* **53** (1929) 553–565. Ibidem **58** (1929) 217–231, **60** (1930) 663–681.

[361] R.Kubo: Path Integration on the Upper Half-Plane. *Prog.Theor.Phys.* **78** (1987) 755–759. Geometry, Heat Equation and Path Integrals on the Poincaré Upper Half-Plane. *Prog.Theor.Phys.* **79** (1988) 217–226.

[362] T.Kubota: Elementary Theory of Eisenstein Series (*John Wiley & Sons*, New York, 1973).

[363] Ю.А.Курочкин и В.С.Отчик: Аналог вектора рунге–ленца и спектр знергий в задаче кеплера на трехмерной цфере. *ДАН БССР* **23** (1979) 987–990. [Yu.A.Kurochkin and V.S.Otchik: Analogue of the Runge-Lenz Vector and Energy Spectrum in the Kepler Problem on the Three-Dimensional Sphere. *DAN BSSR* **23**

(1979) 987–990].

А.А.Богуш, В.С.Отчик и В.М.Редьков: Разделение перемениых в уравнении
шрединтера и нормированные функции состояний для задачи кеплера в
трехмерных пространствах постоянной кривизны. *Вести Акад.Наук.БССР* **3**
(1983) 56–62.

[A.A.Bogush, V.S.Otchik and V.M.Red'kov: Separation of Variables in the Schrödinger Equation and Normalized State Functions for the Kepler Problem in the Three-
Dimensional Spaces of Constant Curvature. *Vesti Akad.Nauk.BSSR* **3** (1983) 56–62].

А.А.Богуш, Ю.А.Курочкин и В.С.Отчик: О квантовомеханической задаче
кеплера в трехмерном пространстве лобачевского. *ДАН БССР* **24** (1980) 19–
22.

[A.A.Bogush, Yu.A.Kurochkin and V.S.Otchik: On the Quantum-Mechanical Kepler
Problem in Three-Dimensional Lobachevskii Space. *DAN BSSR* **24** (1980) 19–22].

[364] V.B.Kuznetsov: Generalized Polyspheroidal Periodic Functions and the Quantum
       Inverse Scattering Method. *J.Math.Phys.* **31** (1990) 1167–1174.

[365] V.B.Kuznetsov: Quadratics on Real Riemannian Spaces of Constant Curvature: Sep-
       aration of Variables and Connection with Gaudin Magnet. *J.Math.Phys.* **33** (1992)
       3240–3254.

[366] V.B.Kuznetsov: Equivalence of Two Graphical Calculi. *J.Phys.A: Math.Gen.* **25**
       (1992) 6005–6026.

[367] G.I.Kusznetsov and Ya.A.Smorodinskiĭ: Integral Representations of Relativistic Am-
       plitudes in the Unphysical Region. *Sov.J.Nucl.Phys.* **3** (1966) 275–283. P.Winternitz,
       Ya.A.Smorodinskiĭ and M.B.Sheftel: Poincaré- and Lorentz-Invariant Expressions of
       Relativistic Amplitudes. *Sov.J.Nucl.Phys.* **7** (1968) 785–792.

[368] M.Lakshmanan and H.Hasegawa: On the Canonical Equivalence of the Kepler Prob-
       lem in Coordinate and Momentum Space. *J.Phys.A: Math.Gen.* **17** (1984) L889–L893.

[369] F.Langouche, D.Roekaerts and E.Tirapegui: Functional Integration and Semiclassical
       Expansion (*Reidel*, Dordrecht, 1982).

[370] S.V.Lawande and K.V.Bhagwat: Derivation of Exact Propagators by Summation of
       Perturbations Series. In "Path Integrals From meV to MeV", Bangkok 1989, 309–
       329. Eds.: V.Sa-yakanit, W.Sritrakool, J.-O. Berananda, M.C.Gutzwiller, A.Inomata,
       S.Lundqvist, J.R.Klauder and L.S.Schulman (*World Scientific*, Singapore, 1989).

[371] H.I.Leemon: Dynamical Symmetries in a Spherical Geometry. *J.Phys.A: Math.Gen.*
       A 12 (1979) 489–501.

[372] S.Levit, K.Möhring, U.Smilanski and T.Dreyfus: Focal Points and the Phase of the
       Semiclassical Propagator. *Ann.Phys.(N.Y.)* **114** (1978) 223–242.

[373] I.Lukàcs: A Complete Set of the Quantum-Mechanical Observables on a Two-Dimen-
       sional Sphere. *Theor.Math.Phys.* **14** (1973) 271–281.

[374] I.Lukàč: Complete Sets of Observables on the Sphere in Four-Dimensional Euclidean Space. *Theor.Math.Phys.* **31** (1977) 457–461.

[375] И.Лукач и М.Надь: Эллиптичкие гармонические функции на сфере в чегырехмерном комплексном пространстве. *ОИЯИ препринт* P2–10736, Дубна 1977, 18pp.
[I.Lukač and M.Nagy: Elliptical Harmonic Functions on the Sphere in Four-Dimensional Complex Space. *JINR Communications*, P2–10736, Dubna 1977, unpublished].

[376] I.Lukàcs and Ya.A.Smorodinskiĭ: Wave Functions for the Asymmetric Top. *Sov.Phys. JETP* **30** (1970) 728–730.

[377] I.Lukàcs and Ya.A.Smorodinskiĭ: Separation of Variables in a Spheroconical Coordinate System and the Schrödinger Equation for a Case of Noncentral Forces. *Theor.Math.Phys.* **14** (1973) 125–131.

[378] M.Luming and E.Predazzi: A Set of Exactly Solvable Potentials. *Nuovo Cimento* **B 44** (1966) 210–212.

[379] И.В.Луценко, Л.Г.Мардоян, Г.С.Погосян и А.Н.Сисакян: Обобщение межбазисного осцилляторного разложения "цилиндр–сфера" в поле кольцеобразного потенциала. *ОИЯИ препринт* P2–89–814, Дубна 1989, 8pp.
[I.V.Lutsenko, L.G.Mardoyan, G.S.Pogosyan and A.N.Sissakian: The Generalization of Interbasis Oscillator Expansions "Cylinder-Sphere" in the Field of a Ring-Shaped Potential. *JINR Communications*, P2-89-814, Dubna, 1989, unpublished].

[380] I.V.Lutsenko, G.S.Pogosyan, A.N.Sisakyan and V.M.Ter-Antonyan: Hydrogen Atom as Indicator of Hidden Symmetry of a Ring-Shaped Potential. *Theor.Math.Phys.* **83** (1990) 633–639.

[381] N.W.MacFadyen and P.Winternitz: Crossing Symmetric Expansions of Physical Scattering Amplitudes. The $O(2, 1)$ Group and Lamé Functions. *J.Math.Phys.* **12** (1971) 281–293.

[382] A.J.MacFarlane: The Quantum Neumann Model with the Potential of Rosochatius. *Nucl.Phys.* **B 386** (1992) 453–467.

[383] H.P.McKean: Selberg's Trace Formula as Applied to a Riemann Surface. *Commun. Pure Appl.Math.* **25** (1972) 225–246. Correction: Ibidem **27** (1974) 134.

[384] D.W.McLaughlin and L.S.Schulman: Path Integrals in Curved Spaces. *J.Math.Phys.* **12** (1971) 2520–2524.

[385] W.Magnus, F.Oberhettinger and R.P.Soni: Formulas and Theorems for the Special Functions of Mathematical Physics (*Springer*, Berlin, 1966).

[386] A.A.Makarov, Ya.A.Smorodinsky, Kh.Valiev and P.Winternitz: A Systematic Search for Nonrelativistic Systems with Dynamical Symmetries. *Nuovo Cimento* **A 52** (1967) 1061–1084.

[387] S.L.Malurkar: Ellipsoidal Wave-Functions. *Indian J.Phys.* **9** (1934) 45–80.

[388] Yu.I.Manin: The Partition Function of the Polyakov String can be Expressed in Terms of Theta Functions. *Phys.Lett.* **B 172** (1986) 184–185.

[389] L.G.Mardoyan, G.S.Pogosyan, A.N.Sissakian and V.M.Ter-Antonyan: Spheroidal Analysis of Hydrogen Atom. *J.Phys.A: Math.Gen.* **16** (1983) 711–718. Spheroidal Corrections to the Spherical and Parabolic Bases of the Hydrogen Atom. *Theor. Math.Phys.* **64** (1985) 762–764. Elliptic Basis for a Circular Oscillator. *Theor. Math.Phys.* **65** (1985) 1113–1122. Interbasis Expansions in a Circular Oscillator. *Nuovo Cimento* **A 86** (1985) 324–336.

[390] L.G.Mardoyan, G.S.Pogosyan, A.N.Sissakian and V.M.Ter-Antonyan: Hidden Symmetry, Separation of Variables and Interbasis Expansions in the Two-Dimensional Hydrogen Atom. *J.Phys.A: Math.Gen.* **18** (1985) 455–466.

[391] L.G.Mardoyan, G.S.Pogosyan, A.N.Sisakyan and V.M.Ter-Antonyan: Two-Dimensional Hydrogen Atom: I. Elliptic Bases. *Theor.Math.Phys.* **61** (1984) 1021–1034. Interbasis Expansion in a Two-Dimensional Hydrogen Atom. *Theor.Math.Phys.* **63** (1985) 597–603.

[392] Мардоян, Л.Г., Погосян, Г.С., Сисакян, А.Н., Тер-Антонян, В.М.: *КВАНТОВЫЕ СИСТЕМЫ СО СКРЫТОЙ СИММЕТРИЕЙ. МЕЖБАЗИСНЫЕ РАЗЛОЖЕНИЯ.* ФИЗМАТЛИТ, Москва, 2006.
[Mardoyan, L.G., Pogosyan, G.S., Sissakian, A.N., Ter-Antonyan, V.M.: *Quantum Systems with Hidden Symmetry. Interbasis Expansions.* FIZMATLIT, Moscow, 2006 (in russian)].

[393] M.S.Marinov and M.V.Terentyev: A Functional Integral on Unitary Groups. *Sov.J. Nucl.Phys.* **28** (1978) 729–737. Dynamics on the Group Manifold and Path Integral. *Fortschr.Phys.* **27** (1979) 511–545.

[394] C.Matthies: Picards Billard: Ein Modell für Arithmetisches Quantenchaos in drei Dimensionen. Dissertation Hamburg 1995, 139 pp.

[395] C.Matthies and F.Steiner: Selberg's Zeta Function and the Quantization of Chaos. *Phys.Rev.* **A 44** (1991) R7877–R7880.

[396] S.Matsumoto, S.Uehara and Y.Yasui: Hadamard Model on the Super Riemann Surface. *Phys.Lett.* **A 134** (1988) 81–86.

[397] S.Matsumoto, S.Uehara and Y.Yasui: A Superparticle on the Super Riemann Surface. *J.Math.Phys.* **31** (1990) 476–501.

[398] I.M.Mayes and J.S.Dowker: Canonical Functional Integrals in General Coordinates. *Proc.Roy.Soc.(London)* **A 327** (1972) 131–135. Hamiltonian Orderings and Functional Integrals. *J.Math.Phys.* **14** (1973) 434–439. The Canonical Quantization of Chiral Dynamics. *Nucl.Phys.* **B 29** (1971) 259–268.

[399] J.Meixner and F.W.Schäfke: Mathieusche Funktionen und Sphäroidfunktionen (*Springer*, Berlin, 1954).

[400] E.Melzer: $N = 2$ Supertori and Their Representations as Algebraic Curves. *J.Math. Phys.* **29** (1988) 1555–1568.

[401] M.L.Metha: Random Matrices (*Academic Press*, San Diego, 1991²).

[402] W.Miller, Jr.: Special Functions and the Complex Euclidean Group in 3−Space. I-III. *J.Math.Phys.* **9** (1968) 1163–1175. Ibidem: 1175–1434, 1434–1444.

[403] W.Miller, Jr.: Lie Theory and Separation of Variables. I: Parabolic Cylinder Coordinates. *SIAM J.Math.Anal.* **5** (1974) 626–642. Lie Theory and Separation of Variables. II: Parabolic Coordinates. *SIAM J.Math.Anal.* **5** (1974) 822–836.

[404] W.Miller, Jr.: Symmetry and Separation of Variables for Linear Partial Differential and Hamilton-Jacobi Equations. In Group Theoretical Methods in Physics, Montreal 1976, 529–548. Eds.: R.T.Sharp and B.Kolman (*Academic Press*, New York, 1977).

[405] W.Miller, Jr.: Symmetry and Separation of Variables (*Addison-Wesley*, Reading, 1977).

[406] W.Miller, Jr., J.Patera and P.Winternitz: Subgroups of Lie Groups and Separation of Variables. *J.Math.Phys.* **22** (1981) 251–260.

[407] S.Minakshisundaram: A Generalization of the Epstein Zeta Functions. *Can.J.Math.* **1** (1949) 320–327.
S.Minakshisundaram and Å.Pleijel: Some Properties of the Eigenfunctions of the Laplace-Operator on Riemannian Manifolds. *Can.J.Math.* **1** (1949) 242–256.

[408] M.M.Mizrahi: The Weyl Correspondence and Path Integrals. *J.Math.Phys.* **16** (1975) 2201–2206.

[409] M.M.Mizrahi: Phase Space Path Integrals, without Limiting Procedure. *J.Math.Phys.* **19** (1978) 298–307.

[410] F.Möglich: Beugungserscheinungen an Körpern von ellipsoidischer Gestalt. *Ann. Phys.* **83** (1927) 609–734.

[411] F.Moon and D.Spencer: Theorems on Separability in Riemannian n-Space. *Proc. Amer.Math.Soc.* **3** (1952) 635-642.

[412] P.Moon and P.E.Spencer: Field Theory Handbook (*Springer*, Berlin, 1961).

[413] G.Moore, P.Nelson and J.Polchinski: Strings and Supermoduli. *Phys.Lett.* **B 169** (1986) 47–53.

[414] C.Morette: On the Definition and Approximation of Feynman's Path Integrals. *Phys.Rev.* **81** (1951) 848–852. Feynman's Path Integral: Definition without Limiting Procedure. *Commun.Math.Phys.* **28** (1972) 47–67.

[415] P.M.Morse: Diatomic Molecules According to the Wave Mechanics II. Vibrational Levels. *Phys.Rev.* **34** (1929) 57–64.

[416] P.M.Morse and H.Feshbach: Methods of Theoretical Physics (*McGraw-Hill*, New York, 1953).

[417] M.Moshinsky, J.Patera, R.T.Sharp and P.Winternitz: Everything You Always Wanted to Know About SU(3) ⊃ O(3). *Ann.Phys.(N.Y.)* **95** (1975) 139–169.

[418] M.A.Namazie and S.Rajeev: On Multiloop Computations in Polyakov's String Theory. *Nucl.Phys.* **B 277** (1986) 332–348.

[419] http://de.wikipedia.org/wiki/Ellipsoid,

   http://www.nasa.gov/mission_pages/hubble/main/index.html

   and http://www.nasa.gov. Note: "NASA Hubble material ... is copyright-free and may be freely used as in the public domain without fee, on the condition that only NASA, STScI, and/or ESA is credited as the source of the material."

[420] E.Nelson: Feynman Integrals and the Schrödinger Equation. *J.Math.Phys.* **5** (1964) 332–343.

[421] C.Neumann: De Problemate Quodam Mechanico, Quod ad Primam Integralium Ultraellipticorum Classem Revocatur. *J.Reine Angew.Math.* **56** (1859) 46–63.

[422] H.Ninnemann: Deformations of Super Riemann Surfaces. *Commun.Math.Phys.* **150** (1992) 267–288.

[423] Y.Nishino: On Quadratic First Integrals in the 'Central Potential' Problem for the Configuration Space of Constant Curvature. *Math.Japonica* **17** (1972) 59–67.

[424] E.Noether: Invariante Variationsprobleme. *Nachr.König.Gesellsch.Wiss.Göttingen* (1918) 235–257.

[425] М.Н.Олевский: Триортогональные системы в пространствах постоянной кривизны, в которых уравнение $\Delta_2 u + \lambda u = 0$ допускает полное разделение переменных. *Мат.Сб.* **27** (1950) 379–426.
   [M.N.Olevskiĭ: Triorthogonal Systems in Spaces of Constant Curvature in which the Equation $\Delta_2 u + \lambda u = 0$ Allows the Complete Separation of Variables. *Math.Sb.* **27** (1950) 379–426].

[426] M.Omote: Point Canonical Transformations and the Path Integral. *Nucl.Phys.* **B 120** (1977) 325–332.

[427] K.Oshima: Contributions of the Discrete Spectrum of the Maass Laplacians to Super Traces of Laplace Operators on Super Riemann Surfaces. *Prog.Theor.Phys.* **82** (1989) 487–492. Completeness Relations for Maass Laplacians and Heat Kernels on the Super Poincaré Upper Half-Plane. *J.Math.Phys.* **31** (1990) 3060–3063. Eigenfunctions and Heat Kernels of Super Maass Laplacians on the Super Poincaré Upper Half-Plane. *J.Math.Phys.* **33** (1992) 1158–1177.

[428] В.С.Отчик и В.М.Редьков: Квантовомеханическая задача кеплера в прос-транствах постоянной кривизны, *Минск Препринт No.298*, 1983, 47pp.
[V.S.Otchik and V.M.Red'kov: Quantum Mechanical Kepler Problem in Space with Constant Curvature. *Minsk Preprint No.298*, 1983, 47pp., unpublished].
А.А.Богуш, В.С.Отчик и В.М.Редьков: Разделение переменных в уравнении шредингера и нормированные функции состояний для задачи кеплера в трех-мерных пространствах постоянной кривизны. *Вести Акад.Наук.БССР* **3** (1983) 56–62.
[A.A.Bogush, V.S.Otchik and V.M.Red'kov: Separation of Variables in the Schrödinger Equation and Normalized State Functions for the Kepler Problem in the Three-Dimensional Spaces of Constant Curvature. *Vesti Akad.Nauk.BSSR* **3** (1983) 56–62].
А.А.Богуш, Ю.А.Курочкин и В.С.Отчик: О квантовомеханической задаче кеплера в трехмерном пространстве лобачевского. *ДАН БССР* **24** (1980) 19–22.
[A.A.Bogush, Yu.A.Kurochkin and V.S.Otchik: On the Quantum-Mechanical Kepler Problem in Three-Dimensional Lobachevskii Space. *DAN BSSR* **24** (1980) 19–22].

[429] N.K.Pak and I.Sökmen: General New-Time Formalism in the Path Integral. *Phys. Rev.* **A 30** (1984) 1629–1635.

[430] J.Patera, R.T.Sharp, P.Winternitz and H.Zassenhaus: Subgroups of the Poincaré Group and Their Invariants. *J.Math.Phys.* **17** (1976) 977–985.

[431] J.Patera, R.T.Sharp, P.Winternitz and H.Zassenhaus: Subgroups of the Similitude Group of Three-Dimensional Minkowski Space. *Can.J.Phys.* **54** (1976) 950–961.

[432] J.Patera and P.Winternitz: A New Basis for the Representations of the Rotation Group. Lamé and Heun Polynomials. *J.Math.Phys.* **14** (1973) 1130–1139.

[433] J.Patera and P.Winternitz: On Bases for Irreducible Representations of O(3) Suitable for Systems with an Arbitrary Finite Symmetry Group. *J.Chem.Phys.* **65** (1976) 2725–2731.

[434] J.Patera, P.Winternitz and H.Zassenhaus: The Maximal Solvable Subgroup of the SU(p,q) Groups and Subgroups of SU$(2,1)$. *J.Math.Phys.* **15** (1974) 1378–1390. The Maximal Solvable Subgroup of the SO(p,q) Groups. *J.Math.Phys.* **15** (1974) 1932–1941. Continuous Subgroups of the Fundamental Groups of Physics I. General Method and the Poincaré Group. *J.Math.Phys.* **16** (1975) 1597–1614. Continuous Subgroups of the Fundamental Groups of Physics II. The Similitude Group. *J.Math.Phys.* **16** (1975) 1615–1624.

[435] J.Patera, P.Winternitz and H.Zassenhaus: Quantum Numbers for Particles in de Sitter Space. *J.Math.Phys.* **17** (1976) 717–728.

[436] W.Paul: Electromagnetic Traps for Charged and Neutral Particles. *Rev.Mod.Phys.* **62** (1990) 531–540.

[437] W.Pauli: Über das Wasserstoffspektrum vom Standpunkt der neuen Quanten-mechanik. *Zeitschr.Phys.* **36** (1926) 336–363, reprinted in [64], 371–399.

[438] W.Pauli: Ausgewählte Kapitel aus der Feldquantisierung (Zürich, 1951).

[439] D.Peak and A.Inomata: Summation Over Feynman Histories in Polar Coordinates. *J.Math.Phys.* **10** (1969) 1422–1428.

[440] A.Pelster and A.Wunderlin: On the Generalization of the Duru-Kleinert-Propagator Transformations. *Zeitschr.Phys.* **B 89** (1992) 373–386.

[441] R.F.Picken: The Propagator for Quantum Mechanics on a Group Manifold from an Infinite Dimensional Analogue of the Duistermaat-Heckman Integration Formula. *J.Phys A: Math.Gen.* **22** (1989) 2285–2297.

[442] B.Podolsky: Quantum-Mechanically Correct Form of Hamiltonian Function for Conservative Systems. *Phys.Rev.* **32** (1928) 812–816.

[443] Lutsenko, I.V., Pogosyan, G.S., Sisakyan, A.N.,TerAntonyan, V.M.: Hydrogen Atom as Indicator of Hidden Symmetry of a Ring-Shaped Potential; *Theor. Math. Phys.* **83** (1990) 633–639.

[444] G.S.Pogosyan, A.N.Sissakian and S.I.Vinitsky: Interbasis "Sphere-Cylinder" Expansions for the Oscillator in the Three-Dimensional Space of Constant Curvature. In: "Frontiers of Fundamental Physics", 429–436. Eds.: M.Barone and F.Selleri (*Plenum Publishing*, New York, 1994).

[445] H.Poincaré: Théorie des Groupes Fuchsiens. *Œvres Complètes* **II** 108–168 (*Gauthiers-Villars*, Paris, 1908).

[446] A.M.Polyakov: Quantum Geometry of Bosonic Strings. *Phys.Lett.* **B 103** (1981) 207–210.

[447] A.M.Polyakov: Quantum Geometry of Fermionic Strings. *Phys.Lett.* **B 103** (1981) 211-213.

[448] C.Quesne: A New Ring-Shaped Potential and its Dynamical Invariance Algebra. *J.Phys.A: Math.Gen.* **21** (1988) 3093–3103.

[449] J.M.Rabin and L.Crane: Global Properties of Supermanifolds. *Commun.Math.Phys.* **100** (1985) 141–160. How Different are the Supermanifolds of Rogers and DeWitt? *Commun.Math.Phys.* **102** (1985) 123–137. Super Riemann Surfaces: Uniformization and Teichmüller Theory. *Commun.Math.Phys.* **113** (1988) 601–623.

[450] B.Randol: Small Eigenvalues of the Laplace Operator on Compact Riemann Surfaces. *Bull.Amer.Math.Soc.* **80** (1974) 996–1000.

[451] D.Ray and I.M.Singer: Analytic Torsion for Complex Manifolds. *Ann.Math.* **98** (1973) 154–177.

[452] B.Riemann: Gesammelte Mathematische Werke (*Teubner*, Leipzig, 1892²).

[453] M.Robnik: Classical Dynamics of a Family of Billiards with Analytic Boundaries. *J.Phys.A: Math.Gen.* **16** (1983) 3971–3986. Quantising a Generic Family of Billiards with Analytic Boundaries. *J.Phys.A: Math.Gen.* **17** (1984) 1049–1074. T.Prosen and M.Robnik: Energy Level Statistics in the Transition Region Between Integrability and Chaos. *J.Phys.A: Math.Gen.* **26** (1993) 2371–2387.

[454] G.Roepstorff: Path Integral Approach to Quantum Physics (*Springer*, Berlin, 1994).

[455] A.Rogers: A Global Theory of Supermanifolds. *J.Math.Phys.* **21** (1980) 1352–1364. On the Existence of Global Integral Forms on Supermanifolds. *J.Math.Phys.* **26** (1985) 2749–2753. Graded Manifolds, Supermanifolds and Infinite-Dimensional Grassmann Algebras. *Commun.Math.Phys.* **105** (1986) 375–384.

[456] O.Rudolph: Sum Over Histories Representation for the Causal Green Function of Free Scalar Field Theory. *Phys.Rev.* **D 51** (1995) 1818–1831.

[457] P.Sarnak: The Arithmetic and Geometry of Some Hyperbolic Three Manifolds. *Acta Mathematica* **151** (1983) 253–295.

[458] P.Sarnak: Determinants of Laplacians. *Commun.Math.Phys.* **110** (1987) 113–120.

[459] L.Schulman: A Path Integral for Spin. *Phys.Rev.* **176** (1968) 1558–1569.

[460] L.S.Schulman: Techniques and Applications of Path Integration (*John Wiley & Sons*, New York, 1981).

[461] R.Schubert: The Trace Formula and the Distribution of Eigenvalues of Schrödinger Operators on Manifolds all of whose Geodesics are Closed. *DESY Report*, DESY 95–090, May 1995, 13pp.

[462] J.H.Schwarz: Superstring Theory. *Phys.Rep.* **89** (1982) 223–322.

[463] J.Sekiguchi: Eigenspaces of the Laplace-Beltrami Operator on a Hyperboloid. *Nagoya Math.J.* **79** (1980) 151–185.

[464] A.Selberg: Harmonic Analysis and Discontinuous Groups in Weakly Symmetric Riemannian Spaces with Application to Dirichlet Series. *J.Indian Math.Soc.* **20** (1956) 47–87.

[465] V.N.Shapovalov: Separation of Variables in Second-Order Linear Differential Equations. *Diff.Eqs.* **16** (1980) 1212–1220.

[466] R.J.Sibner: Symmetric Fuchsian Groups. *Am.J.Math.* **90** (1968) 1237–1259.

[467] M.Sieber: The Hyperbola Billiard: A Model for the Semiclassical Quantization of Chaos. *DESY Report*, DESY 91–030, April 1991 (Dissertation Hamburg 1991), 101pp.

[468] M.Sieber, U.Smilansky, S.C.Craegh and R.G.Littlejohn: Non-Generic Spectral Statistics in the Quantised Stadium Billiard. *J.Phys.A: Math.Gen.* **26** (1993) 6217–6230.

[469] M.Sieber and F.Steiner: Generalized Periodic-Orbit Sum Rules for Strongly Chaotic Systems. *Phys.Lett.* **A 144** (1990) 159–163. Quantum Chaos in the Hyperbola Billiard. *Phys.Lett.* **A 148** (1990) 415–420. Classical and Quantum Mechanics of a Strongly Chaotic Billiard System. *Physica* **D 44** (1990) 248-266. On the Quantization of Chaos. *Phys.Rev.Lett.* **67** (1991) 1941–1944.

[470] B.Simon: Functional Integration and Quantum Physics (*Academic Press*, New York, 1979).

[471] Ya.G.Sinai: On the Foundations of the Ergodic Hypothesis for a Dynamical System of Statistical Mechanics. *Sov.Math.Dokl.* **4** (1963) 1818–1822. Dynamical System with Elastic Reflections. *Russian Math.Surveys* **25** (1970) 137–191.

[472] L.P.Singh and F.Steiner: Fermionic Path Integrals, the Nicolai Map and the Witten Index. *Phys.Lett.* **B 166** (1986) 155–159.

[473] Ya.A.Smorodinsky and I.I.Tugov: On Complete Sets of Observables. *Sov.Phys.JETP* **23** (1966) 434–437.

[474] I.Sökmen: Exact Path-Integral Solution for a Charged Particle in a Coulomb Plus Aharonov-Bohm Potential. *Phys.Lett.* **A 132** (1988) 65–68.

[475] R.D.Spence: Angular Momentum in Sphero-Conal Coordinates. *Amer.J.Phys.* **27** (1959) 329–335.

[476] F.Steiner: Space-Time Transformations in Radial Path Integrals. *Phys.Lett.* **A 106** (1984) 356–362.

[477] F.Steiner: Exact Path Integral Treatment of the Hydrogen Atom. *Phys.Lett.* **A 106** (1984) 363–367.

[478] F.Steiner: Path Integrals in Polar Coordinates From eV to GeV. In the Proceeding of the International Conference on "Path Integrals From meV to MeV", Bielefeld 1985, 335–359. Eds.: M.C.Gutzwiller, A.Inomata, J.R.Klauder and L.Streit (*World Scientific*, Singapore, 1986).

[479] F.Steiner: On Selberg's Zeta-Function for Compact Riemann Surfaces. *Phys.Lett.* **B 188** (1987) 447–454.

[480] F.Steiner: Quantum Chaos. In "Schlaglichter der Forschung. Zum 75. Jahrestag der Universität Hamburg 1994", 543–564. Ed.: R.Ansorge (*Reimer*, Berlin, 1994).

[481] F.Steiner and P.Trillenberg: Refined Asymptotic Expansion of the Heat Kernel for Quantum Billiards in Unbounded Regions. *J.Math.Phys.* **31** (1990) 1670–1676.

[482] W.Stepanov: Sur l'équation de Laplace et certain système triples orthogonaux. *Math.Sb.* **11** (1942) 204–238.

[483] K.Stewartson and R.T.Waechter: On Hearing the Shape of a Drum: Further Results. *Proc.Camb.Phil.Soc.* **69** (1971) 353–363.

[484] S.N.Storchak: Path Reparametrization in a Path Integral on a Finite-Dimensional Manifold. *Theor.Math.Phys.* **75** (1988) 610–618. Rheonomic Homogeneous Point Transformation and Reparametrization in the Path Integral. *Phys.Lett.* **A 135** (1989) 77–85.

[485] R.S.Strichartz: Harmonic Analysis on Hyperboloids. *J.Funct.Anal.* **12** (1973) 341–383.

[486] T.Szeredi and D.A.Goodings: Classical and Quantum Chaos of the Wedge Billiard. I. Classical Mechanics. *Phys.Rev.* **E 48** (1993) 3518–3528. Classical and Quantum Chaos of the Wedge Billiard. II. Quantum Mechanics and Quantization Rules. *Phys.Rev.* **E 48** (1993) 3529–3544.

[487] N.Subia: Formule de Selberg et Formes D'Espace Hyperboliques Compactes. In: "Analyse Harmonique sur les Groupes de Lie", *Lectures Notes in Mathematics* **497**, 674–710. Eds.: P.Eymard, J.Faraut, G.Schiffmann and R.Takahashi (*Springer*, Berlin, 1975).

[488] J.Szmidt: The Selberg Trace Formula for the Picard Group SL(2, $\mathbb{Z}$[i]). *Acta Arithmetica* **42** (1983) 391–424.

[489] R.Takahashi: Sur les Représentations Unitaires des Groupes de Lorentz Généralisés. *Bull. Soc.Math.France* **91** (1963) 289–433.

[490] R.Takahashi: Spherical Functions in $Spin_0(1,d)/Spin(d-1)$ for $d = 2, 4$ and 8. In: "Non-Commutative Harmonic Analysis", *Lecture Notes in Mathematics* **587**, 226–240. Eds.: J.Camora and M.Vergne (*Springer*, Berlin, 1977). Quelques Resultats sur l'Analyse Harmonique Dans l'Espace Symmetrique Non Compact de Rang 1 du Type Exceptionnel. In: "Analyse Harmonique sur les Groupes des Lie II", *Lecture Notes in Mathematics* **739**, 511–567. Eds.: P.Eymard, J.Faraut, G.Schiffmann and R.Takahashi (*Springer*, Berlin, 1977).

[491] R.Takahashi: Fonctions Spheriques dans les Groupes Sp(n, 1). In: "Théorie du Potentiel et Analyse Harmonique", *Lecture Notes in Mathematics* **404**, 218–228. Ed.: J.Faraut (*Springer*, Berlin, 1974).

[492] K.Takase: On Special Values of Selberg Type Zeta Functions on SU(1, q + 1). *J.Math.Soc.Japan* **43** (1991) 631–659.

[493] G.Tanner and D.Wintgen: Quantization of Chaotic Systems. *CHAOS* **2** (1992) 53–59.

[494] E.Teller: Über das Wasserstoffmolekülion. *Zeitschr.Phys.* **61** (1930) 458–480.

[495] A.Terras: Harmonic Analysis on Symmetric Spaces and Applications, Vol.I&II (*Springer*, New York, 1987).

[496] E.Titchmarsh: Eigenfunctions Expansions, Vol.I (*Oxford University Press*, London, 1962).

[497] R.Tomaschitz: On the Calculation of Quantum Mechanical Ground States From Classical Geodesic Motion on Certain Spaces of Constant Curvature. *Physica* **D 34** (1989) 42–89.

[498] Z.Y.Turakulov: Separating Coordinate Systems in the Minkowski Space-Time. *Int.J. Mod.Phys.* **A 6** (1991) 3109–3117.

[499] S.Uehara and Y.Yasui: A Superparticle on the "Super" Poincaré Upper Half Plane. *Phys.Lett.* **B 202** (1988) 530–534. Super-Selberg's Trace Formula from the Chaotic Model. *J.Math.Phys.* **29** (1988) 2486–2490.

[500] K.M.Urwin: Integral Equations for the Paraboloidal Wave Functions (II). *Quart. J.Math.Oxford* **16** (1965) 257–262.

[501] K.M.Urwin and F.M.Arscott: Theory of the Whittaker-Hill Equation. *Proc.Roy.Soc. Edinburgh* **A 69** (1970) 28–44.

[502] A.G.Ushveridze: Special Case of the Separation of Variables in the Multidimensional Schrödinger Equation. *J.Phys.A: Math.Gen.* **21** (1988) 1601–1605.

[503] A.N.Vaidya and H.Boschi-Filho: Algebraic Calculation of the Green's Function of the Hartmann Potential. *J.Math.Phys.* **31** (1990) 1951–1954.

[504] J.H.Van Vleck: The Correspondence Principle in the Statistical Interpretation of Quantum Mechanics. *Proc.Naut.Amer.Soc.* **14** (1928) 178–188.

[505] A.B.Venkov: Expansion in Automorphic Eigenfunctions of the Laplace Operator and the Selberg Trace Formula in the Space $SO_0(n, 1)/SO(n)$. *Soviet Math.Dokl.* **12** (1971) 1363–1366.

[506] A.B.Venkov: Expansion in Automorphic Eigenfunctions of the Laplace-Beltrami Operator in Classical Symmetric Spaces of Rank One, and the Selberg Trace Formula. *Proc.Math.Inst.Steklov* **125** (1973) 1–48.

[507] A.B.Venkov: On an Asymptotic Formula Connected with the Number of Eigenvalues Corresponding to Odd Eigenfunctions of the Laplace-Beltrami Operator on a Fundamental Region of the Modular Group $PSL(2, \mathbb{Z})$. *Dokl.Akad.Nauk SSSR* **233** (1977) 524–526.

[508] A.B.Venkov: Spectral Theory of Automorphic Functions, the Selberg Trace Formula, and Some Problems of Analytic Number Theory and Mathematical Physics. *Russian Math.Surveys* **34** (1979) 79–153.

[509] A.B.Venkov: Spectral Theory of Automorphic Functions. *Proc.Math.Inst.Steklov* **153** (1981) 1–163.

[510] N.Ja.Vilenkin: Special Functions Connected with Class 1 Representations of Groups of Motions in Spaces of Constant Curvature. *Trans.Moscow Math.Soc.* (1963) 209–295.

[511] N.Ya.Vilenkin, G.I.Kuznetsov and Ya.A.Smorodinskiĭ: Eigenfunctions of the Laplace Operator Providing Representations of the $U(2), SU(2), SO(3), U(3)$ and $SU(3)$ Groups and the Symbolic Method. *Sov.J.Nucl.Phys.* **2** (1965) 645–652.

[512] N.Ya.Vilenkin and Ya.A.Smorodinsky: Invariant Expansions of Relativistic Amplitudes. *Sov.Phys.JETP* **19** (1964) 1209–1218.

[513] S.I.Vinitsky, L.G.Mardoyan, G.S.Pogosyan, A.N.Sissakian and T.A.Strizh: A Hydrogen Atom in the Curved Space. Expansion over Free Solutions on the Three-Dimensional Sphere. *Phys.At.Nucl.* **56** (1993) 321–327.

[514] S.I.Vinitskiĭ, V.N.Pervushin, G.S.Pogosyan and A.N.Sisakyan: Equation for Quasiradial Functions in the Momentum Representation on a Three-Dimensional Sphere. *Phys.At.Nucl.* **56** (1993) 1027–1034.

[515] A.Voros: Spectral Functions, Special Functions and the Selberg Zeta Function. *Commun.Math.Phys.* **110** (1987) 439–465.

[516] D.I.Wallace: Terms in the Selberg Trace Formula for $SL(3, \mathbb{Z}) \backslash SL(3, \mathbb{R})/SO(3, \mathbb{R})$ Associated to Eisenstein Series Coming From Maximal Parabolic Subgroups. *Proc. Amer.Math.Soc.* **106** (1989) 875–883.

[517] A.Weil: Sur les 'Formules Explicites' de la Théorie des Nombres Premiere. *Comm. Sém.Math.Lund (Medd.Lunds Univ.Mat.Sem.)* Tome Supplémentaire (1952) 252–265.

[518] H.Weyl: Das asymptotische Verteilungsgesetz der Eigenwerte linearer partieller Differentialgleichungen (mit einer Anwendung auf die Theorie der Hohlraumstrahlung). *Math.Ann.* **71** (1912) 441–479.

[519] F.W.Wiegel: Introduction to Path-Integral Methods in Physics and Polymer Science (*World Scientific*, Singapore, 1986).

[520] N.Wiener: The Average of an Analytic Functional. *Proc.Nat.Acad.Sci.USA* **7** (1921) 253–260. The Average of an Analytic Functional and the Brownian Movement. Ibidem 294–298. Differential Space. *J.Math.and Phys.* **2** (1923) 131–174. The Average Value of a Functional. *Proc.London Math.Soc.* **22** Ser.2 (1924) 454–467. Generalized Harmonic Analysis. *Acta Math.* **55** (1930) 117–258.

[521] http://www.wikipedia.de

[522] D.Wintgen, K.Richter and G.Tanner: The Semiclassical Helium Atom. *CHAOS* **2** (1992) 19–32.

[523] P.Winternitz: Subgroups of Lie Groups and Symmetry Breaking. In Group Theoretical Methods in Physics, Montreal 1976, 549–572. Eds.: R.T.Sharp and B.Kolman (*Academic Press*, New York, 1977).

[524] P.Winternitz and I.Friš: Invariant Expansions of Relativistic Amplitudes and Subgroups of the Proper Lorentz Group. *Sov.J.Nucl.Phys.* **1** (1965) 636–643.

[525] P.Winternitz, I.Lukač and Ya.A.Smorodinskiĭ: Quantum Numbers in the Little Group of the Poincaré Group. *Sov.J.Nucl.Phys.* **7** (1968) 139–145.

[526] P.Winternitz, Ya.A.Smorodinskiĭ and M.Uhlíř: On Relativistic Angular Momentum Theory. *Sov.J.Nucl.Phys.* **1** (1965) 113–119.

[527] P.Winternitz, Ya.A.Smorodinskiĭ, M.Uhlir and I.Fris: Symmetry Groups in Classical and Quantum Mechanics. *Sov.J.Nucl.Phys.* **4** (1967) 444–450.

[528] E.Witten: Constraints on Supersymmetry Breaking. *Nucl.Phys.* **B 202** (1982) 253–316.

[529] J.A.Wolf: Spaces of Constant Curvature (*McGraw-Hill*, New York, 1967).

[530] A.Young and C.DeWitt-Morette: Time Substitution in Stochastic Processes as a Tool in Path Integration. *Ann.Phys.* **169** (1986) 140–166.

[531] B.Zaslow and M.E.Zandler: Two-Dimensional Analog of the Hydrogen Atom. *Amer.J.Phys.* **35** (1967) 1118–1119.

[532] A.S.Zhedanov: The "Higgs Algebra" as a 'Quantum' Deformation of SU(2). *Mod. Phys.Lett.* **A 7** (1992) 507–512.

[533] A.S.Zhedanov: Hidden Symmetry Algebra and Overlap Coefficients for Two Ring-Shaped Potentials. *J.Phys.A: Math.Gen.* **26** (1993) 4633–4641.

[534] J.S.Zmuidzinas: Unitary Representations of the Lorentz Group on 4-Vector Manifolds. *J.Math.Phys.* **7** (1966) 764–780.

Index